Technical
Mathematics I

Technical Mathematics I

THOMAS J. McHALE
PAUL T. WITZKE

ADDISON-WESLEY PUBLISHING COMPANY

Reading, Massachusetts • Menlo Park, California • New York
Don Mills, Ontario • Wokingham, England • Amsterdam
Bonn • Sydney • Singapore • Tokyo • Madrid
Bogotá • Santiago • San Juan

Library of Congress Cataloging-in-Publication Data

McHale, Thomas J., 1931–
    Technical mathematics I.

    Includes index.
    1. Mathematics—1961–   . I. Witzke, Paul T.
II. Title.
QA39.2.M3218 1987   512.1   87-17501
ISBN 0-201-15408-0

Reprinted with corrections, May 1989.

22 23 24 25 CRS 04030201

Preface

Technical Mathematics I and Technical Mathematics II are designed to provide students with the mathematical skills needed for success in technical programs. Each text is written for one semester of a two-semester course. Though the texts presuppose one year of high school algebra or the equivalent, a thorough review of algebra is included. Geometric facts, concepts and relationships are presented or reviewed as needed.

Technical Mathematics I and Technical Mathematics II are an adaptation of the following four texts: Applied Algebra I, Calculation and Calculators, Applied Algebra II, and Applied Trigonometry. In the adaptation, those four texts are reduced to two texts. Most of the chapters of Applied Trigonometry are included unchanged. The chapters of the other three texts have been somewhat condensed, with the major changes occurring for the chapters of Calculation and Calculators. In the revised texts, greater emphasis is given to the solution of applied problems. The treatment of factoring, fractions, and analytic geometry is also expanded.

LEARNABILITY

Twenty years of experience have shown that students achieve a high level of learning using these texts and the accompanying tests. The texts are successful because the content is presented on the basis of a learning task analysis. That is, each learning objective is analyzed to identify the skills, concepts, and procedures needed for mastery, and then those skills, concepts, and procedures are presented by a carefully selected sequence of examples proceeding from the simple to the complex. Worked-out examples are given for the full range of problems covered by each objective so that there are no "gaps" from the student's point of view. The continual testing supports the texts because it enables the student to know whether he/she is succeeding or not, and can serve as a basis for needed remediation.

INTERACTIVE FORMAT

The format of each text is unique because the students have to interact with the content as it is presented. The significant features of the format are described below.

Frames. The content is presented in a step-by-step manner with each step called a frame. Each frame begins with some instruction, including one or more examples, and ends with some problem or problems that the student must do. Since answers for each frame are given to the right of the following frame, the students are given immediate feedback as they proceed from frame to frame.

Assignment Self-Tests. Each chapter is divided into a number of assignments which can ordinarily be covered in one class period. A self-test is provided in the text at the end of each assignment. Students can either use the self-tests as assignment tests or save them for a chapter review after all assignments in the chapter are completed.

Supplementary Problems. Supplementary problems for each assignment are provided at the end of each chapter. They can be assigned as needed by instructors or simply used as further practice by students. Answers for all supplementary problems are given in the back of the text.

DIAGNOSTIC AND CRITERION TESTS

Each text is accompanied by a test book which contains the various diagnostic and criterion tests described below. Answer keys are provided for all tests. The test book is provided only to instructors. Copies for student use must be made by some copying process.

Diagnostic Pre-Test. This test covers all chapters. It can be used either to get a measure of the entry skills of the students or as a basis for prescribing an individualized program.

Assignment Tests. Each chapter is divided into a number of assignments. After the students have completed each assignment and the assignment self-test (in the text), the assignment test (from the test book) can be administered and used as a basis for tutoring or assigning supplementary problems. The assignment tests are simply a diagnostic tool and need not be graded.

Chapter And Multi-Chapter Tests. After the appropriate assignments are completed, either a chapter test or a multi-chapter test can be administered. Ordinarily these tests should be graded. Three parallel forms are provided to facilitate the test administration, including the retesting of students who do not achieve a satisfactory score.

Comprehensive Test. The comprehensive test covers all chapters. Three parallel forms are provided. Since the comprehensive test is a parallel form of the diagnostic pre-test, the difference score can be used as a measure of each student's improvement in the course.

TEACHING MODES

This text can be used in various ways. Some possibilities are described below.

<u>Lecture Class</u>. The text can be used to reinforce lectures and to provide highly structured outside assignments related to the lectures.

<u>Mini-Lecture Class</u>. An instructor can give a brief lecture on difficult points in a completed assignment before the assignment test is administered. An instructor can also give a brief overview of each new assignment before it is begun by students.

<u>No-Lecture Class</u>. The text is well-suited for a no-lecture class that is either paced or self-paced. Class time can then be used to administer tests and to tutor individual students when tutoring is needed.

<u>Learning Laboratory</u>. The text is also well-suited for a learning laboratory where students proceed at their own pace. The instructor can manage the instruction by administering tests and tutoring when necessary.

ACKNOWLEDGMENTS

The authors wish to thank the following people who reviewed the manuscript and made many valuable comments and suggestions: Purl Dietzman, Lakeshore Technical College, Paul Eldersveld, College of DuPage, Roberta Hinkle Gansman, Guilford Technical Community College, and Gerald MacNab, Milwaukee Area Technical College. They also thank Arleen D'Amore who typed the camera-ready copy, Peggy McHale who prepared the drawings and made the corrections, Gail W. Davis who did the final proofreading, and Allan A. Christenson who prepared the index.

The assistance and cooperation of Stuart W. Johnson and others on the Addison-Wesley staff during the design and production of the texts is also appreciated.

ASSIGNMENTS FOR TECHNICAL MATHEMATICS I

Contents

Real Numbers

┌─────────┐
│   1   │
└─────────┘

In this chapter, we will define real numbers and discuss the basic operations with real numbers. We will define powers with integral exponents and discuss the laws of exponents, powers of ten, and scientific notation. We will discuss the proper order of operations for expressions involving more than one operation. We will also translate English phrases to algebraic expressions and do some evaluations with algebraic expressions and formulas.

1-1   REAL NUMBERS

In this section, we will define real numbers and the following subsets of real numbers: natural numbers, whole numbers, integers, rational numbers, and irrational numbers.

1. Any number that can be represented by a point on the number line is called a <u>real</u> <u>number</u>. For example, all of the numbers shown below are real numbers.

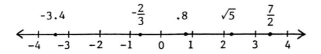

From the number line, you can see these facts:

   1. <u>Positive</u> numbers go to the right of 0.
   2. <u>Negative</u> numbers go to the left of 0.
   3. The number 0 is neither positive nor negative.

Continued on following page.

1. Continued

   Positive and negative numbers are called <u>signed</u> <u>numbers</u>.

   For <u>negative</u> numbers, we always use the - sign.

   "Negative 3" is <u>always</u> written -3.

   For <u>positive</u> numbers, we can use the + sign or no sign, but we usually use no sign.

   "Positive 5" is usually written 5 instead of +5.

   a) Instead of writing +9 for "positive 9", we usually write _____.

   b) Would it make sense to write either +0 or -0? _____

---

2. There are various subsets of the real numbers. Two subsets are:

   The set of <u>natural</u> <u>numbers</u> or <u>counting</u> <u>numbers</u>.

   $$1, 2, 3, 4, 5, 6, 7,...$$

   The set of <u>whole</u> <u>numbers</u>, which includes the natural numbers and 0.

   $$0, 1, 2, 3, 4, 5, 6, 7,...$$

   Which number is a whole number, but not a natural number? _____

   a) 9

   b) No. The number 0 <u>is</u> <u>not</u> a signed number.

---

3. Another subset of the real numbers is the set of <u>integers</u> which is shown below. Integers include the whole numbers plus negative numbers like -1, -2, -3, and so on.

   $$..., -3, -2, -1, 0, 1, 2, 3,...$$

   a) Is 4 both a whole number and an integer? _____

   b) Is 0 both a whole number and an integer? _____

   c) Is -5 both a whole number and an integer?

   0

---

4. Another subset of the real numbers is the set of <u>rational</u> <u>numbers</u>. A rational number is a number that can be expressed in the form $\frac{a}{b}$, where $\underline{a}$ and $\underline{b}$ are integers and $\underline{b}$ is not 0. All of the following are rational numbers because each is a division of integers with a non-zero denominator.

   $$\frac{1}{2} \qquad \frac{7}{6} \qquad \frac{-4}{5} \qquad \frac{9}{-2} \qquad \frac{256}{87}$$

   Any negative fraction is a rational number because it can be written as a division of integers with a non-zero denominator. For example:

   $$-\frac{5}{6} \text{ can be written } \frac{-5}{6} \text{ or } \frac{5}{-6}$$

   $$-\frac{13}{9} \text{ can be written } \underline{\quad} \text{ or } \underline{\quad}$$

   a) Yes

   b) Yes

   c) No. -5 is an integer, but not a whole number.

5. To convert a rational number to a decimal, we divide. Two examples are shown. Notice that we get either a terminating decimal (.625) or a nonterminating decimal with a repeating pattern (.3636...).

$$\frac{5}{8} = 8\overline{)5.000} \quad \begin{array}{r} .625 \\ -4\ 8 \\ \hline 20 \\ -16 \\ \hline 40 \\ -40 \\ \hline \end{array}$$

$$\frac{4}{11} = 11\overline{)4.0000} \quad \begin{array}{r} .3636... \\ -3\ 3 \\ \hline 70 \\ -66 \\ \hline 40 \\ -33 \\ \hline 70 \\ -66 \\ \hline 4 \end{array}$$

For non-terminating decimals with a repeating pattern, the three dots are often replaced by a bar over the repeating part. For example:

Instead of .3636..., we write .$\overline{36}$.

In applied problems, we usually round a repeating pattern to a specific place.

Rounded to thousandths, $\frac{4}{11}$ = _____

6. A fraction is a <u>proper</u> fraction if its numerator is <u>smaller than</u> its denominator.

$\frac{2}{3}$, $\frac{4}{11}$, and $\frac{1}{4}$ are <u>proper</u> fractions.

A fraction is an <u>improper</u> fraction if its numerator <u>is equal to or larger</u> than its denominator.

$\frac{7}{4}$, $\frac{15}{5}$, and $\frac{8}{8}$ are <u>improper</u> fractions.

When an improper fraction converts to a decimal, <u>the decimal is greater than or equal to</u> "1". An example is shown. Complete the other conversion. Round to hundredths.

$$\frac{9}{4} = 4\overline{)9.00} \quad \begin{array}{r} 2.25 \\ -8 \\ \hline 10 \\ -8 \\ \hline 20 \\ -20 \\ \hline \end{array} \qquad \frac{7}{6} =$$

7. Any decimal is a rational number because it can be converted to a fraction. For example:

$2.4 = \frac{24}{10}$ (<u>One</u> decimal place in 2.4; <u>one</u> 0 in 10.)

$.61 = \frac{61}{100}$ (<u>Two</u> decimal places in .61; <u>two</u> 0's in 100.)

Convert each decimal to a fraction.

a) 13.7 = _____   b) 4.19 = _____   c) .854 = _____

Answer column:

$\frac{-13}{9}$ or $\frac{13}{-9}$

.364

1.17

8. Any mixed number is a rational number because it can be converted to an improper fraction. Two examples are shown. (<u>Note</u>: In algebra, we usually use improper fractions instead of mixed numbers.)

$$1\frac{2}{5} = 1 + \frac{2}{5} \qquad\qquad 3\frac{1}{2} = 3 + \frac{1}{2}$$

$$= \frac{5}{5} + \frac{2}{5} \qquad\qquad = \frac{6}{2} + \frac{1}{2}$$

$$= \frac{7}{5} \qquad\qquad\qquad = \frac{7}{2}$$

Convert each mixed number to an improper fraction.

a) $1\frac{3}{4} =$ _____   b) $2\frac{1}{3} =$ _____   c) $4\frac{1}{2} =$ _____

a) $\dfrac{137}{10}$

b) $\dfrac{419}{100}$

c) $\dfrac{854}{1000}$

---

9. Any integer is a rational number because it can be written as a division of itself and "1". For example:

$$8 = \frac{8}{1} \qquad -3 = \frac{-3}{1} \qquad$$

a) 140 = _____   b) -79 = _____

a) $\dfrac{7}{4}$ b) $\dfrac{7}{3}$ c) $\dfrac{9}{2}$

---

10. <u>Rational</u> <u>numbers</u> are one major subset of the real numbers. <u>Irrational</u> <u>numbers</u> are the other major subset of the real numbers. The set of irrational numbers includes all real numbers that are not rational. Any nonterminating decimal <u>without</u> <u>a</u> <u>repeating</u> <u>pattern</u> is an irrational number. For example, the number π is an irrational number.

$$\pi = 3.1415926535...$$

State whether each number is rational or irrational.

a) 0.125 (terminating)

b) 3.424242... (repeating pattern)

c) 4.030030003... (pattern does not repeat)

a) $\dfrac{140}{1}$ b) $\dfrac{-79}{1}$

---

11. When a number is a perfect square, its square root is rational because it is an integer. For example:

$$\sqrt{36} \text{ is rational, because } \sqrt{36} = 6$$

When a number is not a perfect square, its square root is irrational because it is a nonterminating, nonrepeating decimal.

$$\sqrt{15} \text{ is irrational, because } \sqrt{15} = 3.8729833...$$

State whether each number is rational or irrational.

a) $\sqrt{3}$ b) $\sqrt{16}$ c) $\sqrt{49}$ d) $\sqrt{88}$

a) rational

b) rational

c) irrational

12. Some subsets of the real numbers are listed below.

a) irrational

b) rational

c) rational

d) irrational

<u>Subsets</u> <u>Of</u> <u>The</u> <u>Real</u> <u>Numbers</u>

Natural numbers or counting numbers	1,2,3,4,5,6,7,...
Whole numbers	0,1,2,3,4,5,6,7,...
Integers	...,-3,-2,-1,0,1,2,3,...
Rational numbers	Expressed as $\frac{a}{b}$, where <u>a</u> and <u>b</u> are integers, and <u>b</u> is not 0.
Irrational numbers	A real number that is not rational.

The diagram below shows that real numbers are either rational or irrational.

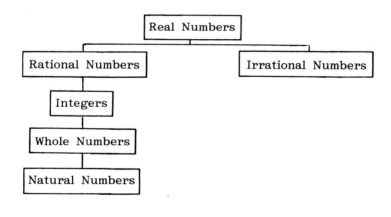

1-2   ORDER AND ABSOLUTE VALUE

In this section, we will define the inequality symbols and absolute value.

13. The symbol = is used for <u>equalities</u>. The symbol ≠ is used for <u>inequalities</u>. The two symbols are defined below. Notice that the slash line in ≠ means "not".

> = means "is equal to".
>
> ≠ means "is not equal to".

Therefore: 5 = 5 is read "5 is equal to 5".

4 ≠ 7 is read "4 is not equal to 7".

Continued on following page.

13.   Continued

In an inequality, one number must be either <u>greater</u> <u>than</u> the other or <u>less</u> <u>than</u> the other. The two symbols > and < are used for "is greater than" and "is less than". That is:

> means "is greater than".
< means "is less than".

Therefore: 7 > 4 is read "7 is greater than 4".

4 < 7 is read "4 is less than 7".

Write either > or < in each blank.

a) 6___9   b) 10___4   c) 1___5

---

14.   The definitions of "is greater than" and "is less than" in terms of the number line are given below.

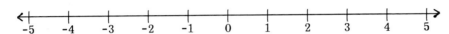

One number <u>is</u> <u>greater</u> <u>than</u> another if it is <u>to</u> <u>the</u> <u>right</u> of the other on the number line. That is:

a > b if <u>a</u> is to the right of <u>b</u>.

Therefore: since 4 is to the right of -1, 4 > -1.

since -2 is to the right of -5, -2 > -5.

One number <u>is</u> <u>less</u> <u>than</u> another if it is <u>to</u> <u>the</u> <u>left</u> of the other on the number line. That is:

a < b if <u>a</u> is to the left of <u>b</u>.

Therefore: since -2 is to the left of 3, -2 < 3.

since -4 is to the left of 0. -4 < 0.

Write either > or < in each blank.

a) 7___-4   d) 0___-7

b) -5___0   e) -6___-2

c) -8___9   f) -4.5___-9.5

a) 6 < 9

b) 10 > 4

c) 1 < 5

a) 7 > -4   d) 0 > -7

b) -5 < 0   e) -6 < -2

c) -8 < 9   f) -4.5 > -9.5

15. The symbols ≥ and ≤ are defined below.

> ≥ means "is greater than or equal to".
>
> ≤ means "is less than or equal to".

Inequalities containing ≥ or ≤ can be either true or false. For example:

   4 ≥ 2 is true, since 4 > 2 is true.

   -3 ≤ -3 is true, since -3 = -3 is true.

   0 ≤ -4 is false, since both 0 < -4 and 0 = -4 are false.

Answer either <u>true</u> or <u>false</u> for these.

   a) 2 ≥ 2   b) 3 ≤ 0   c) -4 ≤ 0   d) -5.2 ≥ -4.6

---

16. The <u>absolute value</u> of a number is its distance from 0 on the number line, with no regard for direction. The symbol | | is used for absolute value.

a) true

b) false

c) true

d) false

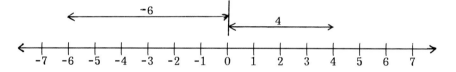

The absolute values of 4 and -6 are shown on the number line above.

   Since 4 is 4 units from 0, |4| = 4

   Since -6 is 6 units from 0, |-6| = 6

Write the absolute value of each number.

   a) |10| = _____   c) |-1| = _____

   b) |$\frac{3}{5}$| = _____   d) |-9.7| = _____

---

17. Since both 3 and -3 are 3 units from 0, they have the same absolute value. That is:

a) 10   c) 1

b) $\frac{3}{5}$ d) 9.7

$$|3| = 3 \text{ and } |-3| = 3$$

The absolute value of 0 is 0. Complete these:

   a) |-12| = _____   c) |0| = _____

   b) |12| = _____   d) |-45| = _____

---

a) 12   c) 0

b) 12   d) 45

1-3 ADDITION AND SUBTRACTION

In this section, we will discuss the procedures for adding and subtracting real numbers. We will also define additive inverses or opposites.

18.   The following two rules are used to add real numbers.

```
         Rules For Adding Real Numbers

  1. When the numbers have the same sign:
     a) Add their absolute values.
     b) Give the sum the same sign as both numbers.

  2. When the numbers have different signs:
     a) Subtract their absolute values.
     b) Give the sum the same sign as the number
        with the larger absolute value.
```

An example of the first rule is shown below.

  $-2 + (-4) = -6$

    ——Sum of absolute values $(2 + 4 = 6)$
    ——Same sign as both numbers.

Two examples of the second rule are shown below.

  $5 + (-2) = +3$

    —— Difference of absolute values $(5 - 2 = 3)$
    —— Sign of number with the larger absolute
      value (5)

  $-9 + 5 = -4$

    —— Difference of absolute values $(9 - 5 = 4)$
    —— Sign of number with the larger absolute
      value (-9)

Use the rules for these:

 a) $-3 + (-7) =$ _____   c) $4 + (-9) =$ _____

 b) $-10 + (-2) =$ _____   d) $-3 + 8 =$ _____

---

19.   Complete: a) $-40 + 10 =$ _____ c) $-20 + (-70) =$ _____   a) -10 c) -5

       b) $80 + (-55) =$ _____ d) $-30 + 90 =$ _____   b) -12 d) 5

---

                     a) -30 c) -90

                     b) 25   d) 60

20. The same rules are used to add fractions. For example:

$$-\frac{2}{5} + \left(-\frac{4}{5}\right) = -\frac{6}{5}$$

$$-\frac{5}{8} + \frac{1}{8} = -\frac{4}{8} = -\frac{1}{2}$$

Add. Reduce each sum to lowest terms.

  a) $-\frac{1}{4} + \left(-\frac{1}{4}\right) = $ _____   b) $\frac{5}{2} + \left(-\frac{1}{2}\right) = $ _____

a) $-\frac{1}{2}$ b) 2

21. To add fractions with unlike denominators, we must get common denominators first. We can use the concept of <u>multiples</u> to find the lowest common denominator (LCD). The multiples of a number can be obtained by counting by the number. For example:

  <u>Multiples</u> <u>of</u> <u>3</u>:   3, 6, 9, 12, 15, 18, 21, 24, 27,...
  <u>Multiples</u> <u>of</u> <u>8</u>:   8, 16, 24, 32, 40, 48, 56, 64, 72,...

When the larger denominator is a multiple of the smaller, the LCD is the larger denominator. For example:

$$\frac{5}{6} + \left(-\frac{1}{3}\right) = \frac{5}{6} + \left(-\frac{2}{6}\right) = \frac{3}{6} = \frac{1}{2}$$

$$-\frac{3}{4} + \frac{3}{8} = -\frac{6}{8} + \frac{3}{8} = -\frac{3}{8}$$

Add. Reduce each sum to lowest terms.

  a) $-\frac{1}{4} + \left(-\frac{1}{2}\right) = $ _____   b) $-\frac{9}{10} + \frac{1}{2} = $ _____

a) $-\frac{3}{4}$ b) $-\frac{2}{5}$

22. To find the LCD when the larger denominator is not a multiple of the smaller, we can <u>test</u> <u>each</u> <u>multiple</u> <u>of</u> <u>the</u> <u>larger</u> <u>denominator</u> until we find the smallest one that is also a multiple of the smaller denominator. An example is discussed.

  For $\frac{5}{6} + \left(-\frac{3}{8}\right)$, we tested the multiples of 8 below.

      Is 8 a multiple of 6?   No.
      Is 16 a multiple of 6?   No.
      Is 24 a multiple of 6?   Yes.

  Therefore, 24 is the LCD. We get:

$$\frac{5}{6} + \left(-\frac{3}{8}\right) = \frac{20}{24} + \left(-\frac{9}{24}\right) = \frac{11}{24}$$

Add. Reduce each sum to lowest terms.

  a) $-\frac{1}{2} + \left(-\frac{1}{5}\right) = $ _____   b) $-\frac{1}{6} + \frac{1}{10} = $ _____

a) $-\frac{7}{10}$ b) $-\frac{1}{15}$

23. To add a fraction and an integer, we convert the integer to a fraction with the same denominator. For example:

$$3 + \left(-\frac{1}{2}\right) = \frac{6}{2} + \left(-\frac{1}{2}\right) = \frac{5}{2}$$

$$-1 + \frac{5}{8} = -\frac{8}{8} + \frac{5}{8} = -\frac{3}{8}$$

Complete these:

a) $1 + \left(-\frac{5}{6}\right) =$ _____   b) $-\frac{4}{3} + (-2) =$ _____

---

24. The same rules are used to add decimals. For example:

$$-10.2 + (-20.5) = -30.7$$
$$-.7 + .55 = -.70 + .55 = -.15$$

Complete these:

a) $-6.6 + (-9.9) =$ _____   c) $-.18 + .129 =$ _____

b) $5.6 + (-3.1) =$ _____   d) $1 + (-.4067) =$ _____

a) $\frac{1}{6}$ b) $-\frac{10}{3}$

---

25. To add three or more numbers, we can add "two at a time" from left to right as we have done below.

$$\underline{3 + (-5)} + 6 + (-8) =$$
$$\underline{-2 \quad + 6} + (-8) =$$
$$4 \quad + (-8) = -4$$

By adding "two at a time", find each sum.

a) $-5 + (-4) + 10 =$ _____   b) $20 + (-10) + (-30) + 5 =$ _____

a) -16.5   c) -.051

b) 2.5   d) .5933

---

26. Two numbers that are the same distance from 0 on the number line, but on opposite sides of 0, are called <u>additive inverses</u> or <u>opposites</u> of each other. For example:

-8 is the additive inverse of 8.

5 is the additive inverse of -5.

The additive inverse of any number <u>a</u> is <u>-a</u>. That is:

The additive inverse of 7 is -7.

The additive inverse of -3 is -(-3) = 3.

Write the additive inverse of each number.

a) -10 _____   b) 17 _____   c) $\frac{3}{4}$ _____   d) -6.5 _____

a) 1   b) -15

27. When two additive inverses are added, their sum is 0. That is:

$$4 + (-4) = 0 \quad \text{a) } -1.75 + 1.75 = \underline{\hspace{1cm}} \quad \text{b) } \frac{7}{8} + \left(-\frac{7}{8}\right) = \underline{\hspace{1cm}}$$

a) 10   c) $-\frac{3}{4}$

b) -17   d) 6.5

---

28. If we subtract 2 from 5, we get 3. That is, 5 - 2 = 3. We also get 3 as an answer if we add -2 to 5. That is, 5 + (-2) = 3. Therefore:

$$5 - 2 = 5 + (-2)$$

Based on the example above, we can define subtraction for any real numbers <u>a</u> and <u>b</u>. Notice that subtracting <u>b</u> from <u>a</u> is the same as adding the additive inverse of <u>b</u> to <u>a</u>.

$$\boxed{a - b = a + (-b)}$$

Therefore, to perform a subtraction, we convert to an equivalent addition and then add. To convert to addition, we <u>ADD</u> <u>THE</u> <u>ADDITIVE</u> <u>INVERSE</u> <u>OF</u> <u>THE</u> <u>SECOND</u> <u>NUMBER</u>. For example:

```
          ┌──── Change - to +
          │  ┌──── Additive inverse of 7
          ↓  ↓
4 - 7 = 4 + (-7) = -3
```

Convert to an equivalent addition and then add.

a) 4 - 10 = \underline{\hspace{4cm}} = \underline{\hspace{2cm}}

b) -6 - 1 = \underline{\hspace{4cm}} = \underline{\hspace{2cm}}

c) -1 - 3 = \underline{\hspace{4cm}} = \underline{\hspace{2cm}}

a) 0   b) 0

---

29. To perform the subtraction below, we also <u>ADDED</u> <u>THE</u> <u>ADDITIVE</u> <u>INVERSE</u> <u>OF</u> <u>THE</u> <u>SECOND</u> <u>NUMBER</u>.

```
          ┌──── Change - to +
          │  ┌──── Additive inverse of -3
          ↓  ↓
6 - (-3) = 6 + 3 = 9
```

Convert to an equivalent addition and then add.

a) 5 - (-2) = \underline{\hspace{4cm}} = \underline{\hspace{2cm}}

b) -1 - (-6) = \underline{\hspace{4cm}} = \underline{\hspace{2cm}}

c) -9 - (-3) = \underline{\hspace{4cm}} = \underline{\hspace{2cm}}

a) 4 + (-10) = -6

b) -6 + (-1) = -7

c) -1 + (-3) = -4

---

a) 5 + 2 = 7

b) -1 + 6 = 5

c) -9 + 3 = -6

30. For each subtraction below, we converted to an equivalent addition and then added.

$$\frac{3}{8} - \frac{7}{8} = \frac{3}{8} + \left(-\frac{7}{8}\right) = -\frac{4}{8} = -\frac{1}{2}$$

$$-\frac{1}{4} - \frac{3}{2} = -\frac{1}{4} + \left(-\frac{3}{2}\right) = -\frac{1}{4} + \left(-\frac{6}{4}\right) = -\frac{7}{4}$$

Use the same method for these.

a) $-\frac{1}{5} - \frac{9}{5} =$ _____

b) $\frac{1}{6} - \frac{1}{2} =$ _____

---

31. For each subtraction below, we converted to an equivalent addition and then added.

$$-\frac{5}{4} - \left(-\frac{1}{4}\right) = -\frac{5}{4} + \frac{1}{4} = -\frac{4}{4} = -1$$

$$\frac{2}{3} - \left(-\frac{1}{5}\right) = \frac{2}{3} + \frac{1}{5} = \frac{10}{15} + \frac{3}{15} = \frac{13}{15}$$

Use the same method for these:

a) $\frac{11}{16} - \left(-\frac{3}{16}\right) =$ _____

b) $-\frac{5}{6} - \left(-\frac{3}{4}\right) =$ _____

a) -2

b) $-\frac{1}{3}$

---

32. Following the examples, complete the other subtractions.

$$\frac{3}{4} - 1 = \frac{3}{4} - \frac{4}{4} = \frac{3}{4} + \left(-\frac{4}{4}\right) = -\frac{1}{4}$$

$$-2 - \left(-\frac{2}{3}\right) = -2 + \frac{2}{3} = -\frac{6}{3} + \frac{2}{3} = -\frac{4}{3}$$

a) $-\frac{2}{5} - 1 =$ _____

b) $3 - \left(-\frac{1}{2}\right) =$ _____

a) $\frac{7}{8}$

b) $-\frac{1}{12}$

---

33. Following the examples, complete the other subtractions.

$$1.4 - 8.5 = 1.4 + (-8.5) = -7.1$$

$$-.25 - (-.32) = -.25 + .32 = .07$$

a) $-3.7 - 9.9 =$ _____

b) $1 - (-.55) =$ _____

a) $-\frac{7}{5}$

b) $\frac{7}{2}$

---

a) -13.6   b) 1.55

rational

SELF-TEST 1 (pages 1-13)

Given this set of numbers: 13, $\sqrt{7}$, -0.6, 0, $\dfrac{9}{7}$, $-\sqrt{5}$, -2, $-\dfrac{3}{8}$, 5.2

1. List the integers. _13, 0, -2_   2. List the irrational numbers. _$\sqrt{7}$, $-\sqrt{5}$_

Write either > or < in each blank. | State whether each is true or false.

3. -8 _<_ -1   4. 0 _>_ -2   5. -7 $\geq$ 3 _F_   6. 8 $\leq$ 0 _F_

Write the absolute values of: 7. |4| _4_   8. $\left|-\dfrac{3}{5}\right|$ _$\dfrac{3}{5}$_

Do these additions.

9. -5 + (-7) = _-12_   11. -2 + 6 + (-1) = _3_   13. $3 + \dfrac{2}{5}$ = $3\dfrac{2}{5} = \dfrac{17}{5}$

10. 8 + (-12) = _-4_   12. $-\dfrac{1}{4} + \dfrac{1}{6}$ = _$-\dfrac{1}{12}$_   14. 1 + (-.27) = _.73_

Write the additive inverse of: 15. -36 _36_   16. $\dfrac{3}{8}$ _$-\dfrac{3}{8}$_

Do these subtractions.   Report fraction answers in lowest terms.

17. 9 - 15 = _-6_   19. $-\dfrac{1}{5} - \dfrac{1}{2}$ = $-\dfrac{7}{10}$ 21. -3.2 - 6.4 = _-9.6_

18. -1 - (-3) = _2_   20. $1 - \left(-\dfrac{2}{3}\right)$ = $\dfrac{5}{3}$ 22. $\dfrac{7}{10} - \dfrac{5}{6}$ = $-\dfrac{2}{15}$

ANSWERS:

1. 13, 0, -2	7. 4	13. $\dfrac{17}{5}$	19. $-\dfrac{7}{10}$
2. $\sqrt{7}$, $-\sqrt{5}$	8. $\dfrac{3}{5}$	14. .73	20. $\dfrac{5}{3}$
3. <	9. -12	15. 36	21. -9.6
4. >	10. -4	16. $-\dfrac{3}{8}$	22. $-\dfrac{2}{15}$
5. False	11. 3	17. -6	
6. False	12. $-\dfrac{1}{12}$	18. 2	

1-4   MULTIPLICATION AND DIVISION

In this section, we will discuss the procedures for multiplying and dividing real numbers. We will also define reciprocals or multiplicative inverses.

34.   In a multiplication, the numbers multiplied are called <u>factors</u>; the answer is called the <u>product</u>.   The following two rules are used to multiply real numbers.

> <u>Rules</u> <u>For</u> <u>Multiplying</u> <u>Real</u> <u>Numbers</u>
>
> 1. When the factors have the <u>same</u> sign, their product is <u>positive</u>.
>
> 2. When the factors have <u>different</u> signs, their product is <u>negative</u>.

Two examples of the first rule are shown below.

$$5(4) = 20 \qquad\qquad (-10)(-7) = 70$$

Two examples of the second rule are shown below.

$$3(-8) = -24 \qquad\qquad (-9)(2) = -18$$

Use the rules for these:

a) $8(-7) =$ _____   c) $(-2)(40) =$ _____

b) $(-6)(-9) =$ _____   d) $(-20)(-5) =$ _____

35.   We used the same rules for the multiplications below.

$$\left(-\frac{2}{3}\right)\left(-\frac{1}{5}\right) = \frac{2}{15} \qquad\qquad 4\left(-\frac{1}{6}\right) = -\frac{4}{6} = -\frac{2}{3}$$

$$-8(-1.2) = 9.6 \qquad\qquad (-.9)(7.5) = -6.75$$

Complete these:

a) $\left(-\frac{5}{2}\right)\left(\frac{2}{7}\right) =$ _____   c) $\left(-\frac{3}{5}\right)(-2) =$ _____

b) $(-7)(-.21) =$ _____   d) $10(-67.5) =$ _____

a) -56   c) -80

b) 54   d) 100

a) $-\frac{5}{7}$ c) $\frac{6}{5}$

b) 1.47   d) -675

See 15 at top.

36. The rules for dividing real numbers are similar to the rules for multiplying real numbers.

> **Rules For Dividing Real Numbers**
>
> 1. When the numbers have the <u>same</u> sign, their quotient is <u>positive</u>.
>
> 2. When the numbers have <u>different</u> signs, their quotient is <u>negative</u>.

Two examples of the first rule are shown below.

$$\frac{35}{7} = 5 \qquad\qquad \frac{-12}{-4} = 3$$

Two examples of the second rule are shown below.

$$\frac{-32}{8} = -4 \qquad\qquad \frac{14}{-2} = -7$$

Complete these:

a) $\frac{-10}{5} =$ _____ b) $\frac{-56}{-7} =$ _____ c) $\frac{28}{-4} =$ _____

---

37. When a division is done correctly, the product of the denominator and the quotient equals the numerator. For example:

$$\frac{12}{3} = 4 \ , \quad \text{since} \quad 3(4) = 12$$

The fact above can be used to justify the rules for dividing real numbers. That is:

$$\frac{-20}{4} = -5, \ \text{since} \ 4(-5) = -20 \qquad \frac{-18}{-6} = 3, \ \text{since} \ -6(3) = \underline{\quad}$$

a) -2

b) 8

c) -7

---

38. Two numbers whose product is "1" are called <u>reciprocals</u> or <u>multiplicative inverses</u> of each other. For example:

-18

Since $(3)\left(\frac{1}{3}\right) = 1$:   The reciprocal of 3 is $\frac{1}{3}$.

The reciprocal of $\frac{1}{3}$ is 3.

Since $\left(-\frac{2}{7}\right)\left(-\frac{7}{2}\right) = 1$:   The reciprocal of $-\frac{2}{7}$ is $-\frac{7}{2}$.

The reciprocal of $-\frac{7}{2}$ is $-\frac{2}{7}$.

Write the reciprocals of these:

a) 10 _____ b) -4 _____ c) $-\frac{1}{6}$ _____ d) $\frac{5}{3}$ _____

---

a) $\frac{1}{10}$ b) $-\frac{1}{4}$ c) -6   d) $\frac{3}{5}$

39.   There are two numbers that are their own reciprocals.

Since $(1)(1) = 1$ the reciprocal of "1" is "1".

Since $(-1)(-1) = 1$, the reciprocal of -1 is -1.

The number 0 has no reciprocal.   To see that fact, answer these:

a) When one factor is 0, the product is always _____.

b) Can we get "1" as a product when one factor is 0? _____

c) Therefore, does 0 have a reciprocal? _____

---

40.   Any division is the same as <u>multiplying</u> <u>the</u> <u>numerator</u> <u>by</u> <u>the</u> <u>reciprocal</u> <u>of</u> <u>the</u> <u>denominator</u>.   Below, for example, we multiplied 12 by $\frac{1}{3}$, the reciprocal of 3.

$$\frac{12}{3} = 12\left(\frac{1}{3}\right) = 4$$

Complete these:

a) $\frac{-16}{8} = -16\left(\frac{1}{8}\right) =$ _____   b) $\frac{-35}{-5} = -35\left(-\frac{1}{5}\right) =$ _____

a) 0

b) No

c) No

---

41.   To divide fractions, we <u>multiply</u> <u>the</u> <u>numerator</u> <u>by</u> <u>the</u> <u>reciprocal</u> <u>of</u> <u>the</u> <u>denominator</u>.   Below, for example, we multiplied $\frac{3}{7}$ by $\frac{9}{4}$, the reciprocal of $\frac{4}{9}$.

$$\frac{\frac{3}{7}}{\frac{4}{9}} = \frac{3}{7}\left(\frac{9}{4}\right) = \frac{27}{28}$$

The same rules for signs apply to divisions of fractions.   That is:

$$\frac{-\frac{2}{3}}{\frac{7}{6}} = \left(-\frac{2}{3}\right)\left(\frac{6}{7}\right) = -\frac{12}{21} = -\frac{4}{7}$$

Complete.   Reduce each quotient to lowest terms.

a) $\dfrac{-\frac{1}{5}}{-\frac{3}{2}} =$ _____   b) $\dfrac{\frac{9}{8}}{-\frac{3}{4}} =$ _____

a) -2   b) 7

---

a) $\frac{2}{15}$ b) $-\frac{3}{2}$

42. In each division below, one term is an integer. In the top division, the denominator is -6. In the bottom division, the numerator is 2.

$$\frac{\frac{3}{4}}{-6} = \frac{3}{4}\left(-\frac{1}{6}\right) = -\frac{3}{24} = -\frac{1}{8}$$

$$\frac{2}{-\frac{3}{7}} = 2\left(-\frac{7}{3}\right) = -\frac{14}{3}$$

Complete. Reduce each quotient to lowest terms.

a) $\dfrac{-\frac{2}{3}}{-8} =$ _____

b) $\dfrac{-4}{\frac{1}{2}} =$ _____

---

43. Two special types of division are given below.

1. When a non-zero number is divided by itself, the quotient is "1".

$$\frac{9}{9} = 1 \qquad\qquad \frac{-\frac{7}{2}}{-\frac{7}{2}} = 1$$

2. When a number is divided by "1", the quotient is <u>identical</u> to the number.

$$\frac{-8}{1} = -8 \qquad\qquad \frac{\frac{3}{5}}{1} = \frac{3}{5}$$

Complete these:

a) $\dfrac{-7}{-7} =$ ____

b) $\dfrac{4}{1} =$ ____

c) $\dfrac{\frac{5}{8}}{\frac{5}{8}} =$ ____

d) $\dfrac{-\frac{12}{11}}{1} =$ ____

a) $\frac{1}{12}$ b) -8

a) 1   b) 4   c) 1   d) $-\frac{12}{11}$

1-5 OPERATIONS WITH ZERO

In this section, we will discuss operations involving the number 0.

44. In each example below, 0 is added to a number. The sum is <u>identical</u> to the other number.

$$8 + 0 = 8 \qquad\qquad 0 + (-5) = -5$$

Using the fact above, complete these:

a) $-7.6 + 0 =$ _____

b) $0 + \dfrac{3}{4} =$ _____

45. In each example below, a number is subtracted from 0.

$$0 - 7 = 0 + (-7) = -7$$
$$0 - (-3) = 0 + 3 = 3$$

In each example below, 0 is subtracted from a number.   The answer is identical to the number.

$$4 - 0 = 4 \qquad\qquad -8 - 0 = -8$$

Following the examples, complete these:

   a) $0 - \dfrac{2}{3} =$ _____   c) $-\dfrac{5}{4} - 0 =$ _____

   b) $17.5 - 0 =$ _____   d) $0 - (-3.96) =$ _____

a) -7.6   b) $\frac{3}{4}$

46. When any number is multiplied by 0, the product is 0.   For example:

$$7(0) = 0 \qquad\qquad 0(-9) = 0$$

Using the fact above, complete these:

   a) $\dfrac{1}{2}(0) =$ _____   b) $0(-6.4) =$ _____

a) $-\frac{2}{3}$ c) $-\frac{5}{4}$
b) 17.5   d) 3.96

47. When a division is done correctly, the product of the denominator and the quotient equals the numerator.   For example:

$$\frac{12}{3} = 4, \quad \text{since} \quad 3(4) = 12$$

When 0 is divided by any other number, the quotient is 0.   We can use the relationship above to see that fact:

$$\frac{0}{5} = 0, \quad \text{since} \quad 5(0) = 0$$

$$\frac{0}{-9} = 0, \quad \text{since} \quad -9(0) = 0$$

Using the fact above, complete these:

   a) $\dfrac{0}{-5.9} =$ _____   b) $\dfrac{0}{\frac{1}{4}} =$ _____

a) 0   b) 0

48. If we divide 5 by 0, we must get a quotient such that 0 times the quotient equals 5.   But 0 times any number is 0, not 5.

$$\frac{5}{0} = ?$$

Therefore, <u>division</u> <u>of</u> <u>any</u> <u>other</u> <u>number</u> <u>by</u> <u>0</u> <u>is</u> IMPOSSIBLE <u>or</u> <u>UNDEFINED</u>.   Perform each possible division below.

   a) $\dfrac{0}{1} =$ _____   b) $\dfrac{1}{0} =$ _____   c) $\dfrac{0}{-1.88} =$ _____   d) $\dfrac{-\frac{3}{4}}{0} =$ _____

a) 0   b) 0

49.   If we divide 0 by 0, we could use any number as the quotient since 0 times any number is 0. For example:

$$\frac{0}{0} = 7, \qquad \text{since} \qquad 0(7) = 0$$

$$\frac{0}{0} = -4, \qquad \text{since} \qquad 0(-4) = 0$$

Since we could use any number as the quotient, we say that division of 0 by 0 <u>cannot</u> <u>be</u> <u>determined</u>. Answer "0", "impossible", or "indeterminate" for these:

a) $\frac{12}{0}$ = _____   b) $\frac{0}{0}$ = _____   c) $\frac{0}{12}$ = _____   d) $\frac{-1}{0}$ = _____

a) 0

b) impossible

c) 0

d) impossible

a) impossible   b) indeterminate   c) 0   d) impossible

1-6   PROPERTIES OF ADDITION AND MULTIPLICATION

In this section, we will discuss the commutative, associative, and identity properties of addition and multiplication.

50.   If we interchange the two numbers in an addition, we get the same sum. That is:

$$10 + (-15) = -15 + 10 \qquad \text{(Both equal -5.)}$$

The property above is called the <u>commutative</u> <u>property</u> <u>of</u> <u>addition</u>. The commutative property is stated for any numbers <u>a</u> and <u>b</u> below.

> Commutative Property Of Addition
>
> a + b = b + a

Using the commutative property, complete these:

a) -8 + 7 = 7 + (-8)   b) $\frac{1}{2} + \left(-\frac{1}{4}\right) = \left(\quad \right) + \frac{1}{2}$

a) 7 + (-8)

b) $\left(-\frac{1}{4}\right) + \frac{1}{2}$

51. If we interchange the two numbers in a multiplication, we get the same product. That is:

$$3(-4) = (-4)(3) \qquad \text{(Both equal -12.)}$$

The property above is called the <u>commutative</u> <u>property</u> <u>of</u> <u>multiplica-</u><u>tion</u>. The commutative property is stated for any numbers <u>a</u> and <u>b</u> below.

> <u>Commutative</u> <u>Property</u> <u>Of</u> <u>Multiplication</u>
>
> $a \cdot b = b \cdot a$

Using the commutative property, complete these:

a) $(-5)(6) = (\quad)(\quad)$ b) $-\frac{2}{3}(-7) = (\quad)\left(\quad\right)$

---

52. Each addition below contains the same three numbers. The parentheses ( ) are grouping symbols which are used to show which two numbers are to be added first. We get the same sum both ways.

$$\underbrace{(5 + 3)}_{} + 6 \qquad\qquad 5 + \underbrace{(3 + 6)}_{}$$
$$8 \quad + 6 = 14 \qquad\qquad 5 + \quad 9 \quad = 14$$

The property above is called the <u>associative</u> <u>property</u> <u>of</u> <u>addition</u>. The associative property is stated for any numbers <u>a</u>, <u>b</u>, and <u>c</u> below.

> <u>Associative</u> <u>Property</u> <u>Of</u> <u>Addition</u>
>
> $(a + b) + c = a + (b + c)$

Using the associative property, complete these:

a) $(-3 + 5) + 9 = -3 + (\underline{\quad} + 9)$

b) $[-6 + (-7)] + (-1) = -6 + [(-7) + (\underline{\quad})]$

---

53. For $4(-2)(5)$, we get the same product whether we multiply 4 and -2 or -2 and 5 first. That is:

$$\underbrace{4(-2)}_{}(5) \qquad\qquad 4\underbrace{(-2)(5)}_{}$$
$$(-8)(5) = -40 \qquad\qquad 4\,(-10) \; = -40$$

The property above is called the <u>associative</u> <u>property</u> <u>of</u> <u>multiplica-</u><u>tion</u>. The associative property is stated for any numbers <u>a</u>, <u>b</u>, and <u>c</u> below.

> <u>Associative</u> <u>Property</u> <u>Of</u> <u>Multiplication</u>
>
> $(a \cdot b) \cdot c = a \cdot (b \cdot c)$

Continued on following page.

Answers (right column):

a) $(6)(-5)$

b) $(-7)\left(-\frac{2}{3}\right)$

a) $-3 + (\underline{5} + 9)$

b) $-6 + [(-7) + (\underline{-1})]$

53. Continued

Using the associative property, complete these:

    a) $(-4 \cdot 5) \cdot 6 = -4 \cdot (5 \cdot \underline{\hphantom{xx}})$

    b) $3 \cdot (-5 \cdot -9) = (3 \cdot \underline{\hphantom{xx}}) \cdot -9$

---

54. When 0 is one term in an addition, the sum is <u>identical</u> to the other term. For example:

$$8 + 0 = 8 \qquad\qquad 0 + (-5) = -5$$

The property above is called the <u>identity</u> <u>property</u> <u>of</u> <u>addition</u>. The number 0 is called the <u>identity</u> <u>element</u> <u>for</u> <u>addition</u>. The identity property is stated for any number <u>a</u> below.

> **Identity <u>Property</u> <u>Of</u> <u>Addition</u>**
>
> $a + 0 = a$ and $0 + a = a$

Using the identity property, complete these:

    a) $6.5 + 0 = \underline{\hphantom{xxxx}}$ b) $0 + \left(-\frac{2}{3}\right) = \underline{\hphantom{xxxx}}$

---

55. When "1" is one factor in a multiplication, the product is <u>identical</u> to the other factor. For example:

$$1(5) = 5 \qquad\qquad (-3)(1) = -3$$

The property above is called the <u>identity</u> <u>property</u> <u>of</u> <u>multiplication</u>. The number "1" is called the <u>identity</u> <u>element</u> <u>for</u> <u>multiplication</u>. The identity property is stated for any number <u>a</u> below.

> **Identity <u>Property</u> <u>Of</u> <u>Multiplication</u>**
>
> $a \cdot 1 = a$ and $1 \cdot a = a$

Using the identity property, complete these:

    a) $1(-3.44) = \underline{\hphantom{xxxx}}$ b) $\frac{5}{3}(1) = \underline{\hphantom{xxxx}}$

Answers (right column):

a) $-4 \cdot (5 \cdot \underline{6})$

b) $(3 \cdot \underline{-5}) \cdot -9$

a) 6.5 b) $-\frac{2}{3}$

a) -3.44 b) $\frac{5}{3}$

56. The properties discussed in this section are summarized below.

> ### Properties Of Real Numbers
>
> For any real numbers a, b, and c:
>
> | Commutative properties | $a + b = b + a$ |
> |---|---|
> | | $ab = ba$ |
> | Associative properties | $(a + b) + c = a + (b + c)$ |
> | | $(ab)c = a(bc)$ |
> | Identity properties | $a + 0 = a$ and $0 + a = a$ |
> | | $a \cdot 1 = a$ and $1 \cdot a = a$ |

Using either "commutative", "associative", or "identity", identify the property.

a) $4(-7) = -7(4)$ _____Com_____

b) $0 + (-1) = -1$ _____Ident_____

c) $(3 \cdot 4)8 = 3(4 \cdot 8)$ _____ass_____

d) $(-5 + 3) + 4 = -5 + (3 + 4)$ _____ass_____

e) $-3 + 2 = 2 + (-3)$ _____Comm_____

f) $1(-7) = -7$ _____Indit_____

a) commutative	c) associative	e) commutative
b) identity	d) associative	f) identity

1-7 INTEGRAL EXPONENTS

In this section, we will define powers whose exponents are integers.

57. Exponential form is a short way of writing a multiplication of identical factors. For example:

$$3 \cdot 3 \cdot 3 \cdot 3 = 3^4$$

In 3^4, 3 is called the base; 4 is called the exponent. The exponent 4 means that 3 appears as a factor 4 times. Two more examples are:

$$8 \cdot 8 \cdot 8 = 8^3$$

$$(-2)(-2)(-2)(-2)(-2)(-2) = (-2)^6$$

Write each multiplication in exponential form.

a) $9 \cdot 9 = $ _____

b) $\left(-\frac{1}{3}\right)\left(-\frac{1}{3}\right)\left(-\frac{1}{3}\right)\left(-\frac{1}{3}\right)\left(-\frac{1}{3}\right) = $ _____

58. Any exponential expression is called a <u>power</u> of the base. The names of some powers are given below.

      5^2 is called "5 <u>squared</u>" or "5 to the <u>second</u> power".

      7^3 is called "7 <u>cubed</u>" or "7 to the <u>third</u> power".

      $(-4)^6$ is called "-4 to the <u>sixth</u> power".

Write the exponential expression for these:

    a) -8.5 cubed = _____   b) 10 to the fifth power = _____

| a) 9^2 b) $\left(-\frac{1}{3}\right)^5$ |

---

59. We evaluated 5^3 below. The exponent 3 tells us to use 5 as a factor three times.

$$5^3 = (5)(5)(5) = 125$$

Evaluate each power.

    a) 3^4 = _____   b) $\left(\frac{1}{2}\right)^5$ = _____

a) $(-8.5)^3$ b) 10^5

---

60. When the base is a negative number, the value of a power is positive for even exponents and negative for odd exponents. For example:

$$(-4)^4 = (-4)(-4)(-4)(-4) = 256$$

$$(-2)^5 = (-2)(-2)(-2)(-2)(-2) = -32$$

Evaluate each power.

    a) $\left(\frac{3}{4}\right)^2$ = _____   b) $(-4)^3$ = _____   c) $(-3)^4$ = _____

a) $(3)(3)(3)(3) = 81$

b) $\left(\frac{1}{2}\right)\left(\frac{1}{2}\right)\left(\frac{1}{2}\right)\left(\frac{1}{2}\right)\left(\frac{1}{2}\right)$ $= \frac{1}{32}$

---

61. When the base is negative, it is <u>always</u> written in parentheses. Therefore, don't confuse $(-5)^2$ with -5^2.

    $(-5)^2$ means $(-5)(-5) = 25$

    -5^2 means $-(5)(5) = -25$

Following the examples, complete these:

    a) $(-2)^4$ = _____   b) -2^4 = _____

a) $\frac{9}{16}$

b) -64

c) 81

---

62. Any positive power of "1" equals "1". Any positive power of 0 equals 0. Any positive power of -1 equals either "1" or "-1". For example:

    $1^3 = (1)(1)(1) = 1$ $(-1)^2 = (-1)(-1) = 1$

    $0^4 = (0)(0)(0)(0) = 0$ $(-1)^3 = (-1)(-1)(-1) = -1$

Complete these:

    a) 1^7 = __ b) 0^{10} = __ c) $(-1)^4$ = __ d) $(-1)^5$ = __

a) 16   b) -16

63. To fit the pattern below, 2^1 must be 2 and 5^1 must be 5.

$2^3 = (2)(2)(2)$ $5^3 = (5)(5)(5)$

$2^2 = (2)(2)$ $5^2 = (5)(5)$

$2^1 = (2)$ or 2 $5^1 = (5)$ or 5

Therefore, we agree to the following definition for any base $\underline{a}$.

Definition: $a^1 = a$

Following the examples, complete these:

$10^1 = 10$ $\left(\dfrac{3}{4}\right)^1 = \dfrac{3}{4}$ a) $6.7^1 =$ _____   b) $(-9)^1 =$ _____

a) 1

b) 0

c) 1

d) -1

---

64. To fit the pattern below, 10^0 must be "1".

$10^3 = 1000$

$10^2 = 100$

$10^1 = 10$

$10^0 = 1$

Therefore, we agree to the following definition for any nonzero base $\underline{a}$. We will show later why the base $\underline{a}$ cannot be 0.

Definition: $a^0 = 1$ $(a \neq 0)$

Following the examples, complete these:

$75^0 = 1$ $\left(\dfrac{5}{2}\right)^0 = 1$ a) $6^0 =$ _____   b) $(-4.5)^0 =$ _____

a) 6.7   b) -9

---

65. To fit the pattern below, 10^{-1} must be $\dfrac{1}{10}$ and 10^{-2} must be $\dfrac{1}{100}$ or $\dfrac{1}{10^2}$.

$10^2 = 100$

$10^1 = 10$

$10^0 = 1$

$10^{-1} = \dfrac{1}{10}$

$10^{-2} = \dfrac{1}{100}$

a) 1   b) 1

Continued on following page.

65. Continued

Therefore, we agree to the following definition, provided that the base $\underline{a}$ is not 0.

> Definition: $a^{-n} = \dfrac{1}{a^n}$ $(a \neq 0)$

Following the examples, complete these:

$5^{-3} = \dfrac{1}{5^3}$ $9^{-6} = \dfrac{1}{9^6}$ a) $7^{-2} = $ _____   b) $(-3)^{-4} = $ _____

---

66. Using the definition, we converted each power below to a fraction.

$7^{-2} = \dfrac{1}{7^2} = \dfrac{1}{49}$ $(-2)^{-3} = \dfrac{1}{(-2)^3} = \dfrac{1}{-8} = -\dfrac{1}{8}$

Convert each power to a fraction.

a) $6^{-2} = $ _____   b) $4^{-3} = $ _____   c) $(-2)^{-5} = $ _____

a) $\dfrac{1}{7^2}$ b) $\dfrac{1}{(-3)^4}$

---

67. We converted each power below to a fraction.

$5^{-1} = \dfrac{1}{5^1} = \dfrac{1}{5}$ $(-3)^{-1} = \dfrac{1}{(-3)^1} = \dfrac{1}{-3} = -\dfrac{1}{3}$

Convert each power to a fraction.

a) $6^{-1} = $ _____   b) $2^{-1} = $ _____   c) $(-9)^{-1} = $ _____

a) $\dfrac{1}{36}$

b) $\dfrac{1}{64}$

c) $-\dfrac{1}{32}$

---

68. The definitions for one, zero, and negative exponents are reviewed below.

> Definitions
>
> Exponent "1" $a^1 = a$
>
> Exponent "0" $a^0 = 1$ $(a \neq 0)$
>
> Negative exponent $a^{-n} = \dfrac{1}{a^n}$ $(a \neq 0)$

Evaluate each of these:

a) $5^1 = $ ___   b) $7^0 = $ ___   c) $4^{-1} = $ ___   d) $(-8)^0 = $ ___

a) $\dfrac{1}{6}$

b) $\dfrac{1}{2}$

c) $-\dfrac{1}{9}$

---

a) 5   b) 1   c) $\dfrac{1}{4}$ d) 1

SELF-TEST 2 (pages 14-26)

Multiply.

1. $(-7)(-8) =$ _____ | 2. $5(-9) =$ _____ | 3. $(0)(-12.4) =$ _____ | 4. $(-1)(3.7) =$ _____

Divide.

5. $\dfrac{72}{-6} =$ _____ | 6. $\dfrac{0}{7} =$ _____ | 7. $\dfrac{-15}{0} =$ _____ | 8. $\dfrac{-1.68}{-1.68} =$ _____ | 9. $\dfrac{0}{0} =$ _____

Do these problems. Report answers in lowest terms.

10. $2\left(-\dfrac{1}{4}\right) =$ _____ | 12. $\dfrac{-\dfrac{1}{6}}{\dfrac{2}{3}} =$ _____ | 14. $\dfrac{-\dfrac{9}{4}}{-6} =$ _____

11. $\dfrac{1}{-\dfrac{1}{5}} =$ _____ | 13. $\left(-\dfrac{3}{2}\right)\left(\dfrac{4}{9}\right) =$ _____ | 15. $(-5)\left(-\dfrac{7}{10}\right) =$ _____

Using either "commutative", "associative", or "identity", identify the property.

16. $\dfrac{2}{5}(1) = \dfrac{2}{5}$ _____ | 18. $-\dfrac{7}{8} + 0 = -\dfrac{7}{8}$ _____

17. $[1 + (-5)] + 3 = 1 + [(-5) + 3]$ _____ | 19. $(-2.7)(1.5) = (1.5)(-2.7)$ _____

Evaluate each power.

20. $5^3 =$ _____ | 22. $2^{-4} =$ _____ | 24. $(7.53)^0 =$ _____

21. $\left(-\dfrac{2}{5}\right)^1 =$ _____ | 23. $\left(-\dfrac{5}{6}\right)^2 =$ _____ | 25. $(-4)^{-1} =$ _____

ANSWERS:
1. 56
2. -45
3. 0
4. -3.7
5. -12
6. 0
7. impossible
8. 1
9. indeterminate
10. $-\dfrac{1}{2}$
11. -5
12. $-\dfrac{1}{4}$
13. $-\dfrac{2}{3}$
14. $\dfrac{3}{8}$
15. $\dfrac{7}{2}$
16. identity
17. associative
18. identity
19. commutative
20. 125
21. $-\dfrac{2}{5}$
22. $\dfrac{1}{16}$
23. $\dfrac{25}{36}$
24. 1
25. $-\dfrac{1}{4}$

1-8 LAWS OF EXPONENTS

In this section, we will define the laws of exponents for multiplication, division, and raising a power to a power.

---

69. We multiplied 4^2 and 4^3 below by substituting $(4)(4)$ for 4^2 and $(4)(4)(4)$ for 4^3. Since there are a total of <u>five</u> 4's, the exponent of the product is 5.

$$4^2 \cdot 4^3 = (4)(4) \cdot (4)(4)(4) = 4^5$$

We can get the exponent of the product <u>by adding the exponents of the factors</u>. That is:

$$4^2 \cdot 4^3 = 4^{2+3} = 4^5$$

Therefore, the general law of exponents for multiplying powers with the same base is:

Multiplication Law
$a^m \cdot a^n = a^{m+n}$

The law above applies to positive, zero, and negative exponents. For example:

$$(-2)^5 \cdot (-2)^4 = (-2)^{5+4} = (-2)^9$$

$$5^{-3} \cdot 5^0 = 5^{-3+0} = 5^{-3}$$

$$7^{-1} \cdot 7^{-5} = 7^{-1+(-5)} = 7^{-6}$$

Using the law, complete these:

a) $8^6 \cdot 8^7 =$ _____   c) $10^0 \cdot 10^4 =$ _____

b) $(-1)^5 \cdot (-1)^{-3} =$ _____   d) $6^{-3} \cdot 6^{-4} =$ _____

---

70. To use the law below, we substituted 3^1 for 3.

$$3 \cdot 3^4 = 3^1 \cdot 3^4 = 3^{1+4} = 3^5$$

Use the same method for these:

a) $5^7 \cdot 5 =$ _____   b) $8 \cdot 8^0 =$ _____   c) $2 \cdot 2^{-3} =$ _____

a) 8^{13} c) 10^4

b) $(-1)^2$ d) 6^{-7}

---

71. The law applies <u>only for powers with the same base</u>. It does not apply to either multiplication below.

$$2^3 \cdot 5^2 \qquad\qquad 6^{-4} \cdot 3^5$$

Use the law if it applies:

a) $6^4 \cdot 6^4 =$ _____   b) $7^{-2} \cdot 8^4 =$ _____   c) $(-3)^{-5} \cdot (-3) =$ _____

a) 5^8

b) $8^1 = 8$

c) 2^{-2}

72. The law also applies to multiplications with more than two factors. For example:

$$4^3 \cdot 4^{-2} \cdot 4^5 = 4^{3 + (-2) + 5} = 4^6$$

Complete these:

a) $7^4 \cdot 7^{-6} \cdot 7^{-3} = $ _____   b) $2^{-3} \cdot 2 \cdot 2^5 \cdot 2^{-1} = $ _____

a) 6^8

b) does not apply

c) $(-3)^{-4}$

---

73. We divided 4^5 by 4^2 below. Notice that $\frac{4 \cdot 4}{4 \cdot 4} = 1$.

$$\frac{4^5}{4^2} = \frac{4 \cdot 4 \cdot 4 \cdot 4 \cdot 4}{4 \cdot 4} = \left(\frac{4 \cdot 4}{4 \cdot 4}\right)(4 \cdot 4 \cdot 4) = 1(4 \cdot 4 \cdot 4) = 4^3$$

We can get the exponent of the quotient by subtracting exponents.

$$\frac{4^5}{4^2} = 4^{5 - 2} = 4^3$$

Therefore, the law of exponents for dividing powers with the same base is:

$$\boxed{\begin{array}{c} \underline{\text{Division Law}} \\[2mm] \dfrac{a^m}{a^n} = a^{m - n} \qquad (a \neq 0) \end{array}}$$

The law applies to positive, zero, and negative exponents. For example:

$$\frac{4^6}{4} = \frac{4^6}{4^1} = 4^{6 - 1} = 4^5 \qquad\qquad \frac{8^{-4}}{8^3} = 8^{-4 - 3} = 8^{-7}$$

Using the law, complete these:

a) $\dfrac{2^9}{2^6} = $ _____   b) $\dfrac{5^0}{5^4} = $ _____   c) $\dfrac{3^{-7}}{3} = $ _____

a) 7^{-5} b) 2^2

---

74. Another division of powers is shown below. Notice how we subtracted the exponents.

$$\frac{8^{-2}}{8^{-5}} = 8^{-2 - (-5)} = 8^{-2 + 5} = 8^3$$

Following the example, complete these:

a) $\dfrac{10^4}{10^{-1}} = $ _____   b) $\dfrac{4^{-9}}{4^{-7}} = $ _____

a) 2^3

b) 5^{-4}

c) 3^{-8}

---

75. The law does not apply to $\dfrac{6^4}{3^2}$ because the powers have different bases. If it applies, use the law for these:

a) $\dfrac{3^8}{4^5} = $ _____   b) $\dfrac{6^{-2}}{6^{-3}} = $ _____   c) $\dfrac{10^{-4}}{12^{-1}} = $ _____

a) 10^5 b) 4^{-2}

76. When a power is divided by itself, the exponent of the quotient is 0. Also, when any nonzero quantity is divided by itself, the quotient is "1". That is:

$$\frac{4^2}{4^2} = 4^{2-2} = 4^0 \qquad\qquad \frac{4^2}{4^2} = 1$$

The facts above confirm the definition we gave earlier for a^0.

> Definition: $a^0 = 1$ $(a \neq 0)$

As you can see from the definition, we do not define 0^0. We do not define it because $0^0 = \frac{0}{0}$ which is indeterminate. That is:

$$0^0 = 0^{1-1} = \frac{0^1}{0^1} = \frac{0}{0}$$

Answer "1" or "indeterminate" for these:

a) $5^0 =$ _____ b) $\frac{0^3}{0^3} =$ _____ c) $\frac{2^{-4}}{2^{-4}} =$ _____ d) $0^0 =$ _____

a) does not apply

b) $6^1 = 6$

c) does not apply

---

77. We raised 5^4 to the third power below by converting to a multiplication.

$$(5^4)^3 = 5^4 \cdot 5^4 \cdot 5^4 = 5^{12}$$

We can get the exponent 12 by multiplying the 4 and 3. That is:

$$(5^4)^3 = 5^{(4)(3)} = 5^{12}$$

Therefore, the law of exponents for raising a power to a power is:

> Power Law: $(a^m)^n = a^{mn}$

The law above applies to both positive and negative exponents. For example:

$$(10^{-5})^2 = 10^{(-5)(2)} = 10^{-10} \qquad (3^{-4})^{-1} = 3^{(-4)(-1)} = 3^4$$

Use the law of exponents for these:

a) $(8^7)^2 =$ _____ b) $(2^{-3})^4 =$ _____ c) $(6^{-2})^{-5} =$ _____

a) 1

b) indeterminate

c) 1

d) indeterminate

---

78. The three basic laws of exponents are summarized below.

Laws Of Exponents	
Multiplication	$a^m \cdot a^n = a^{m+n}$
Division	$\frac{a^m}{a^n} = a^{m-n}$ $(a \neq 0)$
Powers	$(a^m)^n = a^{mn}$

Using the laws, complete these:

a) $4^{-5} \cdot 4 =$ _____ c) $(10^4)^2 =$ _____ e) $\frac{2^{-1}}{2^{-7}} =$ _____

b) $\frac{7^2}{7^4} =$ _____ d) $6^8 \cdot 6^{-3} =$ _____ f) $(3^{-3})^5 =$ _____

a) 8^{14}

b) 2^{-12}

c) 6^{10}

---

a) 4^{-4} b) 7^{-2} c) 10^8 d) 6^5 e) 2^6 f) 3^{-15}

1-9   POWERS OF TEN

In this section, we will discuss powers of ten and the decimal-point-shift method for multiplying by powers of ten.

---

79.   Powers of ten with positive exponents equal whole numbers like 10, 100, 1,000 , and so on. For example:

$$10^1 = 10$$
$$10^2 = 10 \cdot 10 = 100$$
$$10^3 = 10 \cdot 10 \cdot 10 = \underline{\hspace{2cm}}$$

---

80.   The table at the right shows the values of various powers of 10. You can see this fact:

    The <u>exponent</u> equals <u>the</u> <u>number</u> <u>of</u> <u>0's</u> <u>in</u> <u>the</u> <u>whole</u> <u>number</u>.

Using the above fact, convert each of these to a power of ten.

    a) 10,000,000 = _____

    b) 1,000,000,000 = _____

$10^6 =$	1,000,000
$10^5 =$	100,000
$10^4 =$	10,000
$10^3 =$	1,000
$10^2 =$	100
$10^1 =$	10
$10^0 =$	1

1,000

---

81.   Convert each whole number to a power of ten and each power of ten to a whole number.

    a) 10^3 = _____   c) 100 = _____

    b) 1,000,000 = _____   d) 10^5 = _____

a) 10^7

b) 10^9

---

82.   Since $10^1 = 10$, $10^2 = 100$, and $10^3 = 1,000$, we can use the decimal-point-shift method to multiply by those powers. <u>The exponent</u> <u>tells</u> <u>us</u> <u>how</u> <u>many</u> <u>places</u> <u>to</u> <u>shift</u> <u>the</u> <u>decimal</u> <u>point</u> <u>to</u> <u>the</u> <u>right</u>. For example:

    10^1 x .56 = .5 6 = 5.6   (Shifted <u>one</u> place. Exponent is 1.)

    10^2 x .39 = .39 = 39   (Shifted <u>two</u> places. Exponent is 2.)

    10^3 x 9.8 = 9.800 = 9,800 (Shifted <u>three</u> places. Exponent is 3.)

Use the decimal-point-shift method for these:

    a) 10^1 x 470 = ____   b) 10^2 x 6.08 = ____   c) 10^3 x 6.5 = ____

a) 1,000

b) 10^6

c) 10^2

d) 100,000

---

a) 4,700

b) 608

c) 6,500

83. The decimal-point-shift method can also be used to multiply by powers of ten with larger exponents. <u>The</u> <u>number</u> <u>of</u> <u>places</u> <u>shifted</u> <u>depends</u> <u>on</u> <u>the</u> <u>exponent</u>. For example:

10^4 x 2.56 = 2.5600 = 25,600   (Shifted <u>four</u> places.)

10^7 x .0029 = .0029000 = 29,000   (Shifted <u>seven</u> places.)

Use the decimal-point-shift method for these:

a) 10^6 x .00098 = _____   b) 10^8 x .000125 = _____

---

84. The decimal-point-shift method can also be used when the power of ten is the second factor. For example:

775 x 10^3 = 775.000 = 775,000

.094 x 10^5 = .09400 = 9,400

Use the decimal-point-shift method for these:

a) 1.5 x 10^1 = ____   b) .27 x 10^4 = ____   c) .00007 x 10^6 = ____

a) 980   b) 12,500

---

85. Since 10^0 = 1, when multiplying by 10^0, the product is <u>identical</u> to the other factor. That is:

10^0 x 55 = 55   1.39 x 10^0 = 1.39

Therefore, when multiplying by 10^0, we do not shift the decimal point.

a) 10^0 = .009 = _____   b) 27.5 x 10^0 = _____

a) 15

b) 2,700

c) 70

---

86. Powers of ten with negative exponents equal decimal numbers like .1 , .01 , .001 , and so on. For example:

$$10^{-1} = \frac{1}{10^1} = \frac{1}{10} = .1$$

$$10^{-2} = \frac{1}{10^2} = \frac{1}{100} = .01$$

$$10^{-3} = \frac{1}{10^3} = \frac{1}{1,000} = \text{_____}$$

a) .009   b) 27.5

---

87. Some powers of ten with negative exponents are shown in the table at the right. You can see this fact:

<u>The</u> <u>absolute</u> <u>value</u> <u>of</u> <u>the</u> <u>exponent</u> <u>equals</u> <u>the</u> <u>number</u> <u>of</u> <u>decimal</u> <u>places</u> <u>in</u> <u>the</u> <u>decimal</u> <u>number</u>.

Using the above fact, convert each of these to a power of ten.

a) .00000001 = _____

b) .0000000001 = _____

10^{-1}	= .1
10^{-2}	= .01
10^{-3}	= .001
10^{-4}	= .0001
10^{-5}	= .00001
10^{-6}	= .000001

.001

88. Convert each decimal to a power of ten and each power of ten to a decimal.

    a) .001 = _____   c) 10^{-5} = _____

    b) 10^{-2} = _____   d) .000001 = _____

    a) 10^{-8}

    b) 10^{-10}

---

89. Since 10^{-1} = .1 , 10^{-2} = .01 , and 10^{-3} = .001 , we can use the decimal-point-shift method to multiply by those powers. The absolute value of the exponent tells us how many places to shift the decimal point to the left.

    10^{-1} x 3.8 = .3.8 = .38   (Shifted one place. Exponent is -1.)

    10^{-2} x 7,500 = 7,5.00. = 75   (Shifted two places. Exponent is -2.)

    10^{-3} x .91 = .000.91 = .00091 (Shifted three places. Exponent is -3.)

    Use the decimal-point-shift method for these:

    a) 10^{-1} x 250 = ____   b) 10^{-2} x 487 = ____   c) 10^{-3} x 6.4 = ____

    a) 10^{-3} c) .00001

    b) .01   d) 10^{-6}

---

90. The decimal-point-shift method can also be used to multiply by powers of ten with larger negative exponents. The number of places shifted depends on the absolute value of the exponent. For example:

    10^{-4} x 36 = .0036. = .0036   (Shifted four places.)

    10^{-7} x 90,000,000 = 9.0,000,000. = 9   (Shifted seven places.)

    Use the decimal-point-shift method for these:

    a) 10^{-6} x 391 = _____   b) 10^{-8} x 45,000,000 = _____

    a) 25

    b) 4.87

    c) .0064

---

91. The same method can be used when the power of ten is the second factor. For example:

    $$510 \times 10^{-3} = .510. = .51$$

    $$1,200 \times 10^{-5} = .01,200. = .012$$

    Use the decimal-point-shift method for these:

    a) 58 x 10^{-1} = ____   b) 700 x 10^{-4} = ____   c) 2.4 x 10^{-6} = ____

    a) .000391

    b) .45

---

a) 5.8

b) .07

c) .0000024

92. Don't confuse the direction of the decimal-point-shift when multiplying powers of ten.

If the exponent is <u>positive</u>, we shift <u>to the right</u> since we are multiplying by 10, 100, 1000, and so on. That is:

$$5.3 \times 10^3 = 5.300 = 5,300$$

If the exponent is <u>negative</u>, we shift <u>to the left</u> since we are multiplying by .1, .01, .001, and so on. That is:

$$5.3 \times 10^{-3} = 005.3 = .0053$$

Use the decimal-point-shift method for these:

a) $72 \times 10^2 =$ _____   c) $.48 \times 10^4 =$ _____

b) $72 \times 10^{-2} =$ _____   d) $.48 \times 10^{-4} =$ _____

---

a) 7,200   b) .72   c) 4,800   d) .000048

1-10   SCIENTIFIC NOTATION

The very large and very small numbers that occur in science and technology are frequently written in <u>scientific notation</u>. We will discuss <u>scientific notation</u> in this section.

---

93. A number is written in <u>scientific notation</u> when:

1) The first factor is <u>a number between 1 and 10</u>.

2) The second factor is <u>a power of ten</u>.

Some numbers are written in scientific notation below.

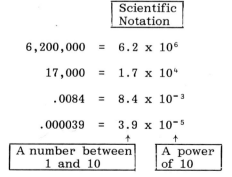

	Scientific Notation
6,200,000 =	6.2×10^6
17,000 =	1.7×10^4
.0084 =	8.4×10^{-3}
.000039 =	3.9×10^{-5}

A number between 1 and 10   A power of 10

Which of the following are written in scientific notation? That is, which are multiplications <u>of a number between 1 and 10</u> and a <u>power of ten</u>? _____

a) 25×10^8 b) 3.4×10^5 c) 9.1×10^{-3} d) $.67 \times 10^{-7}$

Only (b) and (c)

94. To convert a number written in scientific notation to a whole number or decimal, we perform the multiplication by the decimal-point-shift method. For example:

$$2.7 \times 10^3 = 2.700 = 2,700$$

$$9.6 \times 10^{-5} = .00009.6 = .000096$$

Convert to a whole number or decimal.

   a) 3×10^4 = _____   c) 2.54×10^2 = _____

   b) 3×10^{-4} = _____   d) 2.54×10^{-2} = _____

---

95. A number written in scientific notation can stand for a very large or very small number. For example:

$$5.1 \times 10^7 = 5.1000000 = 51,000,000$$

$$3.9 \times 10^{-9} = .000000003.9 = .0000000039$$

Convert to a whole number or decimal.

   a) 7.13×10^9 = _____   b) 5×10^{-10} = _____

a) 30,000   c) 254

b) .0003   d) .0254

---

96. To convert a number larger than "1" to scientific notation, we must find both the first factor and the exponent for the power of ten. Since the first factor must be a number between 1 and 10, we can find it by writing a caret (∧) after the first digit on the left. That is:

   For 7∧3,000 , the first factor is 7.3 .

   For 9∧00,000 , the first factor is 9 .

   For 1∧8.4 , the first factor is 1.84 .

Having found the first factor, we can find the exponent of the power of ten by counting the number of places from the caret to the decimal point. That is:

   7∧3,000. = 7.3×10^4 (Four places to the decimal point)

   8∧00,000. = 8×10^5 (Five places to the decimal point)

   9∧1.7 = 9.17×10^1 (One place to the decimal point)

Write each number in scientific notation.

   a) 47,000 = _____ x _____   c) 88.55 = _____ x _____

   b) 200 = _____ x _____   d) 9,160,000 = _____ x _____

a) 7,130,000,000

b) .0000000005

a) 4.7×10^4

b) 2×10^2

c) 8.855×10^1

d) 9.16×10^6

97. To convert a number smaller than "1" to scientific notation, we must again find both the first factor and the exponent for the power of ten. Since the first factor must be a number between 1 and 10, we can find it by writing a caret after the first non-zero digit in the number. That is:

For $.06_\wedge 8$, the first factor is 6.8 .

For $.1_\wedge 95$, the first factor is 1.95 .

For $.00007_\wedge$, the first factor is 7 .

Having found the first factor, we can find the exponent of the power of ten by counting the number of places from the caret to the decimal point. Since we count to the left, the exponent is negative. That is:

$.06_\wedge 8$ = 6.8×10^{-2} (Two places to the decimal point)

$.1_\wedge 95$ = 1.95×10^{-1} (One place to the decimal point)

$.00007_\wedge$ = 7×10^{-5} (Five places to the decimal point)

Write each number in scientific notation.

a) .00025 = _____ x _____   c) .06718 = _____ x _____

b) .4 = _____ x _____   d) .0000019 = _____ x _____

---

98. In scientific notation, the exponent can be either positive or negative.

If the number is larger than "1", we count to the right from the caret to the decimal point. Therefore, the exponent is positive. For example.

$$9_\wedge 1,300 = 9.13 \times 10^4$$

If the number is smaller than "1", we count to the left from the caret to the decimal point. Therefore, the exponent is negative. For example:

$$.009_\wedge 13 = 9.13 \times 10^{-3}$$

Write each number in scientific notation.

a) 5,600 = _____ x _____   c) 300,000,000 = _____ x _____

b) .056 = _____ x _____   d) .000000003 = _____ x _____

---

a) 2.5×10^{-4}

b) 4×10^{-1}

c) 6.718×10^{-2}

d) 1.9×10^{-6}

---

a) 5.6×10^3

b) 5.6×10^{-2}

c) 3×10^8

d) 3×10^{-9}

99. Scientific notation is used to express the very large and very small numbers that occur in science and technology. For example:

The speed of light is 2.998×10^8 meters per second.

The diameter of a large molecule is 1.7×10^{-7} centimeter.

Using the decimal-point-shift method, convert each measurement above to an ordinary number.

a) 2.998×10^8 meters per second = _____ meters per second.

b) 1.7×10^{-7} centimeter = _____ centimeter

100. Two more measurements are given below.

A light-year (the distance light travels in one year) is 5,870,000,000,000 miles.

One cycle of a television broadcast signal takes .00000000481 second.

Ordinarily measurements of that size are expressed in scientific notation. Convert them to scientific notation below.

a) 5,870,000,000,000 miles = _____ miles

b) .00000000481 second = _____ second

a) 299,800,000

b) .00000017

a) 5.87×10^{12} miles

b) 4.81×10^{-9} second

SELF-TEST 3 (pages 27-37)

Using the laws of exponents, write each answer as a single power.

1. $8^{-5} \cdot 8^2 =$ _____

2. $\dfrac{7^3}{7^{-2}} =$ _____

3. $(10^3)^2 =$ _____

4. $\dfrac{3^{-4}}{3^{-5}} =$ _3′_

5. $(4^{-2})^5 =$ _4³⁰_

6. $\dfrac{2'}{2^2} =$ _2¹_

7. $5^{-6} \cdot 5 \cdot 5^3 =$ _5⁻²_

8. $(6^0)^{-3} =$ _6⁻⁰_

Convert to a power of ten.

9. 10,000,000 _____

10. .0001 _____

Convert to a whole number or decimal.

11. 10^4 _____

12. $10^{-6} =$ _____

Using the decimal-point-shift method, do these multiplications.

13. $10^{-3} \times 640 =$ _____

14. $.0175 \times 10^5 =$ _____

15. $28.7 \times 10^{-6} =$ _____

Write in scientific notation.

16. $53,200 =$ _____

17. $.00928 =$ _____

18. $146,000,000 =$ _____

Convert to a whole number or decimal.

19. $1.2 \times 10^{-5} =$ _____

20. $7.04 \times 10^6 =$ _____

21. $5.23 \times 10^0 =$ _____

22. The total land area of the earth is 5.747×10^7 square miles. Write this area as an ordinary number.

_____ sq mi

23. Light travels 1 kilometer in .00000334 second. Write this time in scientific notation.

_____ sec

ANSWERS:

1. 8^{-3}
2. 7^5
3. 10^6
4. 3^1

5. 4^{-10}
6. 2^{-1}
7. 5^{-2}
8. 6^0

9. 10^7
10. 10^{-4}
11. 10,000
12. .000001

13. .64
14. 1,750
15. .0000287
16. 5.32×10^4

17. 9.28×10^{-3}
18. 1.46×10^8
19. .000012
20. 7,040,000

21. 5.23
22. 57,470,000
23. 3.34×10^{-6}

1-11   ORDER OF OPERATIONS

In this section, we will discuss the proper order of operations for evaluating expressions that contain more than one operation.

101.   When evaluating an expression containing more than one operation, the following order of operation is used:

> ### Order Of Operations
>
> 1. If the expression contains grouping symbols like parentheses ( ) or brackets [ ] or braces { }, <u>do the operations within the grouping symbols first</u> before evaluating the whole expression.
>
> 2. If the expression does not contain grouping symbols, use the following steps.
>
>   a) Evaluate any power.
>
>   b) Do all multiplications and divisions from left to right.
>
>   c) Do all additions and subtractions from left to right.
>
>   d) If the expression contains a fraction bar, do the operations above and below the bar before dividing.

The expression below does not contain grouping symbols.

$$10 - 2^3 + 3(4) - 8$$

To evaluate it, we evaluate the power 2^3, do the multiplication $3(4)$, and then add or subtract from left to right.   We get:

$$10 - 2^3 + 3(4) - 8$$
$$10 - 8\ + 3(4) - 8$$
$$10 - 8\ + 12\ - 8$$
$$2\ \ \ + 12\ - 8$$
$$14\ \ \ \ \ - 8 = 6$$

Following the example, evaluate these:

a) $3^3 - 4(5) - 4$ b) $6(-1) + (-9) - 5(2)$

102. To evaluate the expression below, we evaluated the power before multiplying. Evaluate the other expression.

$2(5)^2 - 30$

$5(-2)^3 + 40$

$2(25) - 30$

$50 - 30 = 20$

a) 3   b) -25

---

103. The expression below contains the grouping (7 - 9).

$$(7 - 9) - 3 + \frac{12}{2} - 4$$

To evaluate it, we simplify (7 - 9), do the division $\frac{12}{2}$, and then add or subtract from left to right. We get:

$$(7 - 9) - 3 + \frac{12}{2} - 4$$

$$-2 \quad - 3 + \frac{12}{2} - 4$$

$$-2 \quad - 3 + 6 - 4$$

$$-5 \quad + 6 - 4$$

$$1 \quad - 4 = -3$$

Following the example, evaluate these:

a) $\frac{-10}{5} + 7 + (6 - 10)$

b) $4 - [3 - (-6)] + \frac{15}{-3}$

0

---

104. Below we simplified (3 + 4) first, then multiplied and subtracted.

$5(3 + 4) - 10$

$5(7) \quad - 10$

$35 \quad - 10 = 25$

Use the same method to evaluate these:

a) $5(-2) - 3(-2 + 4)$

b) $4[-2 - (-3)] + \frac{-40}{-5}$

a) 1   b) -10

---

105. Below we simplified (-1 - 4) and (-3 - 5) before multiplying and then subtracting. Evaluate the other expression.

$(-1 - 4)(-3 - 5) - 10$

$(5 - 7)[3 - (-1)] + 2^3$

$(-5) \quad (-8) \quad - 10$

$40 \quad - 10 = 30$

a) -16   b) 12

40 • REAL NUMBERS

106. To simplify the expression in the brackets below, we began by simplifying $(4 - 7)$. Evaluate the other expression.

$$10 - [9 - (4 - 7)] \qquad\qquad 20 - [8 - 2(3 - 5)]$$
$$10 - [9 - (-3)]$$
$$10 - [9 + 3]$$
$$10 - 12 \qquad\qquad = -2$$

| 0 |

107. To simplify the expression below, we worked from the inside out. That is, first we simplified the parentheses, then the brackets, and then the braces.

$$10 - \{8 - 3[7 - (6 + 3)]\}$$
$$10 - \{8 - 3[7 - 9]\}$$
$$10 - \{8 - 3(-2)\}$$
$$10 - \{8 - (-6)\}$$
$$10 - \{8 + 6\}$$
$$10 - 14 = -4$$

Following the example, simplify this expression.

$$6\{-4 + 2[9 - 5(8 - 7)]\}$$

| 8 |

108. When an expression contains a fraction bar, <u>we</u> <u>do</u> <u>all</u> operations <u>above</u> <u>and</u> <u>below</u> <u>the</u> <u>bar</u> <u>before</u> <u>dividing</u>. For example:

$$\frac{4(3 + 5) - 2}{10 - 4} = \frac{4(8) - 2}{6} = \frac{32 - 2}{6} = \frac{30}{6} = 5$$

Using the same method, evaluate this one:

$$\frac{5(-1 - 9)}{3(4) - 2} =$$

| 24 |

109. Evaluate these:

a) $\dfrac{2(-4) + (-8)(-5)}{-7 + 5 - 2} =$

b) $\dfrac{30 - 2(7 - 4)}{3(-1) + 7} =$

| -5 |

| a) -8 |

| b) 6 |

110. Following the example, evaluate the other expressions.

$$2\left(-\frac{1}{3} - 1\right) = 2\left(-\frac{1}{3} - \frac{3}{3}\right) = 2\left(-\frac{4}{3}\right) = -\frac{8}{3}$$

a) $3\left(\frac{5}{8}\right) - 1 =$

b) $2\left(\frac{1}{2}\right)^2 - \frac{1}{2} + 1 =$

---

111. Following the example, evaluate the other expression.

$$\frac{4\left(-\frac{3}{5}\right) + 1}{5} = \frac{-\frac{12}{5} + \frac{5}{5}}{5} = \frac{-\frac{7}{5}}{5} = \left(-\frac{7}{5}\right)\left(\frac{1}{5}\right) = -\frac{7}{25}$$

$$\frac{2}{7\left(\frac{5}{4} - 1\right)} =$$

a) $\frac{7}{8}$

b) 1

$\frac{8}{7}$

1-12   TRANSLATING PHRASES TO ALGEBRAIC EXPRESSIONS

In this section, we will show how English phrases can be translated to algebraic expressions.

---

112. The table on page 42 gives some translations of English phrases to algebraic expressions. Each letter in an algebraic expression is called a <u>variable</u> because it represents various numbers. Translate each phrase below to an algebraic expression. Use <u>x</u> as the variable.

a) the sum of a number and 10 _____ $x + 10$

b) 25 more than a number _____ $x + 25$

c) 7 less than a number _____ $x - 7$

d) the difference between 100 and a number _____ $100 - x$

---

113. When a variable is multiplied by a number, we write the factors side by side with the number first. For example:

"5 times a number" is written 5x

Translate each phrase to an algebraic expression. Use <u>x</u> as the variable.

a) the product of 10 and a number _____ $10 \cdot x$

b) double a number _____ $2x$

c) 20 multiplied by a number _____ $20 \cdot x$

a) x + 10

b) x + 25

c) x - 7

d) 100 - x

Translating English Phrases To Algebraic Expressions

English Phrase	Algebraic Expression
Addition	
the sum of a number and 3	x + 3
9 more than a number	x + 9
a number plus 7	x + 7
15 added to a number	x + 15
a number increased by 10	x + 10
the sum of two numbers	x + y
Subtraction	
5 less than a number	x - 5
20 minus a number	20 - x
a number decreased by 6	x - 6
a number reduced by 12	x - 12
a number subtracted from 50	50 - x
the difference between 18 and a number	18 - x
the difference between two numbers	x - y
Multiplication	
14 times a number	14x
the product of 4 and a number	4x
a number multiplied by 8	8x
twice a number or double a number	2x
three times a number or triple a number	3x
$\frac{3}{4}$ of a number	$\frac{3}{4}x$
the product of two numbers	xy
Division	
a number divided by 5	$\frac{x}{5}$
the quotient of 7 and some number	$\frac{7}{x}$
the ratio of two numbers or the quotient of two numbers	$\frac{x}{y}$
Powers	
the square of a number	x^2
the cube of a number	x^3

114. Translate each phrase to an algebraic expression. Use <u>x</u> as the variable.

a) 10 divided by a number $\dfrac{10}{x}$

b) the quotient of a number and 5 $\dfrac{x}{5}$

c) the quotient of 5 and a number $\dfrac{5}{x}$

a) 10x

b) 2x

c) 20x

---

115. Since x^2 means $(x)(x)$ and x^3 means $(x)(x)(x)$, we can use x^2 and x^3 for the phrases "the square of a number" and "the cube of a number". Translate each phrase to an algebraic expression. Use <u>x</u> as the variable.

a) 10 more than the square of a number $x^2 + 10$

b) 20 less than the cube of a number $x^3 - 20$

c) triple the square of a number $3x^2$

d) the quotient of the cube of a number and 9 $x^3/9$

a) $\dfrac{10}{x}$

b) $\dfrac{x}{5}$

c) $\dfrac{5}{x}$

---

116. A translation is shown below.

$$\begin{array}{ccc} \underline{10} & \text{plus} & \underline{6 \text{ times a number}} \\ \downarrow & \downarrow & \downarrow \\ 10 & + & 6x \end{array}$$

Translate these to algebraic expressions. Use <u>x</u> as the variable.

a) 20 decreased by double a number $20 - 2x$

b) the sum of triple a number and 5 $3x + 5$

c) 7 more than 8 times the square of a number $8x^2 + 7$

d) 50 reduced by the product of 4 and the cube of a number $50 - 4x^3$

a) $x^2 + 10$

b) $x^3 - 20$

c) $3x^2$

d) $\dfrac{x^3}{9}$

---

117. Two translations are shown below.

triple a number, plus 2   3x + 2

the square of a number, minus 3 $x^2 - 3$

Translate these to algebraic expressions. Use <u>x</u> as the variable.

a) double a number, minus 10 $2x - 10$

b) 4 times the cube of a number, plus 7 $4x^3 + 7$

a) 20 - 2x

b) 3x + 5

c) $8x^2 + 7$

d) $50 - 4x^3$

---

a) 2x - 10

b) $4x^3 + 7$

118. When translating to an algebraic expression, we can use any letter for the variable. Use <u>y</u> as the variable to translate these to algebraic expressions.

   a) 25 divided by double a number   _____

   b) 4 times a number divided by 5   _____

   c) triple the square of a number divided by 10   _____

a) $\dfrac{25}{2y}$

b) $\dfrac{4y}{5}$

c) $\dfrac{3y^2}{10}$

119. When an expression contains two variables, we use two letters. When two variables are multiplied, we write the factors side by side. Usually they are written in alphabetical order. For example:

   "<u>p</u> times <u>q</u>" is written   pq

We used <u>x</u> and <u>y</u> as the variables for the translations below.

the sum of two numbers	x + y
the difference of two numbers	x - y
the product of two numbers	xy
the quotient of two numbers	$\dfrac{x}{y}$

Using the letters <u>x</u> and <u>y</u>, translate these to algebraic expressions.

   a) the sum of two numbers minus 5   _____

   b) 6 more than the difference of two numbers   _____

   c) the product of two numbers, plus "1"   _____

   d) triple the product of two numbers   _____

a) x + y - 5

b) x - y + 6

c) xy + 1

d) 3xy

120. Use the letters <u>a</u> and <u>b</u> to translate these.

   a) the sum of the squares of two numbers   _____

   b) the difference of the cubes of two numbers   _____

   c) the product of the squares of two numbers   _____

   d) the quotient of the cubes of two numbers   _____

a) $a^2 + b^2$

b) $a^3 - b^3$

c) a^2b^2

d) $\dfrac{a^3}{b^3}$

121. Sometimes we need parentheses to translate to algebraic expressions. Two examples are shown.

        10 minus the sum of two numbers   10 - (x + y)

        the product of 2, and the difference
          of a number and 5   2(x - 5)

Using $\underline{x}$ for one variable and $\underline{x}$ and $\underline{y}$ for two variables, translate these.

    a) 7 times the sum of the square of a number and 3

    b) 50 minus the difference of two numbers

    c) the product of 10 and the sum of two numbers

---

122. The phrases below translate to inequalities.

      all numbers greater than -1   x > -1

      5 and all numbers less than 5   x ≤ 5

      all positive numbers   x > 0

Using $\underline{x}$ as the variable, translate these to inequalities.

    a) all numbers less than 7

    b) -1 and all numbers greater than -1

    c) all negative numbers

a) $7(x^2 + 3)$
b) $50 - (x - y)$
c) $10(x + y)$

a) x < 7
b) x ≥ -1
c) x < 0

1-13   EVALUATING ALGEBRAIC EXPRESSIONS AND FORMULAS

In this section, we will do some evaluations with algebraic expressions and formulas.

123. An <u>algebraic expression</u> is a collection of numbers, variables, operation symbols, and grouping symbols. Some examples are:

    $\dfrac{2(x + 1)}{5}$ $5y^2 + 3y - 1$ $\dfrac{a + 3}{b - 2}$

Continued on following page.

123. Continued

To evaluate an algebraic expression, we must substitute some number for each variable. An example is shown. Do the other evaluation.

Evaluate $2x + 5$ when $x = -1$.

$2x + 5 = 2(-1) + 5 = -2 + 5 = 3$

Evaluate $\dfrac{3(y - 1)}{6}$ when $y = 9$.

$\dfrac{3(y - 1)}{6} =$

[handwritten: $3(9-1)$ over 6 $\dfrac{3(8)}{6} = \dfrac{24}{6} = 4$]

---

124. Following the example, do the other evaluation.

Evaluate $2x - 3y$ when $x = 10$ and $y = 2$.

$2x - 3y = 2(10) - 3(2) = 20 - 6 = 14$

Evaluate $\dfrac{x + 1}{y - 1}$ when $x = 9$ and $y = 3$.

$\dfrac{x + 1}{y - 1} =$

[handwritten: $\dfrac{9+1}{3-1} = \dfrac{10}{2} = 5$]

4

---

125. Evaluate each expression when $x = 10$ and $y = -2$.

a) $x - (2y + 5) = $ *[handwritten: $10 - (2(-2) + 5)$]*

b) $2x - (y - 6) = $ *[handwritten: $20 - (-2 - 6)$ $20 + (+8)$ 28]*

5

---

126. Following the example, do the other evaluation.

Evaluate $2xy - 15$ when $x = -8$ and $y = 5$.

$2xy - 15 = 2(-8)(5) - 15 = -80 - 15 = -95$

Evaluate $\dfrac{cd - 8}{10}$ when $c = 6$ and $d = -7$.

$\dfrac{cd - 8}{10} =$

a) 9

b) 28

---

127. We evaluated the first expression when $y = \dfrac{3}{8}$. Evaluate the second expression when $t = \dfrac{5}{4}$.

$$\dfrac{2(y + 1)}{5} = \dfrac{2\left(\dfrac{3}{8} + 1\right)}{5} = \dfrac{2\left(\dfrac{11}{8}\right)}{5} = \dfrac{\dfrac{11}{4}}{5} = \dfrac{11}{4}\left(\dfrac{1}{5}\right) = \dfrac{11}{20}$$

$3t - (1 - t) =$

-5

4

128. Following the example, do the other evaluation.

Evaluate $3x^2 + 2xy$ when $x = -2$ and $y = 4$.

$3x^2 + 2xy = 3(-2)^2 + 2(-2)(4) = 3(4) + 2(-8) = 12 + (-16) = -4$

Evaluate $2y^2 - y + 1$ when $y = -\frac{1}{3}$.

$2y^2 - y + 1 =$

---

129. The relationship stated below contains three variables.

The distance traveled by a moving object can be found by multiplying its average velocity by the time traveled.

The relationship is usually written as a formula with letters as abbreviations for the variables. That is:

$$\boxed{s = vt}$$ where: $s =$ distance traveled

$v =$ average velocity

$t =$ time traveled

By substituting numbers for the variables, we can use the formula to solve applied problems. For example:

If the average velocity is 50 miles per hour and the time traveled is 5 hours, the distance traveled is 250 miles, since:

$s = vt = 50(5) = 250$ miles

If the average velocity is 300 kilometers per hour and the time traveled is 10 hours, the distance traveled is 3,000 kilometers, since:

$s = vt = 300(10) = $ _____ kilometers

$\dfrac{14}{9}$

---

130. When it makes sense to use the same letter for more than one variable in a formula, subscripts are used. Either letters or numbers can be used as the subscripts. For example, the formula below shows the relationship between total force and the three contributing forces in a situation. The letter "T" and the numbers 1, 2, and 3 are used as subscripts.

$$\boxed{F_T = F_1 + F_2 + F_3}$$ where: $F_T =$ total force

$F_1 =$ first force

$F_2 =$ second force

$F_3 =$ third force

Let's use the formula to find F_T when $F_1 = 10$, $F_2 = 20$, and $F_3 = 40$.

$F_T = F_1 + F_2 + F_3 = 10 + 20 + 40 = $

3,000 kilometers

131.   We did one evaluation with a formula below.   Do the other evaluation.

$\qquad$ In $C = \frac{5}{9}(F - 32)$,   find C when   F = 50.

$\qquad$ $C = \frac{5}{9}(F - 32) = \frac{5}{9}(50 - 32) = \frac{5}{9}(18) = 10$

$\qquad$ In $P = 2L + 2W$,   find P when   L = 25   and   W = 10.

$\qquad$ $P = 2L + 2W =$

---

132.   Complete each evaluation.

$\qquad$ a) In $s = \frac{1}{2}gt^2$, find <u>s</u> when   g = 32   and   t = 3.

$\qquad$ $s = \frac{1}{2}gt^2 =$

$\qquad$ b) In $A = \frac{B}{B + 1}$, find A when   B = 99.

$\qquad$ $A = \frac{B}{B + 1} =$

P = 70

---

133.   Complete each evaluation.

$\qquad$ a) In $m = \frac{y_2 - y_1}{x_2 - x_1}$, find m when $y_2 = 18$, $y_1 = 10$, $x_2 = 10$, and $x_1 = 6$.

$\qquad$ $m = \frac{y_2 - y_1}{x_2 - x_1} =$

$\qquad$ b) In $A = P(1 + rt)$, find A when   P = 1000, r = 0.12, and t = 2.

$\qquad$ $A = P(1 + rt) =$

a) s = 144

b) A = .99

---

a) m = 2

b) A = 1240

SELF-TEST 4 (pages 38-49)

Evaluate these expressions.

1. $2(-5)^2 + 3(-5) =$ _65_   2. $(1 - 4)[5 - (-2)] + 2^4 =$ _-5_   3. $6 - [(5 - 8) + 2] =$ _7_

Evaluate these expressions.

4. $\dfrac{4 - 2(3 - 7)}{1 - 5} =$ _-3_   5. $\dfrac{3\left(\frac{2}{5}\right) - 1}{5} =$ _____   6. $6\left(\frac{1}{3}\right)^2 + -5\left(\frac{1}{3}\right) + 1 =$ _____

Translate each phrase to an algebraic expression.   Use x as the variable.

7. 3 plus the product of 5 and a number   _3 + 5x_

9. 15 decreased by double the cube of a number   _15 - 2x³_

8. 1 less than the square of a number   _x² - 1_

10. 6 divided by the sum of a number and 2   _$\frac{6}{x+2}$_

Translate each phrase to an algebraic expression.   Use x and y as the variables.

11. 5 minus the difference of two numbers   _5 - (x - y)_

12. 4 times the sum of the squares of two numbers   _4(x² + y²)_

13. Using x, write as an inequality: all numbers less than -2   _x < -2_

Evaluate each algebraic expression.

14. When $t = \dfrac{1}{2}$,
    $5t + -2(1 - t) =$ _$1\frac{1}{2}$_

15. When $x = 2$ and $y = -1$,
    $x^2 + 2xy + y^2 =$ _1_
    4 + -4 + 1

Evaluate each formula.

16. Find B when A = .9.

$$B = \dfrac{A}{1 - A}$$

B = _9_

17. Find R_t when $R_1 = 20$ and $R_2 = 30$.

$$R_t = \dfrac{R_1 R_2}{R_1 + R_2}$$

$R_t =$ _12_

18. Find F when m = 100, v = 20, and r = 80.

$$F = \dfrac{mv^2}{r}$$

F = _____

ANSWERS: 1. 65   5. $\dfrac{1}{25}$ 9. 15 - 2x³   13. x < -2   16. B = 9

2. -5   6. 0   10. $\dfrac{6}{x + 2}$ 14. $\dfrac{3}{2}$ 17. R_t = 12

3. 7   7. 3 + 5x   11. 5 - (x - y)   15. 1   18. F = 500

4. -3   8. x² - 1   12. 4(x² + y²)

SUPPLEMENTARY PROBLEMS - CHAPTER 1

<u>NOTE</u>:   Answers for all supplementary problems are in the back of the text.

<u>Assignment 1</u>

Given this set of numbers: $\{-5,\ \sqrt{2},\ 30,\ 0,\ -\frac{9}{2},\ -\sqrt{5},\ 6,\ \frac{3}{8},\ -2,\ 1.49\}$

    1. List the natural numbers.   2. List the whole numbers.

    3. List the integers.   4. List the irrational numbers.

Write "true" or "false" for each statement.

    5. -9, -6, and -2 are integers.   6. 0 is a natural number.

    7. All square roots are irrational numbers.   8. All integers are rational numbers.

    9. 1.575757... is an irrational number.   10. $\frac{187}{301}$ is a rational number.

Write either < or > in each blank.

    11. 2___-4   12. -8___-3   13. 0___-1   14. -5___2   15. 8___0

Write the mathematical symbol for each.

    16. "is less than or equal to"   17. "is greater than or equal to"

Find each absolute value.

    18. $|12|$ 19. $\left|-\frac{1}{3}\right|$ 20. $|0|$ 21. $\left|\frac{9}{4}\right|$ 22. $|-6.81|$

Do each addition.   Report answers in lowest terms.

    23. -2 + 7   24. 5 + (-11)   25. -6 + (-6)   26. 0 + (-25)

    27. $\frac{5}{6} + \left(-\frac{1}{6}\right)$ 28. $-\frac{9}{2} + \frac{3}{2}$ 29. $\frac{7}{3} + \left(-\frac{5}{6}\right)$ 30. $-\frac{1}{4} + \left(-\frac{1}{10}\right)$

    31. $-1 + \frac{4}{3}$ 32. $3 + \left(-\frac{5}{4}\right)$ 33. $-\frac{3}{8} + \left(-\frac{1}{6}\right)$ 34. $\frac{7}{5} + \left(-\frac{7}{5}\right)$

    35. 2.3 + (-7.8)   36. 1 + (-.45)   37. -.96 + 5   38. -1 - (6.4)

    39. -2 + 7 + (-5)   40. -45 + 32 + (-15)   41. 5 + (-3) + 9 + (-7) + 0

Write the additive inverse of each number.

    42. -17   43. 3.96   44. $-\frac{5}{2}$ 45. 10

Do each subtraction.   Report answers in lowest terms.

    46. 3 - 8   47. -6 - 1   48. -9 - (-15)   49. 0 - (-50)

    50. -20 - (-20)   51. 1.4 - 8.7   52. -5.14 - 2.35   53. 1 - (-.619)

    54. $-\frac{7}{4} - \frac{3}{4}$ 55. $\frac{3}{10} - \left(-\frac{5}{6}\right)$ 56. $1 - \frac{1}{5}$ 57. $-\frac{2}{9} - (-2)$

Assignment 2

Do each multiplication. Report answers in lowest terms.

1. $8(-7)$
2. $(-5)(4)$
3. $(-9)(-3)$
4. $0(-3)$

5. $(-1)(-10)$
6. $1(-3.9)$
7. $10(-.63)$
8. $(-1.2)(-1.2)$

9. $\left(\frac{1}{2}\right)\left(-\frac{3}{4}\right)$
10. $\left(-\frac{5}{6}\right)\left(-\frac{3}{5}\right)$
11. $(-4)\left(\frac{5}{8}\right)$
12. $\left(-\frac{2}{3}\right)(-9)$

13. $(-7)(-1)(4)$
14. $(-3)(-5)(-2)$
15. $(-9)(0)(-1)(-2)$
16. $(1)(-6)(3)(-7)$

Do each division. Report answers in lowest terms.

17. $\frac{42}{-7}$
18. $\frac{-20}{4}$
19. $\frac{-32}{-4}$
20. $\frac{0}{-6}$

21. $\frac{59}{-1}$
22. $\frac{-17.6}{1}$
23. $\frac{4.4}{0}$
24. $\frac{-8.8}{-1.1}$

25. $\dfrac{\frac{1}{8}}{-\frac{3}{4}}$
26. $\dfrac{-\frac{45}{4}}{-\frac{15}{8}}$
27. $\dfrac{-\frac{8}{7}}{2}$
28. $\dfrac{-15}{-\frac{3}{5}}$

Using either "commutative", "associative", or "identity", identify the property.

29. $0 + (-3) = -3$
32. $5 + (-9) = -9 + 5$
35. $1(10) = 10$

30. $(-9)(1) = -9$
33. $(2 \cdot 3) \cdot 5 = 2 \cdot (3 \cdot 5)$
36. $-1 + 2 = 2 + (-1)$

31. $(-2 + 5) + 3 = -2 + (5 + 3)$
34. $(-1)(-8) = (-8)(-1)$
37. $12 + 0 = 12$

Evaluate each power.

38. 2^3
39. $(-4)^4$
40. 1^8
41. $(-9)^1$
42. 0^5

43. 7^0
44. 8^{-1}
45. $(-7)^{-2}$
46. 6^{-3}
47. $(-2)^{-5}$

48. $(-.5)^3$
49. $(1.5)^2$
50. $\left(\frac{7}{8}\right)^0$
51. $\left(-\frac{9}{5}\right)^1$
52. $\left(\frac{1}{4}\right)^3$

Assignment 3

Using the laws of exponents, write each answer as a single power.

1. $\frac{6^2}{6^3}$
2. $3^4 \cdot 3^{-2}$
3. $(5^2)^3$
4. $8^{-1} \cdot 8^{-5}$

5. $(9^2)^{-4}$
6. $\frac{4^{-1}}{4^{-3}}$
7. $10^3 \cdot 10^0$
8. $(2^{-3})^{-1}$

9. $\frac{10}{10^4}$
10. $2 \cdot 2^{-4} \cdot 2^5$
11. $\frac{7^{-2}}{7^{-2}}$
12. $4^3 \cdot 4^0 \cdot 4^{-7} \cdot 4$

13. $(6^0)^{-3}$
14. $\frac{a^m}{a^n}$
15. $a^m \cdot a^n$
16. $(a^m)^n$

Convert each number to a power of ten.

17. $10,000$
18. $.001$
19. $100,000,000$
20. $.000001$

Convert each power of ten to a whole number or decimal.

21. 10^3
22. 10^{-1}
23. 10^0
24. 10^{-4}

Using the decimal-point-shift method, do these multiplications.

    25. 10^4 x .0168   26. 10^{-2} x 93.7   27. 10^6 x 415   28. 10^{-5} x .62

Write each of the following in scientific notation.

    29. .000714   30. 953   31. 300,000,000   32. .188

Convert each scientific-notation form to a whole number or decimal.

    33. 5×10^3 34. 3.61×10^{-3} 35. 9.82×10^6 36. 7.9×10^{-7}

37. A computer can execute 18,000,000 instructions per second. Write this capability in scientific notation.

38. There are 1.609344×10^3 meters in a mile. Write this length as a decimal number.

39. The wave length of red light is .000065 centimeter. Write this length in scientific notation.

40. The operating frequency of a communications satellite is 4,380,000,000 hertz. Write this frequency in scientific notation.

41. The time for one cycle of a radio wave is 7.25×10^{-7} second. Write this time as a decimal number.

42. The earth's diameter at the equator is 1.276×10^7 meters. Write this diameter as a whole number.

43. One watt equals .00134 horsepower. Write this amount in scientific notation.

Assignment 4

Evaluate these expressions.

    1. $6(-4) + 2(3)^2$ 2. $3(-2)^3 + 4(-2) + 40$ 3. $5(-6) - (-1)(12)$

    4. $9 + 5(3 - 7)$ 5. $4(2 - 5) - 7(3 - 8)$ 6. $4^2 - [1 - (1 - 2)]$

    7. $1 - [3 - (4 - 2)]$ 8. $4 - (-5 + 2)(-6 + 4)$ 9. $20 - [2(-1 - 5) + 8]$

Evaluate these expressions.

    10. $6\left(\frac{1}{3} - 1\right)$ 11. $2\left(1 - \frac{3}{8}\right)$ 12. $\left(\frac{1}{2}\right)^2 - 2\left(\frac{1}{2}\right) + 1$

    13. $\frac{5(1 - 3) + 2}{4 - 6}$ 14. $\frac{20 - 3(10 - 4)}{5(-1) + 7}$ 15. $\frac{4(-7) + (-9)(-6)}{-2(5) + 9}$

    16. $\frac{6\left(\frac{7}{2} - 1\right)}{5}$ 17. $\frac{6\left(-\frac{3}{4}\right) + 2}{8}$ 18. $\frac{1}{1 - \frac{5}{3}}$

Translate each phrase to an algebraic expression. Use x as the variable.

    19. the sum of 20 and twice a number   20. the quotient of 30 and a number

    21. 8 decreased by the product of 5 and a number   22. 15 times the square of a number

    23. 50 divided by the difference of a number and 10   24. 12 times the cube of a number less 6

Translate each phrase to an algebraic expression. Use $\underline{x}$ and $\underline{y}$ as the variables.

25. 18 minus double the product of two numbers

26. the sum of the squares of two numbers divided by the difference of the two numbers

27. one-half the difference of the squares of two numbers

28. 50 decreased by the sum of the cubes of two numbers

29. 9 less than the quotient of the cubes of two numbers

30. the square of the sum of two numbers

Using $\underline{x}$, write each of the following as an inequality.

31. all numbers greater than 5

32. 0 and all numbers less than 0

Do these evaluations.

33. When $x = -1$,

$4x - 3 =$ _____.

34. When $y = 5$,

$\dfrac{4(2 - y)}{3} =$ _____.

35. When $a = 2$ and $b = 4$,

$5a - 2b =$ _____.

36. When $t = \dfrac{3}{2}$,

$t^2 - 2(t - 1) =$ _____.

37. When $r = \dfrac{1}{3}$,

$\dfrac{r}{1 - r} =$ _____.

38. When $s = 3$ and $w = 10$,

$\dfrac{50 - sw}{20} =$ _____.

39. When $y = -1$,

$y^3 - 2y + 3 =$ _____.

40. When $x = 5$ and $y = 3$,

$x^2 - y^2 =$ _____.

41. When $x = 2$ and $y = -3$,

$x^2 - 2xy + 3y =$ _____.

Do these evaluations.

42. In $A = \dfrac{1}{2}bh$, find A when $b = 20$ and $h = 12$.

43. In $F = pA$, find F when $p = 15$ and $A = 40$.

44. In $r = \dfrac{v^2}{a}$, find $\underline{r}$ when $v = 10$ and $a = 5$.

45. In $F = \dfrac{9}{5}C + 32$, find F when $C = 35$.

46. In $M_r = \dfrac{F}{P + S} + 1$, find M_r when $F = 30$, $P = 2$, and $S = 4$.

47. In $v = \dfrac{s_2 - s_1}{t_2 - t_1}$, find $\underline{v}$ when $s_2 = 80$, $s_1 = 20$, $t_2 = 26$, and $t_1 = 14$.

48. In $R_t = \dfrac{R_1 R_2}{R_1 + R_2}$, find R_t when $R_1 = 150$ and $R_2 = 100$.

2 | Solving Equations

In this chapter, we will discuss the algebraic principles and processes that are used to solve non-fractional equations and pure quadratic equations. Some applied problems involving formulas and some other applied problems are included.

___

2-1 THE ADDITION AXIOM

In this section, we will discuss equations and show how the addition axiom is used to solve equations.

___

1. An <u>equation</u> is a statement that two algebraic expressions are equal. Two examples are:

$$x + 4 = 9 \qquad 5y - 3 = 2y + 9$$

An equation contains a left side and a right side connected by an equal sign. That is:

$$\boxed{\text{Left Side}} \quad = \quad \boxed{\text{Right Side}}$$

In $5y - 3 = 2y + 9$: the left side is $5y - 3$.

the right side is _____.

2.   An equation is neither true nor false until we substitute a number for the variable.   Then it becomes either true or false.   For example:

    If we substitute 3 for <u>x</u>,   x + 4 = 9 is false.

    If we substitute 5 for <u>x</u>,   x + 4 = 9 is true.

The substitution that makes the equation true is called the <u>solution</u> or <u>root</u> of the equation.   The solution or root of   x + 4 = 9 is 5.

    a) Is 9 the solution or root of   y + 10 = 19?   _____

    b) Is 7 the solution or root of   3 + d = 9?   _____

> 2y + 9

---

3.   The solution of x + 4 = 9 is 5.   Instead of saying "the solution is 5", we simply say x = 5.

    The solution of   2 + y = 8   is 6.   We say: y = 6

    The solution of   9 - t = 6   is 3.   We say: _____

> a) Yes
>
> b) No

---

4.   To solve an equation means to find its solution or root.   One principle used to solve equations is the <u>addition</u> <u>axiom</u> <u>for</u> <u>equations</u>.   It is stated below.

> t = 3

+---+
| |
| The <u>Addition</u> <u>Axiom</u> <u>For</u> <u>Equations</u> |
| |
| IF WE ADD THE SAME QUANTITY TO BOTH SIDES |
| OF AN EQUATION THAT IS TRUE, THE NEW |
| EQUATION IS ALSO TRUE.   That is: |
| |
| If: A = B |
| |
| Then: A + C = B + C |
| |
+---+

To solve an equation we must get the variable alone on one side. We used the addition axiom to solve each equation below.

To get <u>x</u> alone, we added   To get <u>y</u> alone, we added
-2 to both sides.   -2 is   3 to both sides.   3 is
the inverse of 2.   the inverse of -3.

$$x + 2 = 7 \qquad\qquad y - 3 = 6$$

$$x + \underbrace{2 + (-2)}_{\downarrow} = 7 + (-2) \qquad y \underbrace{- 3 + 3}_{\downarrow} = 6 + 3$$

$$x + \quad 0 \quad = 5 \qquad\qquad y + \quad 0 \quad = 9$$

$$x = 5 \qquad\qquad\qquad y = 9$$

To check a solution, we substitute it in the original equation to see that it satisfies the equation.   We checked 5 as the solution of x + 2 = 7.   Check the other solution.

   <u>Check</u>   <u>Check</u>

$$\begin{aligned} x + 2 &= 7 \\ 5 + 2 &= 7 \\ 7 &= 7 \end{aligned}$$

5. To solve the equation below, we added -10 to both sides. Solve the other equation by adding 3 to both sides.

$$7 = 10 + d \qquad\qquad -9 = -3 + t$$

$$-10 + 7 = \underline{-10 + 10} + d$$
$$\qquad\qquad \downarrow$$
$$-3 = \quad 0 \quad + d$$
$$-3 = d$$
$$d = -3$$

y - 3 = 6
9 - 3 = 6
6 = 6

6. What number would we add to both sides to solve each equation below?

a) 12 + x = 4 _____

b) 20 = p + 30 _____

c) V - 4 = -5 _____

d) 2 = -9 + y _____

-9 = -3 + t
3 - 9 = $\underline{3 - 3}$ + t
$\downarrow$
-6 = 0 + t
-6 = t
t = -6

7. Use the addition axiom to solve each equation.

a) y - 2 = -3 b) -20 = Q + 13

a) -12	c) 4
b) -30	d) 9

8. To solve the equation below, we added 2 to both sides.

$$x - 2 = 0$$

$$x \underline{- 2 + 2} = 0 + 2$$
$$\qquad \downarrow$$
$$x + \quad 0 \quad = 2$$
$$x = 2$$

Solve each of these equations:

a) m + 5 = 0 b) 0 = F - 7

a) y = -1
b) Q = -33

9. To solve the equation below, we added -1.2 to both sides. Solve and check the other equation.

$$x + 1.2 = 3.7 \qquad\qquad 2.4 = m - 8.1$$

$$x + \underline{1.2 + (-1.2)} = 3.7 + (-1.2)$$
$$\qquad \downarrow$$
$$x + \quad 0 \quad = 2.5$$
$$x = 2.5$$

<u>Check</u> <u>Check</u>

$$x + 1.2 = 3.7$$
$$2.5 + 1.2 = 3.7$$
$$3.7 = 3.7$$

a) m = -5
b) F = 7

10. To solve the equation below, we added $-\frac{2}{5}$ to both sides. Solve and check the other equation.

$$x + \frac{2}{5} = \frac{3}{5} \qquad\qquad y + \frac{1}{4} = \frac{3}{4}$$

$$x + \frac{2}{5} + \left(-\frac{2}{5}\right) = \frac{3}{5} + \left(-\frac{2}{5}\right)$$

$$x + \quad 0 \quad = \frac{1}{5}$$

$$x = \frac{1}{5}$$

Check $\qquad\qquad$ Check

$$x + \frac{2}{5} = \frac{3}{5}$$

$$\frac{1}{5} + \frac{2}{5} = \frac{3}{5}$$

$$\frac{3}{5} = \frac{3}{5}$$

| $m = 10.5$ |
| Check |
| $2.4 = m - 8.1$ |
| $2.4 = 10.5 - 8.1$ |
| $2.4 = 2.4$ |

---

11. Following the example, solve the other equation.

$$m - \frac{2}{3} = \frac{1}{6} \qquad\qquad d - \frac{1}{4} = \frac{5}{8}$$

$$m - \frac{2}{3} + \frac{2}{3} = \frac{1}{6} + \frac{2}{3}$$

$$m + \quad 0 \quad = \frac{1}{6} + \frac{4}{6}$$

$$m = \frac{5}{6}$$

| $y = \frac{1}{2}$ |
| Check |
| $y + \frac{1}{4} = \frac{3}{4}$ |
| $\frac{1}{2} + \frac{1}{4} = \frac{3}{4}$ |
| $\frac{3}{4} = \frac{3}{4}$ |

---

12. Following the example, solve the other equation.

$$x + 1 = \frac{2}{5} \qquad\qquad y - 1 = \frac{7}{8}$$

$$x + 1 + (-1) = \frac{2}{5} + (-1)$$

$$x + \quad 0 \quad = \frac{2}{5} + \left(-\frac{5}{5}\right)$$

$$x = -\frac{3}{5}$$

| $d = \frac{7}{8}$ |

---

13. Following the example, solve the other equation.

$$y - \frac{1}{2} = 0 \qquad\qquad x + 3.5 = 0$$

$$y - \frac{1}{2} + \frac{1}{2} = 0 + \frac{1}{2}$$

$$y + \quad 0 \quad = \frac{1}{2}$$

$$y = \frac{1}{2}$$

| $y = \frac{15}{8}$ |

| $x = -3.5$ |

2-2 THE MULTIPLICATION AXIOM

In this section, we will show how the multiplication axiom for equations is used to solve equations.

---

14. Expressions like 3x or -5y indicate a multiplication of a number and a variable. That is:

3x means "3 times x".

-5y means "-5 times y".

In expressions like those above, the numerical factor is called the coefficient of the variable. That is:

In 3x, 3 is the coefficient of x.

In -5y, _____ is the coefficient of y.

---

15. A second principle used to solve equations is the multiplication axiom for equations. It is stated below. | -5

> **The Multiplication Axiom For Equations**
>
> IF WE MULTIPLY BOTH SIDES OF AN EQUATION THAT IS TRUE BY THE SAME QUANTITY, THE NEW EQUATION IS ALSO TRUE. That is:
>
> If: B = C
>
> Then: AB = AC

We used the multiplication axiom to solve each equation below. Notice that we multiplied both sides by the reciprocal of the coefficient of the variable.

We multiplied both sides by $\frac{1}{5}$.

$$5x = 20$$
$$\frac{1}{5}(5x) = \frac{1}{5}(20)$$
$$1\,x = 4$$
$$x = 4$$

We multiplied both sides by $-\frac{1}{3}$.

$$-3y = 18$$
$$-\frac{1}{3}(-3y) = -\frac{1}{3}(18)$$
$$1\,y = -6$$
$$y = -6$$

We checked one solution below. Check the other solution.

Check

$$5x = 20$$
$$5(4) = 20$$
$$20 = 20$$

Check

16.   We used the multiplication axiom to solve two equations below.

$$6x = 24 \qquad\qquad -5y = 35$$

$$\tfrac{1}{6}(6x) = \tfrac{1}{6}(24) \qquad\qquad -\tfrac{1}{5}(-5y) = -\tfrac{1}{5}(35)$$

$$1\,x = 4 \qquad\qquad\qquad 1\ y = -7$$
$$x = 4 \qquad\qquad\qquad\quad y = -7$$

You can see that using this axiom is the same as dividing the number by the coefficient of the variable. Therefore, we can use that shorter method to solve the same two equations.

$$6x = 24 \qquad\qquad -5y = 35$$
$$x = \frac{24}{6} \qquad\qquad\quad y = \frac{35}{-5}$$
$$x = 4 \qquad\qquad\qquad y = -7$$

Use the shorter method to solve these:

   a) $7m = 56$ b) $-2y = 20$ c) $-10F = -40$

$-3y = 18$
$-3(-6) = 18$
$18 = 18$

---

17.   In each solution below, we also divided by the coefficient of the variable.

$$-16 = 2t \qquad\qquad -30 = -5V$$
$$\frac{-16}{2} = t \qquad\qquad \frac{-30}{-5} = V$$
$$t = -8 \qquad\qquad\quad V = 6$$

Use the shorter method to solve these:

   a) $4 = 4x$ b) $-30 = 3t$ c) $-48 = -6m$

a) $m = 8$

b) $y = -10$

c) $F = 4$

---

18.   When the number is 0, the solution is 0. Two examples are shown.

$$5x = 0 \qquad\qquad 0 = -9y$$

$$x = \frac{0}{5} \qquad\qquad \frac{0}{-9} = y$$
$$x = 0 \qquad\qquad\quad y = 0$$

Solve these:

   a) $-4t = 0$ b) $0 = 7d$ c) $-8 = 8R$

a) $x = 1$

b) $t = -10$

c) $m = 8$

19. When the solution is a fraction, the fraction is always reduced to lowest terms if possible. For example:

$$5x = 4 \qquad\qquad 15 = 9m$$

$$x = \frac{4}{5} \qquad\qquad \frac{15}{9} = m$$

$$\frac{5}{3} = m$$

Solve. Reduce to lowest terms.

   a) $11R = 13$ b) $2 = 6y$ c) $30p = 45$

| a) $t = 0$ |
| b) $d = 0$ |
| c) $R = -1$ |

---

20. When a fractional solution is negative, we write the negative sign in front of the fraction. For example:

$$7y = -3 \qquad\qquad 8 = -6d$$

$$y = \frac{-3}{7} \qquad\qquad \frac{8}{-6} = d$$

$$y = -\frac{3}{7} \qquad\qquad -\frac{4}{3} = d$$

Solve. Reduce to lowest terms.

   a) $-6x = 1$ b) $-2 = 4R$ c) $-8h = 10$

| a) $R = \frac{13}{11}$ |
| b) $y = \frac{1}{3}$ |
| c) $p = \frac{3}{2}$ |

---

21. When both terms in the division are negative, the solution is positive. For example:

$$-3y = -1 \qquad\qquad -4 = -6m$$

$$y = \frac{-1}{-3} \qquad\qquad \frac{-4}{-6} = m$$

$$y = \frac{1}{3} \qquad\qquad \frac{2}{3} = m$$

Solve. Reduce to lowest terms.

   a) $-9x = -2$ b) $-1 = -5p$ c) $-12d = -15$

| a) $x = -\frac{1}{6}$ |
| b) $R = -\frac{1}{2}$ |
| c) $h = -\frac{5}{4}$ |

| a) $x = \frac{2}{9}$ |
| b) $p = \frac{1}{5}$ |
| c) $d = \frac{5}{4}$ |

22. Any variable with a - in front of it has a coefficient of -1. That is:

$$-x = -1x \qquad\qquad -y = -1y$$

To solve each equation below, we began by writing the -1 coefficient explicitly.

$$-x = 10 \qquad\qquad -4 = -y$$
$$-1x = 10 \qquad\qquad -4 = -1y$$
$$x = \frac{10}{-1} \qquad\qquad \frac{-4}{-1} = y$$
$$x = -10 \qquad\qquad y = 4$$

Using the same method, solve these:

    a) -t = 7   b) -p = -9   c) 10 = -V

---

23. We used both the multiplication axiom and the shorter method to solve 1.2x = 4.8 below.

$$1.2x = 4.8 \qquad\qquad\qquad 1.2x = 4.8$$
$$\frac{1}{1.2}(1.2x) = \frac{1}{1.2}(4.8) \qquad\qquad x = \frac{4.8}{1.2}$$
$$1 \quad x = \frac{4.8}{1.2} \qquad\qquad\qquad x = 4$$
$$x = 4$$

When decimals are involved, we usually use the shorter method. Use that method to solve these:

    a) -1.5y = 7.5   b) .06 = .3d

a) t = -7

b) p = 9

c) V = -10

---

24. We used both the multiplication axiom and the shorter method to solve $\frac{3}{5}x = \frac{1}{2}$ below.

$$\frac{3}{5}x = \frac{1}{2} \qquad\qquad\qquad \frac{3}{5}x = \frac{1}{2}$$
$$\frac{5}{3}\left(\frac{3}{5}x\right) = \frac{5}{3}\left(\frac{1}{2}\right) \qquad\qquad x = \frac{\frac{1}{2}}{\frac{3}{5}}$$
$$1\,x = \frac{5}{6}$$
$$x = \frac{5}{6} \qquad\qquad\qquad x = \frac{1}{2}\left(\frac{5}{3}\right) = \frac{5}{6}$$

a) y = -5

b) d = .2

Continued on following page.

24. Continued

Ordinarily we use the axiom for such solutions to avoid the division of fractions. We used the axiom below. Notice that we multiplied both sides by $\frac{3}{2}$, the reciprocal of $\frac{2}{3}$. Solve the other equation by multiplying both sides by $\frac{2}{7}$.

$$\frac{2}{3}x = 5 \qquad\qquad\qquad -3 = \frac{7}{2}y$$

$$\frac{3}{2}\left(\frac{2}{3}x\right) = \frac{3}{2}(5)$$
$$1\ x = \frac{15}{2}$$
$$x = \frac{15}{2}$$

---

25. To solve the equation below, we multiplied both sides by $\frac{7}{5}$. Solve the other equation.

$$\frac{5}{7}p = \frac{3}{4} \qquad\qquad\qquad \frac{1}{5} = \frac{2}{3}t$$

$$\frac{7}{5}\left(\frac{5}{7}p\right) = \frac{7}{5}\left(\frac{3}{4}\right)$$
$$1\ p = \frac{21}{20}$$
$$p = \frac{21}{20}$$

$y = -\frac{6}{7}$

---

26. Following the example, solve the other equation.

$$4t = \frac{1}{2} \qquad\qquad\qquad \frac{7}{8} = 3m$$

$$\frac{1}{4}(4t) = \frac{1}{4}\left(\frac{1}{2}\right)$$
$$1\ t = \frac{1}{8}$$
$$t = \frac{1}{8}$$

$t = \frac{3}{10}$

---

27. Following the example, solve the other equation.

$$\frac{1}{3}x = 4 \qquad\qquad\qquad \frac{1}{2}y = 1$$

$$3\left(\frac{1}{3}x\right) = 3(4)$$
$$1\ x = 12$$
$$x = 12$$

$m = \frac{7}{24}$

42. A common error is forgetting to multiply the second term of the addition by the first factor. Two examples are shown.

	Error (see arrow)	Correct

Error (see arrow)

$$3(x + 8) = 3x + 8$$

$$7(1 + 6m) = 7 + 6m$$

Correct

$$3(x + 8) = 3x + 24$$

$$7(1 + 6m) = 7 + 42m$$

Avoiding the common error, complete these:

a) $2(x + 7) =$ _____   c) $5(6 + y) =$ _____

b) $7(5b + 1) =$ _____   d) $8(5 + 7m) =$ _____

43. There is also a distributive principle for subtraction. It is stated below.

---

a) 2x + 14

b) 35b + 7

c) 30 + 5y

d) 40 + 56m

> The Distributive Principle Of Multiplication Over Subtraction
>
> $$a(b - c) = ab - ac$$

Two examples of multiplying by the distributive principle over subtraction are shown below.

$$9(x - 3) = 9(x) - 9(3) = 9x - 27$$

$$4(6 - 2y) = 4(6) - 4(2y) = 24 - 8y$$

Usually we write the final product in one step. For example:

$$9(x - 3) = 9x - 27$$

$$4(6 - 2y) = 24 - 8y$$

Complete these. Remember that the terms are subtracted, not added.

a) $6(m - 1) =$ _____   c) $3(5x - 2) =$ _____

b) $5(4 - h) =$ _____   d) $10(1 - 4y) =$ _____

44. Be careful with the signs for these.

a) $7(2 - x) =$ _____   c) $4(2a + 1) =$ _____

b) $6(y + 9) =$ _____   d) $9(7 - 8b) =$ _____

---

a) 6m - 6

b) 20 - 5h

c) 15x - 6

d) 10 - 40y

---

45. In each multiplication below, the first factor is negative. Notice how we wrote the final products to minimize the number of signs between terms. Both -2x - 6 and -4t + 20 have only one sign between terms.

$$-2(x + 3) = (-2)(x) + (-2)(3) = -2x + (-6) = -2x - 6$$

$$-4(t - 5) = (-4)(t) - (-4)(5) = -4t - (-20) = -4t + 20$$

Continued on following page.

---

a) 14 - 7x

b) 6y + 54

c) 8a + 4

d) 63 - 72b

45.   Continued

Following the examples, do these:

    a) -3(y + 7) = _____

    b) -5(2d + 1) = _____

    c) -1(m - 9) = _____

    d) -10(4 - 2d) = _____

a) -3y - 21   b) -10d - 5   c) -1m + 9   d) -40 + 20d

2-5 COMBINING LIKE TERMS

In this section, we will show how like terms can be combined by factoring by the distributive principle. We will then use that process to solve some equations.

46.   We can also multiply by the distributive principle when the addition or subtraction is the first factor. For example:

$$(2 + 3)x = 2x + 3x$$

$$(8 - 5)y = 8y - 5y$$

By reversing the process, we can break up each product into the original factors. Doing so is called factoring by the distributive principle. That is:

    2x + 3x can be factored to get (2 + 3)x

    8y - 5y can be factored to get _____

47.   In an algebraic expression, the parts separated by a plus sign or a minus sign are called terms. The signs go with the terms. For example:

    In   x + 2y - 3 , the terms are x, 2y, and -3.

    In   -3x - 4y + 5 , the terms are -3x, -4y, and 5.

Though the definition will be stated more precisely later, for now we will define like terms as terms with the same variable. For example:

        3x and 9x are like terms.

        7y and -5y are like terms.

Like terms can be combined by factoring by the distributive principle. That is:

        3x + 9x = (3 + 9)x = 12x

    7y + (-5y) = [7 + (-5)]y = _____

(8 - 5)y

48. As you can see, we can combine like terms by simply adding their coefficients. That is:

$$7x + 5x = 12x, \quad \text{since} \quad 7 + 5 = 12$$

$$-8y + 3y = -5y, \quad \text{since} \quad -8 + 3 = -5$$

By simply adding coefficients, complete these:

a) $11m + 3m =$ _____   c) $6t + (-9t) =$ _____

b) $-2V + 7V =$ _____   d) $-5x + (-5x) =$ _____

2y

---

49. Two multiplications by the distributive principle over subtraction are shown below.

$$(8 - 4)x = 8x - 4x$$

$$(2 - 7)y = 2y - 7y$$

We can also factor by the distributive principle to combine the like terms above. That is:

$$8x - 4x = (8 - 4)x = 4x$$

$$2y - 7y = (2 - 7)y = -5y$$

By simply subtracting coefficients, combine the like terms below:

a) $10R - 2R =$ _____   c) $-2x - 5x =$ _____

b) $5t - 15t =$ _____   d) $9y - 9y =$ _____

a) 14m   c) -3t

b) 5V   d) -10x

---

50. Any variable without a coefficient has a coefficient of "1". That is:

$$x = 1x \qquad\qquad y = 1y$$

To combine like terms below, we wrote the coefficient "1" explicitly.

$$7x + x = 7x + 1x = 8x$$

$$y - 3y = 1y - 3y = -2y$$

Complete these:

a) $x + 3x =$ _____   c) $10d - d =$ _____

b) $m + m =$ _____   d) $V - 4V =$ _____

a) 8R   c) -7x

b) -10t   d) 0

---

51. Just as $x = 1x$ and $y = 1y$, $-x = -1x$ and $-y = -1y$. It is helpful to write the -1 coefficient explicitly in additions and subtractions like those below.

$$-x + 3x = -1x + 3x = 2x$$

$$4y + (-y) = 4y + (-1y) = 3y$$

$$-t - 6t = -1t - 6t = -7t$$

Continued on following page.

a) 4x   c) 9d

b) 2m   d) -3V

51. Continued

Following the examples, complete these:

a) -m + 10m = _____   c) 6d + (-d) = _____

b) -R - 2R = _____   d) -p - p = _____

---

52. Since 0x means "0 times x", 0x = 0. Therefore, each sum below is 0.

$$5x + (-5x) = 0x = 0$$

$$x + (-x) = 1x + (-1x) = 0x = 0$$

Two quantities are called <u>additive</u> <u>inverses</u> if their sum is 0. Two like terms are additive inverses <u>if</u> <u>their</u> <u>coefficients</u> are <u>additive</u> <u>inverses</u>. That is:

5x and -5x are additive inverses, since 5 + (-5) = 0

x and -x are additive inverses, since 1 + (-1) = 0

Write the additive inverse of each term.

a) 8y _____   b) d _____   c) -12t _____   d) -V _____

a) 9m	c) 5d
b) -3R	d) -2p

---

53. To solve the equations below, we combined like terms and then used the multiplication axiom.

$$5x + 2x = 21$$ $$20 = y - 5y$$
$$7x = 21$$ $$20 = -4y$$
$$x = \frac{21}{7}$$ $$\frac{20}{-4} = y$$
$$x = 3$$ $$y = -5$$

Using the same steps, solve these:

a) 2t + t = 27   b) 20 = 8d - 3d

a) -8y	c) 12t
b) -d	d) V

---

54. Solve: a) 45 = -6t + (-3t)   b) 9x - x = 16

a) t = 9	b) d = 4

---

55. Solve: a) y + 4y = 3   b) -d - 2d = 7

a) t = -5	b) x = 2

56. Solve. Reduce each solution to lowest terms.

    a) $-12 = 10V - 20V$ b) $-x + 7x = 3$

a) $y = \frac{3}{5}$ b) $d = -\frac{7}{3}$

57. The solution of the equation below is 0. Solve the other equation.

    $3x + 4x = 0$ $y - 2y = 0$

       $7x = 0$

        $x = \frac{0}{7}$

        $x = 0$

a) $V = \frac{6}{5}$ b) $x = \frac{1}{2}$

58. Following the example, solve the other equation.

    $1.5x + 3.1x = 13.8$ $6.8y - 2.4y = -88$

       $4.6x = 13.8$

         $x = \frac{13.8}{4.6}$

         $x = 3$

$y = 0$

59. Following the example, solve the other equation.

    $\frac{5}{6}x - \frac{2}{3}x = 5$ $\frac{1}{2}y + \frac{1}{4}y = 2$

    $\frac{5}{6}x - \frac{4}{6}x = 5$

       $\frac{1}{6}x = 5$

    $6\left(\frac{1}{6}x\right) = 6(5)$

        $x = 30$

$y = -20$

60. Following the example, solve the other equation.

    $x - \frac{1}{3}x = 5$ $y + \frac{1}{4}y = 10$

      $\frac{2}{3}x = 5$

    $\frac{3}{2}\left(\frac{2}{3}x\right) = \frac{3}{2}(5)$

       $x = \frac{15}{2}$

$y = \frac{8}{3}$

$y = 8$

61.　Some expressions containing more than two terms can be simplified by combining numbers or combining like terms.　Two examples are shown.

$$7 + 3x - 5 = 2 + 3x$$

$$-5d - 4 - d = -6d - 4$$

Following the examples, simplify these:

a) $10 + 7x + 5 =$ _____　　c) $-6y + 4 + 2y =$ _____

b) $y + 9 + 3y =$ _____　　d) $5 - t - 4 =$ _____

62.　We simplified two more expressions below.

$$4 - 6x - 7 = -3 - 6x$$

$$-y - 3y + 2 = -4y + 2$$

Following the examples, simplify these:

a) $10x - 7 - x =$ _____　　c) $x - 7x + 4 =$ _____

b) $3 + 5y - 9 =$ _____　　d) $-t + 6t - 7 =$ _____

a) $15 + 7x$

b) $4y + 9$

c) $-4y + 4$

d) $1 - t$

63.　To solve the equation below, we began by simplifying the left side. Solve the other equation.

$$7 + 3x - 9 = 6$$

$$-2 + 3x = 6$$

$$2 - 2 + 3x = 6 + 2$$

$$3x = 8$$

$$x = \frac{8}{3}$$

$$4 - 5y - 3 = 2$$

a) $9x - 7$

b) $-6 + 5y$

c) $-6x + 4$

d) $5t - 7$

64.　To solve the equation below, we began by simplifying the right side. Solve the other equation.

$$10 = x - 5x + 4$$

$$10 = -4x + 4$$

$$-4 + 10 = -4x + 4 + (-4)$$

$$6 = -4x$$

$$\frac{6}{-4} = x$$

$$x = -\frac{3}{2}$$

$$9 = -y + 10 - y$$

$y = -\frac{1}{5}$

$y = \frac{1}{2}$

2-6 EQUATIONS WITH LIKE TERMS ON OPPOSITE SIDES

In this section, we will use both axioms to solve equations with like terms on opposite sides.

---

65. The equation $7x = 2x + 20$ has like terms on opposite sides. To solve it, we use both axioms.

The <u>addition</u> <u>axiom</u> is used to get both like terms on one side so that they can be combined.

The <u>multiplication</u> <u>axiom</u> is then used to solve the resulting equation.

The steps used to solve the equation are described below.

1) To get both like terms on the same side so that they can be combined, we add -2x to both sides.

$$7x = 2x + 20$$
$$-2x + 7x = -2x + 2x + 20$$
$$5x = 0 + 20$$

2) Then we use the multiplication axiom to solve $5x = 20$.

$$5x = 20$$
$$x = \frac{20}{5} = 4$$

Complete the checking of 4 as the solution of the original equation at the right.

$$7x = 2x + 20$$
$$7(4) = 2(4) + 20$$
$$\underline{\quad} = \underline{\quad} + 20$$
$$\underline{\quad} = \underline{\quad}$$

---

66. For equations like those above, we use the addition axiom to move the like term <u>on</u> <u>the</u> <u>same</u> <u>side</u> <u>as</u> <u>the</u> <u>number</u> to the other side. To do so, we add its <u>opposite</u> to both sides.

For $5y + 36 = 9y$, we move 5y by adding -5y to both sides.

For $7d = 8d - 60$, we move 8d by adding -8d to both sides.

Using the facts above, solve these equations.

a) $5y + 36 = 9y$ b) $7d = 8d - 60$

Answer column:

$28 = 8 + 20$
$28 = 28$

a) $y = 9$ b) $d = 60$

67.   Following the example, solve the other equation.

$$7x = 40 - 3x$$ $$11 - 4y = 4y$$

$$3x + 7x = 40 \underline{\;- 3x + 3x}$$
                          $$\downarrow$$
$$10x = 40 + \quad 0$$

$$10x = 40$$

$$x = \frac{40}{10}$$

$$x = 4$$

---

68.   Following the example, solve the other equation. $$y = \frac{11}{8}$$

$$y = 14 + 3y$$ $$4 - x = 4x$$

$$-3y + y = 14 + 3y + (-3y)$$

$$-2y = 14$$

$$y = \frac{14}{-2}$$

$$y = -7$$

---

69.   Following the example, solve the other equation. $$x = \frac{4}{5}$$

$$\frac{4}{9}x - \frac{2}{9} = \frac{5}{9}x$$ $$\frac{1}{2}y + \frac{1}{4} = \frac{3}{4}y$$

$$-\frac{4}{9}x + \frac{4}{9}x - \frac{2}{9} = \frac{5}{9}x + \left(-\frac{4}{9}x\right)$$

$$-\frac{2}{9} = \frac{1}{9}x$$

$$9\left(-\frac{2}{9}\right) = 9\left(\frac{1}{9}x\right)$$

$$-2 = x$$

---

70.   Following the example, solve the other equation. $$y = 1$$

$$5.2x - 4.5 = 3.7x$$ $$3.6t = 1.6t + 5$$

$$-5.2x + 5.2x - 4.5 = 3.7x + (-5.2x)$$

$$-4.5 = -1.5x$$

$$x = \frac{-4.5}{-1.5}$$

$$x = 3$$

$$t = 2.5$$

71.   The equation $7x + 9 = 4x + 21$ has a variable term and a number on each side. To solve it, we must get both like terms on one side and both numbers on the other side so that they can be combined. To do so, we must move a variable term from one side and a number from the other side. We have a choice of two pairs:

   1) moving 4x and 9   or   2) moving 7x and 21

With either choice, <u>we</u> <u>must</u> <u>use</u> <u>the</u> <u>addition</u> <u>axiom</u> <u>twice</u>. We solved below by moving 4x and 9.

   To move 4x, we add -4x to both sides.

$$\mathbf{7x + 9} = 4x + 21$$
$$-4x + 7x + 9 = -4x + 4x + 21$$
$$3x + 9 = 21$$

   To move 9, we add -9 to both sides.

$$3x + 9 + (-9) = 21 + (-9)$$
$$3x = 12$$
$$x = \frac{12}{3} = 4$$

Check the solution at the right. $7x + 9 = 4x + 21$

---

72.   By moving either possible pair, solve these:

   a) $3y + 8 = 9y + 2$ b) $8t + 1 = 4t - 9$

$7(4) + 9 = 4(4) + 21$
$28 + 9 = 16 + 21$
$37 = 37$

---

73.   Following the example, solve the other equation.

$$30 - 5d = 20 - 9d$$ $15 - 2h = 8 - h$
$$30 - 5d + 5d = 20 - 9d + 5d$$
$$30 = 20 - 4d$$
$$-20 + 30 = -20 + 20 - 4d$$
$$10 = -4d$$
$$d = -\frac{10}{4}$$
$$d = -\frac{5}{2}$$

a) $y = 1$

b) $t = -\dfrac{5}{2}$

$h = 7$

74. Following the example, solve the other equation.

$$\frac{7}{8}t - \frac{1}{4} = \frac{1}{2}t + \frac{5}{8}$$

$$\frac{7}{8}t - \frac{1}{4} + \frac{1}{4} = \frac{1}{2}t + \frac{5}{8} + \frac{1}{4}$$

$$\frac{7}{8}t = \frac{1}{2}t + \frac{7}{8}$$

$$-\frac{1}{2}t + \frac{7}{8}t = -\frac{1}{2}t + \frac{1}{2}t + \frac{7}{8}$$

$$\frac{3}{8}t = \frac{7}{8}$$

$$\frac{8}{3}\left(\frac{3}{8}t\right) = \frac{8}{3}\left(\frac{7}{8}\right)$$

$$t = \frac{7}{3}$$

$$x + \frac{1}{3} = \frac{2}{3}x + 1$$

x = 2

SELF-TEST 6 (pages 68-78)

Multiply:

1. 4(1 - 2d) =

2. -3(x + 5) =

Combine like terms:

3. 4t - t =

4. -m + 7 - 6m =

Solve each equation. Report solutions in lowest terms.

5. -2d + 9d = 0

6. 2 = r - 5r

7. 7t + 8 = 3t

8. w = 10 - 5w

9. 3.3p - 4.8 = 2.1p

10. $\frac{1}{3}x + \frac{1}{2} = x$

11. 11 - 2P = P + 11

12. 6h - 13 = 22 - 9h

ANSWERS:

1. 4 - 8d

2. -3x - 15

3. 3t

4. -7m + 7

5. d = 0

6. r = $-\frac{1}{2}$

7. t = -2

8. w = $\frac{5}{3}$

9. p = 4

10. x = $\frac{3}{4}$

11. P = 0

12. h = $\frac{7}{3}$

2-7   REMOVING GROUPING SYMBOLS

In this section, we will discuss the procedures for removing grouping symbols to simplify expressions.

---

75.   If a grouping is preceded by a + sign or no sign, we can simply drop the grouping symbols. For example:

$$+ (x - 4) \quad \text{can be written} \quad x - 4$$

$$(3y + 1) \quad \text{can be written} \quad 3y + 1$$

Using the fact above, we simplified the expressions below. Simplify the other expressions.

$$7 + (x - 4) = 7 + x - 4 = 3 + x$$

$$(3y + 1) - y = 3y + 1 - y = 2y + 1$$

a) $4t + (3t - 8) =$ _____

b) $10 + (7 - 2d) =$ _____

---

76.   If a grouping is preceded by a - sign, we <u>cannot</u> simply drop the grouping symbols. To get rid of the grouping symbols, we can substitute -1 for the - sign and multiply by the distributive principle. That is:

$$-(3y + 2) = -1(3y + 2) = (-1)(3y) + (-1)(2) = -3y + (-2) = -3y - 2$$

$$-(6x - 4) = -1(6x - 4) = (-1)(6x) - (-1)(4) = -6x - (-4) = -6x + 4$$

As you can see, we can remove the grouping symbols above <u>by simply changing the sign of each term</u>. That is:

$$-(3y + 2) = -3y - 2$$

$$-(6x - 4) = -6x + 4$$

Remove the grouping symbols from these <u>by simply changing the sign of each term</u>.

a) $-(7x + 1) =$ _____   c) $-(t + 10) =$ _____

b) $-(4y - 5) =$ _____   d) $-(8 - d) =$ _____

Answer column:

a) $7t - 8$

b) $17 - 2d$

---

77.   To remove the grouping symbols below, we changed the sign of each term in $(3y + 2)$. We then simplified.

$$5 - (3y + 2) = 5 - 3y - 2 = 3 - 3y$$

Following the example, remove the grouping symbols and simplify.

a) $2d - (4d + 7) =$ _____ = _____

b) $7 - (p + 1) =$ _____ = _____

Answer column:

a) $-7x - 1$

b) $-4y + 5$

c) $-t - 10$

d) $-8 + d$

78. To remove the grouping symbols below, we changed the sign of each term in (6x - 4). We then simplified.

$$7 - (6x - 4) = 7 - 6x + 4 = 11 - 6x$$

Following the example, remove the grouping symbols and simplify.

   a) 5 - (x - 3) = _____ = _____

   b) 6y - (5 - 2y) = _____ = _____

a) 2d - 4d - 7 = -2d - 7

b) 7 - p - 1 = 6 - p

---

79. 3(x + 5) and 4(2y - 6) are instances of the distributive principle. To remove the grouping symbols, we multiply.

$$3(x + 5) = 3x + 15$$

$$4(2y - 6) = 8y - 24$$

Remove the grouping symbols in these.

   a) 5(4d + 1) = _____   b) 7(5 - p) = _____

a) 5 - x + 3 = 8 - x

b) 6y - 5 + 2y = 8y - 5

---

80. When an instance of the distributive principle is preceded by a + sign or no sign, we can remove the grouping symbols by multiplying. For example:

$$3(x + 4) + 7 = 3x + 12 + 7 = 3x + 19$$

$$10 + 6(2y - 3) = 10 + 12y - 18 = -8 + 12y$$

Following the examples, remove the grouping symbols and simplify.

   a) 2(4d - 6) - 3d = _____

   b) 12 + 4(1 - 3x) = _____

a) 20d + 5

b) 35 - 7p

---

81. When an instance of the distributive principle is preceded by a - sign, we multiply first and then remove the grouping symbols in the usual way. It <u>is</u> <u>very</u> <u>helpful</u> <u>to</u> <u>draw</u> <u>brackets</u> <u>around</u> <u>the</u> <u>instance</u> <u>of</u> <u>the</u> <u>distributive</u> <u>principle</u> <u>before</u> <u>multiplying</u>. For example:

$$10 - 2(x + 3) = 10 - [2(x + 3)]$$ Drawing brackets

$$= 10 - [2x + 6]$$ Multiplying

$$= 10 - 2x - 6$$ Removing the grouping symbols

$$= 4 - 2x$$ Simplifying

Following the example, simplify these. <u>Begin</u> <u>by</u> <u>drawing</u> <u>brackets</u> <u>around</u> <u>the</u> <u>instance</u> <u>of</u> <u>the</u> <u>distributive</u> <u>principle</u>.

   a) 5 - 3(d + 1)   b) t - 2(4t + 5)

a) 5d - 12

b) 16 - 12x

82. To simplify the expression below, we began by drawing brackets around 5(x - 1).

$$6 - 5(x - 1) = 6 - [5(x - 1)]$$
$$= 6 - [5x - 5]$$
$$= 6 - 5x + 5$$
$$= 11 - 5x$$

Following the example, simplify these. Draw brackets.

    a) 20 - 4(d - 2)   b) 10y - 3(2y - 9)

a) 2 - 3d

b) -7t - 10

83. -2(x + 6) and -5(3y - 4) are instances of the distributive principle. To remove the grouping symbols, we multiply.

$$-2(x + 6) = -2x - 12$$
$$-5(3y - 4) = -15y + 20$$

Remove the grouping symbols by multiplying.

    a) -3(7p + 1) = _____   b) -4(10 - t) = _____

a) 28 - 4d

b) 4y + 27

84. Following the example, simplify the other expressions.

$$-3(t + 4) + 5 = -3t - 12 + 5 = -3t - 7$$

    a) -2(3y + 1) - 7 = _____

    b) -5(4 - d) - 2d = _____

a) -21p - 3

b) -40 + 4t

a) -6y - 9   b) -20 + 3d

2-8 EQUATIONS WITH GROUPING SYMBOLS

In this section, we will solve equations that contain grouping symbols.

85. To solve the equation below, we removed the grouping symbols, simplified the left side, and then used the axioms. Solve the other equation.

$$3 + (2y + 1) = 9 \qquad (t - 3) - 8 = 0$$
$$3 + 2y + 1 = 9$$
$$4 + 2y = 9$$
$$-4 + 4 + 2y = 9 + (-4)$$
$$2y = 5$$
$$y = \frac{5}{2}$$

86. To solve the equation below, we removed the grouping symbols, simplified the left side, and then used the axioms. Solve the other equation.

$$5x - (2x + 1) = 4$$

$$5x - 2x - 1 = 4$$

$$3x - 1 = 4$$

$$3x - 1 + 1 = 4 + 1$$

$$3x = 5$$

$$x = \frac{5}{3}$$

$$F = 11 - (6F + 7)$$

t = 11

87. To solve the equation below, we removed the grouping symbols, simplified the left side, and then used the axioms. Solve the other equation.

$$6y - (2y - 3) = 1$$

$$6y - 2y + 3 = 1$$

$$4y + 3 = 1$$

$$4y + 3 + (-3) = 1 + (-3)$$

$$4y = -2$$

$$y = -\frac{2}{4}$$

$$y = -\frac{1}{2}$$

$$x = 10 - (5x - 1)$$

$F = \frac{4}{7}$

88. Following the example, solve the other equation.

$$4t - (t + 2) = 0$$

$$4t - t - 2 = 0$$

$$3t - 2 = 0$$

$$3t - 2 + 2 = 0 + 2$$

$$3t = 2$$

$$t = \frac{2}{3}$$

$$8p - (2p - 1) = 0$$

$x = \frac{11}{6}$

89. To solve the equation below, we began by multiplying by the distributive principle. Solve the other equation.

$$3(x + 2) = 21$$

$$3x + 6 = 21$$

$$3x + 6 + (-6) = 21 + (-6)$$

$$3x = 15$$

$$x = 5$$

$$9(2b - 4) = 18$$

$p = -\frac{1}{6}$

90. Following the example, solve the other equation.

$$5(3 + 2y) = 20y \qquad\qquad 3(7 - m) = m$$

$$15 + 10y = 20y$$

$$15 + 10y + (-10y) = 20y + (-10y)$$

$$15 = 10y$$

$$\frac{15}{10} = y$$

$$y = \frac{3}{2}$$

b = 3

---

91. Following the example, solve the other equation.

$$5(x - 1) = 2x - 7 \qquad\qquad 2y + 9 = 3(y + 2)$$

$$5x - 5 = 2x - 7$$

$$-2x + 5x - 5 = -2x + 2x - 7$$

$$3x - 5 = -7$$

$$3x - 5 + 5 = -7 + 5$$

$$3x = -2$$

$$x = -\frac{2}{3}$$

$m = \frac{21}{4}$

---

92. To solve the equation below, we multiplied by the distributive principle on both sides. Solve the other equation.

$$7(d + 2) = 2(3 + d) \qquad\qquad 5(t + 3) = 3(t - 5)$$

$$7d + 14 = 6 + 2d$$

$$-2d + 7d + 14 = 6 + 2d + (-2d)$$

$$5d + 14 = 6$$

$$5d + 14 + (-14) = 6 + (-14)$$

$$5d = -8$$

$$d = -\frac{8}{5}$$

y = 3

---

t = -15

93. To solve the equation below, we simplified the expression on the left side and then used the axioms. Solve the other equation.

$$2(m + 4) - 3 = 10$$

$$2m + 8 - 3 = 10$$

$$2m + 5 = 10$$

$$2m + 5 + (-5) = 10 + (-5)$$

$$2m = 5$$

$$m = \frac{5}{2}$$

$$4t = 5 + 3(2t - 3)$$

---

94. To solve the equation below, we began by simplifying the expression on the left side. Solve the other equation.

$$10 - 3(x + 4) = 16$$

$$10 - [3(x + 4)] = 16$$

$$10 - [3x + 12] = 16$$

$$10 - 3x - 12 = 16$$

$$-2 - 3x = 16$$

$$2 - 2 - 3x = 16 + 2$$

$$-3x = 18$$

$$x = -6$$

$$9y - 7(y + 2) = 4$$

t = 2

---

95. To solve the equation below, we began by simplifying the expression on the right side. Solve the other equation.

$$20 = t - 3(5 - 2t)$$

$$35 = 47 - 6(4 - x)$$

$$35 = 47 - [6(4 - x)]$$

$$35 = 47 - [24 - 6x]$$

$$35 = 47 - 24 + 6x$$

$$35 = 23 + 6x$$

$$-23 + 35 = -23 + 23 + 6x$$

$$12 = 6x$$

$$x = 2$$

y = 9

---

t = 5

2-9   SQUARE ROOTS

In this section, we will define <u>square roots</u> and show that any positive number has two square roots.

---

96.   A <u>square root</u> of a number N is <u>a number whose square is N</u>.

   Any positive number has two square roots, one positive and one negative. For example:

   4 is a square root of 16, since $(4)^2 = 16$.

   -4 is a square root of 16, since $(-4)^2 = 16$.

   The number 0 has only one square root. It is 0, since $(0)^2 = 0$.

   Complete: a) The <u>positive</u> square root of 36 is ___8___ .

   b) The <u>negative</u> square root of 100 is __-18__ .

   c) The two square roots of 49 are __-7__ and __7__ .

   d) The square root of 0 is ___0___ .

---

97.   Though negative square roots are called <u>negative</u> square roots, positive square roots are called <u>principal</u> square roots.

   The symbol $\sqrt{}$, called a <u>radical sign</u>, is used for <u>principal</u> square roots.   That is:

   $\sqrt{25} = 5$ $\sqrt{81} = 9$ $\sqrt{400} = 20$

   The symbol $-\sqrt{}$ is used for <u>negative</u> square roots.   That is:

   $-\sqrt{9} = -3$ $-\sqrt{1} = -1$ $-\sqrt{225} = -15$

   Complete these:

   a) $\sqrt{1} =$ __1__   b) $-\sqrt{64} =$ __-32__   c) $-\sqrt{4} =$ __-2__

a) 6
b) -10
c) 7 and -7
d) 0

---

98.   The symbol ± means "both + and -". It can be written in front of a radical or a number. For example:

   $\pm\sqrt{25}$ means both $\sqrt{25}$ and $-\sqrt{25}$.

   ±5 means both 5 and -5.

   Therefore: $\pm\sqrt{25} = \pm5$ means $\sqrt{25} = 5$ and $-\sqrt{25} = -5$.

   Using the ± symbol, complete these.

   a) $\pm\sqrt{4} =$ _____   b) $\pm\sqrt{1} =$ _____   c) $\pm\sqrt{81} =$ _____

a) 1
b) -8
c) -2

a) ±2
b) ±1
c) ±9

99. Any number that is the square of an integer is called a <u>perfect square</u>. For example:

Since $7^2 = 49$, 49 is a perfect square.

The first ten perfect squares are listed below:

1, 4, 9, 16, 25, 36, 49, 64, 81, 100

Any whole number that is a perfect square has two integers as its square roots. For example:

Since both 12^2 and $(-12)^2 = 144$, $\pm\sqrt{144} = \pm 12$.

Since both 30^2 and $(-30)^2 = 900$, $\pm\sqrt{900} = $ _____ .

---

100. Any fraction that is the square of a fraction is called a <u>perfect square</u>. That is:

Since $\left(\frac{7}{8}\right)^2 = \frac{49}{64}$, $\frac{49}{64}$ is a perfect square.

Since $\left(-\frac{1}{3}\right)^2 = \frac{1}{9}$, $\frac{1}{9}$ is a perfect square.

Any fraction that is a perfect square has two fractions as its square roots. That is:

$$\pm\sqrt{\frac{49}{64}} = \pm\frac{7}{8} \qquad\qquad \pm\sqrt{\frac{1}{9}} = \text{_____}$$

	± 30

---

101. A fraction is a perfect square only if both of its terms are perfect squares. That is:

$\frac{36}{25}$ is a perfect square, since both 36 and 25 are perfect squares.

$\frac{1}{81}$ is a perfect square, since both "1" and 81 are perfect squares.

To find the square root of perfect-square fractions, we find the square roots of both terms. Therefore:

$$\pm\sqrt{\frac{36}{25}} = \pm\frac{6}{5} \qquad\qquad \pm\sqrt{\frac{1}{81}} = \pm\frac{1}{9}$$

Following the examples, complete these:

a) $\sqrt{\frac{4}{9}} = $ _____ b) $-\sqrt{\frac{64}{49}} = $ _____ c) $\sqrt{\frac{1}{16}} = $ _____

$\pm\frac{1}{3}$

---

a) $\frac{2}{3}$

b) $-\frac{8}{7}$

c) $\frac{1}{4}$

102. Any decimal that is the square of a decimal is also called a <u>perfect square</u>. For example:

Since $(1.4)^2 = 1.96$, 1.96 is a perfect square.

Since $(-.8)^2 = .64$, .64 is a perfect square.

Any decimal that is a perfect squre has two decimals as its square roots. That is:

Since $(1.2)^2$ and $(-1.2)^2 = 1.44$, $\pm\sqrt{1.44} = \pm1.2$

Since $(.04)^2$ and $(-.04)^2 = .0016$, $\pm\sqrt{.0016} = \underline{\pm.04}$

103. Except for perfect squares, the square roots of numbers are non-terminating decimal numbers. We can use a calculator to find square roots. Using a calculator, we get:

$$\sqrt{19} = 4.3588989...$$

$$\sqrt{84} = 9.1651514...$$

A table of square roots of numbers from 1 to 100 is given on page 88. In the table, the square roots of numbers that are not perfect squares are rounded to <u>two decimal places</u>. For example, the table has these entries.

N	$\sqrt{N}$
19	4.36
84	9.17

Though the entries are not exact, we will treat them as if they were exact. That is:

We will say: $\sqrt{19} = 4.36$ $\sqrt{84} = 9.17$

Using the table, complete these:

a) $\sqrt{28} =$ _____   b) $-\sqrt{41} =$ _____   c) $\sqrt{97} =$ _____

104. Negative numbers do not have real number square roots because the square of any real number is positive. For example, -16 has no real number square root.

$$\sqrt{-16} \neq 4, \quad \text{because} \quad (4)^2 = 16$$

$$\sqrt{-16} \neq -4, \quad \text{because} \quad (-4)^2 = 16$$

Complete these:

a) $\sqrt{-81} =$ _____   b) $-\sqrt{81} =$ _____   c) $\sqrt{\frac{1}{4}} =$ _____

Answers column (103): $\pm.04$

Answers column (104):
a) 5.29

b) -6.40

c) 9.85

---

a) -81 has no real number square root.   b) -9   c) $\frac{1}{2}$

SQUARE ROOT APPROXIMATIONS

N	$\sqrt{N}$	N	$\sqrt{N}$	N	$\sqrt{N}$	N	$\sqrt{N}$
1	1	26	5.10	51	7.14	76	8.72
2	1.41	27	5.20	52	7.21	77	8.77
3	1.73	28	5.29	53	7.28	78	8.83
4	2	29	5.39	54	7.35	79	8.89
5	2.24	30	5.48	55	7.42	80	8.94
6	2.45	31	5.57	56	7.48	81	9
7	2.65	32	5.66	57	7.55	82	9.06
8	2.83	33	5.74	58	7.62	83	9.11
9	3	34	5.83	59	7.68	84	9.17
10	3.16	35	5.92	60	7.75	85	9.22
11	3.32	36	6	61	7.81	86	9.27
12	3.46	37	6.08	62	7.87	87	9.33
13	3.61	38	6.16	63	7.94	88	9.38
14	3.74	39	6.24	64	8	89	9.43
15	3.87	40	6.32	65	8.06	90	9.49
16	4	41	6.40	66	8.12	91	9.54
17	4.12	42	6.48	67	8.19	92	9.59
18	4.24	43	6.56	68	8.25	93	9.64
19	4.36	44	6.63	69	8.31	94	9.70
20	4.47	45	6.71	70	8.37	95	9.75
21	4.58	46	6.78	71	8.43	96	9.80
22	4.69	47	6.86	72	8.49	97	9.85
23	4.80	48	6.93	73	8.54	98	9.90
24	4.90	49	7	74	8.60	99	9.95
25	5	50	7.07	75	8.66	100	10

SELF-TEST 7 (pages 79-89)

Solve each equation. Report solutions in lowest terms.

1. $2h + 1(7h + 3) = 0$

$2h + 7h + 3 = 0$

$9h = -3$

$h = -\frac{1}{3}$

2. $5 = 2E + 1(4E + 1)$

$5 = 2E + -4E + -1$

$6 = -2E$

$E = -3$

3. $1 + 1(2 - 3t) = t$

$1 + -2 + 3t = t$

$-1 = -2t$

$\frac{1}{2} = t$

4. $2w + 7 = 3(5 - 4w)$

$12w + 2w + 7 = 15 - 12w + 12w$

$14w = 8$

$w = \frac{8}{14} = \boxed{\frac{4}{7}}$

5. $2t = 9 - 5(2t + 3)$

$2t = 9t - 10t - 15$

$2t = -6$

$12t = -6$

$\boxed{-\frac{1}{2}}$

6. $5x - 2(3x - 1) = 1$

$5x - 6x + 2 = 1$

$-x = -1$

$x = 1$

Complete: **7.** $\sqrt{64} = \underline{8}$ **8.** $-\sqrt{36} = \underline{-6}$ **9.** $\pm\sqrt{\frac{1}{25}} = \underline{\pm\frac{1}{5}}$

10. Which of these are perfect squares?   7, 16, 59, $\frac{1}{4}$, $\frac{3}{7}$, $\frac{4}{9}$

Using the table, complete these: **11.** $\sqrt{48} = \underline{6.93}$ **12.** $-\sqrt{75} = \underline{-8.66}$

ANSWERS: **1.** $h = -\frac{1}{3}$ **4.** $w = \frac{4}{7}$ **7.** 8   **10.** 16, $\frac{1}{4}$, $\frac{4}{9}$

2. $E = -3$ **5.** $t = -\frac{1}{2}$ **8.** -6   **11.** 6.93

3. $t = \frac{1}{2}$ **6.** $x = 1$ **9.** $\pm\frac{1}{5}$ **12.** -8.66

2-10   PURE QUADRATIC EQUATIONS

A quadratic equation has two solutions or roots.   Equations of the form $x^2 = k$ are called pure quadratic equations.   We will use the square root method to solve equations of that type in this section.

---

105.   Equations of the form $x^2 = k$ are called <u>pure quadratic equations</u>.
We can use the <u>square root principle</u> to solve equations of that type.
The principle is stated below.

> <u>The Square Root Principle For Equations</u>
>
> $$x^2 = k$$
>
> $$\sqrt{x^2} = \pm\sqrt{k}$$
>
> $$x = \pm\sqrt{k}$$
>
> <u>Note</u>: $x = \pm\sqrt{k}$ means: $x = \sqrt{k}$ and $x = -\sqrt{k}$.

We used the square root principle to solve $x^2 = 64$ below.   Use the same method to solve the other equation.

$$x^2 = 64 \qquad\qquad\qquad y^2 = 81$$

$$\sqrt{x^2} = \pm\sqrt{64}$$

$$x = \pm\sqrt{64}$$

$$x = \sqrt{64} \text{ and } -\sqrt{64}$$

$$x = 8 \text{ and } -8$$

---

106.   To solve the equation below, we used the addition axiom to get t^2 on one side and 100 on the other side.   Solve the other equation.

$$t^2 - 100 = 0 \qquad\qquad\qquad m^2 - 1 = 0$$

$$t^2 = 100$$

$$t = \pm\sqrt{100}$$

$$t = 10 \text{ and } -10$$

y = 9 and -9

---

107.   In each equation below, the number is a perfect-square fraction.   The two solutions are <u>the two square roots of the fraction</u>.   That is:

$$x^2 = \frac{36}{25} \qquad\qquad\qquad y^2 = \frac{1}{4}$$

$$x = \pm\sqrt{\frac{36}{25}} \qquad\qquad\qquad y = \pm\sqrt{\frac{1}{4}}$$

$$x = \frac{6}{5} \text{ and } -\frac{6}{5} \qquad\qquad\qquad y = \underline{\hspace{1cm}} \text{ and } \underline{\hspace{1cm}}$$

m = 1 and -1

108. To solve the equation below, we used the multiplication axiom to isolate t^2 and then found both square roots of $\frac{9}{16}$. Solve the other equation.

$$16t^2 = 9 \qquad\qquad\qquad 49d^2 = 100$$

$$t^2 = \frac{9}{16}$$

$$t = \pm\sqrt{\frac{9}{16}}$$

$$t = \frac{3}{4} \text{ and } -\frac{3}{4}$$

$\frac{1}{2}$ and $-\frac{1}{2}$

109. To solve the equation below, we used both axioms to isolate x^2 and then found both square roots of $\frac{81}{64}$. Solve the other equation.

$$64x^2 - 81 = 0 \qquad\qquad 9b^2 - 1 = 0$$

$$64x^2 = 81$$

$$x^2 = \frac{81}{64}$$

$$x = \frac{9}{8} \text{ and } -\frac{9}{8}$$

$d = \frac{10}{7}$ and $-\frac{10}{7}$

110. When the number is not a perfect square, we can use the square root table. Using the table, we get:

$$x^2 = 69$$
$$x = \pm\sqrt{69}$$
$$x = \pm 8.31$$

Use the square root table to solve these:

a) $t^2 - 94 = 0$ \qquad\qquad b) $m^2 - 80 = 0$

$b = \frac{1}{3}$ and $-\frac{1}{3}$

111. Following the example, use the square root table to solve the other equation.

$$2p^2 = 30 \qquad\qquad\qquad 5d^2 - 475 = 0$$

$$p^2 = \frac{30}{2}$$

$$p^2 = 15$$

$$p = \pm 3.87$$

a) $t = \pm\sqrt{94} = \pm 9.70$

b) $m = \pm\sqrt{80} = \pm 8.94$

$d = \pm 9.75$

112. Since there is no real number that is the square root of a negative number, the equation below has no real number solution.

$$y^2 = -100$$
$$y = \pm\sqrt{-100}$$

Solve each equation if possible.

a) $x^2 - 64 = 0$ b) $x^2 + 64 = 0$

---

a) $x = 8$ and -8 b) No real number solution.

2-11   APPLIED PROBLEMS INVOLVING FORMULAS

In this section, we will solve some applied problems involving formulas.

---

113. We used the given formula to solve one problem below. Notice how we had to solve an equation. Solve the other problem.

The Celsius (C) and Kelvin (K) scales for measuring temperature are related by the formula: $C = K - 273^\circ$. If water freezes at $0^\circ C$, at what Kelvin temperature does it freeze?

$$C = K - 273^\circ$$

$$0^\circ = K - 273^\circ$$

$$K = 273^\circ$$

Therefore, water freezes at $273^\circ K$.

In a series electric circuit, total resistance (R_t) in ohms and three resistors (R_1, R_2, and R_3) in ohms are related by the formula: $R_t = R_1 + R_2 + R_3$. Find R_2 when $R_t = 105$ ohms, $R_1 = 27$ ohms, and $R_3 = 48$ ohms.

---

114. Solve each problem.

a) The area (A), length (L), and width (W) of a rectangle are related by the formula: $A = LW$. Find the width of a rectangle whose area is 240 in^2 if its length is 20 in.

b) In an electric circuit, voltage (V) in volts, current (I) in amperes, and resistance (R) in ohms are related by the formula: $V = IR$. Find the current in a circuit whose voltage is 36 volts if the resistance is 4 ohms.

$R_2 = 30$ ohms

115. Solve each problem.

a) The volume (V), length (L), width (W), and height (H) of a rectangular solid are related by the formula: V = LWH. Find the length of a rectangular solid whose volume is 40 ft³ if its width is 2 ft and its height is 4 ft.

b) If the pressure of a confined gas changes from P_1 to P_2, its volume changes from V_1 to V_2. The pressures and volumes are related by the formula: $P_1 V_1 = P_2 V_2$. A confined gas has a volume of 240 in³ and a pressure of 60 lbs/in². Find its volume when the pressure is increased to 300 lbs/in².

a) W = 12 in

b) I = 9 amperes

116. Solve each problem.

a) The perimeter (P), length (L), and width (W) of a rectangle are related by the formula: P = 2L + 2W. If the perimeter of a rectangle is 120 cm and its width is 24 cm, find its length.

b) Steel expands when heated. The elongation (e) of a steel rail L inches long when its temperature is increased $\underline{t}$ degrees Fahrenheit is: e = (0.000006)Lt. What temperature increase will produce a 0.09 inch elongation in a 300 inch steel rail?

a) L = 5 ft, from:

40 = 8L

b) V_2 = 48 in³

117. In the figure below, the beam balances when the product of distance d_1 and weight W_1 on the left side equals the product of distance d_2 and weight W_2 on the right side. The formula is: $d_1 W_1 = d_2 W_2$.

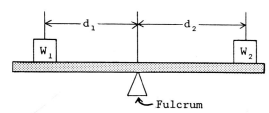

Continued on following page.

a) L = 36 cm

b) an increase of 50°F

117. Continued

a) If 6 lbs is placed 20 in from the fulcrum on the left side, how far from the fulcrum on the right side must 4 lbs be placed to balance the beam?

b) If 150 kilograms is placed 3 meters from the fulcrum on the right side, what weight must be placed 2 meters from the fulcrum on the left side to balance the beam?

---

118. To solve the problem below, we must solve a quadratic equation.

Find the side of a square if its area is 144 cm².

$$A = s^2$$

$$144 = s^2$$

$$s = \pm \sqrt{144} = \pm 12 \text{ cm}$$

Since a negative length does not make sense, the side of the square is _____.

a) d_2 = 30 in

b) W_1 = 225 kilograms

---

119. When solving for a variable that is squared in a formula, we report only the positive root because the negative root does not ordinarily make sense. An example is shown. Solve the other problem.

If an object is dropped, the approximate distance d it falls in t seconds is given by the formula: $d = 16t^2$. How long would it take an object to fall 80 ft?

Velocity v, acceleration a, and distance s are related by the formula: $v^2 = 2as$. Find the velocity when the acceleration is 3 meters/sec² and the distance is 15 meters.

$$d = 16t^2$$

$$80 = 16t^2$$

$$t^2 = \frac{80}{16}$$

$$t^2 = 5$$

$$t = \sqrt{5}$$

$$t = 2.24 \text{ seconds}$$

12 cm

v = 9.49 meters/sec

120. Solve each problem.

a) The area (A) and radius (r) of a circle are related by the formula: $A = \pi r^2$. Find the radius when the area is 314 cm^2. Use 3.14 for π.

b) In an electric circuit, power (P) in watts, current (I) in amperes, and resistance (R) in ohms are related by the formula: $P = I^2R$. Find the current in a circuit when the power is 200 watts and the resistance is 8 ohms.

a) r = 10 cm   b) I = 5 amperes

2-12 OTHER APPLIED PROBLEMS

In this section, we will solve some number problems, some part-whole problems, and some motion problems.

121. The following steps are useful for solving word problems.

Four Steps For Solving Word Problems

1. Represent the unknown or unknowns with a letter or algebraic expression.

2. Translate the problem to an equation.

3. Solve the equation.

4. Check the solution in the original words of the problem.

The four steps are used to solve the problem below.

Problem: If we subtract 5 from double a number, we get 11. Find the number.

Step 1:   Represent the unknown with a letter.

Let x equal the unknown number.

Step 2:   Translate the problem to an equation.

2x - 5 = 11

Step 3:   Solve the equation.

2x - 5 = 11

2x = 16

x = 8

Continued on following page.

121. Continued

Step 4:   Check the solution in the original words of the problem.
Double 8 is 16. If we subtract 5 from 16, we get 11. The solution checks.

Using the same steps, solve these:

a) Five times a number, plus 10 equals 25. Find the number.

b) $\frac{3}{4}$ of a number is 36. Find the number.

---

122. The problem below involves dividing a whole into two parts.

Problem: A wire 50 centimeters long is cut into two parts. The larger part is 10 centimeters longer than the smaller part. How long is each part?

We drew a diagram for the problem below. In the diagram, we let x equal the length of the smaller part. Therefore, the length of the larger part is x + 10.

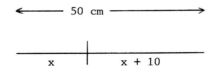

We translated to an equation, solved the equation, and checked the solution below.

$$\begin{array}{ccccc} \text{smaller} & + & \text{larger} & = & 50 \\ \downarrow & \downarrow & \downarrow & \downarrow & \downarrow \\ x & + & (x + 10) & = & 50 \end{array}$$

$$2x + 10 = 50$$
$$2x = 40$$
$$x = 20 \quad (\text{and} \quad x + 10 = 30)$$

Check: The smaller part is 20 cm and the larger part is 30 cm. 30 is 10 longer than 20 and 30 + 20 equals 50. Therefore, the solution checks.

Continued on following page.

---

a) 3, from:

$$5x + 10 = 25$$

b) 48, from:

$$\frac{3}{4}x = 36$$

122. Continued

Following the example, solve these:

a) A rope 90 feet long is cut into two parts. One part is 14 feet longer than the other. Find the lengths of the two parts.

b) A man wants to cut a 48 inch board into two pieces so that one piece will be 8 inches longer than the other. How long should the shorter piece be?

---

123. To solve the problem below, we used x̲ for the first part and 2x̲ for the second part. Solve the other problem.

A 120 meter cable must be cut into two parts so that one part is twice as long as the other. Find the length of each part.

A 68 cm rod is cut into two parts so that one part is three times as long as the other. Find the length of the larger part.

x + 2x = 120

3x = 120

x = 40

Therefore, the two lengths are 40 meters and 80 meters.

a) 38 ft and 52 ft

b) 20 in

---

124. To solve the problem below, we used x̲ for the first piece, 2x̲ for the second piece, and 3(2x) = 6x̲ for the third piece.

A 270 cm wire is cut into three pieces. The second piece is twice as long as the first. The third piece is three times as long as the second. How long is each piece?

A 350 meter rope is cut into three pieces. The second piece is twice as long as the first. The third piece is four times as long as the first. How long is each piece?

x + 2x + 3(2x) = 270

x + 2x + 6x = 270

9x = 270

x = 30

The three pieces are 30 cm, 60 cm, and 180 cm.

51 cm

125. To solve the problem below, we used <u>x</u>, <u>2x</u>, and <u>x - 5</u> for the three branches. Solve the other problem.

An electric current of 83 amperes is branched off into three circuits. The second branch carries twice as much current as the first. The third branch carries 5 amperes less current than the first. Find the amount of current in each branch.

The flow capacity of a pipeline is 800 liters per minute. The pipeline separates into three branches. The second branch has 3 times the capacity of the first branch. The third branch has a capacity of 50 liters more than the first. Find the capacity of each branch.

$$x + 2x + (x - 5) = 83$$

$$4x - 5 = 83$$

$$4x = 88$$

$$x = 22$$

Therefore, the amount of current in the three branches is 22 amperes, 44 amperes, and 17 amperes.

126. The formula $d = rt$ shows the relationship among distance traveled (d), rate (r), and time traveled (t). The formula can be used to solve many applied problems involving motion. Here is an example:

Problem: A car drove from city A to city B in 2.4 hours at a rate (or speed) of 55 mph. The return trip was made in 2 hours. Find the speed for the return trip.

To solve a problem of this type, it is helpful to make a diagram as we have done below. You can see that the two distances (from A to B and from B to A) are equal.

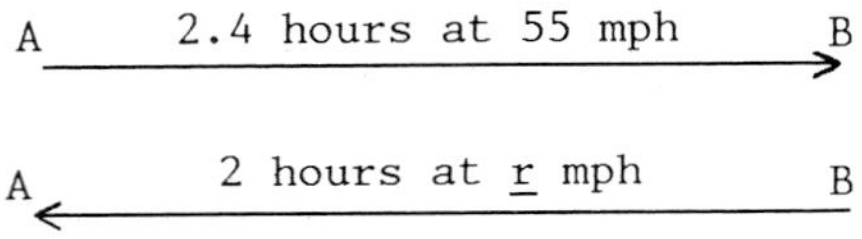

A 2.4 hours at 55 mph → B

A ← 2 hours at <u>r</u> mph B

We can organize what we know in the chart below. We used $d = rt$ to get $d_2 = 2r$.

	Distance	Rate	Time
A to B	$d_1 = (2.4)(55)$	55	2.4
B to A	$d_2 = 2r$	r	2

Continued on the following page.

Answer column:

50 meters, 100 meters, and 200 meters

150 liters, 450 liters, and 200 liters.

126. Continued

We know that the two distances (d_1 and d_2) are equal.

$$d_1 = d_2$$

We also know that $d_1 = (2.4)(55)$ and $d_2 = 2r$. Substituting, we get:

$$(2.4)(55) = 2r$$

Solving the equation, we get:

$$(2.4)(55) = 2r$$
$$132 = 2r$$
$$r = 66 \text{ mph}$$

The solution checks in the original problem because $(2.4)(55) = 132$ and $2(66) = 132$.

Following the example, draw a diagram and make up a chart to solve this one.

Problem: A train travels from city C to city D in 3.6 hours at a speed of 85 mph. The return trip was made in 4 hours. Find the speed for the return trip.

---

127. We used the same general method for the problem below. In this problem, the distance traveled is also the same for both cars.

Problem: One car leaves a motel traveling 42 mph. A second car leaves the same motel one hour later traveling 56 mph on the same highway. How long does it take the second car to catch up to the first car?

We diagrammed the problem below. The two distances are equal. Since we want to find the time traveled by the second car, we used t for the time of the second car.

Car 1

42 mph   (t + 1) hour →

Car 2

56 mph   t hours →

76.5 mph

Continued on following page.

127. Continued

We organized the data in the chart below. We used d = rt to get 42(t + 1) and 56t.

	Distance	Rate	Time
Car 1	d_1 = 42(t + 1)	42	t + 1
Car 2	d_2 = 56t	56	t

From the diagram, we can see that the two distances are equal. Therefore, we can set up and solve the following equation.

$$d_1 = d_2$$

$$42(t + 1) = 56t$$

$$42t + 42 = 56t$$

$$42 = 14t$$

$$t = 3 \text{ hours (and } t + 1 = 4 \text{ hours)}$$

Therefore, it takes the second car 3 hours to catch up to the first car (which travels 4 hours). This checks since 4(42) = 168 and 3(56) = 168.

Solve this one. Use a diagram and a chart.

Problem: A small plane leaves an airport and flies due north at 180 km/hr. A jet leaves the same airport two hours later and flies due north at 900 km/hr. How long will it take the jet to catch up to the small plane?

$\frac{1}{2}$ hour, from:

180(t + 2) = 900t

128. The formula can also be used for the problem below in which the two distances are not equal.

> <u>Problem</u>: A car and a bus leave a city at the same time traveling in opposite directions. The car travels 50 mph and the bus travels 60 mph. In how many hours will they be 275 miles apart?

A diagram of the problem is given below.

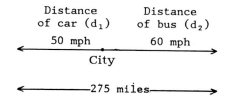

We can see that the two distances (d_1 and d_2) are not equal because the car and bus traveled at different rates for the same period of time. However, the time <u>t</u> is equal for both. We can organize what we know in the chart below.

	Distance	Rate	Time
Car	$d_1 = 50t$	50	t
Bus	$d_2 = 60t$	60	t

From the diagram, we know that the sum of the distance traveled by the car and bus equals 275 miles. That is:

$$d_1 + d_2 = 275$$

But based on d = rt, $d_1 = 50t$ and $d_2 = 60t$, as we can see from the chart. Substituting, we get:

$$50t + 60t = 275$$

Now solving the equation, we get:

$$50t + 60t = 275$$

$$110t = 275$$

$$t = \frac{275}{110} = \frac{5}{2} = 2\frac{1}{2} \text{ hours}$$

The solution checks in the original problem because $2\frac{1}{2}$ x 50 = 125, $2\frac{1}{2}$ x 60 = 150, and 125 + 150 = 275.

Continued on following page.

128. Continued

To solve the problem below, draw a diagram and make up a chart like those in the example problem.

> Problem: A car and a bus leave a city at the same time traveling in opposite directions. The car travels 45 mph and the bus travels 57 mph. In how many hours will they be 170 miles apart?

$\frac{5}{3} = 1\frac{2}{3}$ hours, from:

$$45t + 57t = 170$$

---

129. The formula can also be used for this problem.

> Problem: A 385 mile I-road connects two cities. A car leaves each city at noon driving towards the other city. If their speeds are 52 mph and 58 mph, at what time will they meet?

A diagram of the problem is given below.

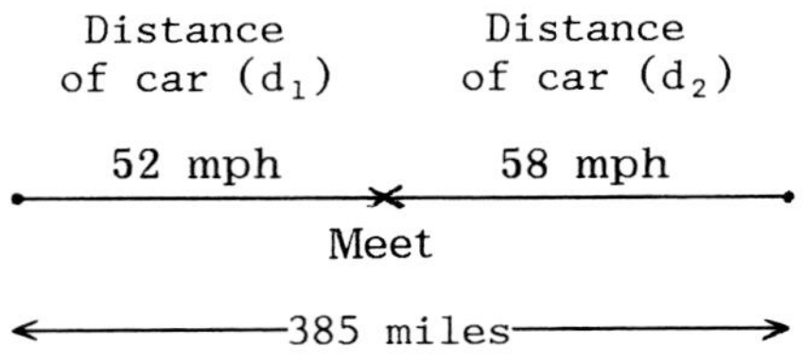

```
   Distance          Distance
 of car (d₁)        of car (d₂)

    52 mph            58 mph
  •————————————✕————————————•
              Meet

  ←—————————385 miles—————————→
```

We organized what we know in the chart below.

	Distance	Rate	Time
Car 1	$d_1 = 52t$	52	t
Car 2	$d_2 = 58t$	58	t

Continued on following page.

129. Continued

From the diagram, we know that the sum of the distances traveled by the cars is 385 miles. That is:

$$d_1 + d_2 = 385$$

Substituting 52t for d_1 and 58t for d_2, we get:

$$52t + 58t = 385$$

Solving the equation, we get:

$$52t + 58t = 385$$

$$110t = 385$$

$$t = \frac{385}{110} = \frac{7}{2} = 3\frac{1}{2} \text{ hours}$$

Therefore, the two cars meet at 3:30 p.m.

Solve this one. Use a diagram and a chart.

Problem: A 425 mile railroad track connects two cities. A train leaves each city at 7:00 a.m. for the other city. If their speeds are 80 mph and 90 mph, at what time will they meet?

9:30 a.m., since

$$t = 2\frac{1}{2} \text{ hours}$$

SELF-TEST 8 (pages 90-104)

Solve each equation.

1. $9x^2 = 25$

$\sqrt{x^2} = \sqrt{\dfrac{25}{9}}$ $x = \pm\dfrac{5}{3}$

2. $4r^2 = 1$

$\sqrt{r^2} = \sqrt{\dfrac{1}{4}}$ $r = \pm\dfrac{1}{2}$

3. $3y^2 = 60$

$\sqrt{x^2} = \sqrt{\dfrac{60}{3}}$ $y = \pm 4.47$

4. In an electric circuit, power (P) in watts, current (I) in amperes, and resistance (R) in ohms are related by the formula: $P = I^2R$. Find the current in a circuit when the power is 100 watts and the resistance is 25 ohms.

$100 = I^2 \cdot 25$

$\sqrt{\dfrac{100}{25}} = \sqrt{I^2}$ $I = 2$ Amps

5. A force (F) equals the product of pressure (P) and area (A). The formula is: $F = PA$. Find the pressure when a 600 lb force is applied to a 30 in^2 area.

$600 = P \cdot 30$

$20 \ lb/in^2$

6. The perimeter (P), length (L), and width (W) of a rectangle are related by the formula: $P = 2L + 2W$. If the perimeter of a rectangle is 256 cm and its length is 76 cm, find its width.

$256 = 152 + 2W$

$52^{cm} = W$

7. A reflected radar pulse takes 0.002 second to travel from a satellite to an earth station. If its speed is 186,000 miles per second, find the distance from the earth station to the satellite.

$\dfrac{186\,000}{1} = \dfrac{x}{.002}$

372

8. A 60 foot cable must be cut into three pieces so that the second piece is 6 feet longer than the first and the third piece is 6 feet longer than the second. Find the lengths of the three pieces.

$x + 6 + x + 12 + x = 60$

$3x + 18 = 60$

$14, 20, 26$

9. A 500 mile I-road connects two cities. A truck leaves each city at 6:00 a.m. driving towards the other city. If their speeds are 60 mph and 65 mph, at what time will they meet?

ANSWERS:

1. $x = \dfrac{5}{3}$ and $-\dfrac{5}{3}$

2. $r = \dfrac{1}{2}$ and $-\dfrac{1}{2}$

3. $y = 4.47$ and -4.47

4. $I = 2$ amperes

5. $P = 20$ lb/in^2

6. $W = 52$ cm

7. 372 miles

8. 14 ft, 20 ft, 26 ft

9. 10:00 a.m.

SUPPLEMENTARY PROBLEMS - CHAPTER 2

Assignment 5

Solve each equation. Report solutions in lowest terms.

1. $y + 9 = 5$

2. $t - 3 = 8$

3. $29 = x - 12$

4. $54 = N + 1$

5. $G - 1 = 0$

6. $5 + d = 0$

7. $x - 2.5 = 4.1$

8. $7.5 = p + 5.7$

9. $d + \frac{1}{6} = \frac{5}{6}$

10. $x - \frac{3}{4} = \frac{1}{8}$

11. $V + 1 = \frac{4}{7}$

12. $m + \frac{2}{3} = 0$

13. $9p = 4$

14. $-8R = -15$

15. $7F = -7$

16. $2 = 14y$

17. $-35 = 21w$

18. $-3h = 0$

19. $-x = 8$

20. $36.6 = 12.2y$

21. $\frac{3}{4}p = 1$

22. $2x = -\frac{4}{5}$

23. $\frac{2}{3}y = \frac{5}{6}$

24. $\frac{7}{8}H = 0$

25. $3x + 5 = 17$

26. $2r + 15 = 9$

27. $11 - 15m = 21$

28. $7 - 12s = 4$

29. $3 = 9 + 4H$

30. $-1 = 1 + 2A$

31. $4y + 7 = 0$

32. $4x - 1 = 3$

33. $3c - 5 = -6$

34. $2 - 11p = 2$

35. $5 - t = 9$

36. $4.5 - m = 0$

37. $.7A - 1.4 = 5.6$

38. $\frac{1}{3}y + \frac{1}{4} = 1$

39. $2x - \frac{1}{4} = \frac{1}{8}$

40. $\frac{2}{3} - s = \frac{2}{3}$

41. $1 - \frac{7}{2}w = 0$

42. $\frac{1}{2} = \frac{1}{4}P + 1$

Assignment 6

Do each multiplication.

1. $3(x + 7)$

2. $9(2 + 3y)$

3. $10(5R + 1)$

4. $7(x - 2)$

5. $5(3 - 4y)$

6. $8(1 - 5d)$

7. $-4(m + 3)$

8. $-1(7 - 6P)$

Combine like terms.

9. $x + 5x$

10. $7y + (-y)$

11. $-5m + (-2m)$

12. $9d - 4d$

13. $r - 7r$

14. $2 + 5p - 3$

15. $6V - 7 - V$

16. $-a - 9 - 2a$

Solve each equation. Report solutions in lowest terms.

17. $4R + (-6R) = 8$

18. $13 = -3y + (-2y)$

19. $9P - 3P = 2$

20. $25 = 2x - 12x$

21. $5w - 2w = 0$

22. $7h - h = 20$

23. $6 = F - 9F$

24. $.4x + .8x = 6$

25. $\frac{3}{4}y - \frac{3}{8}y = \frac{1}{2}$

26. $6r - 5 = 2r$

27. $b = 3b + 4$

28. $5x - 4 = 11x$

29. $a = 1 - a$

30. $10 - V = 9V$

31. $1.9t - 2.4 = 1.1t$

32. $\frac{1}{8}x + \frac{1}{4} = \frac{7}{8}x$

33. $5N + 1 = 2N + 2$

34. $2c + 7 = 7 + 9c$

35. $E + 1 = 3 - E$

36. $2h - 7 = 3h + 5$

37. $2 + 3y = 10 - y$

38. $15 - 6w = 25 - 2w$

39. $5 - k = 1 - 5k$

40. $\frac{2}{3}d - \frac{1}{2} = \frac{1}{6}d - \frac{1}{3}$

Assignment 7

Solve each equation.   Report solutions in lowest terms.

1. $a + (3a + 2) = 1$

2. $4R = (5 - 2R) + 3$

3. $18 - (2s + 7) = 12$

4. $x = 1 - (5x + 2)$

5. $6 - (2 + 5w) = 4$

6. $3b - (b + 5) = 0$

7. $G - (2G - 3) = 5$

8. $9 - (2 - 5h) = 3h$

9. $2r - (7r - 3) = 1$

10. $4t - (1 - t) = 0$

11. $4(x + 3) = 5$

10. $7y = 5(4 + y)$

13. $2 = 4(P - 1)$

14. $5(3F - 4) = 7F$

15. $2(d + 4) = 3(d + 2)$

16. $E + 11 = 4(5 + 4E)$

17. $2(x - 4) = 5x - 2$

18. $3(V + 1) = 2(V - 1)$

19. $3w + 5(w + 2) = 50$

20. $7 + 4(3d - 1) = 0$

21. $25 - 2(4x + 5) = 45$

22. $8y - 3(2y + 7) = 9$

23. $3 = 9 - 2(1 - 4E)$

24. $4 - 5(2r - 7) = 3r$

Find these square roots.   Where necessary, use the table.

25. $\sqrt{4}$

26. $-\sqrt{1}$

27. $\pm\sqrt{100}$

28. $-\sqrt{9}$

29. $\sqrt{2}$

30. $\pm\sqrt{37}$

31. $-\sqrt{75}$

32. $\sqrt{99}$

33. $\sqrt{\frac{25}{16}}$

34. $\pm\sqrt{\frac{1}{9}}$

35. $-\sqrt{\frac{36}{64}}$

36. $\pm\sqrt{\frac{1}{100}}$

37. Which of these are perfect squares?

   $1, 7, 18, 25, 40, 48, 64, 72, 81$

38. Which of these are perfect squares?

   $\frac{1}{2}, \frac{1}{100}, \frac{9}{20}, \frac{16}{49}, \frac{81}{60}, \frac{100}{81}, \frac{1}{8}, \frac{25}{64}$

Assignment 8

Solve each equation.

1. $x^2 = 49$

2. $y^2 = \frac{1}{100}$

3. $p^2 = \frac{81}{64}$

4. $2t^2 = 72$

5. $r^2 - 4 = 0$

6. $16a^2 = 1$

7. $4x^2 - 25 = 0$

8. $3m^2 - 48 = 0$

9. $V^2 = 80$

10. $4h^2 = 48$

11. $5x^2 - 340 = 0$

12. $10 - 5s^2 = 0$

Solve each problem.

13. The area (A), length (L), and width (W) of a rectangle are related by the formula: $A = LW$. Find the length of a rectangle whose area is 320 cm² if its width is 16 cm.

14. In an electric circuit, total resistance (R_t) in ohms and the resistances of three series resistors (R_1, R_2, and R_3) in ohms are related by the formula: $R_t = R_1 + R_2 + R_3$.   Find R_1 when $R_t = 210$ ohms, $R_2 = 50$ ohms, and $R_3 = 70$ ohms.

15. The Kelvin (K) and Celsius (C) temperature scales are related by the formula: $K = C + 273$. Water boils at 373° Kelvin. Find the corresponding Celsius temperature.

16. Steel expands when heated. The elongation e of a steel bar L inches long when its temperature is increased t degrees Fahrenheit is: $e = (0.000006)Lt$. What temperature increase will produce a .024 inch elongation in a 200 inch steel rail?

17. If the pressure of a confined gas changes from P_1 to P_2, its volume changes from V_1 to V_2. The pressures and volumes are related by the formula: $P_1V_1 = P_2V_2$. If a confined gas has a volume of 300 in^3 and a pressure of 80 lbs/in^2, find its volume when the pressure is decreased to 48 lbs/in^2.

18. In an electric circuit, voltage (V) in volts, current (I) in amperes, and resistance (R) in ohms are related by the formula: $V = IR$. Find the resistance of a circuit whose voltage is 12 volts and its current is 0.3 ampere.

19. If an object is dropped, the approximate distance d in feet that it falls in t seconds is given by the formula: $d = 16t^2$. How long would it take a dropped object to fall 400 feet?

20. A gear which has N_1 teeth and a rotational speed of R_1 rpm drives a gear which has N_2 teeth and a rotational speed of R_2 rpm. The teeth and speeds are related by the formula: $N_1R_1 = N_2R_2$. If a 72-tooth gear whose rotational speed is 40 rpm drives a 48-tooth gear, find the rotational speed of the driven gear.

21. When a beam is balanced on a fulcrum, the product of distance d_1 and weight W_1 on the left side equals the product of distance d_2 and weight W_2 on the right side. The formula is: $d_1W_1 = d_2W_2$. If 40 kilograms is placed 1.5 meters from the fulcrum on the left side, how far from the fulcrum on the right side must 20 kilograms be placed to balance the beam?

22. In an electric circuit, power (P) in watts, current (I) in amperes, and resistance (R) in ohms are related by the formula: $P = I^2R$. Find the current in a circuit when the power is 80 watts and the resistance is 16 ohms.

23. Velocity v, acceleration a, and distance s are related by the formula: $v^2 = 2as$. Find the acceleration when the velocity is 8 meters/sec and the distance is 16 meters.

24. A force (F) in kilograms applied to a spring produces a stretch (s) in centimeters according to the formula: $F = 16s$. Find the stretch in the spring when a force of 40 kilograms is applied.

25. An electrical transformer has an input voltage V_1, an input winding of N_1 turns, an output voltage V_2, and an output winding of N_2 turns. The variables are related by the formula: $V_1N_2 = V_2N_1$. For an input voltage of 20 volts, find the output voltage of a transformer whose input and output windings are 400 and 1,000 turns, respectively.

Solve each problem.

26. If 3 is subtracted from four times a number, the difference is 5. Find the number.

27. If 8 is added to a number, the sum is three times the number. Find the number.

28. A steel bar 120 cm long must be cut into two parts so that one part is three times as long as the other. Find the length of each part.

29. An electric current of 12 amperes branches into two circuits. One branch carries 4 amperes more than the other. Find the current in each branch.

30. A 96-inch board is to be cut into three parts so that the second part equals the first and the third part is 24 inches longer than the first. Find the lengths of the three parts.

31. A water pipeline carrying 2,000 gal/min separates into three branches. The capacity of the second branch is 150 gal/min more than the first. The capacity of the third branch is 250 gal/min less than the first. Find the capacity of each branch.

32. How long does it take an airplane to travel 3,300 miles when its speed is 600 mph?

33. An orbiting space station traveled 9,000 miles in 30 minutes. Find its speed in miles per hour.

34. A train traveled from city A to city B in 4.5 hours at a speed of 72 mph. The return trip was made in 4 hours. Find the speed for the return trip.

35. A small airplane flies due west from an airport at 160 km/hr. A jet airplane leaves the same airport three hours later and flies due west at 880 km/hr. How long will it take the jet to catch up to the small plane?

36. A bus and a truck leave the same terminal at noon traveling in opposite directions. The bus travels 55 mph and the truck travels 45 mph. In how many hours will they be 150 miles apart?

37. A 300 mile railroad track connects two cities. A train leaves each city at 2:00 p.m. for the other city. If their speeds are 45 mph and 75 mph, at what time will they meet?

Fractional Equations

3

A fractional equation is an equation that contains one or more algebraic fractions. That is, it contains a fraction that has a variable in its numerator or denominator or both. In this chapter, we will discuss a method for solving fractional equations. Some applied problems involving fractional equations are included.

3-1 EQUATIONS CONTAINING ONE FRACTION

To solve equations containing one algebraic fraction, we begin by clearing the fraction. We will discuss the method in this section.

1. An <u>algebraic fraction</u> is a fraction that contains a variable in either its numerator or denominator or both. Some examples are:

$$\frac{x}{3} \qquad \frac{5}{2y} \qquad \frac{m+1}{4} \qquad \frac{t}{t-1}$$

Just like numerical fractions, algebraic fractions stand for a division. That is:

$\frac{x}{3}$ means $x \div 3$ $\qquad$ $\frac{t}{t-1}$ means _____ ÷ _____

$t \div (t - 1)$

2. Some substitutions in algebraic fractions do not make sense because they lead to a division by 0, and division by 0 is not possible. For example:

If we substitute 0 for $\underline{x}$ below, we get a division by 0.

$$\frac{10}{x} = \frac{10}{0} \qquad \underline{\text{Note}}: \quad \frac{10}{0} \text{ is not possible.}$$

If we substitute 3 for $\underline{x}$ below, we get a division by 0.

$$\frac{x + 5}{x - 3} = \frac{3 + 5}{3 - 3} = \frac{8}{0} \qquad \underline{\text{Note}}: \quad \frac{8}{0} \text{ is not possible.}$$

To avoid the problem above, we will do the following in this text:

> We will avoid substitutions in algebraic fractions that do not make sense. That is, we will assume that a variable does not take on a value that leads to a division by 0.

---

3. When the denominator of an algebraic fraction is "1", the fraction equals its numerator. That is:

$$\frac{x}{1} = x \qquad\qquad \frac{4y}{1} = 4y \qquad\qquad \frac{t + 7}{1} = \underline{\hspace{2cm}}$$

---

4. When the numerator and denominator of an algebraic fraction are identical, the fraction equals "1". That is:

$$\frac{x}{x} = 1 \qquad\qquad \frac{5y}{5y} = 1 \qquad\qquad \frac{m - 3}{m - 3} = \underline{\hspace{1.5cm}}$$

t + 7

---

5. In each example below, we multiplied an algebraic fraction by its denominator. Notice that each product is the numerator of the original fraction.

$$3\left(\frac{x}{3}\right) = \frac{(3)(x)}{3} = \left(\frac{3}{3}\right)(x) = (1)(x) = x$$

$$y\left(\frac{5}{y}\right) = \frac{(y)(5)}{y} = \left(\frac{y}{y}\right)(5) = (1)(5) = 5$$

The "cancelling" shortcut below can be used for the same multiplications.

$$\overset{1}{\cancel{3}}\left(\frac{x}{\cancel{3}}\right) = \frac{x}{1} = x \qquad\qquad \cancel{y}\left(\frac{5}{\cancel{y}}\right) = \frac{5}{1} = 5$$

Use the shortcut for these:

a) $7\left(\dfrac{d}{7}\right) = \underline{\hspace{1.5cm}}$ \qquad b) $m\left(\dfrac{8}{m}\right) = \underline{\hspace{1.5cm}}$

1

---

a) d   b) 8

6. In each example below, we used the "cancelling" shortcut to multiply an algebraic fraction by its denominator. Notice that each product is <u>the numerator of the original fraction</u>.

$$\overset{1}{\cancel{5}}\left(\frac{4y}{\cancel{5}_1}\right) = \frac{4y}{1} = 4y \qquad \overset{1}{\cancel{2x}}\left(\frac{9}{\cancel{2x}_1}\right) = \frac{9}{1} = 9$$

Use the shortcut for these:

a) $10\left(\frac{3t}{10}\right) = $ _____

b) $7m\left(\frac{6}{7m}\right) = $ _____

---

7. In each example below, we multiplied an algebraic fraction by its denominator. Notice again that each product is <u>the numerator of the original fraction</u>.

$$\overset{1}{\cancel{9}}\left(\frac{x + 5}{\cancel{9}_1}\right) = \frac{x + 5}{1} = x + 5 \qquad (\overset{1}{\cancel{t - 5}})\left(\frac{t}{\cancel{t - 5}_1}\right) = \frac{t}{1} = t$$

Use the same method for these:

a) $y\left(\frac{y - 3}{y}\right) = $ _____

b) $(x + 8)\left(\frac{4}{x + 8}\right) = $ _____

a) 3t   b) 6

---

8. Whenever an algebraic fraction is multiplied by its denominator, the product is identical to its numerator. That is:

$$7x\left(\frac{1}{7x}\right) = 1 \qquad V\left(\frac{V - 7}{V}\right) = V - 7 \qquad (x + 4)\left(\frac{5}{x + 4}\right) = 5$$

Using the above fact, complete these:

a) $6\left(\frac{5R}{6}\right) = $ ___ b) $5\left(\frac{1 + h}{5}\right) = $ ___ c) $(10 - x)\left(\frac{x}{10 - x}\right) = $ ___

a) y - 3   b) 4

---

9. To solve the fractional equation below, we began by <u>clearing the fraction</u>. To do so, we used the multiplication axiom, multiplying both sides by 5, <u>the denominator of the fraction</u>. Solve the other equation by multiplying both sides by 8.

Check

$$\frac{x}{5} = 7 \qquad\qquad \frac{x}{5} = 7 \qquad\qquad 4 = \frac{t}{8}$$

$$\cancel{5}\left(\frac{x}{\cancel{5}}\right) = 5(7) \qquad \frac{35}{5} = 7$$

$$x = 35 \qquad\qquad 7 = 7$$

a) 5R

b) 1 + h

c) x

$$8(4) = 8\left(\frac{t}{8}\right)$$

$$32 = t$$

$$t = 32$$

10. To clear the fraction below, we multiplied both sides by y, the denominator of the fraction. To solve the other equation, begin by multiplying both sides by R.

$$\frac{12}{y} = 3 \qquad\qquad 6 = \frac{42}{R}$$

$$\cancel{y}\left(\frac{12}{\cancel{y}}\right) = y(3)$$

$$12 = 3y$$

$$\frac{12}{3} = y$$

$$y = 4$$

---

11. To solve the equation below, we began by multiplying both sides by 3 to clear the fraction. To solve the other equation, begin by multiplying both sides by 2.

$$R(6) = R\left(\frac{42}{R}\right)$$
$$6R = 42$$
$$R = 7$$

<u>Check</u>

$$\frac{2m}{3} = 6 \qquad \frac{2m}{3} = 6 \qquad\qquad 12 = \frac{3x}{2}$$

$$\cancel{3}\left(\frac{2m}{\cancel{3}}\right) = 3(6) \qquad \frac{2(9)}{3} = 6$$

$$2m = 18 \qquad\qquad \frac{18}{3} = 6$$

$$m = \frac{18}{2} \qquad\qquad 6 = 6$$

$$m = 9$$

---

12. To solve the equation below, we began by multiplying both sides by 3t. To solve the other equation, begin by multiplying both sides by 9q.

$$x = 8$$

$$\frac{7}{3t} = 2 \qquad\qquad 1 = \frac{5}{9q}$$

$$\cancel{3t}\left(\frac{7}{\cancel{3t}}\right) = 3t(2)$$

$$7 = 6t$$

$$t = \frac{7}{6}$$

$$q = \frac{5}{9}$$

13. Following the example, solve the other equation.

$$\frac{x^2}{3} = 12 \qquad\qquad 16 = \frac{9y^2}{4}$$

$$3\left(\frac{x^2}{3}\right) = 3(12)$$

$$x^2 = 36$$

$$x = 6 \text{ and } -6$$

---

14. To clear the fraction below, we multiplied both sides by $4d^2$. Solve the other equation.

$$\frac{25}{4d^2} = 9 \qquad\qquad 81 = \frac{49}{t^2}$$

$$4d^2\left(\frac{25}{4d^2}\right) = 4d^2(9)$$

$$25 = 36d^2$$

$$d^2 = \frac{25}{36}$$

$$d = \frac{5}{6} \text{ and } -\frac{5}{6}$$

$y = \frac{8}{3}$ and $-\frac{8}{3}$

---

15. To solve the equation below, we began by multiplying both sides by 2. Solve the other equation.

$$\frac{x + 3}{2} = 5 \qquad\qquad 1 = \frac{y - 3}{5}$$

$$2\left(\frac{x + 3}{2}\right) = 2(5)$$

$$x + 3 = 10$$

$$x + 3 + (-3) = 10 + (-3)$$

$$x = 7$$

$t = \frac{7}{9}$ and $-\frac{7}{9}$

---

16. To solve the equation below, we began by multiplying both sides by $\underline{x}$. Solve the other equation.

$$\frac{x + 1}{x} = 3 \qquad\qquad 1 = \frac{3y - 5}{2y}$$

$$x\left(\frac{x + 1}{x}\right) = x(3)$$

$$x + 1 = 3x$$

$$-x + x + 1 = 3x + (-x)$$

$$1 = 2x$$

$$x = \frac{1}{2}$$

$y = 8$

17. Following the example, solve the other equation.

$$\frac{2m}{7} = 0 \qquad\qquad \frac{y + 5}{9} = 0$$

$$7\left(\frac{2m}{7}\right) = 7(0)$$

$$2m = 0$$

$$m = \frac{0}{2}$$

$$m = 0$$

$y = 5$

---

18. To solve the equation below, we began by multiplying both sides by $(x + 3)$. Notice that we then had to multiply by the distributive principle on the right side. Solve the other equation.

$$\frac{7}{x + 3} = 5 \qquad\qquad 4 = \frac{y}{y - 1}$$

$$(x + 3)\left(\frac{7}{x + 3}\right) = 5(x + 3)$$

$$7 = 5x + 15$$

$$-15 + 7 = 5x + 15 + (-15)$$

$$-8 = 5x$$

$$x = -\frac{8}{5}$$

$y = -5$

---

19. Following the example, solve the other equation.

$$\frac{2x - 1}{2x + 1} = 3 \qquad\qquad 4 = \frac{w + 3}{2w - 3}$$

$$(2x + 1)\left(\frac{2x - 1}{2x + 1}\right) = 3(2x + 1)$$

$$2x - 1 = 6x + 3$$

$$-2x + 2x - 1 = -2x + 6x + 3$$

$$-1 = 4x + 3$$

$$-1 + (-3) = 4x + 3 + (-3)$$

$$-4 = 4x$$

$$x = \frac{-4}{4} = -1$$

$y = \frac{4}{3}$

$w = \frac{15}{7}$

32. To clear the fraction below, we multiplied both sides by 3t.   Solve the other equation.

$$2 - \frac{45}{3t} = 5$$

$$3t\left(2 - \frac{45}{3t}\right) = 3t(5)$$

$$3t(2) - 3t\left(\frac{45}{3t}\right) = 15t$$

$$6t - 45 = 15t$$

$$-45 = 9t$$

$$t = -5$$

$$7 = 10 - \frac{5}{2m}$$

$y = 2$, from:

$$3y = 4 + y$$

$m = \frac{5}{6}$, from: $14m = 20m - 5$

SELF-TEST 9 (pages 109-119)

Solve each equation.   Report each solution in lowest terms.

1. $\frac{6}{5x} = 2$	2. $3 = \frac{2t - 7}{3t}$	3. $\frac{9}{4d^2} = 25$
4. $7 = \frac{r - 8}{r + 6}$	5. $\frac{4(w - 3)}{9w} = 0$	6. $6 = \frac{3m}{4(m + 2)}$ $4m+8$ $(4m+8)6 = 3m$ $24m + 48 = 3m$ $48 = -21m$ $-\frac{16}{7}$ or $-2\frac{2}{7}$
7. $\frac{5(y + 4)}{2(y + 1)} = 1$	8. $4 = \frac{5}{x} - 6$	9. $3 - \frac{10}{2s} = 9$

ANSWERS: 1. $x = \frac{3}{5}$ 4. $r = -\frac{25}{3}$ 7. $y = -6$

2. $t = -1$ 5. $w = 3$ 8. $x = \frac{1}{2}$

3. $d = \frac{3}{10}$ and $-\frac{3}{10}$ 6. $m = -\frac{16}{7}$ 9. $s = -\frac{5}{6}$

3-4   PROPORTIONS

A <u>proportion</u> is a fractional equation that contains only one fraction on each side. In this section, we will discuss a method for solving proportions.

---

33. To clear the fractions in proportions with numerical denominators, we must do multiplications like those below.

$$10\left(\frac{x}{5}\right) = \frac{(10)(x)}{5} = \left(\frac{10}{5}\right)(x) = 2x$$

$$6\left(\frac{y-1}{2}\right) = \frac{(6)(y-1)}{2} = \left(\frac{6}{2}\right)(y-1) = 3(y-1)$$

We can use cancelling to shorten the multiplications above. We get:

$$\overset{2}{\cancel{10}}\left(\frac{x}{\cancel{5}}\right) = 2x \qquad\qquad \overset{3}{\cancel{6}}\left(\frac{y-1}{\cancel{2}}\right) = 3(y-1)$$

Use cancelling to complete these:

a) $12\left(\frac{m}{3}\right) =$ _____

b) $10\left(\frac{t+3}{2}\right) =$ _____

---

34. In the proportion below, the larger denominator is a multiple of the smaller. We cleared the fractions by multiplying both sides by 8. Solve the other proportion. Begin by multiplying both sides by 9.

| | | a) 4m   b) 5(t + 3) |

$$\frac{x}{4} = \frac{7}{8} \qquad \underline{\text{Check}} \qquad\qquad \frac{1}{3} = \frac{y}{9}$$

$$\overset{2}{\cancel{8}}\left(\frac{x}{\cancel{4}}\right) = \cancel{8}\left(\frac{7}{\cancel{8}}\right) \qquad \frac{\frac{7}{2}}{4} = \frac{7}{8}$$

$$2x = 7 \qquad \frac{7}{2}\left(\frac{1}{4}\right) = \frac{7}{8}$$

$$x = \frac{7}{2} \qquad\qquad \frac{7}{8} = \frac{7}{8}$$

---

35. When the larger denominator is a multiple of the smaller, we can clear the fractions by multiplying both sides <u>by the larger denominator</u>. Following the example, solve the other proportion.

y = 3

$$\frac{5-x}{12} = \frac{1}{4} \qquad\qquad \frac{1}{2} = \frac{5-y}{10}$$

$$\cancel{12}\left(\frac{5-x}{\cancel{12}}\right) = \overset{3}{\cancel{12}}\left(\frac{1}{\cancel{4}}\right)$$

$$5 - 1x = 3$$

$$(-5) + 5 - 1x = 3 + (-5)$$

$$-1x = -2$$

$$x = \frac{-2}{-1} = 2$$

36. Following the example, solve the other proportion.

$$\frac{5}{6} = \frac{m + 2}{3}$$

$$\overset{2}{\cancel{6}}\left(\frac{5}{\cancel{6}}\right) = \overset{2}{\cancel{6}}\left(\frac{m + 2}{\cancel{3}}\right)$$

$$5 = 2(m + 2)$$

$$5 = 2m + 4$$

$$1 = 2m$$

$$m = \frac{1}{2}$$

$$\frac{t + 1}{2} = \frac{1}{6}$$

$y = 0$

---

37. When the larger denominator is not a multiple of the smaller, we can clear the fractions in a proportion by multiplying both sides <u>by the lowest common multiple of both denominators</u>. That is:

For $\frac{4x}{5} = \frac{7}{3}$, we multiply both sides by 15.

For $\frac{1}{4} = \frac{7 - t}{6}$, we multiply both sides by 12.

To clear the fractions below, we multiplied both sides by 12. Solve the other proportion.

$$\frac{x}{3} = \frac{3}{4}$$

$$\overset{4}{\cancel{12}}\left(\frac{x}{\cancel{3}}\right) = \overset{3}{\cancel{12}}\left(\frac{3}{\cancel{4}}\right)$$

$$4x = 9$$

$$x = \frac{9}{4}$$

$$\frac{1}{6} = \frac{y}{8}$$

$t = -\frac{2}{3}$, from:

$$3t + 3 = 1$$

---

38. Solve: a) $\frac{5}{6} = \frac{3y}{10}$

b) $\frac{2F - 1}{5} = \frac{1}{2}$

$y = \frac{4}{3}$, from:

$$4 = 3y$$

a) $y = \frac{25}{9}$, from:

$$25 = 9y$$

b) $F = \frac{7}{4}$, from:

$$2(2F - 1) = 5$$

39. Following the example, solve the other proportion.

$$\frac{x^2}{20} = \frac{5}{2}$$

$$\cancel{20}\left(\frac{x^2}{\cancel{20}}\right) = \cancel{20}^{10}\left(\frac{5}{\cancel{2}}\right)$$

$$x^2 = 50$$

$$x = \pm\sqrt{50} = \pm 7.07$$

$$\frac{5}{6} = \frac{y^2}{12}$$

---

40. To clear the fractions in proportions with a literal denominator, we must do multiplications like those below.

$$8x\left(\frac{7}{8}\right) = \frac{(8)(x)(7)}{8} = \left(\frac{8}{8}\right)(x)(7) = (1)(x)(7) = 7x$$

$$6y\left(\frac{y - 5}{y}\right) = \frac{(6)(y)(y - 5)}{y} = \left(\frac{y}{y}\right)(6)(y - 5) = (1)(6)(y - 5) = 6(y - 5)$$

We can use cancelling for the same multiplications. We get:

$$\cancel{8}x\left(\frac{7}{\cancel{8}}\right) = 7x \qquad 6\cancel{y}\left(\frac{y - 5}{\cancel{y}}\right) = 6(y - 5)$$

Use cancelling for these:

a) $3y\left(\frac{5}{3}\right) = $ _____

b) $4d\left(\frac{d + 7}{d}\right) = $ _____

[answer, right column] $y = \pm\sqrt{10} = \pm 3.16$

---

41. To solve proportions with a literal denominator, we also must do multiplications like those below.

$$6x\left(\frac{5}{3x}\right) = \frac{(6)(x)(5)}{(3)(x)} = \left(\frac{6}{3}\right)\left(\frac{x}{x}\right)(5) = (2)(1)(5) = 10$$

$$12t\left(\frac{t - 2}{4t}\right) = \frac{(12)(t)(t - 2)}{(4)(t)} = \left(\frac{12}{4}\right)\left(\frac{t}{t}\right)(t - 2) = (3)(1)(t - 2) = 3(t - 2)$$

We can use cancelling for the same multiplications. We get:

$$\overset{2}{\cancel{6}}x\left(\frac{5}{\cancel{3}x}\right) = 10 \qquad \overset{3}{\cancel{12}}\cancel{t}\left(\frac{t - 2}{\cancel{4}\cancel{t}}\right) = 3(t - 2)$$

Use cancelling for these:

a) $8m\left(\frac{9}{2m}\right) = $ _____

b) $10d\left(\frac{d + 7}{5d}\right) = $ _____

[answer, right column] a) 5y

b) 4(d + 7)

[answer, right column] a) 36

b) 2(d + 7)

42. To clear the fractions in proportions like those below, we multiply both sides by both denominators at the same time. That is:

For $\frac{6}{5} = \frac{9}{x}$,   we multiply by 5x.

To clear the fraction below, we multiplied both sides by 6x. Solve the other proportion.

$$\frac{5}{6} = \frac{2}{x}$$ $$\frac{4}{y} = \frac{3}{2}$$

$$6x\left(\frac{5}{6}\right) = 6x\left(\frac{2}{x}\right)$$

$$5x = 12$$

$$x = \frac{12}{5}$$

---

43. To clear the fractions below, we multiplied both sides by $18d^2$. Solve the other proportion.

$$\frac{25}{18} = \frac{2}{d^2}$$ $$\frac{3}{y^2} = \frac{4}{3}$$

$$18d^2\left(\frac{25}{18}\right) = 18d^2\left(\frac{2}{d^2}\right)$$

$$25d^2 = 36$$

$$d^2 = \frac{36}{25}$$

$$d = \frac{6}{5} \text{ and } -\frac{6}{5}$$

$y = \frac{8}{3}$, from:

$$8 = 3y$$

---

44. To clear the fractions below, we multiplied both sides by 4t. Solve the other proportion.

$$\frac{t + 8}{t} = \frac{3}{4}$$ $$\frac{4}{3} = \frac{d - 1}{d}$$

$$4t\left(\frac{t + 8}{t}\right) = 4t\left(\frac{3}{4}\right)$$

$$4(t + 8) = 3t$$

$$4t + 32 = 3t$$

$$32 = -1t$$

$$t = \frac{32}{-1} = -32$$

$y = \frac{3}{2}$ and $-\frac{3}{2}$

---

d = -3, from:

$$4d = 3(d - 1)$$

45. To clear the fractions in proportions like those below, we multiply both sides by a term in which the coefficient is the <u>lowest</u> <u>common</u> <u>multiple</u> of the numerical factors. That is:

For $\dfrac{7}{10} = \dfrac{3}{2d}$, we multiply by 10d.

For $\dfrac{t - 2}{4t} = \dfrac{1}{6t}$, we multiply by 12t.

To clear the fractions below, we multiplied both sides by 6x. Solve the other proportion.

$$\frac{5}{3x} = \frac{1}{2} \qquad\qquad \frac{7}{10} = \frac{3}{2d}$$

$$\overset{2}{\cancel{6}}x\left(\frac{5}{\cancel{3x}}\right) = \overset{3}{\cancel{6}}x\left(\frac{1}{\cancel{2}}\right)$$

$$10 = 3x$$

$$x = \frac{10}{3}$$

---

46. Following the example, solve the other proportion.

$$6t \cdot \frac{t - 2}{6t} = \frac{1}{8t} \qquad\qquad \frac{1}{3y} = \frac{y + 1}{4y}$$

$$\overset{4}{\cancel{24t}}\left(\frac{t - 2}{\cancel{6t}}\right) = \overset{3}{\cancel{24t}}\left(\frac{1}{\cancel{8t}}\right) \qquad 8t\,(t-2) = 6t$$

$$4(t - 2) = 3$$

$$4t - 8 = 3$$

$$4t = 11$$

$$t = \frac{11}{4}$$

$d = \dfrac{15}{7}$, from:

$7d = 15$

---

47. To clear the fractions in proportions like the one below, we multiply both sides by both denominators at the same time. That is:

$2x+6 = 40$
$2x = 34$
$x =$

For $\dfrac{2}{5} = \dfrac{8}{x + 3}$, we multiply by $5(x + 3)$.

To clear the fractions above, we must do these multiplications.

$$\frac{2}{5}(5)(x + 3) = \frac{(2)(5)(x + 3)}{5} = \left(\frac{5}{5}\right)(2)(x + 3) = (1)(2)(x + 3) = 2(x + 3)$$

$$5(x + 3)\left(\frac{8}{x + 3}\right) = \frac{(5)(x + 3)(8)}{x + 3} = \left(\frac{x + 3}{x + 3}\right)(5)(8) = (1)(5)(8) = 40$$

$y = \dfrac{1}{3}$, from:

$4 = 3(y + 1)$

Continued on following page.

47. Continued

We can use cancelling for the same multiplications. We get:

$$\frac{2}{3}(3)(x + 3) = 2(x + 3) \qquad 5(x + 3)\left(\frac{8}{x + 3}\right) = 40$$

Use cancelling for these:

a) $3(y - 7)\left(\frac{10}{y - 7}\right) = $ _____ b) $\frac{7}{3}(3)(t - 4) = $ _____

---

48. To clear the fractions below, we multiplied both sides by $5(t + 9)$. Solve the other proportion.

$$\frac{16}{t + 9} = \frac{4}{5} \qquad\qquad \frac{3m}{m - 5} = \frac{3}{2}$$

$$5(t + 9)\left(\frac{16}{t + 9}\right) = \frac{4}{5}(5)(t + 9)$$

$$80 = 4(t + 9)$$

$$80 = 4t + 36$$

$$44 = 4t$$

$$t = 11$$

a) 30

b) $7(t - 4)$

---

49. To clear the fractions below, we multiplied both sides by $x(x - 3)$. Solve the other equation.

$$\frac{4}{x - 3} = \frac{2}{x} \qquad\qquad \frac{5}{y} = \frac{1}{3y + 4}$$

$$x(x - 3)\left(\frac{4}{x - 3}\right) = x(x - 3)\left(\frac{2}{x}\right)$$

$$x(4) = (x - 3)2$$

$$4x = 2x - 6$$

$$2x = -6$$

$$x = -3$$

$m = -5$, from:

$$6m = 3m - 15$$

$y = -\dfrac{10}{7}$, from:

$$15y + 20 = y$$

50. To clear the fractions below, we multiplied both sides by $5(x + 5)$. Solve the other equation.

$$\frac{x - 3}{x + 5} = \frac{3}{5} \qquad\qquad \frac{y + 4}{4y - 1} = \frac{1}{2}$$

$$5(\cancel{x + 5})\left(\frac{x - 3}{\cancel{x + 5}}\right) = \cancel{5}(x + 5)\left(\frac{3}{\cancel{5}}\right)$$

$$5(x - 3) = (x + 5)(3)$$

$$5x - 15 = 3x + 15$$

$$2x = 30$$

$$x = 15$$

51. To clear the fractions below, we multiplied both sides by $(x + 1)(x - 4)$. Solve the other equation.

$$\frac{3}{x + 1} = \frac{2}{x - 4} \qquad\qquad \frac{4}{y - 1} = \frac{5}{y + 3}$$

$$(\cancel{x+1})(x-4)\left(\frac{3}{\cancel{x+1}}\right) = (x+1)(\cancel{x-4})\left(\frac{2}{\cancel{x-4}}\right)$$

$$(x - 4)(3) = (x + 1)(2)$$

$$3x - 12 = 2x + 2$$

$$x = 14$$

$y = \frac{9}{2}$, from:

$$2y + 8 = 4y - 1$$

$y = 17$, from: $4y + 12 = 5y - 5$

3-5   OTHER EQUATIONS CONTAINING TWO OR MORE FRACTIONS

In this section, we will discuss the method for solving other equations that contain two or more fractions.

52. When an equation contains two numerical denominators, we can clear the fractions by multiplying both sides by the lowest common multiple of both denominators. For example, we multiplied by 12 below. Notice that we had to multiply by the distributive principle on the right side. Solve the other equation.

$$\frac{x}{12} = x + \frac{1}{4} \qquad\qquad 1 - \frac{y}{4} = \frac{3y}{8}$$

$$\cancel{12}\left(\frac{x}{\cancel{12}}\right) = 12\left(x + \frac{1}{4}\right)$$

$$x = 12(x) + \overset{3}{\cancel{12}}\left(\frac{1}{\cancel{4}}\right)$$

$$x = 12x + 3$$

$$-11x = 3$$

$$x = -\frac{3}{11}$$

53. To clear the fractions below, we multiplied both sides by 12. Solve the other equation.

$$\frac{y}{4} = \frac{5y}{6} + 2 \qquad\qquad \frac{3x}{2} - 3 = \frac{x}{5}$$

$$\overset{3}{\cancel{12}}\left(\frac{y}{\cancel{4}}\right) = 12\left(\frac{5y}{6} + 2\right)$$

$$3y = \overset{2}{\cancel{12}}\left(\frac{5y}{\cancel{6}}\right) + 12(2)$$

$$3y = 10y + 24$$

$$-7y = 24$$

$$y = -\frac{24}{7}$$

$y = \frac{8}{5}$, from:

$$8 - 2y = 3y$$

---

54. Following the example, solve the other equation.

$$\frac{5x}{3} + \frac{1}{2} = 0 \qquad\qquad 0 = \frac{y}{8} - \frac{1}{2}$$

$$6\left(\frac{5x}{3} + \frac{1}{2}\right) = 6(0)$$

$$\overset{2}{\cancel{6}}\left(\frac{5x}{\cancel{3}}\right) + \overset{3}{\cancel{6}}\left(\frac{1}{\cancel{2}}\right) = 0$$

$$10x + 3 = 0$$

$$10x = -3$$

$$x = -\frac{3}{10}$$

$x = \frac{30}{13}$, from:

$$15x - 30 = 2x$$

---

55. When an equation contains more than two fractions, we also clear the fractions by multiplying both sides by the lowest common multiple of all the denominators. That is:

For $\frac{x}{2} + \frac{x}{5} = \frac{3x}{10} + 4$, we multiply by 10.

For $\frac{5y}{7} - 1 = \frac{y}{2} + \frac{1}{2}$, we multiply by 14.

To clear the fractions below, we multiplied both sides by 8. Solve the other equation.

$$\frac{d}{2} - \frac{d}{4} = \frac{d-1}{8} \qquad\qquad \frac{F}{4} + \frac{3}{5} = 2 - \frac{3F}{5}$$

$$8\left(\frac{d}{2} - \frac{d}{4}\right) = \cancel{8}\left(\frac{d-1}{\cancel{8}}\right)$$

$$\overset{4}{\cancel{8}}\left(\frac{d}{\cancel{2}}\right) - \overset{2}{\cancel{8}}\left(\frac{d}{\cancel{4}}\right) = d - 1$$

$$4d - 2d = d - 1$$

$$2d = d - 1$$

$$d = -1$$

$y = 4$, from:

$$0 = y - 4$$

56. To clear the fractions below, we multiplied both sides by 2x. Solve the other equation.

$$\frac{3}{x} + \frac{7}{2} = 1 \qquad\qquad \frac{1}{3} = \frac{5}{m} - 4$$

$$2x\left(\frac{3}{x} + \frac{7}{2}\right) = 2x(1)$$

$$2x\left(\frac{3}{x}\right) + 2x\left(\frac{7}{2}\right) = 2x$$

$$6 + 7x = 2x$$

$$6 = -5x$$

$$x = -\frac{6}{5}$$

$F = \frac{28}{17}$, from:

$5F + 12 = 40 - 12F$

57. To clear the fractions below, we multiplied both sides by 3F. Solve the other equation.

$$\frac{1}{3F} - 2 = \frac{5}{F} \qquad\qquad \frac{15}{4d} = 1 + \frac{5}{d}$$

$$3F\left(\frac{1}{3F} - 2\right) = 3F\left(\frac{5}{F}\right)$$

$$3F\left(\frac{1}{3F}\right) - 3F(2) = 15$$

$$1 - 6F = 15$$

$$-6F = 14$$

$$F = -\frac{14}{6} = -\frac{7}{3}$$

$m = \frac{15}{13}$, from:

$m = 15 - 12m$

58. To clear the fractions below, we multiplied both sides by 4t. Solve the other equation.

$$\frac{3}{t} + \frac{1}{4} = \frac{5}{4t} \qquad\qquad \frac{3}{5} - \frac{1}{5m} = \frac{4}{m}$$

$$4t\left(\frac{3}{t} + \frac{1}{4}\right) = 4t\left(\frac{5}{4t}\right)$$

$$4t\left(\frac{3}{t}\right) + 4t\left(\frac{1}{4}\right) = 5$$

$$12 + t = 5$$

$$t = -7$$

$d = -\frac{5}{4}$, from:

$15 = 4d + 20$

$m = 7$, from:

$3m - 1 = 20$

59. To clear the fractions in equations like those below, we multiply both sides by a term in which the coefficient is the lowest common multiple of the numerical factors. That is:

For $\dfrac{3}{2x} - 1 = \dfrac{5}{3x}$, we multiply by 6x.

For $\dfrac{1}{4} = \dfrac{1}{8} + \dfrac{1}{y}$, we multiply by 8y.

Solve each equation.

a) $\left(\dfrac{3}{2x} - 4\right) = \left(\dfrac{1}{10x}\right)$

$15 - 40x = 0$

$-40x = -14$

$x = \dfrac{14}{40}$

b) $\dfrac{1}{3} = \left(\dfrac{1}{9} + \dfrac{1}{y}\right)\dfrac{9y}{1}$

$3y = y + 9$

$2x = 7$

$x = \dfrac{9}{2}$

a) $x = \dfrac{7}{20}$, from:

$15 - 40x = 1$

b) $y = \dfrac{9}{2}$, from:

$3y = y + 9$

60. Sometimes we have to multiply by the distributive principle on both sides to clear the fractions. An example is shown. Solve the other equation.

$$\dfrac{1}{m} - \dfrac{3}{4} = \dfrac{1}{4} + \dfrac{5}{m}$$

$$4m\left(\dfrac{1}{m} - \dfrac{3}{4}\right) = 4m\left(\dfrac{1}{4} + \dfrac{5}{m}\right)$$

$$4m\left(\dfrac{1}{m}\right) - 4m\left(\dfrac{3}{4}\right) = 4m\left(\dfrac{1}{4}\right) + 4m\left(\dfrac{5}{m}\right)$$

$$4 - 3m = m + 20$$

$$-3m = m + 16$$

$$-4m = 16$$

$$m = -4$$

$$\dfrac{1}{2} + \dfrac{2}{3y} = \dfrac{1}{y} - \dfrac{1}{3}$$

$3y + 4 = 6 - 2y$

$y = \dfrac{2}{5}$, from:

$3y + 4 = 6 - 2y$

SELF-TEST 10 (pages 120-130)

Solve each equation. Report each solution in lowest terms.

1. $\frac{5x}{3} = \frac{7}{12}$ $60x = 21$ $x = \frac{21}{60}$ $\frac{7}{20}$	2. $\frac{2t - 3}{t} = \frac{5}{2}$ $4t - 6 = 5t$ $-6 = t$	3. $\frac{9}{8} = \frac{2}{w^2}$ $9w^2 = 16$ $w^2 = \frac{16}{9}$ $w = \pm\frac{4}{3}$
4. $\frac{5}{6y} = \frac{y + 2}{2y}$ $\frac{10y}{6y} = 6y(y+2)$ $\frac{10}{6} = \frac{(y+2)6y}{1}$ $10 = 6y + 12$ $\left(-\frac{1}{3}\right)$	5. $\frac{1}{3r - 2} = \frac{3}{r}$ $r = 9r - 6$ $6 = 8r$ $\frac{6}{8} = r$ $\frac{3}{4}$	6. $\frac{d}{2} = 3d + \frac{5}{8}$
7. $\frac{2w}{3} - 1 = \frac{w}{2} + \frac{5}{3}$	8. $\frac{2}{3x} - 5 = \frac{4}{x}$	9. $\frac{3}{5} + \frac{1}{2y} = \frac{2}{y} - \frac{3}{10}$

ANSWERS:

1. $x = \frac{7}{20}$

2. $t = -6$

3. $w = \frac{4}{3}$ and $-\frac{4}{3}$

4. $y = -\frac{1}{3}$

5. $r = \frac{3}{4}$

6. $d = -\frac{1}{4}$

7. $w = 16$

8. $x = -\frac{2}{3}$

9. $y = \frac{5}{3}$

3-6   APPLIED PROBLEMS INVOLVING FORMULAS

In this section, we will solve some applied problems that involve an evaluation with a formula.   All of the evaluations require solving a fractional equation.

61.   Following the example, do the other evaluation.

For an engine, power P in foot-pounds per minute, force F in pounds, distance $\underline{s}$ in feet, and time $\underline{t}$ in minutes are related by the formula: $P = \dfrac{Fs}{t}$.   When the power is 100,000 foot-pounds per minute, distance is 500 feet, and the time is 8 minutes, find the force.

$$P = \frac{Fs}{t}$$

$$100,000 = \frac{F(500)}{8}$$

$$800,000 = 500F$$

$$F = \frac{800,000}{500}$$

$$F = 1,600 \text{ pounds}$$

In an electric circuit, current I in amperes, voltage V in volts, and resistance R in ohms, are related by the formula: $I = \dfrac{V}{R}$.   Find the resistance when the current is 5 amperes and the voltage is 20 volts.

62.   Complete these:

a) The area A of a triangle with base $\underline{b}$ and altitude $\underline{h}$ is given by the formula: $A = \dfrac{bh}{2}$.   Find the altitude of a triangle whose base is 20 cm and whose area is 140 cm^2.

b) Pressure P (in kg/cm^2), force F (in kg), and area A (in cm^2) are related by the formula: $P = \dfrac{F}{A}$.   When the pressure is 25 kg/cm^2 and the force is 500 kg, find the area.

R = 4 ohms

a) h = 14 cm

b) A = 20 cm^2

63. Fahrenheit temperatures F and Celsius temperatures C are related by the formula: $F = \dfrac{9C + 160}{5}$. We used the formula for one conversion below. Do the other conversion.

Convert 86°F to Celsius degrees.   Convert 14°F to Celsius degrees.

$$F = \frac{9C + 160}{5}$$

$$86 = \frac{9C + 160}{5}$$

$$430 = 9C + 160$$

$$270 = 9C$$

$$C = 30°$$

C = -10°

64. Complete these:

a) The acceleration a of an object, its initial velocity v_1, its final velocity v_2, and its time traveled t are related by the formula: $a = \dfrac{v_2 - v_1}{t}$. When the acceleration of an object is 3m/sec², its final velocity is 25m/sec, and its time traveled is 5 seconds, find its initial velocity.

b) The area A of a trapezoid, its parallel sides a and b, and its altitude h are related by the formula: $A = \dfrac{(a + b)h}{2}$. Find parallel side a of a trapezoid when its area is 640 ft², parallel side b is 30 ft, and the altitude is 16 ft.

---

65. Following the example, do the other evaluation.

The maximum height h in feet reached by a ball thrown vertically upward with an initial velocity v in feet/second is given by the formula: $h = \dfrac{v^2}{64}$. If the maximum height reached is 100 feet, find the initial velocity.

In an electric circuit, resistance R in ohms, current I in amperes, and power P in watts are related by the formula: $R = \dfrac{P}{I^2}$. Find the current when the resistance is 5 ohms and the power is 320 watts.

$$h = \frac{v^2}{64}$$

$$100 = \frac{v^2}{64}$$

$$6,400 = v^2$$

$$v = 80 \text{ ft/sec}$$

a) v_1 = 10m/sec

b) a = 50 ft

66. Following the example, do the other evaluation.

For a gas, the initial pressure P_1 and the final pressure P_2 (both in kg/cm^2) and the initial volume V_1 and the final volume V_2 (both in liters) are related by the formula: $\dfrac{P_1}{P_2} = \dfrac{V_2}{V_1}$. When the initial pressure is 4 kg/cm^2, the final pressure is 6 kg/cm^2, and the final volume is 8 liters, find the initial volume.

$$\frac{P_1}{P_2} = \frac{V_2}{V_1}$$

$$\frac{4}{6} = \frac{8}{V_1}$$

$$6V_1\left(\frac{4}{6}\right) = 6V_1\left(\frac{8}{V_1}\right)$$

$$4V_1 = 48$$

$$V_1 = 12 \text{ liters}$$

For a gas, the initial volume V_1 and the final volume V_2 (both in liters) and the initial temperature T_1 and the final temperature T_2 (both in degrees Kelvin) are related by the formula: $\dfrac{V_1}{V_2} = \dfrac{T_1}{T_2}$.

When the initial volume is 40 liters and the initial and final temperatures are 280°K and 350°K respectively, find the final volume.

I = 8 amperes

---

67. Complete these:

a) The distance s in feet that an object falls from rest in t seconds is given by the formula: $s = \dfrac{gt^2}{2}$, where g is the acceleration due to gravity. Find the time needed to fall 400 ft when g = 32 ft/sec^2.

b) The illumination I on an object depends on the square of its distance d from a light source. The relationship between the illumination on two objects and their distances from a light source is given by the formula: $\dfrac{I_1}{I_2} = \dfrac{(d_2)^2}{(d_1)^2}$. If $I_1 = 16$ units, $I_2 = 4$ units, and the second object is 10 feet from the light source, how far is the first object from the light source?

$V_2 = 50$ liters

a) t = 5 seconds

b) $d_1 = 5$ feet

68. Following the example, do the other evaluation.

For a differential pulley, the mechanical advantage M, the wheel diameter D, and the axle diameter d are related by the formula: $M = \dfrac{2D}{D - d}$. When the mechanical advantage is 12 and the axle diameter is 5 inches, find the wheel diameter.

In an electric circuit, the current I in amperes, the voltage V in volts, and two series resistances R_1 and R_2 in ohms are related by the formula: $I = \dfrac{V}{R_1 + R_2}$. When the current is 2 amperes, the voltage is 60 volts, and the first resistance is 12 ohms, find the second resistance.

$$M = \frac{2D}{D - d}$$

$$12 = \frac{2D}{D - 5}$$

$$12(D - 5) = (D - 5)\left(\frac{2D}{D - 5}\right)$$

$$12D - 60 = 2D$$

$$10D = 60$$

$$D = 6 \text{ inches}$$

$R_2 = 18$ ohms

69. For a transistor, the collector current gain α and the emitter current gain β are related by the formula: $\alpha = \dfrac{\beta}{\beta + 1}$. We used the formula for one evaluation below. Do the other evaluation.

Find the emitter current gain when the collector current gain is 0.99 .

Find the emitter current gain when the collector current gain is 0.9 .

$$\alpha = \frac{\beta}{\beta + 1}$$

$$.99 = \frac{\beta}{\beta + 1}$$

$$.99(\beta + 1) = (\beta + 1)\left(\frac{\beta}{\beta + 1}\right)$$

$$.99\beta + .99 = \beta$$

$$.99 = 1\beta - .99\beta$$

$$.99 = .01\beta$$

$$\beta = \frac{.99}{.01} = 99$$

$\beta = 9$

70. Following the example, do the other evaluation.

In an electric circuit, two parallel resistances R_1 and R_2 have a total resistance R_t. The three resistances, all in ohms, are related by the formula:

$R_t = \dfrac{R_1 R_2}{R_1 + R_2}$. When the first parallel resistance is 12 ohms and the total resistance is 3 ohms, find the second parallel resistance.

In an electric circuit, two series capacitances C_1 and C_2 have a total capacitance C_t. The three capacitances, all in microfarads, are related by the formula:

$C_t = \dfrac{C_1 C_2}{C_1 + C_2}$. When the first series capacitance is 40 microfarads and the total capacitance is 8 microfarads, find the second series capacitance.

$$R_t = \frac{R_1 R_2}{R_1 + R_2}$$

$$3 = \frac{12R_2}{12 + R_2}$$

$$3(12 + R_2) = \cancel{(12 + R_2)}\left(\frac{12R_2}{\cancel{12 + R_2}}\right)$$

$$36 + 3R_2 = 12R_2$$

$$36 = 9R_2$$

$$R_2 = 4 \text{ ohms}$$

---

71. Following the example, do the other evaluation.

For a lens, the focal length f, the object distance s_o, and the image distance s_i are related by the formula: $\dfrac{1}{f} = \dfrac{1}{s_o} + \dfrac{1}{s_i}$. When the focal length is 6 cm and the object distance is 10 cm, find the image distance.

In an electric circuit, two parallel resistances R_1 and R_2 have a total resistance R_t. The three resistances, all in ohms, are related by the formula: $\dfrac{1}{R_t} = \dfrac{1}{R_1} + \dfrac{1}{R_2}$. When the two parallel resistances are 20 ohms and 30 ohms, find the total resistance.

$C_2 = 10$ microfarads

$$\frac{1}{f} = \frac{1}{s_o} + \frac{1}{s_i}$$

$$\frac{1}{6} = \frac{1}{10} + \frac{1}{s_i}$$

$$\overset{5}{\cancel{30}} s_i\left(\frac{1}{\cancel{6}}\right) = 30 s_i\left(\frac{1}{10} + \frac{1}{s_i}\right)$$

$$5s_i = \overset{3}{\cancel{30}} s_i\left(\frac{1}{\cancel{10}}\right) + 30\cancel{s_i}\left(\frac{1}{\cancel{s_i}}\right)$$

$$5s_i = 3s_i + 30$$

$$2s_i = 30$$

$$s_i = 15 \text{ cm}$$

72. Complete these:

R_t = 12 ohms

a) When a substance is cooled, its heat loss H (in BTUs), weight $\underline{w}$ (in pounds), initial temperature t_1 (in °F), final temperature t_2 (in °F), and specific heat $\underline{s}$ are related by the formula: $s = \dfrac{H}{w(t_2 - t_1)}$.

Find the heat loss for a 10-pound metal gear cooled from 300°F to 70°F when the specific heat is 0.20 .

b) For a rocket launch vehicle, the mass ratio M, fuel weight F, payload weight P, and structure weight S are related by the formula: $M = \dfrac{F + S + P}{S + P}$.

When the mass ratio is 6, the fuel weight is 200 tons, and structure weight is 36 tons, find the payload weight.

a) H = 460 BTUs   b) P = 4 tons

3-7   OTHER APPLIED PROBLEMS

In this section, we will solve some applied problems that involve fractional equations.

73. To solve the problem below, we set up and solved a fractional equation. We let $\underline{x}$ equal the smaller number and $\underline{x + 9}$ equal the larger number. Solve the other problem.

One number is 9 more than another. The quotient of the larger divided by the smaller is 4. Find the two numbers.

$$\frac{x + 9}{x} = 4$$

$$\cancel{x}\left(\frac{x + 9}{\cancel{x}}\right) = x(4)$$

$$x + 9 = 4x$$

$$9 = 3x$$

$$x = 3$$

Therefore, the two numbers are 3 and 12 which is (x + 9).

One number is 4 more than another. The quotient of the larger divided by the smaller is $\frac{5}{4}$. Find the two numbers.

74. We use the formula d = rt for the problem below. In that formula, d = distance traveled, r = rate (or speed), and t = time traveled.

<u>Problem</u>: One car travels 12 miles an hour faster than another. While one car travels 200 miles, the other travels 260 miles. Find the speed of each car.

We diagrammed the problem below.

Car 1

200 miles   x mph ⟶

Car 2

260 miles   (x + 12) mph ⟶

The data is summarized in the chart below. The time <u>t</u> for each is equal. We used $t = \dfrac{d}{r}$ to get $\dfrac{200}{x}$ and $\dfrac{260}{x + 12}$.

	Distance	Rate	Time
Car 1	200	x	$t = \dfrac{200}{x}$
Car 2	260	x + 12	$t = \dfrac{260}{x + 12}$

Since the two times are equal, we can set up and solve the following fractional equation.

$$\frac{200}{x} = \frac{260}{x + 12}$$

$$\cancel{x}(x + 12)\left(\frac{200}{\cancel{x}}\right) = x\cancel{(x + 12)}\left(\frac{260}{\cancel{x + 12}}\right)$$

$$200x + 2400 = 260x$$

$$2400 = 60x$$

$$x = 40 \text{ mph}$$

Therefore, the speed of Car 1 is 40 mph and the speed of Car 2 is 40 + 12 = 52 mph. These values check since $\dfrac{200}{40}$ = 5 and $\dfrac{260}{52}$ = 5. Therefore, they both travel 5 hours at those speeds.

Using a diagram and a chart, do this one:

<u>Problem</u>: A freight train travels 30 mph slower than a passenger train. While the freight train travels 220 miles, the passenger train travels 385 miles. Find the speed of each train.

smaller is 16, larger is 20, from:

$$\frac{x + 4}{x} = \frac{5}{4}$$

75.   Since the two times are equal, we use a fractional equation to solve the problem below.

Problem: A river has a current of 2 mph.   A boat takes as long to go 18 miles downstream as to go 12 miles upstream. What is the speed of the boat in still water?

We diagrammed the problem below.   The two distances are not equal.

Downstream

18 miles   (x + 2) mph

Upstream

12 miles   (x - 2) mph

We organized the data in the chart below.   The time t for each trip is equal.   We used $t = \dfrac{d}{r}$ to get $\dfrac{18}{x + 2}$ and $\dfrac{12}{x - 2}$.

	Distance	Rate	Time
Downstream	18	x + 2	$t = \dfrac{18}{x + 2}$
Upstream	12	x - 2	$t = \dfrac{12}{x - 2}$

Since the two times are equal, we can set up and solve the equation below.

$$\frac{18}{x + 2} = \frac{12}{x - 2}$$

$$(x + 2)(x - 2)\left(\frac{18}{x + 2}\right) = (x + 2)(x - 2)\left(\frac{12}{x - 2}\right)$$

$$18x - 36 = 12x + 24$$

$$6x = 60$$

$$x = 10 \text{ mph}$$

Therefore, the speed of the boat is 10 mph in still water.   This checks since $\dfrac{18}{12} = \dfrac{3}{2}$ and $\dfrac{12}{8} = \dfrac{3}{2}$.   Therefore, both trips take $1\frac{1}{2}$ hours.

Using a diagram and chart, solve this one.

Problem: An airplane takes as long to fly 425 miles with the wind as it does to fly 375 miles against the wind.   If the wind is blowing at 10 mph, what is the speed of the plane without a wind?

The speed of the freight train is 40 mph; the speed of the passenger train is 70 mph.

From either:

$$\frac{220}{x} = \frac{385}{x + 30}$$

or $\dfrac{385}{x} = \dfrac{220}{x - 30}$

76. We also use a fractional equation to solve the problem below.

Problem: A jet flies at an average speed of 500 mph without a wind. If it takes as long to fly 910 miles with the wind as it does to fly 840 miles against the wind, how strong is the wind?

We diagrammed the problem below. The two distances are not equal.

With the wind

910 miles   (500 + x) mph

Against the wind

840 miles   (500 - x) mph

We organized the data in the chart below.

	Distance	Rate	Time
With the wind	910	500 + x	$t = \dfrac{910}{500 + x}$
Against the wind	840	500 - x	$t = \dfrac{840}{500 - x}$

Since the two times are equal, we can set up and solve this equation where $\underline{x}$ is the speed of the wind.

$$\frac{910}{500 + x} = \frac{840}{500 - x}$$

$$(500 + x)(500 - x)\left(\frac{910}{500 + x}\right) = (500 + x)(500 - x)\left(\frac{840}{500 - x}\right)$$

$$455,000 - 910x = 420,000 + 840x$$

$$35,000 = 1750x$$

$$x = 20 \text{ mph}$$

Therefore, the wind is blowing at 20 mph. This checks since $\frac{910}{520} = \frac{7}{4}$ and $\frac{840}{480} = \frac{7}{4}$. Therefore, both flights take $1\frac{3}{4}$ hours.

Using a diagram and chart, solve this one:

Problem: Joe can row 4 miles per hour in still water. It takes as long to row 15 miles upstream as it does to row 25 miles downstream. How fast is the current?

160 mph, from:

$$\frac{425}{x + 10} = \frac{375}{x - 10}$$

1 mph, from:

$$\frac{15}{4 - x} = \frac{25}{4 + x}$$

SELF-TEST 11 (pages 131-140)

1. A substance's weight W in grams, volume V in cm^3, and density D in grams/cm^3 are related by the formula: $D = \dfrac{W}{V}$.
The density of copper is 9 grams/cm^3. Find the volume of 180 grams of copper.

2. In an electric circuit, power P in watts, voltage V in volts, and resistance R in ohms are related by the formula: $P = \dfrac{V^2}{R}$. Find the voltage when the power is 4 watts and the resistance is 9 ohms.

3. The area A of a triangle with base b and altitude h is given by the formula: $A = \dfrac{bh}{2}$. Find the base of a triangle whose altitude is 36 inches and whose area is 900 square inches.

4. For a gas, the initial pressure P_1 and the final pressure P_2 (both in lb/in^2) and the initial volume V_1 and the final volume V_2 (both in in^3) are related by the formula: $\dfrac{P_1}{P_2} = \dfrac{V_2}{V_1}$.
When the initial volume is 30 in^3, the final volume is 80 in^3, and the initial pressure is 200 lb/in^2, find the final pressure.

5. In an electric circuit, two parallel resistances R_1 and R_2 have a total resistance R_t. The three resistances, all in ohms, are related by the formula: $\dfrac{1}{R_t} = \dfrac{1}{R_1} + \dfrac{1}{R_2}$. When the two parallel resistances are 40 ohms and 60 ohms, find the total resistance.

6. A passenger train travels 30 mph faster than a freight train. While the passenger train travels 320 miles, the freight train travels 200 miles. Find the speed of each train.

7. An airplane takes as long to fly 600 miles with the wind as it does to fly 480 miles against the wind. If the wind is blowing at 30 mph, what is the speed of the plane without a wind?

ANSWERS:
1. V = 20 cm^3

2. V = 6 volts

3. h = 50 inches

4. P_2 = 75 lb/in^2

5. R_t = 24 ohms

6. Passenger train: 80 mph
Freight train: 50 mph

7. 270 mph

SUPPLEMENTARY PROBLEMS - CHAPTER 3

Assignment 9

Solve each equation. Report each solution in lowest terms.

1. $\dfrac{w}{6} = 9$

2. $\dfrac{24}{x} = 8$

3. $15 = \dfrac{3y}{5}$

4. $\dfrac{36}{4R} = 3$

5. $4 = \dfrac{5G}{3}$

6. $\dfrac{1}{2v} = 5$

7. $1 = \dfrac{7}{4d}$

8. $\dfrac{3p}{8} = 1$

9. $\dfrac{5t}{8} = 0$

10. $\dfrac{h + 9}{2} = 3$

11. $x = \dfrac{3x - 1}{4}$

12. $\dfrac{4 - 2y}{3} = 2y$

13. $\dfrac{4w + 3}{6} = 0$

14. $\dfrac{8}{F + 3} = 2$

15. $\dfrac{a}{a - 5} = 6$

16. $\dfrac{2k - 1}{2k + 1} = 5$

17. $2 = \dfrac{1 - x}{3}$

18. $\dfrac{3 - 2t}{t + 4} = 0$

19. $\dfrac{4(x + 5)}{3} = 7$

20. $6 = \dfrac{3(H - 1)}{2}$

21. $\dfrac{3(v - 5)}{4v} = 2$

22. $5 = \dfrac{6(P + 3)}{P}$

23. $\dfrac{2(y + 1)}{5y} = 0$

24. $1 = \dfrac{2}{5(B - 1)}$

25. $\dfrac{t}{3(t + 2)} = 1$

26. $6 = \dfrac{2k}{3(k - 1)}$

27. $\dfrac{3s}{5(s + 4)} = 0$

28. $5 - \dfrac{p}{4} = 7$

29. $1 = \dfrac{4m}{7} + 5$

30. $\dfrac{3}{y} - 6 = 15$

31. $8 = \dfrac{3}{5R} + 2$

32. $1 - \dfrac{3}{4w} = 4$

33. $7 + \dfrac{4}{3h} = 3$

Assignment 10

Solve each equation. Report each solution in lowest terms.

1. $\dfrac{t}{5} = \dfrac{7}{10}$

2. $\dfrac{5y}{9} = \dfrac{1}{3}$

3. $\dfrac{3}{w} = \dfrac{5}{2}$

4. $\dfrac{8}{5} = \dfrac{3}{2x}$

5. $\dfrac{5x - 2}{6} = \dfrac{x}{2}$

6. $\dfrac{1}{5} = \dfrac{w + 8}{10}$

7. $\dfrac{5}{4} = \dfrac{2r}{3}$

8. $\dfrac{d - 1}{2} = \dfrac{3}{5}$

9. $\dfrac{9 + a}{6} = \dfrac{7}{8}$

10. $\dfrac{x^2}{3} = \dfrac{18}{6}$

11. $\dfrac{5}{y^2} = \dfrac{4}{5}$

12. $\dfrac{3}{7} = \dfrac{12}{t^2}$

13. $\dfrac{40}{5} = \dfrac{2a^3}{3}$

14. $\dfrac{h - 1}{6} = \dfrac{3h}{4}$

15. $\dfrac{3P + 5}{10} = \dfrac{1}{4}$

16. $\dfrac{5}{2} = \dfrac{b - 1}{2b}$

17. $\dfrac{2E + 3}{5E} = \dfrac{1}{4E}$

18. $\dfrac{7}{3} = \dfrac{2}{3t - 1}$

19. $\dfrac{3}{2R + 5} = \dfrac{1}{R}$

20. $\dfrac{2}{4y} = \dfrac{3}{y - 5}$

21. $\dfrac{F}{F + 1} = \dfrac{3}{8}$

22. $\dfrac{7}{4} = \dfrac{A - 2}{2A}$

23. $\dfrac{10}{r + 1} = \dfrac{8}{r - 1}$

24. $\dfrac{N + 3}{4} = \dfrac{2N + 5}{10}$

25. $\dfrac{t}{6} = \dfrac{3t}{2} + 4$

26. $\dfrac{h}{3} + 1 = \dfrac{h}{9}$

27. $\dfrac{4p}{3} - \dfrac{3p}{2} = 1$

28. $\dfrac{7w}{5} - \dfrac{3}{4} = 0$

29. $\dfrac{2m}{3} + \dfrac{4}{3} = 2 - \dfrac{m}{3}$

30. $\dfrac{5k}{3} - \dfrac{1}{6} = \dfrac{3k}{2} + 1$

31. $\dfrac{3}{4} + \dfrac{2}{m} = \dfrac{1}{3m}$

32. $5 - \dfrac{2}{h} = \dfrac{3}{2h} + \dfrac{7}{4}$

33. $\dfrac{1}{4y} + \dfrac{5}{6} = \dfrac{2}{y} - \dfrac{5}{2}$

Assignment 11

1. The acceleration $\underline{a}$ of an object, its initial velocity v_1, its final velocity v_2, and its time traveled $\underline{t}$ are related by the formula: $a = \dfrac{v_2 - v_1}{t}$. When the acceleration of an object is 5 m/sec^2, its initial velocity is 12 m/sec, and its time traveled is 8 seconds, find its final velocity.

2. For a gas, the initial volume V_1 and the final volume V_2 (both in liters) and the initial temperature T_1 and the final temperature T_2 (both in degrees Kelvin) are related by the formula: $\dfrac{V_1}{V_2} = \dfrac{T_1}{T_2}$. When the final volume is 100 liters and the initial and final temperatures are 360°K and 300°K respectively, find the initial volume.

3. In an electric circuit, resistance R in ohms, current I in amperes, and power P in watts are related by the formula: $R = \dfrac{P}{I^2}$. Find the current when the power is 100 watts and the resistance is 4 ohms.

4. Fahrenheit and Celsius temperatures are related by the formula: $C = \dfrac{5F - 160}{9}$. Convert 20°C to Fahrenheit degrees.

5. The area A of a trapezoid, its parallel sides $\underline{a}$ and $\underline{b}$, and its altitude $\underline{h}$ are related by the formula: $A = \dfrac{(a + b)h}{2}$. Find parallel side $\underline{b}$ of a trapezoid when its area is 1,000 cm^2, parallel side $\underline{a}$ is 50 cm, and the altitude is 25 cm.

6. The distance $\underline{s}$ in feet that an object falls from rest in $\underline{t}$ seconds is given by the formula: $s = \dfrac{gt^2}{2}$, where $\underline{g}$ is the acceleration due to gravity. Find the time needed to fall 100 feet when $g = 32$ ft/sec^2.

7. In an electric circuit, voltage V in volts, power P in watts, and current I in amperes are related by the formula: $V = \dfrac{P}{I}$. Find the current when the voltage is 6 volts and the power is 30 watts.

8. An organ pipe of length L feet emits sound at a frequency of $\underline{n}$ vibrations per second according to the formula: $n = \dfrac{1,100}{4L}$. What pipe length is required for 50 vibrations per second?

9. In a centrifuge for astronauts, the acceleration $\underline{a}$ in ft/sec^2, the velocity $\underline{v}$ in ft/sec, and the radius $\underline{r}$ in feet are related by the formula: $a = \dfrac{v^2}{r}$. When the velocity is 80 ft/sec and the acceleration is 160 ft/sec^2, what is the radius?

10. In an electric circuit, two series capacitances C_1 and C_2 have a total capacitance C_t. The three capacitances, all in microfarads, are related by the formula: $\dfrac{1}{C_t} = \dfrac{1}{C_1} + \dfrac{1}{C_2}$. When the first series capacitance is 10 microfarads and the total capacitance is 8 microfarads, find the second series capacitance.

11. For a transistor, the collector current gain α and the emitter current gain β are related by the formula: $\beta = \dfrac{\alpha}{1 - \alpha}$. Find the collector current gain when the emitter current gain is 24.

12. For a differential pulley, the mechanical advantage M, the wheel diameter D, and the axle diameter $\underline{d}$ are related by the formula: $M = \dfrac{2D}{D - d}$. When the mechanical advantage is 10 and the wheel diameter is 8 inches, find the axle diameter.

13. A river has a current of 3 mph. A boat takes as long to go 30 miles downstream as to go 15 miles upstream. What is the speed of the boat in still water?

14. One car travels 10 mph faster than another. When one car travels 240 miles, the other travels 290 miles. Find the speed of each car.

15. A jet plane travels 180 mph faster than a prop plane. While the prop plane travels 600 miles, the jet plane travels 1,050 miles. Find the speed of each plane.

16. A military plane flies at an average speed of 500 mph without a wind. If it takes as long to fly 780 miles with the wind as it does to fly 720 miles against the wind, how strong is the wind?

4 Percent, Proportion, Variation

In this chapter, we will review the decimal number system and the rules for rounding and introduce some basic operations on a scientific calculator. We will discuss percents, ratios, proportions, and the following types of variation: direct, inverse, direct square, inverse square, joint, and combined. Applied problems of various types are included.

4-1 PLACE NAMES AND WORD NAMES

In this section, we will review the place-names in whole numbers and decimal numbers and the word names for whole numbers and decimal numbers.

1. Our number system is called the <u>decimal</u> number system because it uses <u>ten</u> symbols called <u>digits</u>. The ten digits are:
 	
        0, 1, 2, 3, 4, 5, 6, 7, 8, 9
 	
Any number is written by using one or more digits. Numbers are often referred to as <u>one</u>-<u>digit</u> numbers, <u>two</u>-<u>digit</u> numbers, and so on. For example:
 	
      5 is a one-digit number.
 	
      603 is a three-digit number.
 	
  188,400 is a _____ number.	

	six-digit

2. The names of the most common places in whole numbers are given in the chart below.

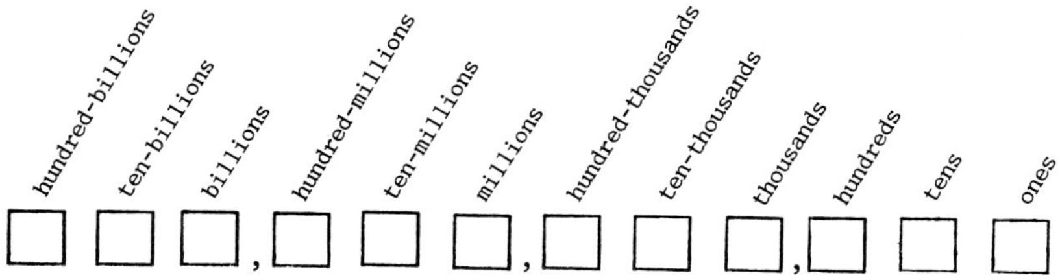

In 817,204,839,561:

        the 6 is in the <u>tens</u> place.

    a) the 9 is in the _____ place.

    b) the 4 is in the _____ place.

    c) the 7 is in the _____ place.

---

3. Name the place in which the 3 appears in each number below.

    a) 4,381 _____   c) 317,690,458 _____

    b) 930,245 _____   d) 439,226,770,000 _____

a) thousands
b) millions
c) billions

---

4. The word names for some two-digit numbers are given below. The word names for 47 and 83 contain a hyphen.

       15 fifteen
       30 thirty
       43 forty-three
       87 eighty-seven

Write the word name for each number.

    a) 12 _____   c) 31 _____

    b) 50 _____   d) 96 _____

a) hundreds
b) ten-thousands
c) hundred-millions
d) ten-billions

---

5. The word names for some three-digit numbers are given below. Notice that the word <u>and</u> is not used.

      400 four hundred
      280 two hundred eighty
      673 six hundred seventy-three

Write the word name for each number.

    a) 509 _____   b) 165 _____

a) twelve
b) fifty
c) thirty-one
d) ninety-six

---

a) five hundred nine
b) one hundred sixty-five

6. Write the number corresponding to each word name.

   a) seventeen _____   c) two hundred seventy _____

   b) sixty-one _____   d) nine hundred forty-four _____

---

7. To make larger numbers easier to read, commas are used every three digits, starting from the right. Each group of three digits is called a period. The most common periods are shown in the chart below.

billions period	millions period	thousands period	ones period
hundred-billions ten-billions billions	hundred-millions ten-millions millions	hundred-thousands ten-thousands thousands	hundreds tens ones

To name a number like 35,641,297,863 , we name the number in each period and then use the period name (except for ones) followed by a comma. The word and is not used. For example:

35,641,297,863   thirty-five billion, six hundred forty-one million, two hundred ninety-seven thousand, eight hundred sixty-three

Write the word name for each number.

   a) 15,906

   b) 144,740,057

   c) 9,200,623,080

a) 17   c) 270

b) 61   d) 944

a) fifteen thousand, nine hundred six

b) one hundred forty-four million, seven hundred forty thousand, fifty-seven

c) nine billion, two hundred million, six hundred twenty-three thousand, eighty

19.   In each example below, we got a 0 in the place rounded to. The 0 is reported so that we know what place was rounded to.

Rounding to <u>tenths</u>, 16.035 rounds to 16.0 (not 16).

Rounding to <u>thousandths</u>, .19967 rounds to .200 (not .2).

Complete: a) Round 5.0966 to <u>hundredths</u>. _5.10_

             b) Round .070449 to <u>thousandths</u>. _.070_

---

20.   Rounding to a definite number of decimal places is the same as rounding to a definite place. For example:

"Round to <u>one</u> decimal place" means "round to <u>tenths</u>".
"Round to <u>two</u> decimal places" means "round to <u>hundredths</u>".
"Round to <u>three</u> decimal places" means "round to <u>thousandths</u>".

a) Round 6.405 to <u>one</u> decimal place. _6.4_

b) Round 1.89526 to <u>two</u> decimal places. _1.90_

c) Round .099853 to <u>three</u> decimal places. _.100_

d) Round .0085044 to <u>five</u> decimal places. _.00850_

a) 5.10

b) .070

---

a) 6.4   b) 1.90   c) .100   d) .00850

4-3 INTRODUCTION TO THE CALCULATOR

In this section, we will show the method for performing addition, subtraction, multiplication, division, squaring, square roots, and reciprocals on a calculator. The operations are extended to negative numbers.

    <u>Note</u>: The instruction in this text is designed for scientific calculators with an <u>algebraic</u> entry method as opposed to an RPN entry method.

---

21.   Turn your calculator on. The display shows "0". To enter a whole number, press the digit keys in order from left to right.

    1. Enter 2,740. The display shows "2740".

    2. Press the <u>clear</u> key (ON/C or C). The display shows "0".

To enter a decimal number, press the digit keys in order from left to right. Press the decimal-point key ⌐·⌐ when needed.

    1. Enter 15.96. The display shows "15.96".

    2. Press the <u>clear</u> key. The display shows "0".

22. The $\boxed{+}$, $\boxed{-}$, $\boxed{\text{x}}$, and $\boxed{\div}$ keys are used for addition, subtraction, multiplication, and division. The $\boxed{=}$ key completes each operation. An example of each operation and the calculator steps for it are given below.

Operation	Calculator Steps
8 + 7	8 $\boxed{+}$ 7 $\boxed{=}$ 15
10 - 4	10 $\boxed{-}$ 4 $\boxed{=}$ 6
6 x 9	6 $\boxed{\text{x}}$ 9 $\boxed{=}$ 54
48 ÷ 6	48 $\boxed{\div}$ 6 $\boxed{=}$ 8

Use the same steps to complete these.

a) 3,658 + 949 = _____   c) 74 x 508 = _____

b) 7.31 - 2.99 = _____   d) 106.95 ÷ 15 = _____

---

23. Any fraction stands for a division. For example:

$$\frac{3}{4} = 3 \div 4 \qquad \frac{97}{20} = 97 \div 20$$

Use a calculator to convert each fraction to a decimal number.

a) $\frac{3}{4}$ = _____   b) $\frac{97}{20}$ = _____   c) $\frac{39.6}{2.4}$ = _____

a) 4,607

b) 4.32

c) 37,592

d) 7.13

---

24. The calculator steps for the operations below are shown.

Operation	Calculator Steps
6 + 9 - 5	6 $\boxed{+}$ 9 $\boxed{-}$ 5 $\boxed{=}$ 10
4 x 3 x 7	4 $\boxed{\text{x}}$ 3 $\boxed{\text{x}}$ 7 $\boxed{=}$ 84
10 x 2 ÷ 4	10 $\boxed{\text{x}}$ 2 $\boxed{\div}$ 4 $\boxed{=}$ 5

Use a calculator for these:

a) 9.75 + 17.6 + 6.34 = _____

b) .075 + .0106 - .0099 = _____

c) 1.5 x 2.8 x 3.2 = _____

d) 7 x 1.2 ÷ .4 = _____

a) .75

b) 4.85

c) 16.5

---

a) 33.69

b) .0757

c) 13.44

d) 21

25. Calculator answers can contain as many as ten digits. For example:

$$.987 \times 5.48 = 5.40876$$
$$44.7 \div 13.3 = 3.360902256$$

Answers like those above are usually rounded to less decimal places. Rounding to hundredths, we get:

$$.987 \times 5.48 = 5.41$$
$$44.7 \div 13.3 = 3.36$$

Complete these. Round each answer to <u>tenths</u>.

a) $69.7 \div 2.6 =$ _____   b) $8.77 \times 5.33 =$ _____

---

26. Convert each fraction to a decimal number.

a) Rounding to <u>two</u> decimal places, $\frac{2}{7} =$ _____

b) Rounding to <u>four</u> decimal places, $\frac{17}{300} =$ _____

a) 26.8   b) 46.7

---

27. The $\boxed{x^2}$ and $\boxed{\sqrt{x}}$ keys are used to square a number or to find the square root of a number. If either function is given above a key, we must press the $\boxed{2nd}$ or $\boxed{INV}$ key first. Some examples are shown.

a) .29

b) .0567

Operation	Calculator Steps
7^2	7 $\boxed{x^2}$ 49
	or 7 $\boxed{2nd}$ $\boxed{x^2}$ 49
$\sqrt{81}$	81 $\boxed{\sqrt{x}}$ 9
	or 81 $\boxed{2nd}$ $\boxed{\sqrt{x}}$ 9

Following the examples, complete these:

a) $256^2 =$ _____   c) $\sqrt{348,100} =$ _____

b) $1.18^2 =$ _____   d) $\sqrt{.000729} =$ _____

---

28. Two numbers are a pair of <u>reciprocals</u> if their product is "1". To find the reciprocal of a number, we divide "1" by that number. For example:

The reciprocal of 25 is $\frac{1}{25} = .04$

The $\boxed{1/x}$ key is used to find the reciprocal of a number. On some calculators, the $\boxed{2nd}$ or $\boxed{INV}$ key must be pressed first. An example is shown.

a) 65,536

b) 1.3924

c) 590

d) .027

Operation	Calculator Steps
Reciprocal of 25	25 $\boxed{1/x}$ $.04$
	or 25 $\boxed{2nd}$ $\boxed{1/x}$ $.04$

Continued on following page.

28.   Continued

Find the reciprocal of each number.

a) 40 _____   b) .25 _____   c) 320 _____

---

29.   Any fraction whose numerator is "1" is the reciprocal of its denominator.   That is:

$$\frac{1}{125} \text{ is the reciprocal of 125.}$$

Therefore, we can convert any fraction of that type to a decimal number by finding the reciprocal of its denominator.   The calculator steps are shown below.

125 $\boxed{1/x}$ .008

or   125 $\boxed{2nd}$ $\boxed{1/x}$ .008

Therefore: $\frac{1}{125}$ = .008

Use the reciprocal key to convert each fraction to a decimal.

a) $\frac{1}{8}$ = _____   b) $\frac{1}{64}$ = _____   c) $\frac{1}{1.25}$ = _____

a) .025

b) 4

c) .003125

---

30.   Round each answer to thousandths.

a) The reciprocal of 13 is _.077_ .   c) $\frac{1}{60}$ = _____

b) $(.859)^2$ = _____   d) $\sqrt{.917}$ = _____

a) .125

b) .015625

c) .8

---

31.   To enter a negative number on a calculator, we enter its absolute value and then use the change sign key $\boxed{+/-}$.   For example, the steps below are used to enter -17.

17 $\boxed{+/-}$ - 17

The calculator steps for some operations involving negative numbers are shown below.

Operation	Calculator Steps
7 + (-10)	7 $\boxed{+}$ 10 $\boxed{+/-}$ $\boxed{=}$ -3
-5 - (-9)	5 $\boxed{+/-}$ $\boxed{-}$ 9 $\boxed{+/-}$ $\boxed{=}$ 4
(8)(-6)	8 $\boxed{x}$ 6 $\boxed{+/-}$ $\boxed{=}$ -48
$\frac{-396}{-12}$	396 $\boxed{+/-}$ $\boxed{÷}$ 12 $\boxed{+/-}$ $\boxed{=}$ 33

Complete these:

a) -2.58 + 1.99 = _____   c) (-2.2)(-8.5) = _____

b) -3.4 - 7.4 = _____   d) $\frac{-42}{168}$ = _____

a) .077   c) .017

b) .738   d) .958

32. The calculator steps for two more operations involving negative numbers are shown below.

Operation	Calculator Steps
$(-7)^2$	7 [+/-] [x^2] 49
$\dfrac{1}{-20}$	20 [+/-] [1/x] -.05

Use a calculator for these.

a) $(-1.8)^2 =$ _____   b) $(-.25)^2 =$ _____   c) $\dfrac{1}{-125} =$ _____

a) -.59

b) -10.8

c) 18.7

d) -.25

---

33. Division by 0, finding the reciprocal of 0, and finding the square root of a negative number are impossible operations. If you try to do them on a calculator, the calculator will show a flashing display, the word "Error", the letter "E", or something else. To confirm this fact, try each of these on your calculator.

$\dfrac{6}{0}$ Find the reciprocal of 0. $\sqrt{-64}$

a) 3.24

b) .0625

c) -.008

SELF-TEST 12 (pages 144-155)

Write the number corresponding to each word name.

1. forty-two million, eighteen thousand, seventy _____

2. two hundred eighty ten-thousandths _____

3. six hundred two and thirty-nine hundredths _____

4. nine hundred fifty-three millionths _____

5. Round to hundred-thousands. 754,593 _____	7. Round to thousandths. 4.12963 _____
6. Round to hundreds. 2,947.883 _____	8. Round to millionths. .00260318 _____

Do Problems 9-16 on a calculator.

9. Round to one decimal place. 4.59 + 17.8 + 1.935 = _____	12. Round to hundred-thousands. $(-7,330)^2 =$ _____
10. Round to hundredths. -2.76 - .4353 = _____	13. Round to ten-thousandths. 4.0615 ÷ 1.7582 = _____
11. Round to thousands. 90.8 x 513 x 7.64 = _____	14. Round to two decimal places. $\sqrt{14.5} =$ _____

15. Convert $\dfrac{41}{65}$ to a decimal number. Round to three decimal places. _____

16. Find the reciprocal of 63. Round to hundred-thousandths. _____

ANSWERS:
1. 42,018,070	5. 800,000	9. 24.3	13. 2.3100
2. .0280	6. 2,900	10. -3.20	14. 3.81
3. 602.39	7. 4.130	11. 356,000	15. .631
4. .000953	8. .002603	12. 53,700,000	16. .01587

4-4 CONVERTING PERCENTS TO FRACTIONS AND DECIMALS

In this section, we will define <u>percents</u> and show the methods for converting percents to fractions and decimals.

---

34. The word <u>percent</u> means <u>per 100</u> or ____/100. That is, 42% means 42 per 100 or 42/100. Therefore, any percent can be written as a <u>hundredths</u> decimal fraction with the percent-value as its numerator. For example:

$$63\% = \frac{63}{100} \qquad\qquad 7.5\% = \frac{7.5}{100} \qquad\qquad 240\% = \frac{240}{100}$$

Convert each percent to a <u>hundredths</u> decimal-fraction.

a) 1% = _____   b) 150% = _____   c) 8.75% = _____

---

35. Some percents can be converted to simple fractions by reducing the decimal fraction to lowest terms. For example:

$$50\% = \frac{50}{100} = \frac{1}{2} \qquad\qquad 80\% = \frac{80}{100} = \frac{4}{5}$$

Convert each percent to a simple fraction.

a) 25% = _____   b) 10% = _____   c) 60% = $\frac{8.75}{100}$

a) $\frac{1}{100}$

b) $\frac{150}{100}$

c) $\frac{8.75}{100}$

---

36. Any multiple of 100% equals a whole number. That is:

$$100\% = \frac{100}{100} = 1 \qquad 200\% = \frac{200}{100} = 2 \qquad 500\% = \frac{500}{100} = 5$$

a) $\frac{1}{4}$ b) $\frac{1}{10}$ c) $\frac{3}{5}$

---

37. To convert each percent below to a decimal, we converted it to a decimal fraction and then divided the numerator by 100. (<u>Note</u>: To divide by 100, we can simply shift the decimal point two places to the left.)

$$83\% = \frac{83}{100} = 83. = .83$$

$$172\% = \frac{172}{100} = 1.72. = 1.72$$

$$6.5\% = \frac{6.5}{100} = 06.5 = .065$$

The same conversions can be made directly <u>by shifting the decimal point two places to the left and dropping the percent sign</u>. That is:

$$83\% = 83. = .83$$

$$172\% = 1.72. = 1.72$$

$$6.5\% = 06.5 = .065$$

Convert each percent to a decimal.

a) 7% = .07   b) 215% = 2.15   c) 12.75% = .1275

5

38. After converting some percents to decimal numbers, we can drop one or two final 0's after the decimal point. For example:

$$30\% = .30. = .30 = .3$$

$$400\% = 4.00. = 4.00 = 4$$

Convert each percent to a decimal number.

a) 90% = __.9__ b) 200% = __2__ c) 130% = __1.3__

a) .07

b) 2.15

c) .1275

39. To convert a mixed-number percent to a decimal number, we convert it to a decimal-number percent first. For example:

$$12\tfrac{1}{2}\% = 12.5\% = .125 \qquad 9\tfrac{3}{4}\% = 9.75\% = .0975$$

Convert each percent to a decimal number.

a) $11\tfrac{3}{4}\%$ = __11.75__   b) $5\tfrac{1}{2}\%$ = __5.5__

a) .9

b) 2

c) 1.3

40. Some common percent-fraction-decimal equivalents are given in the table below. Because of their usefulness, they <u>should</u> <u>be</u> <u>memorized</u>.

$25\% = \tfrac{1}{4} = .25$	$33\tfrac{1}{3}\% = \tfrac{1}{3} = .33$	$20\% = \tfrac{1}{5} = .2$	$10\% = \tfrac{1}{10} = .1$
$50\% = \tfrac{1}{2} = .5$	$66\tfrac{2}{3}\% = \tfrac{2}{3} = .67$	$40\% = \tfrac{2}{5} = .4$	$30\% = \tfrac{3}{10} = .3$
$75\% = \tfrac{3}{4} = .75$		$60\% = \tfrac{3}{5} = .6$	$70\% = \tfrac{7}{10} = .7$
		$80\% = \tfrac{4}{5} = .8$	$90\% = \tfrac{9}{10} = .9$

<u>Note</u>: The decimal equivalents for $33\tfrac{1}{3}\%$ and $66\tfrac{2}{3}\%$ are rounded <u>to the nearest hundredth</u>.

From memory, convert each percent to a fraction.

a) 25% = ____ b) 80% = ____ c) $66\tfrac{2}{3}\%$ = __2/3__

d) 90% = ____ e) 50% = ____ f) $33\tfrac{1}{3}\%$ = __1/3__

a) .1175   b) .055

a) $\tfrac{1}{4}$ b) $\tfrac{4}{5}$ c) $\tfrac{2}{3}$

d) $\tfrac{9}{10}$ e) $\tfrac{1}{2}$ f) $\tfrac{1}{3}$

4-5 CONVERTING DECIMALS AND FRACTIONS TO PERCENTS

In this section, we will discuss the methods for converting decimals and fractions to percents.

---

41. Because <u>percent</u> means <u>hundredths</u>, any hundredths decimal fraction can be converted directly to a percent. For example:

$$\frac{9}{100} = 9\% \qquad \frac{60}{100} = 60\% \qquad \frac{225}{100} = \underline{\hspace{2cm}}$$

---

42. We converted the decimals below to percents by converting them to a <u>hundredths</u> decimal fraction first.

$$.05 = \frac{5}{100} = 5\% \qquad\qquad 1.37 = \frac{137}{100} = 137\%$$

$$.4 = \frac{4}{10} = \frac{40}{100} = 40\% \qquad 1.8 = \frac{18}{10} = \frac{180}{100} = 180\%$$

The same conversions can be made directly <u>by shifting the decimal point two places to the right and attaching the percent sign</u>. That is:

$$.05 = .05_\curvearrowright = 5\% \qquad 1.37 = 1.37_\curvearrowright = 137\%$$

$$.4 = .40_\curvearrowright = 40\% \qquad 1.8 = 1.80_\curvearrowright = 180\%$$

Convert to a percent.

a) .83 = _83%_   b) .01 = _1%_   c) 2.96 = _296%_

d) .1 = _10%_   e) .9 = _90%_   f) 2.3 = _230%_

225%

---

43. Any whole number can be converted to a percent by converting it to a <u>hundredths</u> decimal fraction first. For example:

$$1 = \frac{100}{100} = 100\% \qquad\qquad 3 = \frac{300}{100} = 300\%$$

The decimal-point-shift method can be used for the same conversions. That is:

$$1 = 1.00_\curvearrowright = 100\% \qquad 3 = 3.00_\curvearrowright = 300\%$$

Convert to a percent.

a) 2 = _200%_   b) 4 = _400%_   c) 7 = _700%_

a) 83%

b) 1%

c) 296%

d) 10%

e) 90%

f) 230%

---

44. We used the decimal-point-shift method for the conversions below.

$$.497 = .49_\curvearrowright 7 = 49.7\% \qquad .0875 = .08_\curvearrowright 75 = 8.75\%$$

Convert to a percent.

a) .065 = _6.5%_   b) .1325 = _13.25%_   c) .0915 = _9.15%_

a) 200%

b) 400%

c) 700%

---

a) 6.5%

b) 13.25%

c) 9.15%

45. Some fractions can be converted to percents by converting them to equivalent <u>hundredths</u> decimal fractions. For example:

$$\frac{1}{2} = \frac{50}{100} = 50\% \qquad\qquad \frac{3}{4} = \frac{75}{100} = 75\%$$

Using the same method, complete these conversions.

a) $\frac{3}{5} = \frac{\boxed{60}}{100} =$ _____   b) $\frac{1}{10} = \frac{\boxed{}}{100} =$ _____

---

46. We can convert fractions to percents by converting them to decimals first. To do so, we divide. An example is shown. Complete the other conversion.

$$\frac{1}{4} = 4\overline{\smash{)}1.00} \quad \frac{.25 = 25\%}{\begin{array}{r} -8 \\ \hline 20 \\ -20 \end{array}}$$

$$\frac{4}{5} = 5\overline{\smash{)}4.} \quad _____ =$$

a) $\frac{60}{100} = 60\%$

b) $\frac{10}{100} = 10\%$

---

47. The division method is generally used to convert fractions to percents. A calculator can be used for the divisions. Two examples are shown. Use a calculator for the other conversions.

$$\frac{51}{68} = .75 = 75\% \qquad\qquad \frac{3}{16} = .1875 = 18.75\%$$

a) $\frac{34}{40} = \underline{.85} = \underline{85\%}$ b) $\frac{10}{16} = _____ = _____$

$\dfrac{.8 = 80\%}{5\overline{)4.0}}$

---

48. When the quotient is larger than "1", the percent is larger than 100%. An example is shown. Use a calculator for the other conversion.

$$\frac{31}{25} = 1.24 = 124\% \qquad\qquad \frac{47}{20} = \underline{2.35} = \underline{235\%}$$

a) $.85 = 85\%$

b) $.625 = 62.5\%$

---

49. When one or both terms of the fraction is a decimal number, we convert it to a percent in the usual way. An example is shown. Use a calculator for the other conversion.

$$\frac{8.4}{17.5} = .48 = 48\% \qquad\qquad \frac{5.65}{4.52} = \underline{1.25} = \underline{125\%}$$

$2.35 = 235\%$

---

50. When the division is non-terminating, we round to a specific place. For example:

To round <u>to a whole number percent</u>, we round <u>to two decimal places</u> before converting.

$$\frac{211}{375} = .56266667 = .56 = 56\%$$

To round <u>to a tenth of a percent</u>, we round <u>to three decimal places</u> before converting.

$$\frac{39}{550} = .07090909 = .071 = 7.1\%$$

$1.25 = 125\%$

Continued on following page.

50. Continued

a) Convert to a whole number percent.

$\frac{18}{170}$ = ___ 11%

b) Convert to a tenth of a percent.

$\frac{15.9}{17.5}$ = ___ 90.9%

a) 11% b) 90.9%

4-6   THE THREE TYPES OF PERCENT PROBLEMS

In this section, we will discuss the three types of percent problems. We will show how they can be solved by converting to equations and then solving the euations.

> Note: You will find it helpful to use a calculator for some problems in this section and the next section.

---

51. Two examples of the first type of percent problems are shown below. We converted each to an equation.

What is 40% of 200?
↓   ↓   ↓   ↓   ↓
x   = 40% · 200

75% of 96 is what?
↓   ↓   ↓   ↓   ↓
75% · 96 =   x

Translate each of these to an equation.

a) What is 50% of 138?   b) 63% of 837 is what?

_____   _____

---

52. Another example of the first type of percent problems is translated to an equation. Translate the other problem to an equation.

Find 8.5% of $500.
↓   ↓   ↓
8.5% · $500 = x

Find 145% of 925.

_____

a) x = 50% · 138
b) 63% · 837 = x

---

53. To solve an equation involving a percent, we must convert the percent to a fraction or decimal. Below, for example, we solved for x by converting $33\frac{1}{3}$% to a fraction to multiply. Use the same method for the other problem.

Find $33\frac{1}{3}$% of 600.

$x = 33\frac{1}{3}\% \cdot 600$

$x = \frac{1}{3}(600) = 200$

25% of 48 is what?

145% · 925 = x

x = 12, from $\frac{1}{4}$(48)

54. Ordinarily we have to convert <u>to</u> a <u>decimal</u> when multiplying by a percent. An example is shown. Do the other problem.

Find 2.4% of 120. What is 18% of $7,500 ?

$x = 2.4\% \cdot 120$

$x = .024(120)$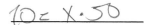

$x = 2.88$

---

55. Two examples of the second type of percent problems are shown below. We translated each to an equation.

17 is <u>what percent</u> of 80?   <u>What percent</u> of 200 is 159 ?
↓ ↓   ↓   ↓ ↓   ↓   ↓ ↓ ↓ ↓
17 =   x   · 80   x   · 200 = 159

Translate each problem to an equation.

a) 20 is what percent of 50?   b) What percent of 900 is 650?

$20 = x \cdot 50$ $x \cdot 900 = 650$

x = $1,350 , from
.18($7,500)

---

56. To solve equations like those in the last frame, we divide the other number by the coefficient of x and then convert the fraction to a percent. An example is shown. Solve the other equation.

$x \cdot 90 = 60$ $20 = x \cdot 50$

$x = \dfrac{60}{90} = \dfrac{2}{3} = 66\dfrac{2}{3}\%$ $\dfrac{}{50}$ $\dfrac{}{50}$

a) 20 = x · 50

b) x · 900 = 650

---

57. Ordinarily we have to use the division method to solve the second type of percent problem. An example is shown. Do the other problem. Round to a whole number percent.

What percent of 70 is 98?   14.9 is what percent of 84.7?

$x \cdot 70 = 98$ $14 = x \cdot 84.7$

$x = \dfrac{98}{70} = 140\%$ $\dfrac{}{84.7}$

40%, from $\dfrac{20}{50}$

---

58. Two examples of the third type of percent problems are shown below.

25 is 12% of what?   80% of what is 62?
↓ ↓ ↓ ↓   ↓   ↓ ↓ ↓ ↓ ↓
25 = 12% ·   x   80% ·   x   = 62

Translate each problem below to an equation.

a) 50% of what is 20?   b) 75 is 140% of what?

$50 \cdot x = 20$ $75 = 140\% x$

x = 18%, from $\dfrac{14.9}{84.7}$

---

a) 50% · x = 20   b) 75 = 140% · x

59. To solve equations like those in the last frame, we can sometimes convert the percent to a fraction. An example is shown. Solve the other equation.

$$30 = 75\% \cdot x$$

$$30 = \frac{3}{4}x$$

$$\frac{4}{3}(30) = \frac{4}{3}\left(\frac{3}{4}x\right)$$

$$40 = x$$

$$50\% \cdot x = 35$$

$$\frac{1}{2}x = 35$$

$$\frac{\frac{1}{2}}{\frac{1}{2}} \quad \frac{35}{\frac{1}{2}}$$

$$x = 70$$

---

60. Usually we have to convert the percent to a decimal to solve the third type of percent problem. An example is shown. Do the other problem. Round to a whole number.

22 is 2.5% of what?

$$22 = 2.5\% \cdot x$$

$$22 = .025x$$

$$x = \frac{22}{.025} = 880$$

12% of what is 200?

$$12\% \cdot x = 200$$

$$.12x = 200$$

$$x = \frac{200}{.12}$$

x = 70

---

61. Do these. a) Find $7\frac{1}{2}\%$ of $600.

$$\$45.00$$

b) 39.6 is what percent of 28.3? (Round to a whole number percent.)

$$\frac{39.6 = 28.3 \times}{28.3}$$

$$\$140$$

x = 1,667, from:
.12x = 200

---

62. Do these. a) What percent of 9,500 is 427? (Round to a tenth of a percent.)

$$9,500 = 427$$

$$4.5\%$$

b) 75 is 45% of what? (Round to a whole number.)

$$75 = 45\% \cdot x$$

$$\frac{432}{167} \quad 45\%$$

a) $45   b) 140%

---

a) 4.5%   b) 167

4-7 APPLIED PROBLEMS INVOLVING PERCENTS

In this section, we will discuss a method for solving applied problems involving percents.

63. To solve the problem below, we reworded it, translated the rewording to an equation, and then solved the equation. Use the same method for the other problem.

How many pounds of zinc are there in 50 pounds of an alloy if the alloy is 30% zinc?

Rewording: What is 30% of 50?

Equation: $x = 30\% \cdot 50$

Solution: $x = .3(50) = 15$

Therefore, there are 15 pounds of zinc in the alloy.

On a 40-item test, a student had 90% of the items correct. How many did she have correct?

Rewording: *What is 90% of 40*

Equation: *$x = 90\% \cdot 40$*

Solution: _____

What is 90% of 40?

$x = 90\% \cdot 40$

$x = .9(40)$
   = 36 items correct

64. We used the rewording-equation method to solve the problem below. Use the same method for the other problem. Round to a tenth of a percent.

Of 268 motors tested, 37 were defective. What percent were defective?

Rewording: 37 is what percent of 268?

Equation: $37 = x \cdot 268$

Solution: $x = \frac{37}{268} = .138 = 13.8\%$

Therefore, 13.8% of the motors were defective.

There are 14 grams of nitrogen in 63 grams of nitric acid. What percent of nitric acid is nitrogen?

Rewording: *What percent of 63 is 14*

Equations: *$63 \cdot x = 14$*

Solution: _____

14 is what percent of 63?

$14 = x \cdot 63$

$x = \frac{14}{63} = 22.2\%$

65. We used the rewording-equation method for the problem below. Use the same method for the other problem.

A stainless steel alloy contains 8% nickel. How many pounds of the alloy can be made from 560 pounds of nickel?

Rewording: 560 is 8% of what?

Equation: 560 = 8% · x

Solution: $x = \frac{560}{.08} = 7,000$

Therefore, 7,000 pounds of the alloy can be made from 560 pounds of nickel.

If silver solder contains 70% silver, how much solder can be made from 420 grams of silver?

Rewording: *what is 70% of 420*   *420 is 70% of what*

Equation: *420 = 70% · x*

Solution: _____

---

66. Use the "rewording" hints to solve the problems in this frame and the next frame.

420 is 70% of what?

420 = 70% · x

$x = \frac{420}{.7} = 600$ grams

a) If a student got 26 items correct on a 32-item test, what percent grade did he receive? (Round to a whole number percent.)

26 is what % of 32

(Hint: 26 is what percent of 32?)

26 = x · 32   *81.5*

b) A cast iron alloy has 3.5% carbon. Find the amount of carbon in 500 pounds of the alloy. *what is 3.5% of steel*

(Hint: What is 3.5% of 500?)

p = 3.5 × 500

17.5

---

67. a) An alloy weighing 32.7 kilograms contains 25.3 kilograms of aluminum. What percent of the alloy is aluminum? (Round to a tenth of a percent.)

(Hint: 25.3 is what percent of 32.7?)

32.7 ÷ 25.3   = 77%

a) 81%

b) 17.5 pounds

b) If iron rust is 70% iron, how much rust can be formed from 40 grams of iron? (Round to a whole number.)

(Hint: 40 is 70% of what?)

40 = 70% · t   *57 grms*

68. Make up your own "rewording" hints for the problems in this frame and the next frame.

    a) 223 kilograms of copper are used to make 260 kilograms of a bronze alloy. What percent of the alloy is copper? (Round to a tenth of a percent.)

    b) How many items would a student have to get correct to get 80% on a 30-item test?

a) 77.4%

b) 57 grams

---

69. a) An alloy contains 18.2% chromium. How much of the alloy can be made from 150 pounds of chromium? (Round to a whole number.)

    b) A sample of polluted lake water contains "87 parts per million" of mercury. What percent is mercury?

a) 85.8%, from:

    223 is what percent of 260?

b) 24 items, from:

    What is 80% of 30?

---

70. The following mixture problem involves percents.

    Problem: How many liters of a 20% acid solution must be mixed with 6 liters of a 70% acid solution to get a 50% acid solution?

    Letting x equal the number of liters of the 20% solution, we can organize the data in the following table. To get the numbers in the right column, we multiplied the strengths and the number of liters.

Strength	Liters of solution	Liters of pure acid
70%	6	.70(6) = 4.2
20%	x	.20x
50%	6 + x	.50(6 + x)

    Since the number of liters of pure acid in the 50% solution is the sum of the number of liters of pure acid in the 70% and 20% solution, we can set up and solve the following equation.

    $$4.2 + .20x = .50(6 + x)$$
    $$4.2 + .20x = 3 + .50x$$
    $$1.2 = .3x$$
    $$x = \frac{1.2}{.3} = 4$$

    Therefore, 4 liters of the 20% solution must be used.

    Continued on following page.

a) 824 pounds, from:

    150 is 18.2% of what?

b) .0087%, from:

    87 is what percent of 1,000,000?

70. Continued

Using the same method, solve this mixture problem.

How many liters of a 10% alcohol solution must be mixed with 20 liters of a 60% alcohol solution to get a 30% solution?

30 liters

SELF-TEST 13 (pages 156-166)

1. Convert 60% to a fraction in lowest terms.

   _____

2. Convert 27.5% to a decimal number.

   _____

3. Convert .0418 to a percent.

   _____

4. Convert $\frac{2}{3}$ to a percent.

   _____

5. Convert $\frac{766}{512}$ to a percent. Round to a whole number percent. _____

6. 150 is what percent of 500?

   _____

7. What is 18.5% of 5,400?

   _____

8. 40% of what is 15?

   _____

9. Find .8% of 42,390. Round to a whole number.

   _____

10. What percent of 24,800 is 672? Round to a hundredth of a percent.

    _____

11. 4,316 is 2.7% of what? Round to thousands.

    _____

12. Find the percent grade for a student who got 27 items correct on a 32-item test. Round to a whole number percent.

    _____

13. A stainless steel alloy contains 11.5% chromium. How much chromium is there in 2,500 pounds of the alloy? Round to tens.

    _____

14. Sea water contains 0.13% magnesium. What weight of sea water contains 50 kilograms of magnesium? Round to hundreds.

    _____

15. A manufacturer made 140,000 metal cans. Inspectors rejected 830 cans. What percent were rejected? Round to a hundredth of a percent.

    _____

ANSWERS TO SELF-TEST 13:

1. $\frac{3}{5}$ 5. 150%   9. 339   13. 290 pounds

2. .275   6. 30%   10. 2.71%   14. 38,500 kilograms

3. 4.18%   7. 999   11. 160,000   15. .59%

4. $66\frac{2}{3}$%   8. 37.5   12. 84%

4-8 RATIO AND PROPORTION

In this section, we will define and discuss ratios and proportions.

---

71. A <u>ratio</u> is a comparison of two quantities by division. A ratio can be expressed in three ways: with a fraction, with the word <u>to</u>, or with a colon. For example, if the length and width of a rectangle are 7 feet and 4 feet, the ratio of length to width can be expressed in each of three ways below.

$$\frac{7 \text{ feet}}{4 \text{ feet}} = \frac{7}{4} \qquad 7 \text{ to } 4 \qquad 7:4$$

A first process takes 11 seconds and a second process takes 20 seconds. Express the time ratio of the first process to the second process in three ways below.

$\frac{11}{20}$ 11:20   11 to 20

---

72. When ratio language is used, the ratio is ordinarily reduced to lowest terms. For example:

The ratio of 25 grams to 15 grams is $\frac{25}{15} = \frac{5}{3}$ or 5 to 3.

When a ratio expressed as a fraction reduces to a whole number, we frequently write "1" as the denominator to make it clear that we are comparing two quantities. For example:

The ratio of 60 meters to 15 meters is $\frac{60}{15} = \frac{4}{1}$ or 4 to 1 .

$\frac{11}{20}$
11 to 20
11:20

---

73. For one problem below, we set up the ratio and then used a calculator to convert it to a decimal number. Do the same for the other problem.

A driving gear with 68 teeth is meshed with a driven gear with 24 teeth. What is the gear ratio? That is, what is the ratio of the number of teeth in the driving gear to the number of teeth in the driven gear? (Round to hundredths.)

$$\frac{68 \text{ teeth}}{24 \text{ teeth}} = 2.83$$

An electrical signal of 0.835 volt is accompanied by a noise voltage of 0.0355 volt. What is the signal-to-noise ratio? (Round to tenths.)

4 to 1

74. When a ratio is expressed as a decimal number, it is helpful to write the decimal over "1" to show that we are comparing two quantities. That is:

    A gear ratio of 2.83 means $\frac{2.83}{1}$ or 2.83 to 1.

    A signal-to-noise ratio of 23.5 means $\frac{23.5}{1}$ or _____ to _____.

| $\frac{0.835 \text{ volt}}{0.0355 \text{ volt}} = 23.5$ |

75. Ratios are frequently used to compare quantities with different units. For example:

    If a spring is stretched 1.3 inches by a force of 21 pounds, we can use a ratio to compare the stretch to the force. We get:

    $\frac{1.3 \text{ inches}}{21 \text{ pounds}}$ or a ratio of 1.3 to 21.

    If there are 55 grams of a gas in a volume of 22.4 liters, we can use a ratio to compare the weight to the volume. We get:

    $\frac{55 \text{ grams}}{22.4 \text{ liters}}$ or a ratio of _____ to _____.

23.5 to 1

76. Two ratios are equal if they both reduce to the same lowest-terms ratio. For example, the two ratios below are equal because they both reduce to 5 to 2.

    50 grams to 20 liters   5 to 2 $\left(\text{from } \frac{50}{20} = \frac{5}{2} \right)$

    100 grams to 40 liters   5 to 2 $\left(\text{from } \frac{100}{40} = \frac{5}{2} \right)$

    By reducing to lowest terms, determine whether each pair of ratios is equal or not.

    a) 8 feet to 6 feet   b) 2 inches to 20 pounds
       40 feet to 30 feet   3 inches to 24 pounds

55 to 22.4

77. Two ratios are also equal if they both convert to the same decimal. For example, the two ratios below are equal because they both convert to 5.5 or 5.5 to 1.

    77 feet to 14 seconds   5.5 to 1 $\left(\text{from } \frac{77}{14} = 5.5 \right)$

    121 feet to 22 seconds   5.5 to 1 $\left(\text{from } \frac{121}{22} = 5.5 \right)$

    By converting to decimals, determine whether each pair of ratios is equal or not.

    a) 735 rpm to 525 rpm   b) 366 miles to 12 gallons
       600 rpm to 375 rpm   427 miles to 14 gallons

a) Equal, since both equal 4 to 3.

b) Not equal.

a) Not equal.

b) Equal, since both equal 30.5 or 30.5 to 1.

78. A <u>proportion</u> is a statement that two ratios are equal. For example:

$$\frac{5 \text{ inches}}{6 \text{ pounds}} = \frac{10 \text{ inches}}{12 \text{ pounds}} \quad \text{is a proportion, since} \quad \frac{5}{6} = \frac{10}{12}$$

Which of the following are proportions? _____

a) $\dfrac{20 \text{ grams}}{10 \text{ liters}} = \dfrac{60 \text{ grams}}{30 \text{ liters}}$ b) $\dfrac{100 \text{ miles}}{2 \text{ hours}} = \dfrac{200 \text{ miles}}{3 \text{ hours}}$

Only (a)

4-9 CROSS-MULTIPLICATION AND PROPORTIONS

Cross-multiplication is a shortcut method that can be used to clear the fractions in a proportion. We will discuss cross-multiplication in this section.

79. To clear the fractions and solve the proportion below, we multiplied both sides by both denominators at the same time.

$$\frac{x}{8} = \frac{3}{4}$$

$$(8)(4)\left(\frac{x}{8}\right) = (8)(4)\left(\frac{3}{4}\right)$$

$$\frac{(8)(4)(x)}{8} = \frac{(8)(4)(3)}{4}$$

$$\left(\frac{8}{8}\right)(4)(x) = \left(\frac{4}{4}\right)(8)(3) \qquad \underline{\text{Check}}: \frac{x}{8} = \frac{3}{4}$$

$$(1)(4x) = (1)(24) \qquad \qquad \frac{6}{8} = \frac{3}{4}$$

$$4x = 24$$

$$x = 6 \qquad \qquad \frac{3}{4} = \frac{3}{4}$$

However, there is a shortcut called <u>cross-multiplication</u> that can be used to clear the fractions in a proportion. The shortcut is shown and discussed below.

$$\frac{x}{8} \diagdown\!\!\!\!\diagup \frac{3}{4}$$

$$(x)(4) = (8)(3)$$

$$4x = 24 \qquad \underline{\text{Note}}: \text{ To get } 4x, \text{ we multiplied } \underline{x} \text{ and } 4.$$

$$x = 6 \qquad \qquad \text{To get } 24, \text{ we multiplied } 8 \text{ and } 3.$$

Did we get the same solution both ways? _____

Yes. We got x = 6.

80. To solve the proportion below, we cross-multiplied to clear the fractions. Use the same steps to solve the other proportion.

$$\frac{x}{6} = \frac{4}{3} \qquad\qquad \frac{2}{5} = \frac{y}{15}$$

$$(x)(3) = (6)(4)$$

$$3x = 24$$

$$x = \frac{24}{3} = 8$$

---

81. To solve the proportion below, we used a calculator to divide 210 by 50. Use a calculator to solve the other proportion.

$$\frac{7}{x} = \frac{50}{30} \qquad\qquad \frac{4}{5} = \frac{10}{y}$$

$$(7)(30) = (x)(50)$$

$$210 = 50x$$

$$x = \frac{210}{50} = 4.2$$

$y = 6$, from:

$$30 = 5y$$

---

82. Following the example, round the other root to hundredths.

$$\frac{3}{7} = \frac{x}{5} \qquad\qquad \frac{9}{8} = \frac{7}{y}$$

$$(3)(5) = (7)(x)$$

$$15 = 7x$$

$$x = \frac{15}{7} = 2.14$$

$y = 12.5$, from:

$$4y = 50$$

---

83. To solve the proportion below, we used a calculator twice: first to multiply 34 and 71.5 and then to divide 2431 by 13.

$$\frac{13}{34} = \frac{71.5}{x}$$

$$(13)(x) = (34)(71.5)$$

$$13x = 2,431$$

$$x = \frac{2,431}{13} = 187$$

$y = 6.22$, from: $\frac{56}{9}$

Continued on following page.

93. Continued

Use the same method to solve this problem.

If the signal-to-noise ratio is 24.6, what is the noise voltage if the signal voltage is 0.925 volt? (Round to four decimal places.)

---

94. The formula for <u>efficiency</u> is shown below. As you can see, efficiency is the ratio of power output to power input.

$$\text{Efficiency} = \frac{\text{Power Output}}{\text{Power Input}}$$

Efficiency is a ratio that is ordinarily stated as a percent. An example is given below.

An electric motor has a power input of 1,500 watts and a power output of 1,420 watts. What is its percent efficiency? (Round to a tenth of a percent.)

$$\text{Efficiency} = \frac{1,420 \text{ watts}}{1,500 \text{ watts}} = .947 = 94.7\%$$

Following the example, solve this problem.

What is the efficiency of a gasoline engine if its power output is 240 horsepower and its power input is 955 horsepower? (Round to a tenth of a percent.)

0.0376 volt, from:

$$24.6 = \frac{0.925 \text{ volt}}{x \text{ volts}}$$

---

95. We used the efficiency formula to solve the problem below.

A gasoline engine has an efficiency of 24%. If its power output is 210 horsepower, find its power input.

$$24\% = \frac{210 \text{ horsepower}}{x \text{ horsepower}} \quad \text{or} \quad .24 = \frac{210}{x}$$

$$.24x = 210$$

$$x = \frac{210}{.24} = 875$$

Therefore, its power input is 875 horsepower.

25.1%, from: $\frac{240}{955}$

Continued on following page.

95. Continued

Following the example, solve this problem.

An electrical transformer has an efficiency of 98.2% and a power input of 67,500 watts. Find its power output in watts. (Round to hundreds.)

66,300 watts, from: $.982 = \dfrac{x}{67,500}$

SELF-TEST 14 (pages 167-176)

1. Write "the ratio of 60 grams to 24 grams" as a fraction in lowest terms.

_____

2. Write "the ratio of 10.2 volts to 2.4 volts" as a decimal number.

_____

Solve each proportion.

3. Round to a whole number.

$$\frac{586}{235} = \frac{x}{179}$$

x = _____

4. Round to tenths.

$$\frac{1.53}{24.7} = \frac{2.81}{y}$$

y = _____

5. Round to thousandths.

$$\frac{0.907}{w} = \frac{3,150}{2,740}$$

w = _____

6. If a 6 inch length on a map represents 80 miles, what length represents 140 miles?

_____

7. If a spring is stretched 2.7 centimeters by a 12 kilogram force, what stretch will be produced by a 20 kilogram force?

_____

8. If 22.4 liters of carbon dioxide gas weigh 44 grams, find the weight of 8 liters. Round to tenths.

_____

9. If a 250 foot length of copper wire weighs 4.95 pounds, find the weight of a 75 foot length. Round to two decimal places.

_____

10. The efficiency of an electric motor is 92.9%. If its power output is 780 watts, find the power input. Round to a whole number.

_____

11. The width-to-height ratio of a TV screen is 1.333 . If the height of a TV screen is 14.4 inches, find its width. Round to tenths.

_____

ANSWERS TO SELF-TEST 14:

1. $\dfrac{5}{2}$ 3. x = 446   6. 10.5 inches   9. 1.49 pounds

2. 4.25   4. y = 45.4   7. 4.5 centimeters   10. 840 watts

5. w = 0.789   8. 15.7 grams   11. 19.2 inches

4-11   DIRECT VARIATION

Any relationship of the form   y = kx   is called <u>direct</u> <u>variation</u>.   We will discuss direct variation in this section.

96.   If a car travels at a constant speed of 50 mph, the ratio of distance traveled (d) to time traveled (t) is a constant.   The constant is 50. That is:

$$\frac{d}{t} = \frac{50}{1} = \frac{100}{2} = \frac{150}{3} = \frac{200}{4} = 50$$

The relationship above can be stated as an equation.

$$\frac{d}{t} = 50 \qquad \text{or} \qquad d = 50t$$

Any relationship of that type is called <u>direct</u> <u>variation</u>.   Above, <u>distance</u> <u>varies</u> <u>directly</u> <u>as</u> <u>time</u>.   When time increases, distance increases.   When time decreases, distance decreases.

The general form for <u>direct</u> <u>variation</u> is given below.

> y = kx   where <u>k</u> is called <u>the</u> <u>constant</u> <u>of</u> <u>variation</u>
> or <u>the</u> <u>constant</u> <u>of</u> <u>proportionality</u>.

Some examples of direct variation are:

$$y = 5x \qquad\qquad d = 50t$$

a) In y = 5x, the constant of variation is _____.

b) In d = 50t, the constant of proportionality is _____.

97.   The following language is used to state a direct variation.

For y = 4x, we say: <u>y</u> varies <u>directly</u> as <u>x</u>.
<u>y</u> is <u>directly</u> <u>proportional</u> to <u>x</u>.

a) For d = 100t, we say: _____ varies directly as _____.

b) For E = 6R, we say: _____ is directly proportional to _____.

---

a) 5

b) 50

---

a) d ... t

b) E ... R

98. If we are given one specific pair of values in a direct variation, we can find $\underline{k}$ by substitution. For example, if $\underline{y}$ varies directly as $\underline{x}$ and $y = 10$ when $x = 2$, we get:

$$y = kx$$
$$10 = k(2)$$
$$k = \frac{10}{2} = 5$$

Since $k = 5$, the direct variation is $y = 5x$. Using that equation, we can find $\underline{y}$ for other values of $\underline{x}$. For example, we found $\underline{y}$ when $x = 3$ below. Find $\underline{y}$ when $x = 10$.

$y = 5x$	$y = 5x$
$y = 5(3)$	
$y = 15$	_____

99. Using the two steps from the last frame, we solved the problem below.

If V is directly proportional to R and $V = 30$ when $R = 3$, find V when $R = 5$.

$V = kR$	$V = 10R$
$30 = k(3)$	$V = 10(5)$
$k = \frac{30}{3} = 10$	$V = 50$

Using the same steps, solve this one.

If $\underline{d}$ varies directly as $\underline{t}$ and $d = 600$ when $t = 10$, find $\underline{d}$ when $t = 7$.

100. Solve these:

a) When the resistance is constant, the amount of current (I) in an electric circuit varies directly as the applied voltage (V). If $I = 30$ amperes when $V = 6$ volts, find I when $V = 24$ volts.

b) The amount of stretch (s) in a spring is directly proportional to the force (F) applied to it. If $s = 6$ centimeters when $F = 40$ kilograms, find $\underline{s}$ when $F = 90$ kilograms.

Answers (right column):

$y = 50$

$d = 420$, from:
$$d = 60t$$

a) 120 amperes
b) 13.5 centimeters

4-12 INVERSE VARIATION

Any relationship of the form $y = \frac{k}{x}$ is called inverse variation. We will discuss inverse variation in this section.

101. Airplanes flying at different rates (or speeds) take different amounts of time to fly 500 miles. At 500 mph, it takes 1 hour. At 250 mph, it takes 2 hours. At 125 mph, it takes 4 hours. In each case, the product of the rate r and the time t is a constant. The constant is 500. That is:

$$rt = (500)(1) = (250)(2) = (125)(4) = 500$$

The relationship above can be stated as an equation.

$$rt = 500 \qquad \text{or} \qquad t = \frac{500}{r}$$

Any relationship of that type is called <u>inverse variation</u>. Above, <u>time</u> <u>varies</u> <u>inversely</u> <u>as</u> <u>rate</u>. When the rate increases, time decreases. When the rate decreases, time increases.

The general form for <u>inverse variation</u> is given below.

$$\boxed{y = \frac{k}{x}} \qquad \text{, where } \underline{k} \text{ is called } \underline{\text{the}} \ \underline{\text{constant}} \ \underline{\text{of}} \ \underline{\text{variation}}$$
$$\text{or } \underline{\text{the}} \ \underline{\text{variation}} \ \underline{\text{constant}}$$

Some examples of inverse variation are:

$$y = \frac{12}{x} \qquad\qquad t = \frac{20}{r} \qquad\qquad V_1 = \frac{8.5}{P_1}$$

a) In $y = \frac{12}{x}$, the constant of variation is _____.

b) In $V_1 = \frac{8.5}{P_1}$, the variation constant is _____.

102. The following language is used to state an inverse variation.

For $y = \frac{12}{x}$, we say: $\underline{y}$ varies <u>inversely</u> as $\underline{x}$

or: $\underline{y}$ is <u>inversely</u> <u>proportional</u> to $\underline{x}$.

a) In $t = \frac{20}{v}$, _____ varies inversely as _____.

b) In $V_1 = \frac{8.5}{P_1}$, _____ is inversely proportional to _____.

a) 12

b) 8.5

a) t ... v

b) V_1 ... P_1

103. If we are given one specific pair of values in an inverse variation, we can find $\underline{k}$ by substitution. For example, if $\underline{y}$ varies inversely as $\underline{x}$ and $y = 50$ when $x = 4$, we get:

$$y = \frac{k}{x}$$

$$50 = \frac{k}{4}$$

$$k = (50)(4) = 200$$

Since $k = 200$, the inverse variation is $y = \frac{200}{x}$. We can use that equation to find $\underline{y}$ for other values of $\underline{x}$.

a) Find $\underline{y}$ when $x = 5$.　　　b) Find $\underline{y}$ when $x = 50$.

$$y = \frac{200}{x}$$　　　　　$$y = \frac{200}{x}$$

$$\frac{50}{4} = \frac{x}{5}$$

104. Using the two steps from the last frame, we solved the problem below.

If I is inversely proportional to R and $I = 20$ when $R = 2$, find I when $R = 8$.

$$I = \frac{k}{R}$$　　　　　$$I = \frac{40}{R}$$

$$20 = \frac{k}{2}$$　　　　　$$I = \frac{40}{8}$$

$$k = 40$$　　　　　$$I = 5$$

Using the same steps, solve this one.

If $\underline{t}$ varies inversely as $\underline{v}$ and $t = 4$ when $v = 100$, find $\underline{t}$ when $v = 50$.

a) y = 40

b) y = 4

t = 8, from:

$$t = \frac{400}{v}$$

105. Solve these:

a) In an electric circuit with fixed power, the current (I) is inversely proportional to the voltage (V). If I = 8 amperes when V = 12 volts, find I when V = 10 volts.

b) The pressure (P) of a gas varies inversely as its volume (V). If P = 10 g/cm^2 when V = 25 cm^3, find P when V = 40 cm^3.

a) I = 9.6 amperes   b) P = 6.25 g/cm^2

4-13 QUADRATIC VARIATION

In this section, we will discuss <u>direct square variation</u> and <u>inverse square variation</u>.

106. Any equation or formula of the form $\boxed{y = kx^2}$ is called <u>direct square variation</u>. Some examples are:

$$y = 5x^2 \qquad P = 20 I^2 \qquad A = .7854d^2$$

In $y = kx^2$, <u>k</u> is called the <u>constant of variation</u> or the <u>variation constant</u>. The following language is used to state a direct square variation.

For $y = 5x^2$, we say: <u>y</u> varies directly as the square of <u>x</u>.

For $P = 20 I^2$, we say: ____ varies directly as the square of ____.

107. If we are given one specific pair of values in a direct square variation, we can find <u>k</u> by substitution. For example, if <u>y</u> varies directly as the square of <u>x</u> and y = 12 when x = 2, we get:

$$y = kx^2$$
$$12 = k(2)^2$$
$$12 = k(4)$$
$$k = 3$$

P ... I

Since k = 3, the direct square variation is $y = 3x^2$. We can use that equation to find <u>y</u> for other values of <u>x</u>.

a) Find <u>y</u> when x = 4.   b) Find <u>y</u> when x = 10.

$$y = 3x^2 \qquad\qquad y = 3x^2$$

108. Using the two steps from the last frame, we solved the problem below.

    If <u>a</u> varies directly as the square of <u>m</u> and a = 100 when m = 5, find <u>a</u> when m = 8.

$a = km^2$	$a = 4m^2$
$100 = k(5)^2$	$a = 4(8)^2$
$100 = k(25)$	$a = 4(64)$
$k = 4$	$a = 256$

    Using the same steps, solve this one.

    If R varies directly as the square of V and R = 500 when V = 10, find R when V = 4.

    a) y = 48

    b) y = 300

109. Solve these:

    a) The kinetic energy (E) of a moving object varies directly as the square of its velocity (v). If E = 400 joules when v = 10 meters per second, find E when v = 8 meters per second.

    b) The power (P) in an electric circuit varies directly as the square of the current (I). If P = 20 watts when I = 2 amperes, find P when I = 3 amperes.

    R = 80, from:

    $R = 5V^2$

110. Any equation or formula of the form $\boxed{y = \dfrac{k}{x^2}}$ is called <u>inverse square variation</u>. Some examples are:

    $$y = \frac{5}{x^2} \qquad F = \frac{65}{d^2} \qquad P = \frac{100}{t^2}$$

    In $y = \dfrac{k}{x^2}$, <u>k</u> is called the <u>constant</u> <u>of</u> <u>variation</u> or the <u>variation constant</u>. The following language is used to state an inverse square variation.

    For $y = \dfrac{5}{x^2}$, we say: <u>y</u> varies inversely as the square of <u>x</u>.

    For $F = \dfrac{65}{d^2}$, we say: ___ varies inversely as the square of ___ .

    a) E = 256 joules

    b) P = 45 watts

F ... d

111. If we are given one specific pair of values in an inverse square variation, we can find k by substitution. For example, if y varies inversely as the square of x and y = 12 when x = 2, we get:

$$y = \frac{k}{x^2}$$

$$12 = \frac{k}{2^2}$$

$$12 = \frac{k}{4}$$

$$k = 48$$

Since k = 48, the inverse square variation is $y = \frac{48}{x^2}$. We can use that equation to find y for other values of x.

    a) Find y when x = 4.   b) Find y when x = 8.

---

112. Using the two steps from the last frame, we solved the problem below.

If H varies inversely as the square of P and H = 4 when P = 5, find H when P = 2.

$$H = \frac{k}{P^2} \qquad\qquad H = \frac{100}{P^2}$$

$$4 = \frac{k}{5^2} \qquad\qquad H = \frac{100}{2^2}$$

$$4 = \frac{k}{25} \qquad\qquad H = \frac{100}{4}$$

$$k = 100 \qquad\qquad H = 25$$

Using the same steps, solve this one.

If m varies inversely as the square of s and m = 4 when s = 6, find m when s = 4.

---

a) y = 3

b) $y = \frac{3}{4}$ or 0.75

---

m = 9, from:

$$m = \frac{144}{s^2}$$

113. Solve these:

a) The intensity (I) of a radio signal varies inversely as the square of its distance (d) from the transmitter. If I = 5 microvolts when d = 10 kilometers, find I when d = 5 kilometers.

b) The gravitational force (F) between two objects varies inversely as the square of the distance (d) between them. If F = 40 dynes when d = 10 centimeters, find F when d = 4 centimeters.

a) I = 20 microvolts   b) F = 250 dynes

4-14   JOINT AND COMBINED VARIATION

In this section, we will discuss joint variation and combined variation. Both types of variation involve two or more independent variables.

114. Any equation or formula of the form $\boxed{y = kxz}$ is called joint variation. Some examples are:

$$y = 4yz \qquad\qquad A = \frac{1}{2}bh \qquad\qquad V = 10hg^2$$

When a formula is a joint variation, the units for the variables are frequently chosen so that k = 1. For example:

$$E = IR \qquad\qquad F = ma \qquad\qquad P = I^2R$$

The following language is used for a joint variation.

   For $y = 4xz$, we say: y varies jointly as x and z.

   For $P = I^2R$, we say: P varies jointly as R and the square of I.

   For $V = IR$, we say: V varies jointly as ____ and ____.

115. To solve the problem below, we found the variation constant and then used $y = 4xz$ to complete the solution.

I and R

   y varies jointly as x and z. If y = 80 when x = 2 and z = 10, find y when x = 3 and z = 6.

y = kxz	y = 4xz
80 = k(2)(10)	y = 4(3)(6)
80 = 20K	y = 12(6)
k = 4	y = 72

Continued on following page.

115. Continued

Use the same method to solve this problem.

P varies jointly as Q and R. If P = 40 when Q = 2 and R = 4, find P when Q = 5 and R = 10.

---

116. We used the same two-step method to solve one problem below. Solve the other problem.

$\underline{b}$ varies jointly as $\underline{c}$ and the square of $\underline{d}$. If b = 200 when c = 4 and d = 5, find $\underline{b}$ when c = 2 and d = 10.

$b = kcd^2$	$b = 2cd^2$
$200 = k(4)(5^2)$	$b = 2(2)(10^2)$
$200 = k(4)(25)$	$b = 2(2)(100)$
$200 = 100k$	$b = 400$
$k = 2$	

$\underline{m}$ varies jointly as $\underline{s}$ and the square of $\underline{t}$. If m = 90 when s = 2 and t = 3, find $\underline{m}$ when s = 8 and t = 5.

---

117. <u>Combined variation</u> includes both direct and inverse variation. Some examples and the language used to describe them are given below.

$y = \dfrac{kx}{z}$ $\underline{y}$ varies directly as $\underline{x}$ and inversely as $\underline{z}$.

$p = \dfrac{kst}{v}$ $\underline{p}$ varies jointly as $\underline{s}$ and $\underline{t}$ and inversely as $\underline{v}$.

For the formulas below, the variation constant $\underline{k}$ equals "1".

$F = \dfrac{mv^2}{r}$ F varies jointly as $\underline{m}$ and the square of $\underline{v}$ and inversely as $\underline{r}$.

$F = \dfrac{m_1 m_2}{rd^2}$ F varies jointly as m_1 and m_2 and inversely as $\underline{r}$ and the square of $\underline{d}$.

Write an equation or formula for each of these.

a) $\underline{y}$ varies directly as $\underline{x}$ and inversely as $\underline{b}$ and the square of $\underline{z}$.   _____

b) P_1 varies jointly as P_2 and V_2 and inversely as V_1, with k = 1.   _____

Answers (right column):

P = 250

m = 1,000

118. We used a two-step process to solve the problem below. Solve the other problem.

$\underline{y}$ varies jointly as $\underline{x}$ and $\underline{z}$ and inversely as $\underline{w}$. If $y = 15$ when $x = 2$, $z = 5$, and $w = 4$, find $\underline{y}$ when $x = 3$, $z = 10$, and $w = 9$.

$$y = \frac{kxz}{w} \qquad\qquad y = \frac{6xz}{w}$$

$$15 = \frac{k(2)(5)}{4} \qquad y = \frac{6(3)(10)}{9}$$

$$15 = \frac{10k}{4} \qquad\qquad y = \frac{180}{9}$$

$$4(15) = 4\left(\frac{10k}{4}\right) \qquad y = 20$$

$$60 = 10k$$

$$k = 6$$

$\underline{b}$ varies directly as $\underline{c}$ and inversely as the square of $\underline{d}$. If $b = 10$ when $c = 20$ and $d = 4$, find $\underline{b}$ when $c = 45$ and $d = 6$.

a) $y = \dfrac{kx}{bz^2}$

b) $P_1 = \dfrac{P_2 V_2}{V_1}$

---

119. Solve each problem:

a) The wind force (F) on a vertical surface varies jointly as the area (A) of the surface and the square of the wind velocity (v). When v = 20 mph and A = 1 sq ft, F = 2 lbs. Find F when v = 30 mph and A = 2 sq ft.

b) The volume (V) of a gas varies directly as the temperature (T) and inversely as the pressure (P). If $V = 200$ cm^3 when $T = 250°$K and $P = 5$ kg/cm^2, find V when $T = 300°$K and $P = 8$ kg/cm^2.

b = 10

a) F = 9 lbs, from:

$$F = .005 \, Av^2$$

b) $V = 150$ cm^3, from:

$$V = \frac{4T}{P}$$

SELF-TEST 15 (pages 177-187)

1. y is directly proportional to x. If $y = 8$ when $x = 2$, find y when $x = 6$.

    y = _____

2. y is inversely proportional to x. If $y = 2$ when $x = 6$, find y when $x = 4$.

    y = _____

3. y varies inversely as the square of x. If $y = 4$ when $x = 3$, find y when $x = 2$.

    y = _____

4. In an electric circuit, power P varies directly as voltage V. If P = 5 watts when V = 20 volts, find P when V = 12 volts.

    P = _____

5. The centripetal force F on an object varies directly as the square of its velocity v. If F = 8 newtons when v = 20 m/sec, find F when v = 40 m/sec.

    F = _____

6. The volume V of a gas varies inversely as the pressure P. If V = 20 cm^3 when P = 15 g/cm^2, find V when P = 30 g/cm^2.

    V = _____

7. The level of illumination E on a surface varies inversely as the square of the distance d from the luminous source. If E = 12 units when d = 2 meters, find E when d = 4 meters.

    E = _____

8. The volume V of a rectangular box varies jointly as its length L and its width W. If V = 200 in^3 when L = 10 in and W = 4 in, find V when L = 12 in and W = 8 in.

    V = _____

9. The electrical resistance R of a wire is directly proportional to its length L and inversely proportional to its cross-sectional area A. If R = 1 ohm when L = 400 meters and A = 10 mm^2, find R when L = 800 meters and A = 5 mm^2.

    R = _____

ANSWERS:

1. y = 24	4. P = 3 watts	7. E = 3 units
2. y = 3	5. F = 32 newtons	8. V = 480 in^3
3. y = 9	6. V = 10 cm^3	9. R = 4 ohms

SUPPLEMENTARY PROBLEMS - CHAPTER 4

Assignment 12

Name the place in which the 8 appears in each number below.

   1. 786,509   2. 92.803   3. 5,802   4. .06813   5. .0015287

Write the number corresponding to each word name.

   6. seven hundred three thousand, fifteen

   7. five billion, two hundred million, eighty thousand

   8. six and nine thousandths

   9. thirty and five tenths

  10. one hundred sixty ten-thousandths

  11. three hundred millionths

Round each number as directed.

  12. 569,417 to thousands   13. 109,817,000 to millions

  14. 74.486 to a whole number   15. .00036049 to millionths

  16. 35.09581 to two decimal places   17. 863.172 to tens

  18. 60,344,815 to hundred thousands   19. .017098 to hundred-thousandths

  20. .059973 to ten-thousandths   21. .600128 to three decimal places

Do these problems on a calculator. Report the entire calculator answer.

  22. $(-5.4)(-3.5)$ 23. $-85.1 - 29.7$ 24. $2.25 \div (-.375)$

  25. $(-.715)^2$ 26. $\dfrac{1}{-62.5}$ 27. $\sqrt{-64}$

Do these problems on a calculator. Round as directed.

  28. Round to hundreds.   29. Round to two decimal places.

      $7,853 + 4,741 + 9,075$ $.8596 \times 23.57$

  30. Round to millions.   31. Round to ten-thousands.

      $30,600 \div .00512$ $(555)^2$

  32. Round to tenths.   33. Round to tens.

      $\sqrt{6,090.2}$ $401.3 - 157.6$

  34. Round to hundred-thousands.   35. Round to thousandths.

      $(2,154)^2$ $\sqrt{.572}$

  36. Round to millionths.   37. Round to thousands.

      $31.7 \div 914$ $7.19 \times 26,800$

  38. Round to hundredths.   39. Round to millionths.

      $\sqrt{5}$ $(.0792)^2$

  40. Round to three decimal places.   41. Round to hundreds.

      $720 \times .000152 \times 3.09$ $673.8 \times 2.915 \div .0491$

  42. Convert $\dfrac{37}{79}$ to a decimal number. Round to ten-thousandths.

  43. Find the reciprocal of 68. Round to thousandths.

Assignment 13

Convert each percent to a fraction in lowest terms.

    1. 75%   2. 20%   3. $33\frac{1}{3}$%   4. 90%   5. 25%

Convert each percent to a decimal number or whole number.

    6. 19%   7. 3.9%   8. .65%   9. 147%   10. 300%

Convert each decimal number or whole number to a percent.

    11. .814   12. .0425   13. .006   14. 1.7   15. 5

Convert each fraction to a percent.

    16. $\frac{1}{2}$ 17. $\frac{4}{5}$ 18. $\frac{3}{10}$ 19. $\frac{2}{3}$ 20. $\frac{59}{100}$

Convert to a whole number percent.   Convert to a hundredth of a percent.

21. $\frac{14}{40}$ 22. $\frac{3.7}{21.9}$ 23. $\frac{243}{185}$ 24. $\frac{76}{1,830}$ 25. $\frac{2.4}{506}$ 26. $\frac{1.79}{4.26}$

Solve the following problems.

    27. 20 is what percent of 50?   28. What is 75% of 400?   29. 30% of what is 15?

    30. What percent of 180 is 60?   31. Find 8.5% of $250.   32. 56 is 8% of what?

    33. .25% of 800 is what?   34. 6% of what is 24?   35. 12 is what percent of 300?

Solve these applied problems involving percents.

    36. A cast iron alloy has 3.5% carbon. Find the amount of carbon in 500 pounds of the alloy.

    37. If a student got 19 problems correct on a 23-problem test, what was her percent grade? Round to a whole number percent.

    38. An ore contains 0.8% uranium. How much uranium is there in 2,000 kilograms of the ore?

    39. If 58.45 grams of sodium chloride contain 35.46 grams of chlorine, what percent is chlorine? Round to a hundredth of a percent.

    40. If silver solder contains 75% silver, how much solder can be made from 500 grams of silver? Round to a whole number.

    41. In making 260 kilograms of a bronze alloy, 223 kilograms of copper are used. What percent of the alloy is copper? Round to a tenth of a percent.

    42. Sea water contains "380 parts per million" of potassium. What percent is potassium?

    43. An ore contains 1.37% copper. How much ore is needed to get 100 pounds of copper? Round to hundreds.

    44. How many liters of a 20% alcohol solution must be mixed with 12 liters of a 50% alcohol solution to get a 30% alcohol solution?

    45. How many liters of water must be mixed with 10 liters of a 40% acid solution to get a 25% acid solution?

Assignment 14

1. Write "the ratio of 30 liters to 72 liters" as a fraction in lowest terms.

2. Write "the ratio of 66.3 kilograms to 20.4 kilograms" as a decimal number.

Solve each proportion.   Round as directed.

3. Round to thousandths.   4. Round to thousands.   5. Round to hundredths.

$$\frac{.261}{303} = \frac{t}{589}$$
$$\frac{1.284}{x} = \frac{.03162}{72,880}$$
$$\frac{5.09}{2.47} = \frac{9.35}{w}$$

6. If a car used 7.3 gallons of gas to travel 235 miles, how far can it travel on 20 gallons of gas? Round to a whole number.

7. If 1 pound equals 453.6 grams, how many pounds are there in 1,000 grams?   Round to one decimal place.

8. If 14.4 grams of carbon are burned, 52.8 grams of carbon dioxide are produced.   How much carbon must be burned to produce 800 grams of carbon dioxide?   Round to a whole number.

9. A 2.5-centimeter length on a drawing represents 3 meters.   What length represents 10 meters? Round to tenths.

10. What force is needed to stretch a spring 2.4 inches if a 6-pound force stretches it 1.8 inches?

11. If 15 metal washers weigh 68 grams, how many washers are there in a lot weighing 971 grams?

12. If the driver-to-driven gear ratio is 2.40 and if the driving gear has 96 teeth, how many teeth are there in the driven gear?

13. If the length-to-width ratio of a rectangle is 7 to 4, find its length if its width is 12.4 centimeters.

14. An electrical transformer's efficiency is 97.8%.   If its power input is 60 kilowatts, find its power output.   Round to tenths.

15. If 500 feet of aluminum wire weighs 15.2 pounds, find the weight of 180 feet of this wire. Round to hundredths.

16. One liter equals 1.0567 quarts.   How many liters are there in 4 quarts?   Round to thousandths.

17. If 1,000 feet of #14 copper wire has an electrical resistance of 2.525 ohms, what length of wire will have a resistance of 1 ohm?   Round to a whole number.

18. One hundred liters of sea water contain 6 milligrams of iodine.   How much iodine is there in 350 liters of sea water?

19. If 1 kilometer equals 0.62137 miles, how many kilometers are there in 200 miles?   Round to a whole number.

20. The "mass ratio" of a rocket vehicle is the ratio of its takeoff weight to its burnout weight. If the mass ratio is 7.82 and the takeoff weight is 648 tons, find the burnout weight.   Round to tenths.

Assignment 15

1. $\underline{w}$ is inversely proportional to $\underline{t}$.   If w = 9 when t = 4, find $\underline{w}$ when t = 2.

2. $\underline{d}$ varies directly as the square of $\underline{r}$.   If d = 8 when r = 4, find $\underline{d}$ when r = 8.

3. $\underline{p}$ varies jointly as $\underline{h}$ and $\underline{s}$. If p = 24 when h = 6 and s = 2, find $\underline{p}$ when h = 5 and s = 7.

4. F varies inversely as the square of $\underline{s}$. If F = 36 when s = 2, find F when s = 6.

5. $\underline{v}$ is directly proportional to $\underline{t}$. If v = 20 when t = 40, find $\underline{v}$ when t = 30.

6. R varies jointly as D and L and inversely as A. If k = 1, find R when D = 3, L = 7, and A = 0.5 .

7. The distance $\underline{d}$ traveled by a car is directly proportional to time $\underline{t}$. If d = 170 kilometers when t = 2 hours, find $\underline{d}$ when t = 5 hours.

8. The pressure P of a gas varies inversely as the volume V. If P = 4 kg/cm^2 when V = 15 liters, find P when V = 10 liters.

9. In an alternating current circuit, inductive reactance X_L is directly proportional to frequency $\underline{f}$. If X_L = 50 ohms when f = 20 hertz, find X_L when f = 48 hertz.

10. The intensity level I of a sound varies inversely as the square of the distance $\underline{d}$ from the source. If I = 12 microwatts when d = 10 meters, find I when d = 20 meters.

11. The kinetic energy E of a moving object varies directly as the square of its velocity $\underline{v}$. If E = 32 joules when v = 4 meters per second, find E when v = 10 meters per second.

12. In an alternating current circuit, capacitive reactance X_C is inversely proportional to frequency $\underline{f}$. If X_C = 80 ohms when f = 400 hertz, find X_C when f = 1,000 hertz.

13. The electrostatic force F between two charged bodies varies inversely as the square of the distance $\underline{d}$ between them. If F = 5 newtons when d = 2 meters, find F when d = 1 meter.

14. A metal bar of original length L has an elongation $\underline{e}$ when its temperature is increased t degrees. $\underline{e}$ varies jointly as L and t. If e = 0.03 inch when L = 60 inches and t = 50°F, find $\underline{e}$ when L = 80 inches and t = 100°F.

15. The weight W of a substance is directly proportional to its volume V. If W = 180 grams when V = 30 cubic centimeters, find W when V = 100 cubic centimeters.

16. The power P in an electric circuit varies jointly as its resistance R and the square of its current I. If P = 180 watts when I = 3 amperes and R = 20 ohms, find P when I = 5 amperes and R = 10 ohms.

17. The centripetal acceleration $\underline{a}$ on an object varies directly as the square of its rotational speed $\underline{v}$ and inversely as the radius $\underline{r}$. For k = 1, find $\underline{a}$ when v = 15 ft/sec and r = 10 ft.

18. The time $\underline{t}$ needed to fill a tank varies inversely as the pumping rate $\underline{r}$. If t = 90 minutes when r = 80 gallons per minute, find $\underline{t}$ when r = 100 gallons per minute.

19. When its brakes are applied, a car's stopping distance $\underline{s}$ varies directly as the square of its velocity $\underline{v}$. If s = 125 ft when v = 50 mph, find $\underline{s}$ when v = 80 mph.

5 Calculator Operations

One of the real powers of the calculator is the ability to do combined operations in one process. In this chapter, we will discuss the calculator steps used to do the types of combined operations needed for formula evaluation. Operations with numbers in scientific notation are included.

5-1 ADDITION, SUBTRACTION, MULTIPLICATION

In this section, we will show how a calculator can be used to evaluate formulas involving addition, subtraction, and multiplication.

1. Some formula evaluations require only an addition, a subtraction, or a multiplication. Use a calculator for these:

   a) In $A = B + C$, find A when B = 1.78 and C = 4.66.

   $$A = \underline{\hspace{1.5cm}}$$

   b) In $X = X_L - X_C$, find X when X_L = 27.8 and X_C = 14.9 .

   $$X = \underline{\hspace{1.5cm}}$$

   c) In $P = 0.433h$, find P when h = 50. $P = \underline{\hspace{1.5cm}}$

2. The evaluation below involves a three-factor multiplication. The calculator steps are shown.

In $V = LWH$, find V when L = 8, W = 6, and H = 10.

Calculator Steps: 8 $\boxed{\text{x}}$ 6 $\boxed{\text{x}}$ 10 $\boxed{=}$ 480

Use a calculator for this evaluation.

In $V = LWH$, when L = 7.5, W = 4.6, and H = 6.9,

$V = $ _____

a) A = 6.44

b) X = 12.9

c) P = 21.65

3. The evaluation below involves a multiplication and an addition. The calculator steps are shown.

In $a = bc + d$, find $\underline{a}$ when b = 3, c = 4, and d = 20.

Calculator Steps: 3 $\boxed{\text{x}}$ 4 $\boxed{+}$ 20 $\boxed{=}$ 32

Use the same steps for this evaluation.

In $V = RS + H$, when R = 2.5, S = 8.4, and H = 9.9,

$V = $ _____

V = 238.05

4. The evaluation below involves a multiplication and a subtraction. The calculator steps are shown.

In $D = K - PQ$, find D when K = 50, P = 5, and Q = 8.

Calculator Steps: 50 $\boxed{-}$ 5 $\boxed{\text{x}}$ 8 $\boxed{=}$ 10

Use the same steps for this one.

In $v_1 = v_2 - at$, when v_2 = 91.6, a = 10.4, and t = 3,

$v_1 = $ _____

V = 30.9

5. The evaluation below involves two multiplications and an addition. The calculator steps are shown.

In $P = 2L + 2W$, find P when L = 7 and W = 5.

Calculator Steps: 2 $\boxed{\text{x}}$ 7 $\boxed{+}$ 2 $\boxed{\text{x}}$ 5 $\boxed{=}$ 24

Use the same steps for this one.

In $E = IR_1 + IR_2$, when I = 8, R_1 = 12.5, and R_2 = 17.5,

$E = $ _____

v_1 = 60.4

E = 240

6. Do these. Round to tenths.

    a) In V = LWH , when L = 12.4, W = 9.3, and H = 10.7,

    V = _____

    b) In h = 1.5v + s , when v = 6.45 and s = 9.03,

    h = _____

    c) In H = E + PV , when E = 7.58, P = 1.44, and V = 3.89,

    H = _____

    d) In Q = cd + ef , when c = 2.33, d = 8.09, e = 5.47, and

    f = 7.91, Q = _____

    ---

    a) V = 1,233.9   b) h = 18.7   c) H = 13.2   d) Q = 62.1

5-2 SINGLE DIVISIONS

In this section, we will show how a calculator can be used to evaluate formulas involving a single division.

7. The evaluations below involve only a simple division. Use a calculator for them. Round to hundredths.

    a) In $L = \dfrac{A}{W}$, when A = 9.68 and W = 1.43, L = _____

    b) In $h = \dfrac{P}{0.433}$, when P = 2.77, h = _____

8. To do the first evaluation below on a calculator, we can either divide 1 by 4 or enter 4 and press the reciprocal key $\boxed{1/x}$. Use either method for the other two evaluations.

    In $t = \dfrac{1}{f}$: When f = 4, $t = \dfrac{1}{4}$ = .25

    a) When f = 10, t = _____

    b) When f = 25, t = _____

a) L = 6.77

b) h = 6.40

9. The evaluation below involves a multiplication (in the numerator) and then a division. The calculator steps are shown.

    In $b = \dfrac{2A}{h}$, find b when A = 20 and h = 5.

    Calculator Steps: 2 $\boxed{\text{x}}$ 20 $\boxed{\div}$ 5 $\boxed{=}$ 8

    Use the same steps for this one.

    In $v = \dfrac{Ftg}{m}$, when F = 20, t = 15, g = 32, and m = 25,

    v = _____

a) t = .1

b) t = .04

v = 384

10. To perform evaluations with formulas like $A = \frac{1}{2}bh$ or $A = \frac{1}{3}Bh$, we can rewrite them to get a division on the right side. That is:

Instead of $A = \frac{1}{2}bh$, we can use $A = \frac{bh}{2}$.

Instead of $A = \frac{1}{3}Bh$, we can use $A = \frac{Bh}{3}$.

a) In $A = \frac{1}{2}bh$, when b = 6.4 and h = 8.5 , A = _____

b) In $A = \frac{1}{3}Bh$, when B = 30 and h = 6 , A = _____

11. The evaluation below involves a division and an addition. The calculator steps are shown.

In $L = \frac{M}{P} + X$, find L when M = 8, P = 2, and X = 10.

Calculator Steps: 8 ÷ 2 + 10 = 14

Use the same steps for this one.

In $F = \frac{9C}{5} + 32°$, when C = 55° , F = _____

12. The evaluation below involves a subtraction and a division. The calculator steps are shown.

In $X = L - \frac{M}{P}$, find X when L = 20, M = 10, and P = 2.

Calculator Steps: 20 − 10 ÷ 2 = 15

Use the same steps for this one.

In $r = R - \frac{2RF}{w}$, when R = 100, F = 25, and w = 80,

r = _____

13. Use a calculator for these. Round to one decimal place.

a) In $T_1 = \frac{V_1 T_2}{V_2}$, when V_1 = 10.4, T_2 = 15, and V_2 = 12.7,

T_1 = _____

b) In $A = \frac{1}{3}Bh$, when B = 18.7 and h = 5.5, A = _____

c) In $X = L - \frac{M}{P}$, when L = 75.9, M = 48.7, and P = 12.3,

X = _____

a) A = 27.2

b) A = 60

F = 131°

r = 37.5

a) T_1 = 12.3

b) A = 34.3

c) X = 71.9

14. The evaluation below involves an addition and a division.

In $R = \dfrac{p + q}{2}$, find R when p = 8 and q = 10.

$$R = \frac{p + q}{2} = \frac{8 + 10}{2} = \frac{18}{2} = 9$$

If we simply perform the addition and then divide immediately on a calculator, we get 13 instead of 9 because the calculator performs the evaluation below. Try it.

$$R = p + \frac{q}{2} = 8 + \frac{10}{2} = 8 + 5 = 13$$

Therefore, to perform the top evaluation correctly on a calculator, we must press $\boxed{=}$ to complete the addition <u>before</u> dividing. The steps are:

Calculator Steps: 8 $\boxed{+}$ 10 $\boxed{=}$ $\boxed{\div}$ 2 $\boxed{=}$ 9

Use the same steps for this evaluation.

In $a = \dfrac{b + c}{d}$, when b = 154, c = 26, and d = 45, a = _____

---

15. The evaluation below involves a subtraction and a division.

$a = 4$

In $P = \dfrac{H - E}{V}$, find P when H = 30, E = 12, and V = 3.

$$P = \frac{H - E}{V} = \frac{30 - 12}{3} = \frac{18}{3} = 6$$

If we simply perform the subtraction and then divide immediately on a calculator, we get 26 instead of 6 because the calculator performs the evaluation below. Try it.

$$P = H - \frac{E}{V} = 30 - \frac{12}{3} = 30 - 4 = 26$$

To perform the top evaluation correctly, we must press $\boxed{=}$ to complete the subtraction <u>before</u> dividing. The steps are:

Calculator Steps: 30 $\boxed{-}$ 12 $\boxed{=}$ $\boxed{\div}$ 3 $\boxed{=}$ 6

Use the same steps for this evaluation.

In $I = \dfrac{E - e}{r}$, when E = 16.7, e = 12.5, and r = 1.2,

I = _____

---

16. Do these on a calculator. Be sure to press $\boxed{=}$ before dividing. Round to two decimal places.

$I = 3.5$

a) In $V = \dfrac{S + T}{M}$, when S = 3.66, T = 7.59, and M = 4.18,

V = _____

b) In $I_a = \dfrac{E_1 - E_0}{R_a}$, when E_1 = 27.5, E_0 = 15.8, and R_a = 10.5,

I_a = _____

17. The evaluation below involves a multiplication, a subtraction, and a division. The calculator steps are shown. Notice again that we pressed $=$ to complete the subtraction before dividing.

In $X = \dfrac{PL - M}{P}$, find X when P = 2, L = 10, and M = 6.

Calculator Steps: 2 $\boxed{x}$ 10 $\boxed{-}$ 6 $\boxed{=}$ $\boxed{\div}$ 2 $\boxed{=}$ 7

Use the same steps for this evaluation.

In $C = \dfrac{5F - 160°}{9}$, when F = 68°, C = _____

a) V = 2.69

b) I_a = 1.11

18. Do these on a calculator. Press $=$ before dividing. Round to tenths.

a) In $L = \dfrac{M + PX}{P}$, when M = 88.7, P = 9.5, and X = 14.1,

L = _____

b) In $c_2 = \dfrac{K - c_1 a}{b}$, when K = 100, c_1 = 2.5, a = 24.8, and

b = 3.5, c_2 = _____

C = 20°

19. To perform the evaluation below, we evaluate the fraction and then add the value of <u>h</u>. The calculator steps are shown.

In $T = \dfrac{a - b}{c} + h$, find T when a = 10, b = 4, c = 2, and h = 5.

Calculator Steps: 10 $\boxed{-}$ 4 $\boxed{=}$ $\boxed{\div}$ 2 $\boxed{+}$ 5 $\boxed{=}$ 8

Use the same steps for this evaluation. Round to hundredths.

In $h = \dfrac{P - p_1}{w} + h_1$, find <u>h</u> when P = 35.9, p_1 = 27.6, w = 2.46, and h_1 = 1.59, h = _____

a) L = 23.4

b) c_2 = 10.9

h = 4.96

5-3 MULTIPLE DIVISIONS

When some evaluations are done on a calculator, more than one division is required. We will discuss divisions of that type in this section.

20. To perform the evaluation below, we can perform the multiplication in the denominator and then divide.

In $a = \dfrac{b}{cd}$, find <u>a</u> when b = 40, c = 2, and d = 5.

$$a = \dfrac{b}{cd} = \dfrac{40}{(2)(5)} = \dfrac{40}{10} = 4$$

Continued on following page.

20. Continued

However, we can perform the same evaluation by dividing 40 by 2 and then dividing 20 by 5.

$$a = \frac{b}{cd} = \frac{\overset{20}{\cancel{40}}}{\underset{1}{(\cancel{2})(5)}} = \frac{20}{5} = 4$$

The second method is used to perform the evaluation on a calculator. That is, we perform 40 ÷ 2 ÷ 5 which requires two divisions. The steps are shown.

Calculator Steps: 40 $\boxed{\div}$ 2 $\boxed{\div}$ 5 $\boxed{=}$ 4

Use the same steps for this one.

In $H = \dfrac{B}{VT}$, when B = 200, V = 4, and T = 10,   H = _____

---

21. Do these on a calculator.   Round to two decimal places.

a) In $v = \dfrac{s}{mp}$, when s = 17.8, m = 2.41, and p = 1.49,   v = _____

b) In $B = \dfrac{C}{DF}$, when C = 2.49, D = 1.33, and F = 0.87,   B = _____

H = 5

---

22. To perform the evaluation below, we can multiply in the numerator and get 48, divide 48 by 4, and then divide 12 by 3.

In $k = \dfrac{cd}{pq}$, find $\underline{k}$ when c = 8, d = 6, p = 4, and q = 3.

$$k = \frac{cd}{pq} = \frac{(8)(6)}{(4)(3)} = \frac{\overset{12}{\cancel{48}}}{\underset{1}{(\cancel{4})(3)}} = \frac{12}{3} = 4$$

To do it on a calculator, we perform (8)(6) ÷ 4 ÷ 3 which requires two divisions.   The steps are shown.

Calculator Steps: 8 $\boxed{x}$ 6 $\boxed{\div}$ 4 $\boxed{\div}$ 3 $\boxed{=}$ 4

Use the same steps for this one.

In $e = \dfrac{FL}{AY}$, when F = 50, L = 4, A = 2, and Y = 5,   e = _____

a) v = 4.96

b) B = 2.15

---

e = 20

23. When a formula contains many letters (or variables), it is helpful to write the formula with the numbers substituted before doing the calculator operations. Do the evaluations below. Round to tenths.

a) In $F = \dfrac{mv}{gt}$, when m = 60.9, v = 87.5, g = 32.2, and t = 10.8,

$$F = \frac{(60.9)(87.5)}{(32.2)(10.8)} = \underline{\hspace{2cm}}$$

b) In $P_1 = \dfrac{P_2 V_2 T_1}{V_1 T_2}$, when P_2 = 40.7, V_2 = 20.5, T_1 = 300, V_1 = 34.9, and T_2 = 550,

$$P_1 = \frac{(40.7)(20.5)(300)}{(34.9)(550)} = \underline{\hspace{2cm}}$$

---

24. To perform the evaluation below, we can simplify the numerator and then divide by 2, 3, and 5.

In $V = \dfrac{PQ}{RST}$, find V when P = 6, Q = 50, R = 2, S = 3, and T = 5 .

$$V = \frac{(6)(50)}{(2)(3)(5)} = \frac{\overset{150}{\cancel{300}}}{(\cancel{2})(3)(5)} = \frac{\overset{50}{\cancel{150}}}{(\cancel{3})(5)} = \frac{50}{5} = 10$$

On a calculator, we do (6)(50) ÷ 2 ÷ 3 ÷ 5 which requires three divisions. The steps are shown.

Calculator Steps: 6 $\boxed{\text{x}}$ 50 $\boxed{÷}$ 2 $\boxed{÷}$ 3 $\boxed{÷}$ 5 $\boxed{=}$ 10

Use the same steps for this one.

In $H = \dfrac{r}{xyz}$, when r = 1,600, x = 40, y = 2, and z = 5,

$$H = \underline{\hspace{2cm}}$$

a) F = 15.3

b) P_1 = 13.0

---

25. Do these on a calculator. Round to thousandths.

a) In $t = \dfrac{LM}{3.14SV}$, when L = 5.07, M = 2.99, S = 4.75, and V = 3.24, t = $\underline{\hspace{2cm}}$

b) In $F = \dfrac{33,000H}{abT}$, when H = 0.025, a = 59.6, b = 7.14, and T = 2.63, F = $\underline{\hspace{2cm}}$

H = 4

---

a) t = 0.314

b) F = 0.737

26. To do the following evaluation on a calculator, two methods are described below.

    In $t = \dfrac{1}{av}$, find $\underline{t}$ when a = 4 and v = 20.

    1. We can enter "1" and then divide by both 4 and 20.

       Calculator Steps: 1 $\boxed{\div}$ 4 $\boxed{\div}$ 20 $\boxed{=}$ 0.0125

    2. We can multiply 4 and 20 and then press the reciprocal key $\boxed{1/x}$ after pressing $\boxed{=}$.

       Calculator Steps: 4 $\boxed{x}$ 20 $\boxed{=}$ $\boxed{1/x}$ 0.0125

    Use either method for this one.

    In $m = \dfrac{1}{2cd}$, when c = 4 and d = 5, m = _____

---

27. To do the evaluation below on a calculator, we evaluate the fraction and then add the value of X. The steps are shown.

    In $L = \dfrac{2M}{WX} + X$, find L when M = 24, W = 4, and X = 3.

    Calculator Steps: 2 $\boxed{x}$ 24 $\boxed{\div}$ 4 $\boxed{\div}$ 3 $\boxed{+}$ 3 $\boxed{=}$ 7

    Use the same steps for this one.

    In $t = \dfrac{ab}{cd} - c$, when a = 16, b = 10, c = 2 and d = 5,

    t = _____

---

28. Do these on a calculator. Round to tenths.

    a) In $V = \dfrac{S}{BT} - M$, when S = 187, B = 5.4, T = 2.3, and M = 4.9,   V = _____

    b) In $H = \dfrac{CD}{ABE} + B$, when C = 17.9, D = 48.3, A = 5.6, B = 10.4, and E = 2.7,   H = _____

---

m = 0.025

t = 14

a) V = 10.2

b) H = 15.9

29. To perform the evaluation below on a calculator, we enter the value for D, press $\boxed{-}$, and then evaluate the fraction in the usual way. The steps are shown.

In $F = D - \dfrac{H}{ms}$, find F when D = 100, H = 60, m = 2, and s = 3.

Calculator Steps: 100 $\boxed{-}$ 60 $\boxed{\div}$ 2 $\boxed{\div}$ 3 $\boxed{=}$ 90

Use the same steps for this one.

In $v = h + \dfrac{3I}{ab}$, when h = 75, I = 16, a = 4, and b = 2,

$v =$ _____

---

30. Do these on a calculator. Round to tenths.

a) In $t_1 = t_2 - \dfrac{H}{ms}$, when $t_2 = 50$, H = 9.68, m = 1.5, and s = 2.03, $t_1 =$ _____

b) In $k = p + \dfrac{q}{4st}$, when p = 41.3, q = 81.6, s = 2.6, and t = 1.9, $k =$ _____

---

31. To do the evaluation below, we evaluate the numerator, press $\boxed{=}$ before dividing, and then divide by the values of C and D. The steps are shown.

In $K = \dfrac{CD - FG}{CD}$, find K when C = 4, D = 25, F = 6, and G = 7.

Calculator Steps: 4 $\boxed{x}$ 25 $\boxed{-}$ 6 $\boxed{x}$ 7 $\boxed{=}$ $\boxed{\div}$ 4 $\boxed{\div}$

25 $\boxed{=}$ 0.58

Use the same steps for this one.

In $V = \dfrac{R + ab}{at}$, when R = 20, a = 2, b = 15, and t = 5,

$V =$ _____

---

32. Use a calculator for these. Round to two decimal places.

a) In $t_1 = \dfrac{mst_2 - H}{ms}$, when m = 50, s = 19.7, $t_2 = 10$, and H = 65.5, $t_1 =$ _____

b) In $H = \dfrac{CK + E}{CKD}$, when C = 12.7, K = 21.9, E = 40.4, and D = 6.4, H = _____

Answers (right column):

v = 81

a) t_1 = 46.8

b) k = 45.4

V = 5

a) t_1 = 9.93   b) H = 0.18

SELF-TEST 16 (pages 192-202)

1. Find H when E = 6.75, P = 9.48, and V = 7.39 . Round to tenths.

$$H = E + PV$$

H = _____

2. Find $\underline{b}$ when A = 150 and h = 12.5.

$$b = \frac{2A}{h}$$

b = _____

3. Find F when C = 27.

$$F = \frac{9C}{5} + 32$$

F = _____

4. Find $\underline{v}$ when d_2 = 32.6, d_1 = 17.9, and t = 8.34 . Round to hundredths.

$$v = \frac{d_2 - d_1}{t}$$

v = _____

5. Find L when M = 780, P = 42, and X = 54. Round to one decimal place.

$$L = \frac{M + PX}{P}$$

L = _____

6. Find $\underline{w}$ when a = 2.18 and d = 1.37 . Round to thousandths.

$$w = \frac{1}{4ad}$$

w = _____

7. Find G when d = 236, w = 505, and b = 1.12 . Round to hundredths.

$$G = \frac{d + w}{bw}$$

G = _____

8. Find F when D = 21.48, H = 159, m = 4.45 and s = 2.63 . Round to two decimal places.

$$F = D - \frac{H}{ms}$$

F = _____

9. Find K when A = 3,720 , P_1 = 81.9, and P_2 = 17.6 . Round to tenths.

$$K = \frac{AP_1 - AP_2}{2P_1P_2}$$

K = _____

ANSWERS: 1. H = 76.8   3. F = 80.6   5. L = 72.6   7. G = 1.31   9. K = 83.0

2. b = 24   4. v = 1.76   6. w = 0.084   8. F = 7.89

5-4 SQUARES

In this section, we will show how a calculator can be used to evaluate formulas containing squares.

33. To do the evaluations below, we use the $\boxed{x^2}$ key to square the value of s. In $A = s^2$: a) When $s = 1.5$, $A =$ _____ b) When $s = 10.4$, $A =$ _____	
34. To do the evaluation below, we square the value of I before multiplying. The steps are shown. In $P = I^2R$, find P when $I = 3$ and $R = 10$. Calculator Steps: 3 $\boxed{x^2}$ $\boxed{x}$ 10 $\boxed{=}$ 90 For this evaluation, press $\boxed{x^2}$ after entering the value of w and then press $\boxed{=}$ to get the final answer. In $d = 4w^2$, when $w = 5$, $d =$ _____	a) $A = 2.25$ b) $A = 108.16$
35. To do the evaluation below, we press $\boxed{x^2}$ immediately after entering the values of x and y. The steps are shown. In $m = x^2y^2$, find m when $x = 5$ and $y = 3$. Calculator Steps: 5 $\boxed{x^2}$ $\boxed{x}$ 3 $\boxed{x^2}$ $\boxed{=}$ 225 Use the same steps for this one. In $a = b^2c^2$, when $b = 9$ and $c = 11$, $a =$ _____	$d = 100$
36. In an evaluation containing a squared letter, the same steps are used except that $\boxed{x^2}$ is pressed immediately after entering each value that is to be squared. Another example is given below. In $G = A - B^2H^2$, find G when $A = 100$, $B = 2$, and $H = 3$. Calculator Steps: 100 $\boxed{-}$ 2 $\boxed{x^2}$ $\boxed{x}$ 3 $\boxed{x^2}$ $\boxed{=}$ 64 Do this one. Round to tenths. In $b = 4c^2F^2 - r$, when $c = 0.15$, $F = 17.1$, and $r = 11.2$, $b =$ _____	$a = 9,801$
	$b = 15.1$

37. Do these. Round to two decimal places.

    a) In $A = 0.7854d^2$, when $d = 2.4$, $A =$ _____

    b) In $v = h^2q^2$, when $h = 1.26$ and $q = 2.09$, $v =$ _____

    c) In $I_c = I_a - d^2m$, when $I_a = 87.5$, $d = 1.24$, and $m = 12$,

                                              $I_c =$ _____

---

38. In the calculator steps below, notice that we pressed $\boxed{x^2}$ immediately after entering the value for $\underline{t}$.

    In $s = \dfrac{at^2}{2}$, find $\underline{s}$ when $a = 40$ and $t = 3$.

    Calculator Steps: $40\ \boxed{\times}\ 3\ \boxed{x^2}\ \boxed{\div}\ 2\ \boxed{=}\ \ \ 180$

Do this one. Round to tenths.

    In $H = \dfrac{D^2N}{2.5}$, when $D = 3.58$ and $N = 5$, $H =$ _____

a) $A = 4.52$

b) $v = 6.93$

c) $I_c = 69.05$

---

39. Notice again in the steps below that we pressed $\boxed{x^2}$ immediately after entering the value for $\underline{d}$.

    In $P = \dfrac{pL}{d^2}$, find P when $p = 10$, $L = 16$, and $d = 2$.

    Calculator Steps: $10\ \boxed{\times}\ 16\ \boxed{\div}\ 2\ \boxed{x^2}\ \boxed{=}\ \ \ 40$

Do this one. Press $\boxed{x^2}$ immediately after entering the values for d_2 and d_1.

    In $I_1 = \dfrac{I_2(d_2)^2}{(d_1)^2}$, when $I_2 = 15$, $d_2 = 12$, and $d_1 = 20$,

                                              $I_1 =$ _____

$H = 25.6$

---

40. Notice in the steps below that we divided by both F^2 and $\underline{t}$.

    In $h = \dfrac{W^2}{F^2t}$, find $\underline{h}$ when $W = 12$, $F = 2$, and $t = 9$.

    Calculator Steps: $12\ \boxed{x^2}\ \boxed{\div}\ 2\ \boxed{x^2}\ \boxed{\div}\ 9\ \boxed{=}\ \ \ 4$

Do this one. Round to tenths.

    In $s = \dfrac{v^2}{2g}$, when $v = 49.7$ and $g = 32$, $s =$ _____

$I_1 = 5.4$

$s = 38.6$

41. Do these. Round to tenths.

    a) In $P = \dfrac{E^2}{R}$, when E = 61.8 and R = 40,   P = _____

    b) In $t = \dfrac{a^2 T}{v^2}$, when a = 21.9, T = 45, and v = 36.8,

                                         t = _____

    c) In $F = \dfrac{m_1 m_2}{rd^2}$, when m_1 = 27.5, m_2 = 23.5, r = 10.8,

                          and d = 2.4,   F = _____

a) P = 95.5

b) t = 15.9

c) F = 10.4

42. To do the evaluation below on a calculator, two methods are described.

    In $m = \dfrac{1}{b^2 t^2}$, find $\underline{m}$ when b = 2 and t = 10.

       1) We can enter "1" and then divide by both 2^2 and 10^2.

       2) We can multiply 2^2 and 10^2 and then press the reciprocal key $\boxed{1/x}$.

Using either method, complete the evaluation.   m = _____

43. Notice in the steps below that we simply subtracted the value for $\underline{d}$ after evaluating the fraction.

m = 0.0025

    In $m = \dfrac{ab}{c^2} - d$, find $\underline{m}$ when a = 6, b = 10, c = 2, and d = 8.

       Calculator Steps: 6 $\boxed{x}$ 10 $\boxed{\div}$ 2 $\boxed{x^2}$ $\boxed{-}$ 8 $\boxed{=}$ 7

Do this one.   Round to a whole number.

    In $P = \dfrac{mv^2}{r} - mg$, when m = 6.24, v = 30.9, r = 12.5, and

                         g = 32.2,   P = _____

44. Notice in the steps below that we pressed $\boxed{=}$ to complete the subtraction in the numerator before dividing.

P = 276

    In $r = \dfrac{s^2 d^2 - h}{2}$, find $\underline{r}$ when s = 3, d = 8, and h = 54.

       Calculator Steps: 3 $\boxed{x^2}$ $\boxed{x}$ 8 $\boxed{x^2}$ $\boxed{-}$ 54 $\boxed{=}$ $\boxed{\div}$ 2 $\boxed{=}$ 261

Do these.   Round to hundredths.

    a) In $d = \dfrac{ab^2 - 4s^2}{b^2}$, when a = 7, b = 14, and s = 9,

                             d = _____

    b) In $s = \dfrac{(v_f)^2 - (v_o)^2}{2g}$, when v_f = 40, v_o = 32, and g = 32.2,

                             s = _____

a) d = 5.35

b) s = 8.94

45. In the evaluation below, we must simplify the expression within parentheses <u>before</u> squaring. Notice in the steps that we press $\boxed{=}$ to complete the division <u>before</u> squaring.

In $a = \left(\dfrac{b}{c}\right)^2$, find <u>a</u> when b = 20 and c = 4.

Calculator Steps: 20 $\boxed{\div}$ 4 $\boxed{=}$ $\boxed{x^2}$ 25

Do these. Press $\boxed{=}$ to complete the operation within parentheses <u>before</u> squaring.

a) In $b = (a - c)^2$, when a = 10 and c = 3, b = _____

b) In $x = \left(\dfrac{b - t}{c}\right)^2$, when b = 270, t = 150, and c = 10,

x = _____

a) b = 49   b) x = 144

5-5   SQUARE ROOTS

In this section, we will show how a calculator can be used to evaluate formulas containing square-root radicals.

46. To do the evaluations below, we use the $\boxed{\sqrt{x}}$ key to find the square root of A.

In $S = \sqrt{A}$:   a) when A = 1.96, s = _____

b) when A = 462.25, s = _____

47. To do the evaluation below, we multiply the values of P and R before finding the square root.

In $E = \sqrt{PR}$, find E when P = 4 and R = 9.

$E = \sqrt{PR} = \sqrt{(4)(9)} = \sqrt{36} = 6$

To do the same evaluation on a calculator, we press $\boxed{=}$ to complete the multiplication before pressing $\boxed{\sqrt{x}}$. The steps are:

Calculator Steps: 4 $\boxed{x}$ 9 $\boxed{=}$ $\boxed{\sqrt{x}}$ 6

Do these. Be sure to press $\boxed{=}$ before pressing $\boxed{\sqrt{x}}$.

a) In $V = \sqrt{64h}$, when h = 9, V = _____

b) In $v = \sqrt{2as}$, when a = 50 and s = 121, v = _____

a) s = 1.4

b) s = 21.5

a) V = 24

b) v = 110

48. Whenever an evaluation involves a square-root radical, the operation within the radical is performed before finding the square root. For example:

In $c = \sqrt{a^2 + b^2}$, find $\underline{c}$ when a = 3 and b = 4.

$$c = \sqrt{a^2 + b^2} = \sqrt{3^2 + 4^2} = \sqrt{9 + 16} = \sqrt{25} = 5$$

When doing the same evaluation on a calculator, we press $\boxed{=}$ to complete the operation within the radical before pressing $\boxed{\sqrt{x}}$. The steps are:

Calculator Steps: 3 $\boxed{x^2}$ + 4 $\boxed{x^2}$ $\boxed{=}$ $\boxed{\sqrt{x}}$ 5

Do these. Round to two decimal places.

a) In $v = \sqrt{w - e}$, when w = 79.6 and e = 37.5, v = _____

b) In $v_o = \sqrt{(v_f)^2 - 2gs}$, when v_f = 37.6, g = 32.2 and

s = 21, v_o = _____

49. We pressed $\boxed{=}$ before pressing $\boxed{\sqrt{x}}$ to complete the evaluation below.

In $t = \sqrt{\dfrac{2s}{a}}$, find $\underline{t}$ when s = 72 and a = 4.

Calculator Steps: 2 $\boxed{x}$ 72 $\boxed{\div}$ 4 $\boxed{=}$ $\boxed{\sqrt{x}}$ 6

Do these. Round to hundredths.

a) In $d = \sqrt{\dfrac{A}{0.7854}}$, when A = 60, d = _____

b) In $d = \sqrt{\dfrac{pL}{R}}$, when p = 14.5, L = 27.2, and R = 18.8,

d = _____

a) v = 6.49

b) v_o = 7.83

50. Notice again that we pressed $\boxed{=}$ before pressing $\boxed{\sqrt{x}}$ to complete the evaluation below.

In $C = \sqrt{1 - \dfrac{H}{h}}$, find C when H = 32 and h = 50.

Calculator Steps: 1 $\boxed{-}$ 32 $\boxed{\div}$ 50 $\boxed{=}$ $\boxed{\sqrt{x}}$ 0.6

Do these. Round to two decimal places.

a) In $d = \sqrt{\dfrac{m}{t} + 1}$, when m = 82.7 and t = 12.5, d = _____

b) In $V = \sqrt{100 - \dfrac{b^2}{F}}$, when b = 37.5 and F = 68.5,

V = _____

a) d = 8.74

b) d = 4.58

a) d = 2.76

b) V = 8.91

51. Do these. Round to hundredths.

a) In $P = \sqrt{Q^2 + R^2}$, when Q = 5.66 and R = 7.59,

P = _____

b) In $t = \sqrt{\dfrac{rw}{2P}}$, when r = 313, w = 36.8 and P = 898,

t = _____

c) In $t = \sqrt{\dfrac{a^2}{r^2} - 1}$, when a = 6.48 and r = 1.59, t = _____

---

52. To do the evaluation below, we pressed $\boxed{=}$ to complete the subtraction in the numerator before dividing and then we pressed $\boxed{=}$ before pressing $\boxed{\sqrt{x}}$.

In $a = \sqrt{\dfrac{b - c}{c}}$, find a when b = 100 and c = 10.

Calculator Steps: 100 $\boxed{-}$ 10 $\boxed{=}$ $\boxed{\div}$ 10 $\boxed{=}$ $\boxed{\sqrt{x}}$ 3

Do these. Round to tenths.

a) In $d = \sqrt{\dfrac{I_A - I_B}{m}}$, when I_A = 395, I_B = 70, and m = 3,

d = _____

b) In $t = \sqrt{\dfrac{1.5s^2 + V}{1.5}}$, when s = 40 and V = 15, t = _____

a) P = 9.47

b) t = 2.53

c) t = 3.95

---

53. To do the evaluation below, we find $\sqrt{100}$ before multiplying.

In $r = a\sqrt{w}$, find r when a = 7 and w = 100.

$r = a\sqrt{w} = 7\sqrt{100} = 7(10) = 70$

To do the same evaluation on a calculator, we simply press $\boxed{\sqrt{x}}$ after entering 100. The steps are:

Calculator Steps: 7 $\boxed{x}$ 100 $\boxed{\sqrt{x}}$ $\boxed{=}$ 70

To do these, press $\boxed{\sqrt{x}}$ after entering the value of the radicand.

a) In $R = P + \sqrt{Q}$, when P = 17 and Q = 81, R = _____

b) In $b = \dfrac{\sqrt{d}}{c}$, when d = 64 and c = 2, b = _____

a) d = 10.4

b) t = 40.1

---

a) R = 26

b) b = 4

54. Whenever a radical containing a variable alone is part of an operation, we do the operation in the usual way. However, we press $\boxed{\sqrt{x}}$ after entering the value of the radicand. For example:

In $b = \dfrac{a\sqrt{x}}{c}$, find b when $a = 8$, $x = 144$, and $c = 3$.

Calculator Steps: 8 $\boxed{x}$ 144 $\boxed{\sqrt{x}}$ $\boxed{\div}$ 3 $\boxed{=}$ 32

Do these. Round to tenths.

a) In $p = m + r\sqrt{t}$, when $m = 40.8$, $r = 11.6$, and $t = 19.9$,

p = _____

b) In $F = \dfrac{D - S}{\sqrt{G}}$, when $D = 477$, $S = 199$, and $G = 30.3$,

F = _____

---

55. Two methods for the evaluation below are described.

In $t = \dfrac{1}{b\sqrt{m}}$, find t when $b = 5$ and $m = 100$.

1) We can enter "1" and then divide by both 5 and $\sqrt{100}$.

2) We can multiply 5 and $\sqrt{100}$, press $\boxed{=}$, and then press the reciprocal key $\boxed{1/x}$.

Using either method, complete the evaluation $t =$ _____

a) p = 92.5

b) F = 50.5

---

56. To do the evaluation below, we evaluate the radical in the numerator and then divide by 4. The steps are shown:

In $s = \dfrac{\sqrt{h + 2r}}{d}$, find s when $h = 50$, $r = 7$, and $d = 4$.

Calculator Steps: 50 $\boxed{+}$ 2 $\boxed{x}$ 7 $\boxed{=}$ $\boxed{\sqrt{x}}$ $\boxed{\div}$ 4 $\boxed{=}$ 2

Do these. Round to hundredths.

a) In $a = \dfrac{\sqrt{bc}}{d}$, when $b = 9.17$, $c = 6.88$, and $d = 1.29$,

a = _____

b) In $M = \dfrac{\sqrt{P^2 - Q^2}}{3R}$, when $P = 97.8$, $Q = 31.4$, and $R = 7.66$,

M = _____

t = 0.02

---

a) a = 6.16

b) M = 4.03

57. When a radical containing an operation is a factor in a multiplication, we must evaluate the radical first on a calculator. That is:

Instead of $a = b\sqrt{cd}$, we must evaluate $a = (\sqrt{cd})(b)$

Instead of $T = \dfrac{m}{b}\sqrt{\dfrac{d}{c}}$, we must evaluate $T = \sqrt{\dfrac{d}{c}}\left(\dfrac{m}{b}\right)$

The calculator steps for the evaluation below are shown. Notice that we evaluated the radical first and then multiplied by 10.

In $v = s\sqrt{\dfrac{p}{q}}$, find $\underline{v}$ when s = 10, p = 48, and q = 3.

Calculator Steps: 48 $\boxed{\div}$ 3 $\boxed{=}$ $\boxed{\sqrt{x}}$ $\boxed{x}$ 10 $\boxed{=}$ 40

Do these by evaluating the radical first. Round to tenths.

a) In $G = H\sqrt{ST}$, when H = 2.78, S = 5.19, and T = 8.77,

$$G = \underline{\hspace{2cm}}$$

b) In $w = 2g\sqrt{\dfrac{a}{r}}$, when g = 9.01, a = 68.9, and r = 12.6,

$$w = \underline{\hspace{2cm}}$$

---

58. The calculator steps for the evaluation below are shown. Notice again that we evaluated the radical before multiplying by 24 and dividing by 3.

In $T = \dfrac{m}{b}\sqrt{\dfrac{d}{c}}$, find T when m = 24, b = 3, d = 18, and c = 2.

Calculator Steps: 18 $\boxed{\div}$ 2 $\boxed{=}$ $\boxed{\sqrt{x}}$ $\boxed{x}$ 24 $\boxed{\div}$ 3 $\boxed{=}$ 24

Do this one. Round to three decimal places.

In $V = \dfrac{1.5}{H}\sqrt{\dfrac{P}{Q}}$, when H = 6.45, P = 7.03, and Q = 1.98,

$$V = \underline{\hspace{2cm}}$$

---

59. When a radical involving an operation is the denominator of a fraction, we evaluate the radical first and put that value in storage (or memory) by pressing $\boxed{STO}$, enter the numerator, and then divide using $\boxed{RCL}$ to recall the value of the denominator. For example:

In $K = \dfrac{L}{\sqrt{MT}}$, find K when L = 100, M = 2, and T = 8.

Calculator Steps: 2 $\boxed{x}$ 8 $\boxed{=}$ $\boxed{\sqrt{x}}$ $\boxed{STO}$ 100 $\boxed{\div}$ $\boxed{RCL}$ $\boxed{=}$ 25

Continued on following page.

Answers (right column):

a) G = 18.8

b) w = 42.1

V = 0.438

59. Continued

Using the same general method, do these. Round to hundredths.

a) In $F = \dfrac{W}{\sqrt{ht}}$, when W = 87.6, h = 21.7, and t = 3.98,

F = _____

b) In $a = \dfrac{b}{\sqrt{\dfrac{c}{d}}}$, when b = 9.47, c = 48.6, and d = 17.9,

a = _____

---

60. Two methods for the evaluation below are described.

In $a = \dfrac{1}{\sqrt{p^2 + w^2}}$, find a when p = 9 and w = 12.

1) We can evaluate the denominator, press $\boxed{\text{STO}}$, enter "1", and then press $\boxed{\div}$ $\boxed{\text{RCL}}$ to divide.

2) We can evaluate the denominator and then press the reciprocal key $\boxed{1/x}$.

Using either method, do the evaluation. Round to thousandths.

a = _____

a) F = 9.43

b) a = 5.75

---

a = 0.067

5-6 CUBES AND CUBE ROOTS

In this section, we will show the calculator procedure for cubing a number and finding the cube root of a number. Some evaluations with formulas containing a cube or cube root are included.

61. To cube a number, we use the number as a factor <u>three</u> times. For example:

$$5^3 = (5)(5)(5) = 125$$

To cube a number on a calculator, we use the <u>power</u> key $\boxed{y^x}$ or $\boxed{x^y}$. We enter the number, press the power key, enter the exponent 3, and then press $\boxed{=}$. The steps for 5^3 are:

Calculator Steps: 5 $\boxed{y^x}$ 3 $\boxed{=}$ 125

Use the power key for these:

a) $(4.5)^3 =$ _____   b) $(115)^3 =$ _____

a) 91.125

b) 1,520,875

62. Do these. Round to three decimal places.

    a) $(0.82)^3$ = _____   b) $(0.659)^3$ = _____

---

63. To do the evaluations below, we cube the value of <u>s</u>.

    In $V = s^3$:   a) When s = 8.5, V = _____

                       b) When s = 47, V = _____

a) 0.551

b) 0.286

---

64. The <u>cube</u> <u>root</u> of a number N is <u>the</u> <u>number</u> <u>whose</u> <u>cube</u> <u>is</u> <u>N</u>. That is:

    Since 3^3 = 27, the cube root of 27 is 3.
    Since 5^3 = 125, the cube root of 125 is 5.

    Instead of saying <u>the</u> <u>cube</u> <u>root</u>, we use the symbol $\sqrt[3]{}$. That is:

    Since 2^3 = 8, $\sqrt[3]{8}$ = 2   Since 6^3 = 216, $\sqrt[3]{216}$ = _____

a) V = 614.125

b) V = 103,823

---

65. There are various methods for finding $\sqrt[3]{64}$ on a calculator.

    1. If there is a <u>cube</u> <u>root</u> key $\boxed{\sqrt[3]{}}$, we use that key.
       The steps are:

       <u>Calculator</u> <u>Steps</u>: 64 $\boxed{\sqrt[3]{}}$ 4

    2. If there is a <u>root</u> key $\boxed{\sqrt[x]{y}}$, we use that key. The steps
       are shown. Notice how we entered 3 for <u>x</u>.

       <u>Calculator</u> <u>Steps</u>: 64 $\boxed{\sqrt[x]{y}}$ 3 $\boxed{=}$ 4

    3. If there is neither a <u>cube</u> <u>root</u> key nor a <u>root</u> key, we use
       the <u>inverse</u> key $\boxed{2nd}$ or $\boxed{INV}$ together with the <u>power</u> key
       $\boxed{y^x}$ or $\boxed{x^y}$. The steps are:

       <u>Calculator</u> <u>Steps</u>: 64 $\boxed{2nd}$ $\boxed{y^x}$ 3 $\boxed{=}$ 4

    Use one of the methods above to complete these:

    a) $\sqrt[3]{512}$ = _____   b) $\sqrt[3]{19,683}$ = _____

6

---

66. Complete: a) Rounding to tenths, $\sqrt[3]{19,400}$ = _____

             b) Rounding to hundredths, $\sqrt[3]{47.5}$ = _____

             c) Rounding to thousandths, $\sqrt[3]{0.646}$ = _____

a) 8   b) 27

---

67. To do the evaluations below, we find the cube root of V.

    In $s = \sqrt[3]{V}$:   a) When V = 2.744, s = _____

                      b) When V = 27,000, s = _____

a) 26.9

b) 3.62

c) 0.864

a) s = 1.4   b) s = 30

SELF-TEST 17 (pages 203-213)

1. Find A when d = 23.8. Round to a whole number.

   $$A = 0.7854d^2$$

   A = _____

2. Find R when p = 0.870, L = 295, and a = 7.34 . Round to hundredths.

   $$R = \frac{pL}{a^2}$$

   R = _____

3. Find $\underline{m}$ when t = 15.4, w = 22.8, and v = 6.13 . Round to one decimal place.

   $$m = \frac{t^2 + w^2}{2v}$$

   m = _____

4. Find Z when R = 5,700 and X = 8,100. Round to hundreds.

   $$Z = \sqrt{R^2 + X^2}$$

   Z = _____

5. Find $\underline{v}$ when F = 3.96 and D = 1.09 . Round to thousandths.

   $$v = \sqrt{\frac{F - D}{F}}$$

   v = _____

6. Find f_1 when f_2 = 256, D_1 = 7.28, and D_2 = 9.47 . Round to a whole number.

   $$f_1 = f_2 \sqrt{\frac{D_2}{D_1}}$$

   f_1 = _____

7. Find P when a = 768, c = 53.9, and r = 8.84 . Round to two decimal places.

   $$P = \frac{a}{c\sqrt{r}}$$

   P = _____

8. Find V when s = 86.8 . Round to thousands.

   $$V = s^3$$

   V = _____

9. Find $\underline{s}$ when V = 2,840. Round to tenths.

   $$s = \sqrt[3]{V}$$

   s = _____

ANSWERS: 1. A = 445   3. m = 61.7   5. v = 0.851   7. P = 4.79   9. s = 14.2

   2. R = 4.76   4. Z = 9,900   6. f_1 = 292   8. V = 654,000

5-7 FORMULAS IN PROPORTION FORM

In this section, we will discuss evaluations with formulas that are written in proportion form.

---

68. The formulas below are written in proportion form.

$$\frac{P_1}{P_2} = \frac{V_2}{V_1} \qquad\qquad \frac{I_1}{I_2} = \frac{(d_2)^2}{(d_1)^2}$$

To perform evaluations with formulas of that type, we substitute and then solve the proportion. An example is shown. Use a calculator to complete the other evaluation.

Find d_1 when $F_1 = 12$, $F_2 = 20$, and $d_2 = 40$.

Find V_1 when $V_2 = 120$, $T_1 = 12$, and $T_2 = 18$.

$$\frac{F_1}{F_2} = \frac{d_1}{d_2} \qquad\qquad \frac{V_1}{V_2} = \frac{T_1}{T_2}$$

$$\frac{12}{20} = \frac{d_1}{40}$$

$$(12)(40) = (20)(d_1)$$

$$d_1 = \frac{(12)(40)}{20}$$

$$d_1 = 24$$

---

69. We did one evaluation with the formula below. Do the other evaluation.

$V_1 = 80$, from:

$$\frac{(120)(12)}{18}$$

Find T_2 when $P_1 = 10$, $V_1 = 16$, $T_1 = 4$, $P_2 = 20$, and $V_2 = 11$.

Find P_2 when $P_1 = 12$, $V_1 = 25$, $T_1 = 3$, $V_2 = 50$, and $T_2 = 7$.

$$\frac{P_1 V_1}{T_1} = \frac{P_2 V_2}{T_2} \qquad\qquad \frac{P_1 V_1}{T_1} = \frac{P_2 V_2}{T_2}$$

$$\frac{(10)(16)}{4} = \frac{(20)(11)}{T_2}$$

$$(10)(16)(T_2) = (4)(20)(11)$$

$$T_2 = \frac{(4)(20)(11)}{(10)(16)}$$

$$T_2 = 5.5$$

$P_2 = 14$, from:

$$P_2 = \frac{(12)(25)(7)}{(3)(50)}$$

70. We did one evaluation with the formula below. Do the other evaluation.

Find I_1 when $I_2 = 10$, $d_2 = 4$, and $d_1 = 8$.

$$\frac{I_1}{I_2} = \frac{(d_2)^2}{(d_1)^2}$$

$$\frac{I_1}{10} = \frac{(4)^2}{(8)^2}$$

$$(I_1)(8^2) = (10)(4^2)$$

$$I_1 = \frac{(10)(4^2)}{8^2}$$

$$I_1 = 2.5$$

Find I_2 when $I_1 = 18$, $d_2 = 6$, and $d_1 = 3$.

$$\frac{I_1}{I_2} = \frac{(d_2)^2}{(d_1)^2}$$

$I_2 = 4.5$, from:

$$\frac{(18)(3^2)}{6^2}$$

71: Do these. Round to tenths.

a) Find d_2 when $F_1 = 14.7$, $F_2 = 40.7$, and $d_1 = 21.6$.

$$\frac{F_1}{F_2} = \frac{d_1}{d_2}$$

b) Find V_1 when $P_1 = 10.7$, $T_1 = 55$, $P_2 = 25.8$, $V_2 = 63.9$, and $T_2 = 90$.

$$\frac{P_1 V_1}{T_1} = \frac{P_2 V_2}{T_2}$$

a) $d_2 = 59.8$ b) $V_1 = 94.2$

5-8   CALCULATOR ANSWERS IN SCIENTIFIC NOTATION

Very large and very small answers are given in scientific notation on a calculator. We will examine answers of that type in this section.

72. As we saw in an earlier chapter, when a number is written in scientific notation, the first factor is a number between 1 and 10 and the second factor is a power of ten. Two examples are:

$$4.16 \times 10^7 \qquad\qquad 9.58 \times 10^{-11}$$

Continued on following page.

72. Continued

    When a calculator displays a number in scientific notation, <u>only the exponent</u> of the power of ten is shown. The exponent appears on the right side of the display. For example:

    The display 2.54 12 means 2.54×10^{12}

    The display 6.81 -09 means 6.81×10^{-9}

    The display 4.05 07 means _____

---

73. When the number of digits in a calculator answer is more than the capacity of the display, the answer is displayed in scientific notation. Two examples are given.

    1. Try 8,300,000 x 250,000 on a calculator.

       Calculator Steps: 8,300,000 $\boxed{\text{x}}$ 250,000 $\boxed{=}$ 2.075 12

    2. Try $\dfrac{.00115}{18,400,000}$ on a calculator.

       Calculator Steps: .00115 $\boxed{\div}$ 18,400,000 $\boxed{=}$ 6.25 -11

    The answers above are $\underline{2.075 \times 10^{12}}$ and $\underline{6.25 \times 10^{-11}}$. To convert them to a whole number and a decimal number, we perform the multiplication by the decimal-point-shift method. That is:

    2.075×10^{12} = 2.075000000000 = 2,075,000,000,000

    6.25×10^{-11} = 00000000006.25 = _____

> 4.05 x 10⁷ → 4.05×10^7

---

74. Complete. Report each answer in scientific notation.

    a) 2,600,000 x 14,500,000 = _____

    b) $\dfrac{.00098}{14,000,000}$ = _____

    c) $(900,000)^2$ = _____

    d) $(.0000085)^3$ = _____

> .0000000000625

---

75. In the last frame, we got $\underline{6.14125 \times 10^{-16}}$ as an answer. The first factor contains five decimal places. We frequently round the first factor <u>to two</u> decimal places. For example:

    6.14125×10^{-16} is rounded to 6.14×10^{-16}

    Do these. Round the first factor to two decimal places.

    a) $(45,600)^3$ = _____

    b) .000686 x .000000199 = _____

> a) 3.77×10^{13}
>
> b) 7×10^{-11}
>
> c) 8.1×10^{11}
>
> d) 6.14125×10^{-16}

> a) 9.48×10^{13} b) 1.37×10^{-10}

5-9   OPERATIONS WITH NUMBERS IN SCIENTIFIC NOTATION

In this section, we will discuss the calculator procedures for multiplying, dividing, squaring, and finding the square root of numbers in scientific notation.   A few applied problems are included.

76. To multiply a number in scientific notation by 4 below, we multiplied the first factor 2 by 4. Do the other multiplication. $\qquad 4 \times (2 \times 10^5) = (4 \times 2) \times 10^5 = 8 \times 10^5$ $\qquad 3 \times (3 \times 10^{-8}) = \underline{\qquad\qquad}$	
77. To multiply the two numbers in scientific notation below, we multiplied the two first factors and the two powers of ten. Do the other multiplication. $\qquad (3 \times 10^4) \times (2 \times 10^5) = (3 \times 2) \times (10^4 \times 10^5) = 6 \times 10^9$ $\qquad (2 \times 10^{-3}) \times (4 \times 10^{-7}) = \underline{\qquad\qquad}$	9×10^{-8}
78. A calculator can be used to multiply numbers in scientific notation. To enter a number in scientific notation, we use the $\boxed{\text{EXP}}$ key or the $\boxed{\text{EE}}$ key. Following the steps below, enter 6.5×10^{10} and 1.9×10^{-11} . $\qquad$ Calculator Steps: 6.5 $\boxed{\text{EXP}}$ 10 $\qquad\qquad\qquad\qquad\qquad$ 1.9 $\boxed{\text{EXP}}$ 11 $\boxed{+/-}$ The calculator steps for two multiplications are shown below. $\qquad 3 \times (2.4 \times 10^{10}) = 7.2 \times 10^{10}$ $\qquad\quad$ Calculator Steps: 3 $\boxed{\text{x}}$ 2.4 $\boxed{\text{EXP}}$ 10 $\boxed{=}$ 7.2 10 $\qquad (3.2 \times 10^{-5}) \times (1.4 \times 10^{-7}) = 4.48 \times 10^{-12}$ $\qquad\quad$ Calculator Steps: 3.2 $\boxed{\text{EXP}}$ 5 $\boxed{+/-}$ $\boxed{\text{x}}$ 1.4 $\boxed{\text{EXP}}$ 7 $\qquad\qquad\qquad\qquad\qquad\qquad\qquad$ $\boxed{+/-}$ $\boxed{=}$ 4.48 -12 Do this one: $(8.4 \times 10^{-15}) \times (2.5 \times 10^4) = \underline{\qquad\qquad}$	8×10^{-10}
79. In the multiplication below, the product is not in scientific notation because 56 is not a number between 1 and 10. We converted the product to scientific notation by writing 56 in scientific notation and then multiplying the two powers of ten. $\qquad (8 \times 10^9) \times (7 \times 10^6) = 56 \times 10^{15} = (5.6 \times 10^1) \times 10^{15}$ $\qquad\qquad\qquad\qquad\qquad\qquad\qquad\qquad\qquad\qquad = 5.6 \times 10^{16}$ However, when the same multiplication is done on a calculator, <u>the product is automatically reported in scientific notation</u>. The steps are: $\qquad$ Calculator Steps: 8 $\boxed{\text{EXP}}$ 9 $\boxed{\text{x}}$ 7 $\boxed{\text{EXP}}$ 6 $\boxed{=}$ 5.6 16 Continued on following page.	2.1×10^{-10}

79. Continued

Do these. Round the first factor to two decimal places.

a) $27.6 \times (1.44 \times 10^{-12}) = $ _____

b) $(7.14 \times 10^{10}) \times (6.76 \times 10^{13}) = $ _____

---

80. For the multiplications below, the number of digits in the product does not exceed the capacity of the display. Some calculators still report the products in scientific notation while others report them as a whole number or decimal number. Try them on your calculator.

a) $7 \times (4.6 \times 10^5) = $ _____

b) $(6.5 \times 10^{-13}) \times (1.4 \times 10^{10}) = $ _____

a) 3.97×10^{-11}

b) 4.83×10^{24}

---

81. To divide a number in scientific notation by 2 below, we divided the first factor 6 by 2. Do the other division.

$$\frac{6 \times 10^5}{2} = \frac{6}{2} \times 10^5 = 3 \times 10^5$$

$$\frac{8 \times 10^{-9}}{4} = \text{_____}$$

a) 3.22×10^6 or
   $3,220,000$

b) 9.1×10^{-3} or
   $.0091$

---

82. To divide the two numbers in scientific notation below, we divided the two first factors and the two powers of ten. Do the other division.

$$\frac{8 \times 10^9}{2 \times 10^4} = \frac{8}{2} \times \frac{10^9}{10^4} = 4 \times 10^5$$

$$\frac{9 \times 10^{-8}}{3 \times 10^{-6}} = \text{_____}$$

2×10^{-9}

---

83. The calculator steps for two divisions are shown below.

$$\frac{6.48 \times 10^{12}}{20} = 3.24 \times 10^{11}$$

Calculator Steps: 6.48 $\boxed{\text{EXP}}$ 12 $\boxed{\div}$ 20 $\boxed{=}$ 3.24 11

$$\frac{1.24 \times 10^{-9}}{4.96 \times 10^4} = 2.5 \times 10^{-14}$$

Calculator Steps: 1.24 $\boxed{\text{EXP}}$ 9 $\boxed{+/-}$ $\boxed{\div}$ 4.96 $\boxed{\text{EXP}}$ 4
$\boxed{=}$ 2.5   -14

Do this one: $\dfrac{6.3 \times 10^7}{4.5 \times 10^{-5}} = $ _____

3×10^{-2}

1.4×10^{12}

84. Do these. Round to two decimal places.

   a) $\dfrac{5.18 \times 10^8}{9.06 \times 10^{-6}}$ = _____ b) $\dfrac{256}{4.18 \times 10^{12}}$ = _____

---

85. Do these. Your calculator may not give the quotient in scientific notation.

   a) $\dfrac{5.19 \times 10^{-6}}{6.92}$ = _____ b) $\dfrac{8.01 \times 10^{18}}{4.45 \times 10^{15}}$ = _____

a) 5.72×10^{13}

b) 6.12×10^{-11}

---

86. A first electrical impulse has a duration of 3.72×10^{-9} second. A second electrical impulse has a duration of 2.27×10^{-6} second. State the ratio of the first impulse to the second impulse as a decimal number rounded to hundred-thousandths.

a) 7.5×10^{-7} or .00000075

b) 1.8×10^3 or 1,800

---

87. To solve the problem below, we set up and solved a proportion involving numbers in scientific notation.

   If it takes 1 second for light to travel 1.86×10^5 miles, how long does it take light to travel from the sun to the earth if that distance is 9.30×10^7 miles?

   $$\frac{1 \text{ sec}}{1.86 \times 10^5 \text{ mi}} = \frac{x \text{ sec}}{9.30 \times 10^7 \text{ mi}} \quad \text{or} \quad \frac{1}{1.86 \times 10^5} = \frac{x}{9.30 \times 10^7}$$

   $$9.30 \times 10^7 = (1.86 \times 10^5)(x)$$

   $$x = \frac{9.30 \times 10^7}{1.86 \times 10^5}$$

   $$x = 5 \times 10^2 = 500$$

   Therefore, it takes light 500 seconds to travel from the sun to the earth.

   Following the example above, use a proportion to solve the following problem.

   Current in a circuit is directly proportional to the applied voltage. If a current of 5.6×10^{-4} ampere is obtained when 1 volt is applied to a circuit, how many volts must be applied to get a current of 1.4×10^{-3} ampere?

.00164

2.5 volts, from:

$$\frac{5.6 \times 10^{-4}}{1} = \frac{1.4 \times 10^{-3}}{x}$$

88. To solve the problem below, we also set up and solved a proportion.

If there are 6.02×10^{23} molecules in 22.4 liters of gas, how many molecules are there in 1 liter of the gas? (Round the first factor to hundredths.)

$$\frac{6.02 \times 10^{23} \text{ molecules}}{22.4 \text{ liters}} = \frac{x \text{ molecules}}{1 \text{ liter}} \quad \text{or} \quad \frac{6.02 \times 10^{23}}{22.4} = \frac{x}{1}$$

$$x = \frac{6.02 \times 10^{23}}{22.4}$$

$$x = 2.69 \times 10^{22}$$

Therefore, there are 2.69×10^{22} molecules in 1 liter of the gas.

Use a proportion to solve the following problem.

The ratio of the speed of light in a vacuum to the speed of light in water is 4 to 3. If the speed of light in a vacuum is 3×10^8 meters per second, what is the speed of light in water?

---

89. To square the number in scientific notation below, we squared the first factor and the power of ten. Square the other number.

$$(2 \times 10^8)^2 = (2)^2 \times (10^8)^2 = 4 \times 10^{16}$$

$$(3 \times 10^{-5})^2 = \underline{\hspace{4cm}}$$

> 2.25×10^8 meters
>   per second
>
> or
>
> $225,000,000$ meters
>   per second
>
> from:
>
> $\frac{4}{3} = \frac{3 \times 10^8}{x}$

---

90. The calculator steps for two squarings are shown below. Notice that we enter the number in the usual way and then press $\boxed{x^2}$.

$$(6.8 \times 10^6)^2 = 4.624 \times 10^{13}$$

Calculator Steps: 6.8 $\boxed{\text{EXP}}$ 6 $\boxed{x^2}$ 4.624 13

$$(5.1 \times 10^{-11})^2 = 2.601 \times 10^{-21}$$

Calculator Steps: 5.1 $\boxed{\text{EXP}}$ 11 $\boxed{+/-}$ $\boxed{x^2}$ 2.601 -21

Do this one: $(4.8 \times 10^{-6})^2 = \underline{\hspace{3cm}}$

> 9×10^{-10}

---

91. The square roots of two powers of ten are given below.

$$\sqrt{10^8} = 10^4 \text{ , since } (10^4)^2 = 10^8$$

$$\sqrt{10^{-6}} = 10^{-3} \text{ , since } (10^{-3})^2 = 10^{-6}$$

To find the square root of a power of ten, we can divide its exponent by 2. That is:

$$\sqrt{10^8} = 10^{\frac{8}{2}} = 10^4 \qquad \sqrt{10^{-6}} = 10^{\frac{-6}{2}} = 10^{-3}$$

Continued on following page.

> 2.304×10^{-11}

91. Continued Use the preceding fact for these. a) $\sqrt{10^4}$ = _____ b) $\sqrt{10^{14}}$ = _____ c) $\sqrt{10^{-12}}$ = _____	
92. To find the square root of a number in scientific notation, we find the square root of the first factor and the square root of the power of ten. For example: $\sqrt{4 \times 10^{10}} = \sqrt{4} \times \sqrt{10^{10}} = 2 \times 10^5$ $\sqrt{9 \times 10^{-8}}$ = _____	a) 10^2 b) 10^7 c) 10^{-6}
93. The calculator steps for finding two square roots are shown below. Notice that we enter the number in the usual way and then press $\boxed{\sqrt{x}}$. $\sqrt{1.44 \times 10^{22}} = 1.2 \times 10^{11}$ Calculator Steps: 1.44 $\boxed{\text{EXP}}$ 22 $\boxed{\sqrt{x}}$ 1.2 11 $\sqrt{5.76 \times 10^{-20}} = 2.4 \times 10^{-10}$ Calculator Steps: 5.76 $\boxed{\text{EXP}}$ 20 $\boxed{+/-}$ $\boxed{\sqrt{x}}$ 2.4 -10 Do this one: $\sqrt{9.61 \times 10^{-24}}$ = _____	3×10^{-4}
94. Do these. Round to two decimal places. a) $(9.55 \times 10^9)^2$ = _____ b) $\sqrt{7.38 \times 10^{-21}}$ = _____	3.1×10^{-12}
95. Do these. Your calculator may not give the answers in scientific notation. a) $(1.7 \times 10^{-3})^2$ = _____ b) $\sqrt{3.61 \times 10^{14}}$ = _____	a) 9.12×10^{19} b) 8.59×10^{-11}

a) 2.89×10^{-6} or .00000289	b) 1.9×10^7 or 19,000,000

5-10 EVALUATIONS INVOLVING NUMBERS IN SCIENTIFIC NOTATION

In this section, we will discuss formula evaluations involving numbers in scientific notation.

96. The steps for the evaluation below are shown. In Q = CE , find Q when C = 8.55×10^{-13} and E = 60. Calculator Steps: 8.55 $\boxed{\text{EXP}}$ 13 $\boxed{+/-}$ $\boxed{\times}$ 60 $\boxed{=}$ 5.13 -11 Therefore: Q = 5.13×10^{-11} Continued on following page.	

96. Continued

Do this one. Round the first factor to two decimal places.

In $E = hf$, when $h = 6.63 \times 10^{-34}$ and $f = 3.25 \times 10^{18}$,

$$E = \underline{\hspace{2cm}}$$

---

97. The steps for the evaluation below are shown.

In $E = mc^2$, find E when $m = 50$ and $c = 3 \times 10^8$.

Calculator Steps: 50 $\boxed{x}$ 3 $\boxed{EXP}$ 8 $\boxed{x^2}$ $\boxed{=}$ 4.5   18

Therefore: $E = 4.5 \times 10^{18}$

In $E = mc^2$, when $m = 690$ and $c = 3 \times 10^8$, $E = \underline{\hspace{2cm}}$

$E = 2.15 \times 10^{-15}$

---

98. Do these. For (a), round the first factor to two decimal places. For (b), your calculator may not give the answer in scientific notation.

a) In $M = \dfrac{gs^2}{G}$, when $g = 9.80$, $s = 6.37 \times 10^6$, and $G = 6.67 \times 10^{-11}$, $M = \underline{\hspace{2cm}}$

b) In $F = \dfrac{kQ_1 Q_2}{s^2}$, when $k = 9 \times 10^9$, $Q_1 = 3 \times 10^{-4}$, $Q_2 = 5 \times 10^{-5}$, and $s = .5$, $$F = \underline{\hspace{2cm}}$$

$E = 6.21 \times 10^{19}$

---

99. The steps for the evaluation below are shown. Notice that we pressed $\boxed{=}$ to complete the operation within the radical before pressing $\boxed{\sqrt{x}}$. We rounded the first factor to two decimal places.

In $Z = \sqrt{R^2 + X^2}$, find Z when $R = 8.5 \times 10^{10}$ and $X = 3.4 \times 10^{10}$.

Calculator Steps: 8.5 $\boxed{EXP}$ 10 $\boxed{x^2}$ $\boxed{+}$ 3.4 $\boxed{EXP}$ 10

$\boxed{x^2}$ $\boxed{=}$ $\boxed{\sqrt{x}}$ 9.15478   10

Therefore: $Z = 9.15 \times 10^{10}$

Do this one. Round the first factor to two decimal places.

In $Z = \sqrt{R^2 + X^2}$, when $R = 9.15 \times 10^{12}$ and $X = 5.86 \times 10^{12}$,

$$Z = \underline{\hspace{2cm}}$$

a) $M = 5.96 \times 10^{24}$

b) $F = 5.4 \times 10^2$
(or 540)

$Z = 1.09 \times 10^{13}$

SELF-TEST 18 (pages 214-223)

1. Find r_1 when $F_1 = 46.2$, $F_2 = 90.7$, and $r_2 = 3.54$. Round to hundredths.

$$\frac{F_1}{F_2} = \frac{r_1}{r_2}$$

$r_1 = $ _____

2. Find I_2 when $I_1 = 46.9$, $d_1 = 3.17$, and $d_2 = 5.48$. Round to tenths.

$$\frac{I_1}{I_2} = \frac{(d_2)^2}{(d_1)^2}$$

$I_2 = $ _____

In Problems 3-12, report each answer in scientific notation with the first factor rounded to two decimal places.

3. $\dfrac{4,700,000}{.0000816}$

_____

4. $(.00000828)^2$

_____

5. $(9.27 \times 10^{-7}) \times (2.14 \times 10^{-5})$

_____

6. $\dfrac{3.19 \times 10^6}{9.76 \times 10^{-6}}$

_____

7. $\sqrt{1.852 \times 10^{24}}$

_____

8. $(5.04 \times 10^{-8})^2$

_____

9. The sun's mass is 1.96×10^{27} tons and the earth's mass is 5.89×10^{21} tons. Find the ratio of the sun's mass to the earth's mass.

_____

10. If it takes 1 second for a radio wave to travel 2.998×10^5 kilometers, how long will it take the wave to travel 1 kilometer?

$$\frac{1}{2.998 \times 10^5} = \frac{x}{1}$$

$$1 = n \ x$$

11. Find V when $Q = 3.53 \times 10^{-7}$ and $C = 1.57 \times 10^{-10}$.

$$V = \frac{Q}{C}$$

$V = $ _____

12. Find $\underline{f}$ when $L = 8.6 \times 10^{-3}$ and $C = 6.2 \times 10^{-11}$.

$$f = \frac{1}{6.28\sqrt{LC}}$$

$f = $ _____

ANSWERS:

1. $r_1 = 1.80$
2. $I_2 = 15.7$
3. 5.76×10^{10}
4. 6.86×10^{-11}

5. 1.98×10^{-11}
6. 3.27×10^{11}
7. 1.36×10^{12}
8. 2.54×10^{-15}

9. 3.33×10^5
10. 3.34×10^{-6} second
11. $V = 2.25 \times 10^3$
12. $f = 2.18 \times 10^5$

5-11 FORMULAS CONTAINING PARENTHESES

In this section, we will show how the parentheses symbols on a calculator can be used for evaluations with formulas containing parentheses.

---

100. To do the evaluation below on a calculator, the parentheses symbols $\boxed{(}$ and $\boxed{)}$ are used. The steps are shown. Notice that we pressed $\boxed{x}$ before pressing $\boxed{(}$.

In $D = P(Q + R)$, find D when P = 5, Q = 4, and R = 3.

Calculator Steps: 5 $\boxed{x}$ $\boxed{(}$ 4 $\boxed{+}$ 3 $\boxed{)}$ $\boxed{=}$ 35

Do these. Don't forget to press $\boxed{x}$ before $\boxed{(}$.

a) In $M = P(L - X)$, when P = 20, L = 44.5, and X = 29.7,

M = _____

b) In $Q = C(T_1 - T_2)$, when C = 2.5, T_1 = 90, and T_2 = 55,

Q = _____

---

101. The steps for the evaluation below are shown. Notice that we press $\boxed{(}$ and $\boxed{)}$ where they appear in the formula.

In $H = ms(t_2 - t_1)$, find H when m = 4, s = 3, t_2 = 10, and t_1 = 5.

Calculator Steps: 4 $\boxed{x}$ 3 $\boxed{x}$ $\boxed{(}$ 10 $\boxed{-}$ 5 $\boxed{)}$ $\boxed{=}$ 60

Do these. Don't forget to press $\boxed{x}$ before $\boxed{(}$.

a) In $K = mt(x - r)$, when m = 25.7, t = 40, x = 19.7, and

r = 12.2, K = _____

b) In $V = SR(b_1 + b_2)$, when S = 3.7, R = 4.2, b_1 = 25, and

b_2 = 75, V = _____

a) M = 296

b) Q = 87.5

---

102. Notice in the steps below that we press $\boxed{x^2}$ after entering 4 and 3.

In $d = a(b^2 - c^2)$, find d when a = 10, b = 4, and c = 3.

Calculator Steps: 10 $\boxed{x}$ $\boxed{(}$ 4 $\boxed{x^2}$ $\boxed{-}$ 3 $\boxed{x^2}$ $\boxed{)}$ $\boxed{=}$ 70

Do these. Round to tens in (a) and hundredths in (b).

a) In $A = 3.14(R^2 - r^2)$, when R = 25.5 and r = 13.7,

A = _____

b) In $A = 0.785(D^2 - d^2)$, when D = 2.54 and d = 1.89,

A = _____

a) K = 7,710

b) V = 1,554

---

a) A = 1,450

b) A = 2.26

103. Notice in the steps below that we began by pressing $\boxed{(}$. Also, we pressed $\boxed{x}$ after $\boxed{)}$.

In $H = (1 - C^2)h$, find H when C = 0.6 and h = 50.

Calculator Steps: $\boxed{(}$ 1 $\boxed{-}$ .6 $\boxed{x^2}$ $\boxed{)}$ $\boxed{x}$ 50 $\boxed{=}$ 32

Do these. Round to hundredths.

a) In $d = (1 - t^2)R$, when t = 0.75 and R = 15, d = _____

b) In $m = (a^2 + b^2)T$, when a = 1.25, b = 2.46, and T = 3.04,

m = _____

---

104. For the evaluation below, we evaluated the numerator and then divided by 2.

In $A = \dfrac{h(b_1 + b_2)}{2}$, find A when h = 20, b_1 = 4, and b_2 = 10.

Calculator Steps: 20 $\boxed{x}$ $\boxed{(}$ 4 $\boxed{+}$ 10 $\boxed{)}$ $\boxed{\div}$ 2 $\boxed{=}$ 140

Do these. Round to hundredths in (b).

a) In $r = \dfrac{R(w - 2F)}{w}$, when R = 12, w = 18, and F = 6,

r = _____

b) In $H = \dfrac{3.14dR(F_1 - F_2)}{33,000}$, when d = 100, R = 15, F_1 = 35.8,

and F_2 = 27.4, H = _____

a) d = 6.56

b) m = 23.15

---

105. For the evaluation below, we evaluated the numerator and then divided by 2 and 3.

In $a = \dfrac{c(d - f)}{2b}$, find a when c = 4, d = 20, f = 8, and b = 3.

Calculator Steps: 4 $\boxed{x}$ $\boxed{(}$ 20 $\boxed{-}$ 8 $\boxed{)}$ $\boxed{\div}$ 2 $\boxed{\div}$ 3 $\boxed{=}$ 8

Do this one. Round to hundredths.

In $F = \dfrac{W(R - r)}{2R}$, when W = 12.4, R = 75.1, and r = 57.3,

F = _____

a) r = 4

b) H = 1.20

---

106. Do these. Round to tenths in (b).

a) In $E = \dfrac{(m_1 + m_2)v^2}{2}$, when m_1 = 35.5, m_2 = 55.5, and

v = 40, E = _____

b) In $H = \dfrac{M[(V_1)^2 - (V_2)^2]}{1,100gt}$, when M = 50, V_1 = 225, V_2 = 110,

g = 16, and t = 5, H = _____

F = 1.47

107. The evaluations below involve numbers in scientific notation. Your calculator may not give the answers in scientific notation.

    a) In $\Delta L = \alpha L(T - T_0)$, when $\alpha = 1.1 \times 10^{-5}$, $L = 60$, $T = 80$,

        and $T_0 = 0$, $\Delta L = $ _____

    b) In $\Delta V = \beta V(T - T_0)$, when $\beta = 9.6 \times 10^{-4}$, $V = 100$, $T = 30$,

        and $T_0 = 0$, $\Delta V = $ _____

a) E = 72,800

b) H = 21.9

---

108. Use parentheses when necessary for these.

    a) In $A = 180 - (B + C)$, when $B = 47$ and $C = 29$, $A = $ _____

    b) In $P = p_1 + w(h - h_1)$, when $p_1 = 20$, $w = 25$, $h = 75$,

        and $h_1 = 40$, $P = $ _____

a) $\Delta L = 5.28 \times 10^{-2}$
   or .0528

b) $\Delta V = 2.88 \times 10^{0}$
   or 2.88

---

109. For the evaluation below, be sure to press $\boxed{=}$ to complete the subtraction in the numerator before dividing.

$$I_b = \frac{I_c - (\beta + 1)I_{co}}{\beta} \text{ , when } I_c = 87.5, \ \beta = 0.5,$$

    and $I_{co} = 42.6$, $I_b = $ _____

a) A = 104

b) P = 895

---

110. The evaluation below contains two sets of parentheses. Notice that we pressed $\boxed{x}$ between the two sets of parentheses.

    In $F = (a + 3)(b + 4)$, find F when $a = 5$ and $b = 2$.

        Calculator Steps: $\boxed{(}$ 5 $\boxed{+}$ 3 $\boxed{)}$ $\boxed{x}$ $\boxed{(}$ 2 $\boxed{+}$ 4 $\boxed{)}$ $\boxed{=}$ 48

Complete these.

    a) In $m = (C - D)(T - S)$, when $C = 35$, $D = 25$, $T = 87$,

        and $S = 51$, $m = $ _____

    b) In $V = 2t(b - t)(a - t_1)$, when $t = 15$, $b = 45$, $a = 37$,

        and $t_1 = 20$, $V = $ _____

$I_b = 47.2$

a) m = 360   b) V = 15,300

5-12 USING PARENTHESES FOR DIVISIONS

In this section, we will show how parentheses can be used for divisions in which the denominator is an addition or subtraction or contains a set of parentheses.

111. To do the evaluation below, we treat $R_1 + R_2$ as if there were parentheses around it. The steps are shown.

If $I = \dfrac{E}{R_1 + R_2}$, find I when E = 100, R_1 = 15, and R_2 = 5.

Calculator Steps: 100 $\boxed{\div}$ $\boxed{(}$ 15 $\boxed{+}$ 5 $\boxed{)}$ $\boxed{=}$ 5

Use the same method for these. Round to hundredths.

a) In $A = \dfrac{B}{B + 1}$, when B = 28.3, A = _____

b) In $P = \dfrac{W_s}{V_1 - V_2}$, when W_s = 9.75, V_1 = 8.64, and V_2 = 5.73,

P = _____

112. Use parentheses around the denominator for these. Round to two decimal places.

a) In $R_t = \dfrac{R_1 R_2}{R_1 + R_2}$, when R_1 = 15 and R_2 = 20,

R_t = _____

b) In $W = \dfrac{2RF}{R - r}$, when R = 37.5, F = 4.60, and r = 2.90,

W = _____

| a) A = 0.97 |
| b) P = 3.35 |

113. Use parentheses around the denominator for these. Round to tenths in (b).

a) In $I = \dfrac{E}{R_1 + R_2 + R_3}$, when E = 200, R_1 = 15, R_2 = 25,

and R_3 = 40, I = _____

b) In $T = \dfrac{BV_1(P_2 - P_1)}{V_1 - V_2}$, when B = 2.5, V_1 = 80, P_2 = 30,

P_1 = 20, and V_2 = 50, T = _____

| a) R_t = 8.57 |
| b) W = 9.97 |

114. Complete these.

a) In $t = \dfrac{h^2}{1 - h}$, when h = 0.75, t = _____

b) In $m = \dfrac{M}{F^2 - 1}$, when M = 150 and F = 11, m = _____

| a) I = 2.5 |
| b) T = 66.7 |

| a) t = 2.25 |
| b) m = 1.25 |

115. In the evaluations below, be sure to press $\boxed{=}$ to complete the subtraction in the numerator before using parentheses to divide. Round to hundredths in (b).

   a) In $v = \dfrac{s_2 - s_1}{t_2 - t_1}$, when $s_2 = 100$, $s_1 = 10$, $t_2 = 60$,

   and $t_1 = 20$, $v =$ _____

   b) In $w = \dfrac{P - p_1}{h - h_1}$, when $P = 45.5$, $p_1 = 28.7$, $h = 40$,

   and $h_1 = 25$, $w =$ _____

---

116. In the evaluation below, we must divide H by both $\underline{m}$ and $(t_2 - t_1)$. Parentheses can be used for the second division. The steps are shown.

   In $s = \dfrac{H}{m(t_2 - t_1)}$, find $\underline{s}$ when $H = 90$, $m = 1.5$, $t_2 = 50$,

   and $t_1 = 20$.

   Calculator Steps: $90 \boxed{\div} 1.5 \boxed{\div} \boxed{(}\ 50 \boxed{-} 20 \boxed{)}\ \boxed{=}$ 2

   Use the same steps for this one. Round to hundredths.

   In $A = \dfrac{3T}{G(F - G)}$, when $T = 50.5$, $G = 12.8$, and $F = 15.1$,

   $A =$ _____

---

117. Do these. In (a), be sure to press $\boxed{=}$ to complete the subtraction in the numerator. Round to thousandths.

   a) In $\alpha = \dfrac{L_2 - L_1}{L_1(t_2 - t_1)}$, when $L_2 = 75$, $L_1 = 40$, $t_2 = 30$,

   and $t_1 = 15$, $\alpha =$ _____

   b) In $\beta = \dfrac{T(V_1 - V_2)}{V_1(P_2 - P_1)}$, when $T = 20$, $V_1 = 90$, $V_2 = 55$,

   $P_2 = 80$, and $P_1 = 65$, $\beta =$ _____

---

118. To do these, we have to divide each numerator by the three factors in the denominator. Round to two decimal places.

   a) In $T = \dfrac{HL}{AK(t_2 - t_1)}$, when $H = 33.8$, $L = 200$, $A = 10$,

   $K = 22.5$, $t_2 = 30$, and $t_1 = 25$,

   $T =$ _____

   b) In $R = \dfrac{33,000H}{3.14d(F_1 - F_2)}$, when $H = 1.5$, $d = 25$, $F_1 = 150$,

   and $F_2 = 75$, $R =$ _____

---

a) $v = 2.25$

b) $w = 1.12$

A = 5.15

a) $\alpha = 0.058$

b) $\beta = 0.519$

a) $T = 6.01$

b) $R = 8.41$

119. In the evaluation below, there are four factors in the denominator. Parentheses must be used for two of the divisions.

$$\text{In} \quad S = \frac{T}{2t(a - t)(b - t_1)} \text{ , when } T = 500, \, t = 25, \, a = 35,$$
$$b = 40, \text{ and } t_1 = 20, \quad S = \underline{\hspace{2cm}}$$

$S = 0.05$

5-13 USING MEMORY FOR DIVISIONS

In this section, we will show how memory can be used instead of parentheses to do the same divisions we did in the last section.

120. Both $\boxed{\text{STO}}$ and $\boxed{\text{RCL}}$ are memory keys. We used both keys for the evaluation below. We began by evaluating the denominator $R_1 + R_2$ and using $\boxed{\text{STO}}$ to put that value in storage. Then we entered the value of E and used $\boxed{\text{RCL}}$ to recall the stored value for the division.

$$\text{In} \quad I = \frac{E}{R_1 + R_2} \text{ , find I when } E = 100, \, R_1 = 15, \text{ and } R_2 = 10.$$

Calculator Steps: $15 \boxed{+} 10 \boxed{=} \boxed{\text{STO}} 100 \boxed{\div} \boxed{\text{RCL}} \boxed{=} 4$

Use the same method for these. Round to hundredths.

a) In $A = \dfrac{B}{B + 1}$, when $B = 28.3$, $A = \underline{\hspace{2cm}}$

b) In $P = \dfrac{W_s}{V_1 - V_2}$, when $W_s = 9.75$, $V_1 = 8.64$, and $V_2 = 5.73$,

$$P = \underline{\hspace{2cm}}$$

121. Do these by evaluating the denominator first and storing that value, then evaluating the numerator and dividing by the stored value. Round to hundredths.

a) In $R_t = \dfrac{R_1 R_2}{R_1 + R_2}$, when $R_1 = 15$ and $R_2 = 20$, $R_t = \underline{\hspace{2cm}}$

b) In $W = \dfrac{2RF}{R - r}$, when $R = 37.5$, $F = 4.6$, and $r = 2.9$,

$$W = \underline{\hspace{2cm}}$$

a) $A = 0.97$

b) $P = 3.35$

a) $R_t = 8.57$

b) $W = 9.97$

122. Use memory for these. In (b), be sure to evaluate $(P_2 - P_1)$ first and press $\boxed{=}$ to complete that subtraction when evaluating the numerator. Round to tenths in (b).

a) In $I = \dfrac{E}{R_1 + R_2 + R_3}$, when $E = 200$, $R_1 = 15$, $R_2 = 25$, and $R_3 = 40$, $I = \underline{\hspace{2cm}}$

b) In $T = \dfrac{BV_1(P_2 - P_1)}{V_1 - V_2}$, when $B = 2.5$, $V_1 = 80$, $P_2 = 30$, $P_1 = 20$, and $V_2 = 50$, $T = \underline{\hspace{2cm}}$

---

123. Use memory for these.

a) In $t = \dfrac{h^2}{1 - h}$, when $h = 0.75$, $t = \underline{\hspace{2cm}}$

b) In $m = \dfrac{M}{F^2 - 1}$, when $M = 150$ and $F = 11$, $m = \underline{\hspace{2cm}}$

a) $I = 2.5$

b) $T = 66.7$

---

124. In the evaluations below, be sure to press $\boxed{=}$ to complete the subtraction in the numerator before dividing by the stored value.

a) In $v = \dfrac{s_2 - s_1}{t_2 - t_1}$, when $s_2 = 100$, $s_1 = 10$, $t_2 = 60$, and $t_1 = 20$, $v = \underline{\hspace{2cm}}$

b) In $w = \dfrac{P - p_1}{h - h_1}$, when $P = 45.5$, $p_1 = 28.7$, $h = 40$, and $h_1 = 25$, $w = \underline{\hspace{2cm}}$

a) $t = 2.25$

b) $m = 1.25$

---

125. In the evaluation below, we must evaluate the denominator first and store that value. To evaluate the denominator, we evaluate $(t_2 - t_1)$ first and press $\boxed{=}$ to complete the subtraction. The steps are shown.

In $s = \dfrac{H}{m(t_2 - t_1)}$, find $\underline{s}$ when $H = 90$, $m = 1.5$, $t_2 = 50$, and $t_1 = 20$.

Calculator Steps: $50 \boxed{-} 20 \boxed{=} \boxed{\times} 1.5 \boxed{=} \boxed{STO} \; 90$
$\boxed{\div} \boxed{RCL} \boxed{=} \; 2$

Use the same steps for this one. Round to hundredths.

In $A = \dfrac{3T}{G(F - G)}$, when $T = 50.5$, $G = 12.8$, and $F = 15.1$, $A = \underline{\hspace{2cm}}$

a) $v = 2.25$

b) $w = 1.12$

$A = 5.15$

126. Do these. In (a), be sure to press $\boxed{=}$ to complete the subtraction in the numerator before dividing by the stored value. Round to thousandths.

a) In $\alpha = \dfrac{L_2 - L_1}{L_1(t_2 - t_1)}$, when $L_2 = 75$, $L_1 = 40$, $t_2 = 30$, and $t_1 = 15$, $\alpha = $ _____

b) In $\beta = \dfrac{T(V_1 - V_2)}{V_1(P_2 - P_1)}$, when $T = 20$, $V_1 = 90$, $V_2 = 55$, $P_2 = 80$, and $P_1 = 65$, $\beta = $ _____

127. To evaluate the denominators below, we begin by performing the subtraction in the parentheses and press $\boxed{=}$ to complete that subtraction. Round to two decimal places.

a) In $T = \dfrac{HL}{AK(t_2 - t_1)}$, when $H = 33.8$, $L = 200$, $A = 10$, $K = 22.5$, $t_2 = 30$, and $t_1 = 25$, $T = $ _____

b) In $R = \dfrac{33,000H}{3.14d(F_1 - F_2)}$, when $H = 1.5$, $d = 25$, $F_1 = 150$, and $F_2 = 75$, $R = $ _____

128. To evaluate the denominator below, we must evaluate $2t(a - t)$ and store that value, then evaluate $(b - t_1)$ and multiply by the stored value. After placing the denominator in storage, we divide T by the value of the denominator. The steps are shown.

In $S = \dfrac{T}{2t(a - t)(b - t_1)}$, find S when $T = 500$, $t = 25$, $a = 35$, $b = 40$, and $t_1 = 20$.

Calculator Steps: 35 $\boxed{-}$ 25 $\boxed{=}$ $\boxed{x}$ 2 $\boxed{x}$ 25 $\boxed{=}$ $\boxed{\text{STO}}$

40 $\boxed{-}$ 20 $\boxed{=}$ $\boxed{x}$ $\boxed{\text{RCL}}$ $\boxed{=}$ $\boxed{\text{STO}}$

500 $\boxed{\div}$ $\boxed{\text{RCL}}$ $\boxed{=}$ 0.05

---

a) $\alpha = 0.058$

b) $\beta = 0.519$

a) $T = 6.01$

b) $R = 8.41$

S = 0.05

SELF-TEST 19 (pages 224-232)

1. Find B when A = 109 and C = 27.

$$B = 180 - (A + C)$$

B = _____

2. Find A when D = 8.17 and d = 7.82 . Round to hundredths.

$$A = 0.7854(D^2 - d^2)$$

A = _____

3. Find F when W = 456, R = 27.3, and r = 13.9 . Round to a whole number.

$$F = \frac{W(R - r)}{2R}$$

F = _____

4. Find C_t when C_1 = 37.5 and C_2 = 81.5 . Round to tenths.

$$C_t = \frac{C_1 C_2}{C_1 + C_2}$$

C_t = _____

5. Find $\underline{v}$ when d_2 = 8,950 , d_1 = 3,470 , t_2 = 4.62, and t_1 = 3.85. Round to tens.

$$v = \frac{d_2 - d_1}{t_2 - t_1}$$

v = _____

6. Find $\underline{s}$ when H = 76.4, m = 23.9, w_1 = 8.55, and w_2 = 3.37 . Round to thousandths.

$$s = \frac{H}{m(w_1 - w_2)}$$

s = _____

7. Find R when H = 5.75, d = 82.9, F_1 = 712, and F_2 = 267. Round to two decimal places.

$$R = \frac{33,000H}{3.14d(F_1 - F_2)}$$

R = _____

ANSWERS: 1. B = 44   3. F = 112   5. v = 7,120   7. R = 1.64

2. A = 4.40   4. C_t = 25.7   6. s = 0.617

SUPPLEMENTARY PROBLEMS - CHAPTER 5

Assignment 16

1. In $R = V - T$, find R when V = 512.7 and T = 179.8 .

2. In $V = LWH$, find V when L = 188, W = 96, and H = 0.355 . Round to hundreds.

3. In $Q = br - hs$, find Q when b = 6.29, r = 31.3, h = 2.46, and s = 57.8 . Round to tenths.

4. In $L = \dfrac{A}{W}$, find L when A = 9,410 and W = 58. Round to a whole number.

5. In $t = \dfrac{1}{f}$, find $\underline{t}$ when f = 576. Round to five decimal places.

6. In $v = \dfrac{Ftg}{m}$, find $\underline{v}$ when F = 450, t = 0.836, g = 32.2, m = 1,790. Round to hundredths.

7. In $X = L - \dfrac{M}{P}$, find X when L = 58.7, M = 263, and P = 9.14 . Round to tenths.

8. In $P = \dfrac{H - E}{V}$, find P when H = 850, E = 570, and V = 390. Round to thousandths.

9. In $C = \dfrac{5F - 160}{9}$, find C when F = -40.

10. In $T = \dfrac{a - b}{c} + h$, find T when a = 12.47, b = 8.93, c = 0.0152, and h = 68.3 .
    Round to a whole number.

11. In $H = \dfrac{B}{VT}$, find H when B = 138, V = 0.515, and T = 143. Round to two decimal places.

12. In $m = \dfrac{1}{4ap}$, find $\underline{m}$ when a = 22.4 and p = 0.128 . Round to four decimal places.

13. In $E = \dfrac{PL}{Ae}$, find E when P = 2,000 , L = 15, A = 0.0205, and e = 0.047 . Round to millions.

14. In $F = D - \dfrac{H}{ms}$, find F when D = 20.7, H = 590, m = 8.22, and s = 11.6 .
    Round to one decimal place.

15. In $P = \dfrac{RT - GW}{RT}$, find P when R = 852, T = 548, G = 175, and W = 319.
    Round to thousandths.

Assignment 17

1. In $P = I^2R$, find P when I = 4.8 and R = 35. Round to a whole number.

2. In $A = 0.7854d^2$, find A when d = 0.942 . Round to three decimal places.

3. In $R = \dfrac{pL}{d^2}$, find R when p = 10.4, L = 750, and d = 64.1. Round to hundredths.

4. In $s = \dfrac{v^2}{2g}$, find $\underline{s}$ when v = 120 and g = 32.2 . Round to a whole number.

5. In $a = (b - c)^2$, find $\underline{a}$ when b = 5.42 and c = 1.86 . Round to one decimal place.

6. In $p = m + r\sqrt{t}$, find $\underline{p}$ when m = 2.94, r = 1.37, and t = 20.6 .
Round to two decimal places.

7. In $E = \sqrt{PR}$, find E when P = 80 and R = 180.

8. In $b = \sqrt{c^2 - a^2}$, find $\underline{b}$ when c = 510 and a = 170. Round to tens.

9. In $w = 2g\sqrt{\dfrac{a}{r}}$, find $\underline{w}$ when g = 47, a = 53, and r = 79. Round to a whole number.

10. In $F = \dfrac{W}{\sqrt{ht}}$, find F when W = 4,860 , h = 12.6, and t = 250. Round to tenths.

11. In $t = \dfrac{1}{b\sqrt{m}}$, find $\underline{t}$ when b = 8 and m = 5.32 . Round to thousandths.

12. In $C = \sqrt{1 - \dfrac{H}{h}}$, find C when H = 75.3 and h = 91.8 . Round to three decimal places.

13. In $V = s^3$, find V when s = 89.2 . Round to thousands.

14. In $s = \sqrt[3]{V}$, find $\underline{s}$ when V = 260. Round to hundredths.

Assignment 18

1. In $\dfrac{V_1}{V_2} = \dfrac{T_1}{T_2}$, find T_1 when V_1 = 22.4, V_2 = 24.3, and T_2 = 320. Round to a whole number.

2. In $\dfrac{A_1}{A_2} = \dfrac{(r_1)^2}{(r_2)^2}$, find A_2 when A_1 = 12.8, r_1 = 2.02, and r_2 = 3.75. Round to tenths.

3. In $\dfrac{P_1 V_1}{T_1} = \dfrac{P_2 V_2}{T_2}$, find P_1 when P_2 = 7.38, V_1 = 51.2, V_2 = 64.7, T_1 = 315, and T_2 = 295.
Round to hundredths.

For problems 4-11, report the answer in scientific notation with the first factor rounded to two decimal places.

4. $(2.78 \times 10^{-15}) \times (7.36 \times 10^4)$

5. $\dfrac{1.68 \times 10^6}{4.07 \times 10^{-9}}$

6. $(5.477 \times 10^{-6})^2$

7. $\dfrac{2.76 \times 10^{-8}}{4,580}$

8. $\sqrt{6.64 \times 10^{21}}$

9. $3,130 \times (8.62 \times 10^9)$

10. For $f = \dfrac{v}{L}$, find $\underline{f}$ when v = 2.998×10^{10} and L = 4.50×10^{-5}.

11. For $R = \sqrt{Z^2 - X^2}$, find R when Z = 2.37×10^6 and X = 1.19×10^6.

Do these.

12. Two successive laser pulses had duration times of 3.37×10^{-6} second and 5.84×10^{-3} second. Write the ratio of the first pulse to the second pulse as a decimal number rounded to millionths.

13. Density is mass divided by volume. Find the density of the earth, in grams per cubic centimeter, if its mass is 5.976×10^{27} grams and its volume is 1.083×10^{27} cubic centimeters. Round to hundredths.

14. One acre equals 4.356×10^4 square feet. How many acres are there in 9,000 square feet? Round to thousandths.

15. A radar pulse travels one mile in 5.38×10^{-6} second. How far will it travel in 0.002 second? Round to a whole number.

16. One milligram equals 3.5274 x 10^{-5} ounce. How many milligrams are there in one ounce? Round to tens.

17. The resistance of one foot of #0000 aluminum wire is 8.04 x 10^{-5} ohm. What length of the wire has one ohm resistance? Round to hundreds.

Assignment 19

1. In $M = P(L - K)$, find M when P = 160, L = 48.3, and K = 17.5 . Round to hundreds.

2. In $A = 0.7854(D^2 - d^2)$, find A when D = 3.28 and d = 2.93 . Round to hundredths.

3. In $r = \dfrac{R(w - 2F)}{w}$, find $\underline{r}$ when R = 59.2, w = 435, and F = 127. Round to tenths.

4. In $d = (1 - t^2)K$, find $\underline{d}$ when t = 0.28 and K = 6,100. Round to hundreds.

5. In $E = \dfrac{(m_1 + m_2)v^2}{2}$, find E when m_1 = 375, m_2 = 540, and v = 81.9 . Round to hundred-thousands.

6. In $A = 180 - (B + C)$, find A when B = 39 and C = 48.

7. In $\Delta L = \alpha L(T - T_0)$, find ΔL when α = 1.3 x 10^{-5}, L = 60, T = 100, and T_0 = 20.

8. In $I = \dfrac{E}{R_1 + R_2}$, find I when E = 12, R_1 = 220, and R_2 = 470. Round to thousandths.

9. In $R = \dfrac{wr}{w - 2F}$, find R when w = 783, r = 2.08, and F = 195. Round to two decimal places.

10. In $v = \dfrac{s_2 - s_1}{t_2 - t_1}$, find $\underline{v}$ when s_2 = 81.4, s_1 = 28.7, t_2 = 3.26, and t_1 = 1.19 . Round to one decimal place.

11. In $m = \dfrac{M}{F^2 - 1}$, find $\underline{m}$ when M = 32,600 and F = 1.82 . Round to hundreds.

12. In $R = \dfrac{33,000H}{3.14d(F_1 - F_2)}$, find R when H = 46.3, d = 5.72, F_1 = 6,180 , and F_2 = 4,930. Round to tenths.

13. In $R_1 = \dfrac{R_2 R_t}{R_2 - R_t}$, find R_1 when R_t = 258 and R_2 = 412. Round to a whole number.

14. In $w = \dfrac{2N}{r(h - r)(p - r)}$, find $\underline{w}$ when N = 860, r = 18, h = 43, and p = 56. Round to hundredths.

6 Measurements

In this chapter, we will discuss the units of length, weight, time, liquid measures, temperature, and rates in both the English System and the Metric System. Conversions of units within each system and between the two systems are discussed. The unity-fraction method is shown for all conversions. The decimal-point-shift method is shown for conversions within the Metric System. We will also discuss the following properties of measurements: precision, lower and upper limits, absolute error, significant digits, relative error, and accuracy. The rules for reporting answers for operations with measurements are introduced and used.

6-1 ENGLISH SYSTEM

In this section, we will give the common units of length, weight (or mass), time, and liquid measures in the English System and show how the unity-fraction method can be used to convert from one unit to another.

Continued on following page.

1. The common equivalent English units of length, weight (or mass), time, and liquid measures are given below.

Length

1 foot (ft) = 12 inches (in)

1 yard (yd) = 3 feet (ft)

1 mile (mi) = 5,280 feet (ft) or 1,760 yards (yd)

*Weight (or Mass)

1 pound (lb) = 16 ounces (oz)

1 ton (T) = 2,000 pounds (lb)

Time

1 minute (min) = 60 seconds (sec)

1 hour (hr) = 60 minutes (min)

1 hour (hr) = 3,600 seconds (sec)

*Liquid Measures

1 pint (pt) = 16 ounces (oz)

1 quart (qt) = 2 pints (pt) = 32 ounces (oz)

1 gallon (gal) = 4 quarts (qt) = 8 pints (pt)

*The unit ounce that is used as a measure of weight is not the same as the unit ounce that is used for liquid measures.

Using the fact that 1 foot = 12 inches, we set up two fractions below. Each fraction equals "1" because the numerator and denominator of each are equal. Because the fractions equal "1", they are called <u>unity</u> fractions.

$$\frac{1 \text{ ft}}{12 \text{ in}} = 1 \qquad\qquad \frac{12 \text{ in}}{1 \text{ ft}} = 1$$

Using the fact that 1 hour = 3,600 seconds, set up two unity fractions below.

_____ and _____

$$\frac{1 \text{ hr}}{3,600 \text{ sec}} \text{ and } \frac{3,600 \text{ sec}}{1 \text{ hr}}$$

2. Using a unity fraction based on 1 yard = 3 feet, we converted 5 yards to feet below. Notice how we cancelled the yd's.

$$5 \text{ yd} = 5 \text{ yd}\left(\frac{3 \text{ ft}}{1 \text{ yd}}\right) = 5(3 \text{ ft}) = 15 \text{ ft}$$

a) Complete the following conversion of 6 pounds to ounces.

$$6 \text{ lb} = 6 \text{ lb}\left(\frac{16 \text{ oz}}{1 \text{ lb}}\right) = \underline{\hspace{3cm}} \text{ oz}$$

b) Complete the following conversion of 3 miles to feet.

$$3 \text{ mi} = 3 \text{ mi}\left(\frac{5,280 \text{ ft}}{1 \text{ mi}}\right) = \underline{\hspace{3cm}} \text{ ft}$$

---

3. Using a unity fraction based on 1 minute = 60 seconds, we converted 480 seconds to minutes below.

$$480 \text{ sec} = 480 \text{ sec}\left(\frac{1 \text{ min}}{60 \text{ sec}}\right) = \frac{480}{60} \text{ min} = 8 \text{ min}$$

a) Complete the following conversion of 48 quarts to gallons.

$$48 \text{ qt} = 48 \text{ qt}\left(\frac{1 \text{ gal}}{4 \text{ qt}}\right) = \underline{\hspace{3cm}} \text{ gal}$$

b) Complete the following conversion of 8,800 yards to miles.

$$8,800 \text{ yd} = 8,800 \text{ yd}\left(\frac{1 \text{ mi}}{1,760 \text{ yd}}\right) = \underline{\hspace{3cm}} \text{ mi}$$

---

4. The steps in the unity-fraction method for the conversion below are described.
$$3.5 \text{ mi} = \underline{\hspace{2cm}} \text{ ft}$$

1. Identify the conversion fact relating miles and feet.
   It is: 1 mi = 5,280 ft

2. Of the two possible unity fractions, use the one with miles in the denominator <u>so</u> <u>that</u> <u>miles</u> <u>can</u> <u>be</u> <u>cancelled</u>.

Therefore: $3.5 \text{ mi} = 3.5 \text{ mi}\left(\frac{5,280 \text{ ft}}{1 \text{ mi}}\right) = 3.5(5,280 \text{ ft}) = 18,480 \text{ ft}$

Use the unity-fraction method for these conversions.

a) 5 qt = _____ oz   b) 7.5 hr = _____ min

---

a) 96 oz

b) 15,840 ft

a) 12 gal

b) 5 mi

a) 160 oz, from:

$$5 \text{ qt}\left(\frac{32 \text{ oz}}{1 \text{ qt}}\right)$$

b) 450 min, from:

$$7.5 \text{ hr}\left(\frac{60 \text{ min}}{1 \text{ hr}}\right)$$

5. The steps in the unity-fraction method for the conversion below are described.

$$192 \text{ oz} = \underline{\hspace{2cm}} \text{ lb}$$

1. Identify the conversion fact relating ounces and pounds. It is: 1 lb = 16 oz

2. Of the two possible unity fractions, use the one with ounces in the denominator <u>so</u> <u>that</u> <u>ounces</u> <u>can</u> <u>be</u> <u>cancelled</u>.

Therefore: $192 \text{ oz} = 192 \text{ }\cancel{oz}\left(\dfrac{1 \text{ lb}}{16 \text{ }\cancel{oz}}\right) = \dfrac{192}{16} \text{ lb} = 12 \text{ lb}$

Use the unity-fraction method for each conversion.

a) 54 ft = _____ yd   b) 18,000 sec = _____ hr

---

6. Do these. Use a calculator.

a) 8.2 ft = _____ in   c) 7.5 lb = _____ oz

b) 210 min = _____ hr   d) 7,744 yd = _____ mi

a) 18 yd, from:

$$54 \text{ ft}\left(\dfrac{1 \text{ yd}}{3 \text{ ft}}\right)$$

b) 5 hr, from:

$$18,000\left(\dfrac{1 \text{ hr}}{3,600 \text{sec}}\right)$$

---

7. When converting 49 ft to yards below, we rounded to tenths.

$$49 \text{ ft} = 49 \text{ }\cancel{ft}\left(\dfrac{1 \text{ yd}}{3 \text{ }\cancel{ft}}\right) = \dfrac{49}{3} \text{ yd} = 16.3 \text{ yd}$$

Do these. Round to tenths.

a) 327 sec = _____ min   b) 500 oz = _____ qt

a) 98.4 in

b) 3.5 hr

c) 120 oz

d) 4.4 mi

---

a) 5.5 min

b) 15.6 qt

6-2 METRIC SYSTEM

The Metric System is the only system of measurement used in all major countries other than the United States. Its use in the United States is increasing. In this section, we will discuss conversions within the Metric System. The decimal-point-shift method for those conversions is emphasized.

---

8. The basic units of length, weight (or mass), time, and liquid measures in the Metric System are a <u>meter</u>, a <u>gram</u>, a <u>second</u>, and a <u>liter</u>. As you can see in the table below, the other units are defined in terms of those basic units.

Lengths

1 <u>kilo</u>meter (km) = 1,000 meters (m)

1 <u>hecto</u>meter (hm) = 100 meters (m)

1 <u>deka</u>meter (dam) = 10 meters (m)

1 meter (m) = 1 meter (m)

1 <u>deci</u>meter (dm) = $\frac{1}{10}$ meter (m)

1 <u>centi</u>meter (cm) = $\frac{1}{100}$ meter (m)

1 <u>milli</u>meter (mm) = $\frac{1}{1,000}$ meter (m)

Weight (or Mass)

1 metric ton (MT or t) = 1,000 kilograms (kg)

1 <u>kilo</u>gram (kg) = 1,000 grams (g)

1 <u>hecto</u>gram (hg) = 100 grams (g)

1 <u>deka</u>gram (dag) = 10 grams (g)

1 gram (g) = 1 gram (g)

1 <u>deci</u>gram (dg) = $\frac{1}{10}$ gram (g)

1 <u>centi</u>gram (cg) = $\frac{1}{100}$ gram (g)

1 <u>milli</u>gram (mg) = $\frac{1}{1,000}$ gram (g)

Time

1 millisecond (msec) = $\frac{1}{1,000}$ second (sec)

Liquid Measure

1 milliliter (mℓ) = $\frac{1}{1,000}$ liter (ℓ)

Continued on following page.

8. Continued

The names for the other units contain a prefix followed by the word for the basic unit. <u>Since the same prefixes and numerical factors are used in all types of metric measures, they should be memorized.</u> For example:

kilo- means 1,000

deka- means 10

centi- means $\frac{1}{100}$

Write the numerical factor that goes with each prefix.

    a) hecto- _____   b) deci- _____   c) milli- _____

---

9. Write the prefix that goes with each numerical factor.

    a) 1,000 _____   b) 10 _____   c) $\frac{1}{100}$ _____

a) 100   b) $\frac{1}{10}$

c) $\frac{1}{1,000}$

---

10. The abbreviations for basic units and other units are given in the table in frame 8. <u>They should be memorized.</u> Some examples are:

liter = ℓ

kilogram = kg

decimeter = dm

millisecond = msec

Write the abbreviation for each unit.

    a) dekagram = _____   c) kilometer = _____

    b) centimeter = _____   d) milliliter = _____

a) kilo-

b) deka-

c) centi-

---

11. Write the unit that goes with each abbreviation.

    a) dam = _____   c) hm = _____

    b) dg = _____   d) msec = _____

a) dag   c) km

b) cm   d) mℓ

---

12. We can use unity fractions for conversions within the Metric System.

Using the fact: 1 dam = 10 m, we can convert 2.9 dam to meters.

$$2.9 \text{ dam} = 2.9 \cancel{\text{dam}} \left(\frac{10 \text{ m}}{1 \cancel{\text{dam}}} \right) = 2.9(10 \text{ m}) = 29 \text{ m}$$

Using the fact: 1 m = 100 cm, we can convert 6.4 m to cm.

$$6.4 \text{ m} = 6.4 \cancel{\text{m}} \left(\frac{100 \text{ cm}}{1 \cancel{\text{m}}} \right) = 6.4(100 \text{ cm}) = 640 \text{ cm}$$

Using the fact: 1 hg = 1,000 dg, we can convert 7.3 hg to dg.

$$7.3 \text{ hg} = 7.3 \text{ hg} \left(\frac{1,000 \text{ dg}}{1 \text{ hg}} \right) = \underline{\hspace{3cm}}$$

a) dekameter

b) decigram

c) hectometer

d) millisecond

13. We did the three conversions below in the last frame. In each one, we converted from a larger unit to a smaller unit.

$$2.9 \text{ dam} = 29 \text{ m}$$

$$6.4 \text{ m} = 640 \text{ cm}$$

$$7.3 \text{ hg} = 7,300 \text{ dg}$$

The units of length and weight in the Metric System are diagrammed below with the length units at the left.

1,000	100	10	1	$\frac{1}{10}$	$\frac{1}{100}$	$\frac{1}{1,000}$
km	hm	dam	m	dm	cm	mm
kg	hg	dag	g	dg	cg	mg

To convert from a larger unit to a smaller unit, we can simply shift the decimal point to the right. The number of places shifted depends on the number of unit-places in the conversion. For example:

To convert 2.9 dam to meters, we shift 1 place to the right because m is 1 unit-place to the right of dam.

$$2.9 \text{ dam} = 2.9_\curvearrowright = 29 \text{ m}$$

To convert 6.4 m to centimeters, we shift 2 places to the right because cm is 2 unit-places to the right of m.

$$6.4 \text{ m} = 6.40_\curvearrowright = 640 \text{ cm}$$

To convert 7.3 hg to decigrams, we shift 3 places to the right because dg is 3 unit-places to the right of hg.

$$7.3 \text{ hg} = \underline{\hspace{2cm}} \text{ dg}$$

7,300 dg

---

14. How many places to the right would we shift the decimal point for each conversion below?

   a) hectometers to meters ____   c) dekagrams to centigrams ____

   b) seconds to milliseconds ____   d) centimeters to millimeters ____

$7.300_\curvearrowright = 7,300$ dg

---

15. Use the decimal-point-shift method for each conversion.

   a) .65 kg = _____ g   c) 7.7 dam = _____ dm

   b) 1.85 ℓ = _____ mℓ   d) 49 cm = _____ mm

a) 2 places

b) 3 places

c) 3 places

d) 1 place

a) 650 g

b) 1,850 mℓ

c) 770 dm

d) 490 mm

16. We can use unity fractions for the conversions below.

Using the fact: 1 m = 10 dm, we can convert 13 dm to meters.

$$13 \text{ dm} = 13 \text{ dm} \left(\frac{1 \text{ m}}{10 \text{ dm}} \right) = \frac{13}{10} \text{ m} = 1.3 \text{ m}$$

Using the fact: 1 hg = 100 g, we can convert 400 g to hg.

$$400 \text{ g} = 400 \text{ g} \left(\frac{1 \text{ hg}}{100 \text{ g}} \right) = \frac{400}{100} \text{ hg} = 4 \text{ hg}$$

Using the fact: 1 dam = 1,000 cm, we can convert 648 cm to dam.

$$648 \text{ cm} = 648 \text{ cm} \left(\frac{1 \text{ dam}}{1,000 \text{ cm}} \right) = \underline{\hspace{2in}}$$

---

17. We did the three conversions below in the last frame. In each one, we converted from a smaller unit to a larger unit.

$$13 \text{ dm} = 1.3 \text{ m}$$
$$400 \text{ g} = 4 \text{ hg}$$
$$648 \text{ cm} = .648 \text{ dam}$$

To convert <u>from a smaller unit to a larger unit</u>, we can simply shift the decimal point <u>to the left</u>. The number of places shifted depends on the number of unit-places in the conversion. For example:

To convert 13 dm to meters, we shift <u>1 place to the left</u> because <u>m</u> is 1 unit-place to the left of <u>dm</u>.

$$13 \text{ dm} = 1\underset{\smile}{.}3 = 1.3 \text{ m}$$

To convert 400 g to hectograms, we shift <u>2 places to the left</u> because <u>hg</u> is 2 unit-places to the left of <u>g</u>.

$$400 \text{ g} = 4\underset{\smile}{.}00. = 4.00 \text{ hg}$$

To convert 648 cm to dekameters, we shift <u>3 places to the left</u> because <u>dam</u> is 3 units-places to the left of <u>cm</u>.

$$648 \text{ cm} = \underline{\hspace{1.5in}} \text{ dam}$$

[margin answer: .648 dam]

---

18. How many places to the left would we shift the decimal point for each conversion below?

a) milliliters to liters ____

b) grams to kilograms ____

c) decimeters to dekameters ____

d) millimeters to centimeters ____

[margin answer: .648. = .648 dam]

---

19. Use the decimal-point-shift method for these.

a) 775 msec = _____ sec

b) 43 m = _____ dam

c) 940 cg = _____ dg

d) 19 dm = _____ hm

[margin answers:]
a) 3 places
b) 3 places
c) 2 places
d) 1 place

20. When using the decimal-point-shift method, remember:

To convert <u>from</u> <u>a</u> <u>larger</u> <u>to</u> <u>a</u> <u>smaller</u> <u>unit</u>, we shift <u>to</u> <u>the</u> <u>right</u>. Therefore, we get <u>more</u> of the smaller unit. For example:

$$3 \text{ kg} = 3,000 \text{ g} \quad (3,000 \text{ is more than } 3)$$
$$5 \text{ m} = 500 \text{ cm} \quad (500 \text{ is more than } 5)$$

To convert <u>from</u> <u>a</u> <u>smaller</u> <u>to</u> <u>a</u> <u>larger</u> <u>unit</u>, we shift <u>to</u> <u>the</u> <u>left</u>. Therefore, we get <u>less</u> of the larger unit. For example:

$$39 \text{ m} = .39 \text{ hm} \quad (.39 \text{ is less than } 39)$$
$$644 \text{ m}\ell = .644\ell \quad (.644 \text{ is less than } 644)$$

Do these.

a) .48 kg = _____ g   c) 1.8 m = _____ cm

b) 27 m = _____ dam   d) 67 msec = _____ sec

a) .775 sec

b) 4.3 dam

c) 94 dg

d) .019 hm

21. Do these.

a) 285 mm = _____ cm   c) 95 dam = _____ km

b) 6.5 dg = _____ mg   d) 149 kg = _____ hg

a) 480 g

b) 2.7 dam

c) 180 cm

d) .067 msec

a) 28.5 cm   b) 650 mg   c) .95 km   d) 1,490 hg

6-3   ENGLISH-METRIC CONVERSIONS

In this section, we will use the unity-fraction method for English-Metric conversions.

22. The common English-Metric equivalents for length, weight, and liquid are given at the right.

Length

1 meter (m) = 39.37 inches (in)

1 inch (in) = 2.54 centimeters (cm)

1 kilometer (km) = 0.6214 mile (mi)

1 mile (mi) = 1.609 kilometers (km)

Weight (or Mass)

1 ounce (oz) = 28.35 grams (g)

1 kilogram (kg) = 2.205 pounds (lb)

Liquid Measures

1 liter (ℓ) = 1.057 quarts (qt)

1 gallon (gal) = 3.785 liters (ℓ)

Continued on following page.

22. Continued

We used the unity-fraction method to convert 7.5 in to centimeters below.

$$7.5 \text{ in} = 7.5 \text{ in}\left(\frac{2.54 \text{ cm}}{1 \text{ in}}\right) = 7.5(2.54 \text{ cm}) = 19.05 \text{ cm}$$

Do these. Report (a) to tenths and (b) to hundredths.

a) 8 oz = _____ g   b) 12 ℓ = _____ qt

---

23. To convert 40 km to miles, we can use either of these two facts: 1 km = 0.6214 mi or 1 mi = 1.609 km. We used both below, rounding to hundredths.

$$40 \text{ km} = 40 \text{ km}\left(\frac{0.6214 \text{ mi}}{1 \text{ km}}\right) = 40(0.6214 \text{ mi}) = 24.86 \text{ mi}$$

$$40 \text{ km} = 40 \text{ km}\left(\frac{1 \text{ mi}}{1.609 \text{ km}}\right) = \frac{40}{1.609} \text{ mi} = 24.86 \text{ mi}$$

Do these. Round (a) to tenths and (b) to hundredths.

a) 100 lb = _____ kg   b) 25 ℓ = _____ gal

a) 226.8 g

b) 12.68 qt

---

24. To convert meters to feet, we can use a two-step process: first converting meters to inches and then converting inches to feet. For example, we converted 6 meters to feet below.

Step 1:   Converting 6 m to inches

$$6 \text{ m} = 6 \text{ m}\left(\frac{39.37 \text{ in}}{1 \text{ m}}\right) = 6(39.37 \text{ in}) = 236.22 \text{ in}$$

Step 2:   Converting 236.22 in to feet

$$236.22 \text{ in} = 236.22 \text{ in}\left(\frac{1 \text{ ft}}{12 \text{ in}}\right) = \frac{236.22}{12} \text{ ft} = 19.685 \text{ ft}$$

We can do the conversion above in one step by multiplying by both unity fractions at the same time. We get:

$$6 \text{ m} = 6 \text{ m}\left(\frac{39.37 \text{ in}}{1 \text{ m}}\right)\left(\frac{1 \text{ ft}}{12 \text{ in}}\right) = \frac{6(39.37)}{12} \text{ ft} = 19.685 \text{ ft}$$

Use the one-step method for the conversion below. Round to tenths.

25 m = _____ yd

a) 45.4 kg

b) 6.61 gal

25. To convert yards to meters, we can use a two-step process: first converting yards to inches and then converting inches to meters. For example, we converted 50 yards to meters below.

Step 1: Converting 50 yd to inches

$$50 \text{ yd} = 50 \text{ yd}\left(\frac{36 \text{ in}}{1 \text{ yd}}\right) = 50(36 \text{ in}) = 1{,}800 \text{ in}$$

Step 2: Converting 1,800 in to meters

$$1{,}800 \text{ in} = 1{,}800 \text{ in}\left(\frac{1 \text{ m}}{39.37 \text{ in}}\right) = \frac{1800}{39.37} \text{ m} = 45.7 \text{ m}$$

We can do the conversion above in one step by multiplying by both unity fractions at the same time. We get:

$$50 \text{ yd} = 50 \text{ yd}\left(\frac{36 \text{ in}}{1 \text{ yd}}\right)\left(\frac{1 \text{ m}}{39.37 \text{ in}}\right) = \frac{50(36)}{39.37} \text{ m} = 45.7 \text{ m}$$

Use the one-step method for the conversion below. Round to tenths.

80 ft = _____ m

27.3 yd, from:

$$\frac{25(39.37)}{36} \text{ yd}$$

24.4 m, from: $\frac{80(12)}{39.37}$ m

6-4 RATES

In this section, we will discuss measures of rate in both the English and the Metric System.

26. Both velocity (or speed) and flow rate are rate measures. Each is stated as a ratio of a quantity to 1 unit of time. Two examples are discussed.

If a car travels 150 miles in 3 hours, its velocity (or speed) is the ratio of miles traveled to hours traveled. The ratio is reduced so that the velocity (or speed) is stated in terms of 1 hour. That is:

$$\frac{150 \text{ miles}}{3 \text{ hours}} \text{ is reduced to } \frac{50 \text{ miles}}{1 \text{ hour}}$$

If 60 liters of water flow into a tank in 10 minutes, the flow rate is the ratio of liters to minutes. The ratio is also reduced so that the flow rate is stated in terms of 1 minute. That is:

$$\frac{60 \text{ liters}}{10 \text{ minutes}} \text{ is reduced to } \frac{6 \text{ liters}}{1 \text{ minute}}$$

Continued on following page.

38. Very high temperatures are used for processes like melting metals. Do these conversions. Round to a whole number when necessary.

    a) 900°F = _____°C   c) 750°C = _____°F

    b) 1,500°F = _____°C   d) 1,253°C = _____°F

a) F = -22°

b) F = 19.4°

---

39. From the diagram in frame 33, you can see that the following formulas can be used for conversions between degrees-Kelvin (K) and degrees-Celsius (C).

$$K = C + 273° \qquad C = K - 273°$$

Therefore: If C = 50°,   K = 50° + 273° = 323°

           If K = 200°,   C = 200° - 273° = -73°

Do these conversions.

    a) -29°C = _____°K   c) 1,200°C = _____°K

    b) 450°K = _____°C   d) 2,000°K = _____°C

a) 482°C

b) 816°C

c) 1,382°F

d) 2,287°F

---

40. To convert 323°K to degrees-Fahrenheit, we must convert to degrees-Celsius first. Therefore, two steps are needed.

    1. Converting 323°K to degrees-Celsius

       C = K - 273° = 323° - 273° = 50°

    2. Converting 50°C to degrees-Fahrenheit

       $F = \dfrac{9C}{5} + 32° = \dfrac{9(50°)}{5} + 32° = 122°$

Using the same two-step process, convert 253°K to degrees-Fahrenheit.

    253°K = _____°C = _____°F

a) 244°K

b) 177°C

c) 1,473°K

d) 1,727°C

-20°C = -4°F

41. To convert 95°F to degrees-Kelvin, we must also convert to degrees-Celsius first. Therefore, a two-step process is needed.

   1. <u>Converting 95°F to degrees-Celsius</u>

      $$C = \frac{5F - 160°}{9} = \frac{5(95°) - 160°}{9} = 35°$$

   2. <u>Converting 35°C to degrees-Kelvin</u>

      $$K = C + 273° = 35° + 273° = 308°$$

   Using the same two-step process, convert 5°F to degrees-Kelvin.

   5°F = _____ °C = _____ °K

-15°C = 258°K

6-6 THE DECIMAL NUMBER SYSTEM AND POWERS OF TEN

In this section, we will relate the place-names in the decimal number system to powers of ten.

42. In the decimal number system, any number can be written in an expanded form in which each term is a multiplication. For example:

   $$5,617 = 5(1,000) + 6(100) + 1(10) + 7(1)$$

   $$.843 = 8(.1) + 4(.01) + 3(.001)$$

   Powers of ten can be used in the expanded forms above. That is:

   $$5,617 = 5(10^3) + 6(10^2) + 1(10^1) + 7(10^0)$$

   $$.843 = 8(10^{-1}) + 4(10^{-2}) + 3(10^{-3})$$

   Write the ordinary number corresponding to each expanded form.

   a) $3(10^3) + 9(10^2) + 4(10^1) + 8(10^0)$ = _____

   b) $7(10^1) + 5(10^0) + 0(10^{-1}) + 6(10^{-2})$ = _____

   c) $0(10^{-1}) + 0(10^{-2}) + 2(10^{-3}) + 1(10^{-4})$ = _____

a) 3,948

b) 75.06

c) .0021

43. In the diagram below, the names of the places in numbers are related to powers of ten.

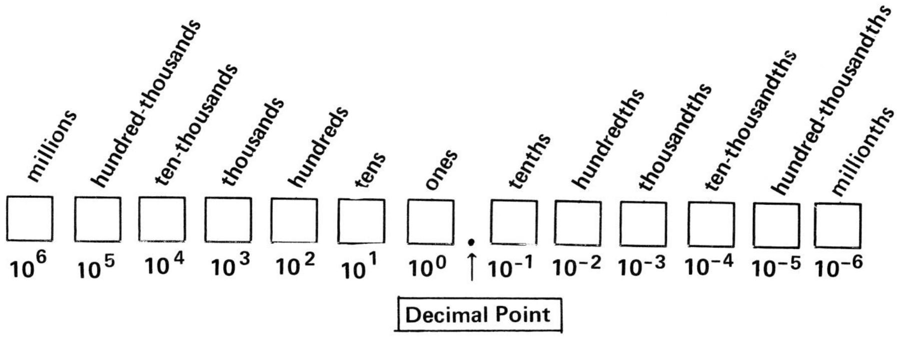

On the left side of the decimal point:

The name of the 10^0 place is <u>ones</u>.

The name of the 10^2 place is <u>hundreds</u>.

a) The name of the 10^3 place is _____ .

b) The name of the 10^6 place is _____ .

44. On the right side of the decimal point.

The name of the 10^{-1} place is <u>tenths</u>.

a) The name of the 10^{-3} place is _____ .

b) The name of the 10^{-6} place is _____ .

a) thousands

b) millions

45. Write the name for each of the following places.

a) 10^1 _____   c) 10^{-2} _____

b) 10^5 _____   d) 10^{-4} _____

a) thousandths

b) millionths

46. Write the power of ten corresponding to each place.

a) hundreds _____   c) thousandths _____

b) millions _____   d) millionths _____

a) tens
b) hundred-thousands

c) hundredths
d) ten-thousandths

a) 10^2 c) 10^{-3}

b) 10^6 d) 10^{-6}

47. Other powers of ten and place names that are sometimes used are given in the table below.

10^9	= billions	10^{-9}	= billionths
10^{12}	= trillions	10^{-12}	= trillionths

Write the name for each place.

a) 10^{12} _____   b) 10^{-9} _____

---

48. Write the power of ten corresponding to each place.

a) billions _____   b) trillionths _____

a) trillions

b) billionths

---

a) 10^9 b) 10^{-12}

6-7   METRIC PREFIXES AND POWERS OF TEN

The numerical factors for metric prefixes are usually stated as powers of ten. In this section, we will show that use of powers of ten and extend the Metric System to include prefixes with larger and smaller numerical factors.

49. The numerical factors for the metric prefixes discussed up to this point are expressed as powers of ten in the table below.

Prefix	Abbreviation	Numerical Factor	Place-Name
kilo-	k	$1000 = 10^3$	thousands
hecto-	h	$100 = 10^2$	hundreds
deka-	da	$10 = 10^1$	tens
Basic Unit →		$1 = 10^0$	ones
deci-	d	$.1 = 10^{-1}$	tenths
centi-	c	$.01 = 10^{-2}$	hundredths
milli-	m	$.001 = 10^{-3}$	thousandths

Write the power of ten related to each prefix.

a) kilo- _____   b) deka- _____   c) centi- _____

a) 10^3

b) 10^1

c) 10^{-2}

50. Write the prefix related to each power of ten.

    a) 10^2 _____ b) 10^{-1} _____ c) 10^{-3} _____

---

51. Write the place-name related to each prefix.

    a) centi- _____ b) kilo- _____

a) hecto-
b) deci-
c) milli-

---

52. Write the prefix related to each place-name.

    a) thousandths _____ b) hundreds _____

a) hundredths
b) thousands

---

53. In the table below, we have extended the Metric System to include prefixes with larger and smaller numerical factors. Because of their size, the numerical factors are expressed only as powers of ten.

a) milli-
b) hecto-

Prefix	Abbreviation	Numerical Factor	Place-Name
tera-	T	10^{12}	trillions
giga-	G	10^9	billions
mega-	M	10^6	millions
kilo-	k	10^3	thousands
hecto-	h	10^2	hundreds
deka-	da	10^1	tens
Basic Unit →		10^0	ones
deci-	d	10^{-1}	tenths
centi-	c	10^{-2}	hundredths
milli-	m	10^{-3}	thousandths
micro-	μ	10^{-6}	millionths
nano-	n	10^{-9}	billionths
pico-	p	10^{-12}	trillionths

Note: 1. Beyond kilo- and milli-, prefixes are only added for every third place.

   2. The abbreviation for micro- is μ, which is the Greek letter for m.

   3. The following pronunciations are used:

   | tera- | (ter' ah) | nano- | (nan' oh) |
   | giga- | (jig' ah) | pico- | (peek' oh) |

Continued on following page.

53. Continued

Write the power of ten related to each prefix.

a) giga- _____   b) micro- _____   c) pico- _____

---

54. Write the prefix related to each power of ten.

a) 10^{12} _____   b) 10^6 _____   c) 10^{-9} _____

a) 10^9
b) 10^{-6}
c) 10^{-12}

---

55. Write the place-name related to each prefix.

a) mega- _____   c) micro- _____

b) giga- _____   d) pico- _____

a) tera-

b) mega-

c) nano-

---

56. Write the prefix related to each place-name.

a) millions _____   c) millionths _____

b) trillions _____   d) billionths _____

a) millions
b) billions
c) millionths
d) trillionths

---

57. We wrote the abbreviations for some metric units below.

megameters = Mm   microseconds = μsec
gigagrams = Gg   picometers = pm

Write the abbreviation for each unit.

a) teragrams = _____   c) micrograms = _____

b) gigameters = _____   d) nanoseconds = _____

a) mega-   c) micro-

b) tera-   d) nano-

---

58. Write the unit that goes with each abbreviation.

a) Mm = _____   c) Tg = _____

b) μg = _____   d) psec = _____

a) Tg   c) μg

b) Gm   d) nsec

---

59. To avoid confusion with the measuring instrument called a <u>micrometer</u>, the word <u>micron</u> is used instead of <u>micrometer</u> in the Metric System. That is:

μm = micron

Write the unit that goes with each abbreviation.

a) Gm = _____   c) μm = _____

b) μsec = _____   d) ng = _____

a) megameter

b) microgram

c) teragram

d) picosecond

60. The Metric System can be extended to prefixes beyond <u>tera-</u> and <u>pico-</u>. Some additional units and their power-of-ten numerical factors are shown below.

Prefix	Abbreviation	Numerical Factor
exa-	E	10^{18}
peta-	P	10^{15}
femto-	f	10^{-15}
atto-	a	10^{-18}

a) gigameter

b) microsecond

c) micron

d) nanogram

6-8 METRIC CONVERSIONS INVOLVING LARGER AND SMALLER NUMBERS

In this section, we will show how the decimal-point-shift method can be used for conversions involving units larger than <u>kilo-</u> and smaller than <u>milli-</u>.

61. The metric prefixes are shown below with the larger ones to the left. The prefixes are similar to the places in the decimal number system.

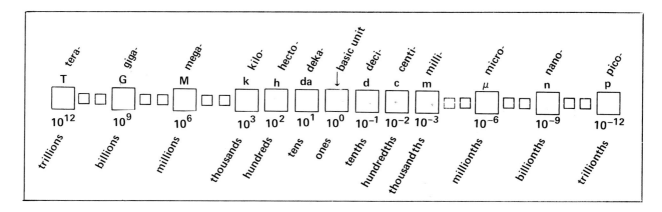

You can see these facts from the diagram.

<u>giga-</u> is larger than <u>mega-</u>

<u>milli-</u> is larger than <u>micro-</u>

Identify the <u>larger</u> prefix in each pair.

a) tera- or giga- _____

b) kilo- or mega- _____

c) micro- or nano- _____

d) pico- or nano- _____

62. You can also see these facts from the same diagram.

       <u>kilo-</u> is smaller than <u>mega-</u>

       <u>pico-</u> is smaller than <u>nano-</u>

   Identify the <u>smaller</u> prefix in each pair.

       a) mega- or giga-   c) nano- or micro-

       b) tera- or giga-   d) milli- or micro-

   a) tera-

   b) mega-

   c) micro-

   d) nano-

---

63. Identify the <u>larger</u> measure in each pair.

       a) km or Mm   b) ng or pg   c) Tm or Mm

   a) mega-
   b) giga-
   c) nano-
   d) micro-

---

64. Identify the <u>smaller</u> measure in each pair.

       a) μsec or nsec   b) Mg or Gg   c) nsec or msec

   a) Mm
   b) ng
   c) Tm

---

65. To convert from a larger unit to a smaller unit, we shift the decimal point <u>to</u> <u>the</u> <u>right</u> and get <u>more</u> of the smaller unit. For example:

   To convert from megameters to kilometers, we shift <u>3 places</u> <u>to the right</u> because <u>km</u> is 3 unit-places to the right of <u>Mm</u>.

   $$1.5 \text{ Mm} = 1.500 = 1,500 \text{ km}$$

   To convert from milliseconds to nanoseconds, we shift <u>6 places</u> <u>to the right</u> because <u>nsec</u> is 6 unit-places to the right of <u>msec</u>.

   $$38 \text{ msec} = \underline{\hspace{2cm}} \text{ nsec}$$

   a) nsec

   b) Mg

   c) nsec

---

66. Do these.

       a) 9.7 Mm = \underline{\hspace{2cm}} km

       b) 14.6 μsec = \underline{\hspace{2cm}} nsec

       c) .58 Tg = \underline{\hspace{2cm}} Mg

       d) 2.3 mm = \underline{\hspace{2cm}} nm

   38,000,000 nsec,

   from: 38.000000

---

   a) 9,700 km

   b) 14,600 nsec

   c) 580,000 Mg

   d) 2,300,000 nm

67. To convert from a smaller unit to a larger unit, we shift the decimal point to the left and get less of the larger unit. For example:

To convert from megameters to gigameters, we shift 3 places to the left because Gm is 3 unit-places to the left of Mm.

$$8.4 \text{ Mm} = {}_{\curvearrowleft}008.4 = .0084 \text{ Gm}$$

To convert from picoseconds to microseconds, we shift 6 places to the left because μsec is 6 unit-places to the left of psec.

475 psec = _____ μsec

---

68. Do these.

a) 255 Mm = _____ Gm   c) 1,400 kg = _____ Gg

b) 88 μsec = _____ msec   d) 980 pm = _____ μm

.000475 μsec,

from: ${}_{\curvearrowleft}000475{}_{\curvearrowright}$

---

69. When using the decimal-point-shift method, remember these facts.

    1. To convert from a larger to a smaller unit, we shift to the right.

    2. To convert from a smaller to a larger unit, we shift to the left.

Do these.

a) 7.5 nsec = _____ psec   c) 66 ng = _____ mg

b) 480 km = _____ Mm   d) .029 Gm = _____ km

a) .255 Gm

b) .088 msec

c) .0014 Gg

d) .00098 μm

---

70. The same method can be used for conversions involving a basic unit. For example:

To convert from seconds to microseconds, we shift 6 places to the right.

$$75 \text{ sec} = 75.000000_{\curvearrowright} = 75,000,000 \text{ μsec}$$

To convert from grams to megagrams, we shift 6 places to the left.

$$93 \text{ g} = {}_{\curvearrowleft}000093. = .000093 \text{ Mg}$$

Do these.

a) .63 Mm = _____ m   c) 45 sec = _____ msec

b) 2.9 g = _____ kg   d) 3,500 μsec = _____ sec

a) 7,500 psec

b) .48 Mm

c) .000066 mg

d) 29,000 km

---

a) 630,000 m   b) .0029 kg   c) 45,000 msec   d) .0035 sec

SELF–TEST 21 (pages 251–262)

1. Room temperature is 68°F. Convert to degrees-Celsius.

_____

2. Gold melts at 1,064°C. Convert to degrees-Fahrenheit. Round to a whole number.

_____

3. Nitrogen liquifies at -210°C. Convert to degrees-Kelvin.

_____

Write the <u>metric</u> <u>prefix</u> related to each power of ten.

4. 10^3 _____

5. 10^{-3} _____

6. 10^{-9} _____

Write the <u>metric</u> <u>prefix</u> related to each place-name.

7. millions _____

8. hundredths _____

9. millionths _____

Do these conversions. Use the diagram inside the front cover of the text.

10. 29 mg = _____ g

11. 7.14 km = _____ m

12. 3,180 nsec = _____ μsec

13. .37 Tm = _____ Mm

14. 85.8 μg = _____ mg

15. .045 mm = _____ μm

16. 8 μsec = _____ psec

17. 1.5 Mm = _____ Gm

18. 640 cg = _____ dag

ANSWERS:

1. 20°C
2. 1,947°F
3. 63°K
4. kilo-
5. milli-

6. nano-
7. mega-
8. centi-
9. micro-
10. .029 g

11. 7,140 m
12. 3.18 μsec
13. 370,000 Mm
14. .0858 mg
15. 45 μm

16. 8,000,000 psec
17. .0015 Gm
18. .64 dag

6-9   PRECISION, LIMITS, ABSOLUTE ERROR

In this section, we will discuss the precision, lower and upper limits, and absolute error of measurements.

---

71.   Counting is a measurement process that leads to <u>exact</u> numbers. For example:

        If we count the number of students in a class and get 29, the number 29 is exact. We know the number is not 28 or 30.

However, the typical measurement process in science and technology leads to only <u>approximate</u> numbers because measuring instruments are used.   For example:

        If we measure the length of a steel bar and get 9.3 in, the number 9.3 is approximate. Using better and more expensive instruments, we could get any of the following values for the same length.

            9.28 in
            9.287 in
            9.2874 in
            9.28746 in
            9.287461 in

But regardless of the instrument used, the measurement is only approximate because we can, at least in principle, use an instrument that allows us to report one more decimal place.

If we measure the weight of an object and report 5.67 kg, is the 5.67 <u>exact</u> or <u>approximate</u>? _____

---

72.   When we use the word <u>measurement</u> in this text, we will always mean an approximate number given by a measuring instrument. The <u>precision</u> of a measurement is determined by the place of the last meaningful digit in the measurement.   For example:

        15 ft is precise to <u>ones</u>.

        12.8 ft is precise to <u>tenths</u>.

        6.04 ft is precise to <u>hundredths</u>.

State the precision of each measurement.

    a) 5.7 ℓ _____   c) 0.0149 in _____

    b) 0.128 g _____   d) 0.005624 sec _____

---

approximate

---

a) tenths

b) thousandths

c) ten-thousandths

d) millionths

73. When a final 0 or 0's are given after the decimal point in a measurement, they indicate the precision of the measurement. For example:

    10.0 sec is precise to <u>tenths</u>.

    6.50 sec is precise to <u>hundredths</u>.

    0.7500 sec is precise to <u>ten-thousandths</u>.

    Since the final 0 or 0's above indicate the precision of the measurement, they cannot be dropped. State the precision of each measurement.

    a) 7.0 in _____   c) 25.00 ℓ _____

    b) 5.070 g _____   d) 0.0800 sec _____

---

74. To indicate a precision to one of the final 0's before the decimal point, we will underline that 0. If no 0 is underlined, the precision is the place of the last non-zero digit. For example:

    34,000 mi is precise to <u>thousands</u>.

    34,<u>0</u>00 mi is precise to <u>hundreds</u>.

    34,0<u>0</u>0 mi is precise to <u>tens</u>.

    State the precision of each measurement.

    a) 1,250,000 mi _____   c) 9,7<u>0</u>0,000 kg _____

    b) 675,<u>0</u>00 km _____   d) 7,1<u>0</u>0 mi _____

a) tenths

b) thousandths

c) hundredths

d) ten-thousandths

---

75. Measurements of the same quantity can be compared in terms of their precision. The farther to the right the place of the final digit, the more precise the measurement. For example:

    6.75 ft is more precise than 12.8 ft,
      because 6.75 is precise to <u>hundredths</u>.

    15 sec is less precise than 5.40 sec,
      because 15 sec is only precise to <u>ones</u>.

    a) Which measurement has the <u>greatest</u> precision? _____

       20.5 ft   0.088 ft   13.60 ft

    b) Which measurement has the <u>least</u> precision? _____

       7.0 sec   127 sec   42.06 sec

a) ten-thousands

b) hundreds

c) thousands

d) tens

---

a) 0.088 ft

b) 127 sec

76. a) Which measurement has the greatest precision? _____

   675,000 mi   24,000 mi   109,000 mi

   b) Which measurement has the least precision? _____

   540,000 kg   600,000 kg   480,000 kg

a) 24,000 mi

b) 480,000 kg

77. Though a measurement is not exact, it does tell us something about the actual value of the measurement. For example:

   If a weight measurement is reported as 27 g, we know that the actual value lies between 26.5 g and 27.5 g.

   If a length measurement is reported as 43.8 cm, we know that the actual value lies between 43.75 cm and 43.85 cm.

26.5 g and 27.5 g are called the lower limit and the upper limit of the weight measurement. The largest possible error that can occur is called the absolute error. The absolute error is 0.5 g, since:

   26.5 g is 0.5 g less than 27 g.

   27.5 g is 0.5 g more than 27 g.

For the length measurement of 43.8 cm:

   a) the lower limit is _____

   b) the upper limit is _____

   c) the absolute error is _____

78. The only non-zero digit in the absolute error of a measurement is a 5. In measurements precise to places to the right of the decimal point, the 5 is always in the place immediately to the right of the final digit of the measurement. That is:

Measurement	Absolute Error	
25 g	0.5	(5 in tenths place)
9.7 sec	0.05	(5 in hundredths place)
1.84 ℓ	0.005	(5 in thousandths place)

To find the lower and upper limits for 9.7 sec, we subtract and add the absolute error to the measurement. We get:

   Lower limit = 9.7 - 0.05 = 9.65 sec

   Upper limit = 9.7 + 0.05 = 9.75 sec

Find the limits for the other two measurements.

Measurement	Lower Limit	Upper Limit
a) 25 g	_____	_____
b) 1.84 ℓ	_____	_____

a) 43.75 cm

b) 43.85 cm

c) 0.05 cm

79. The absolute errors for some more-precise measurements are given below. Notice again that the 5 appears in the place immediately to the right of the final digit of the measurement.

Measurement	Absolute Error	
0.425 cm	0.0005	(5 in <u>ten-thousandths</u> place)
0.0926 oz	0.00005	(5 in <u>hundred-thousandths</u> place)
0.00282 sec	0.000005	(5 in <u>millionths</u> place)

To find the lower and upper limits for 0.0926 oz, we subtract and add 0.00005 to the measurement. We get:

Lower limit = 0.0926 - 0.00005 = 0.09255 oz

Upper limit = 0.0926 + 0.00005 = 0.09265 oz

Find the limits of the other two measurements.

Measurement	Lower Limit	Upper Limit
a) 0.425 cm	_____	_____
b) 0.00282 cm	_____	_____

a) 24.5   25.5

b) 1.835   1.845

---

80. The absolute errors for some measurements precise to places to the left of the decimal point are given below. Notice the place of the 5.

Measurements	Absolute Error	
17,600 kg	50	(5 in <u>tens</u> place)
283,000 mi	500	(5 in <u>hundreds</u> place)

We found the lower and upper limits for 17,600 kg below.

Lower limit = 17,600 - 50 = 17,550 kg

Upper limit = 17,600 + 50 = 17,650 kg

Find the lower and upper limits for 283,000 mi.

a) Lower limit = _____

b) Upper limit = _____

a) 0.4245 0.4255

b) 0.002815 0.002825

---

a) 282,500 mi

b) 283,500 mi

81. Measurements are sometimes reported with their absolute error attached as we have done below. The ± symbol means that the absolute error is both added and subtracted to find the limits.

$$250 \pm 5 \text{ mi}$$

$$17 \pm 0.5 \text{ mi}$$

$$7.9 \pm 0.05 \text{ g}$$

$$5.33 \pm 0.005 \text{ lb}$$

Insert the absolute error in each measurement below.

a) 13 ± _____ oz   c) 17,800 ± _____ mi

b) 0.69 ± _____ ℓ   d) 0.00205 ± _____ sec

a) ±0.5   b) ±0.005   c) ±50   d) ±0.000005

6-10   SIGNIFICANT DIGITS

In this section, we will discuss the meaning of <u>significant</u> <u>digits</u> in measurements. We will also discuss rounding to a certain number of significant digits. To simplify the instruction, we will not use measuring units in this section.

82. The <u>significant</u> <u>digits</u> in a measurement are the <u>meaningful</u> <u>digits</u> in the measurement. All non-zero digits in a measurement are significant. For example:

17.4 contains <u>three</u> significant digits.

35.96 contains <u>four</u> significant digits.

State the number of significant digits in each measurement.

a) 47 _____   b) 1.35 _____   c) 639.6 _____

83. Any 0 between non-zero digits in a measurement is significant. For example:

20.6 contains <u>three</u> significant digits.

9.005 contains <u>four</u> significant digits.

State the number of significant digits in each measurement.

a) 307 _____   b) 1.08 _____   c) 40.03 _____

a) two

b) three

c) four

a) three

b) three

c) four

84. Any 0 before the first non-zero digit in a measurement is not significant. For example:

   0.75 contains <u>two</u> significant digits.

   0.0813 contains <u>three</u> significant digits.

   State the number of significant digits in these.

   a) 0.419 _____   b) 0.053 _____   c) 0.00808 _____

---

85. Final 0's after the decimal point are included to indicate the precision of the measurement. Therefore, they are significant. For example:

   3.0 contains <u>two</u> significant digits.

   5.900 contains <u>four</u> significant digits.

   State the number of significant digits in these.

   a) 15.0 _____   b) 0.060 _____   c) 0.001500 _____

   a) three
   b) two
   c) three

---

86. Final 0's before the decimal point may or may not be significant. It depends on the precision of the measurement. For example:

   350,000 contains <u>two</u> significant digits.

   35<u>0</u>,000 contains <u>three</u> significant digits.

   350,<u>0</u>00 contains <u>four</u> significant digits.

   State the number of significant digits in these.

   a) 45,000 _____   b) 609,<u>0</u>00 _____   c) 40,<u>0</u>00 _____

   a) three
   b) two
   c) four

---

87. When a measurement has a 0 or 0's after the decimal point, all 0's immediately to the left of the decimal point are significant. For example:

   40.0 contains <u>three</u> significant digits.

   900.00 contains <u>five</u> significant digits.

   State the number of significant digits in these.

   a) 170.0 _____   b) 80.00 _____   c) 50.0 _____

   a) two
   b) four
   c) three

---

   a) four
   b) four
   c) three

88. The statements applying to significant digits in measurements are summarized below.

Significant Digits

1. All non-zero digits are significant.

   Example: 17.6 has three significant digits.

2. 0's lying between non-zero digits are significant.

   Example: 50.08 has four significant digits.

3. 0's before the first non-zero digit are not significant.

   Example: 0.0624 has three significant digits.

4. Final 0's after the decimal point are significant.

   Example: 9.100 has four significant digits.

5. Final 0's before the decimal point may or may not be significant, depending on the precision of the measurement.

   Example: 36,000 has two significant digits.
   36,<u>0</u>00 has three significant digits.

6. If there is a 0 or 0's after the decimal point, all 0's immediately to the left of the decimal point are significant.

   Example: 90.00 has four significant digits.

State the number of significant digits in these.

a) 0.83 _____   e) 65.000 _____

b) 5.70 _____   f) 0.0903 _____

c) 0.0025 _____   g) 176,000 _____

d) 160.0 _____   h) 9,00<u>0</u>,000 _____

89. As we shall see in a later section, some answers are rounded to a certain number of significant digits. To round to <u>three</u> significant digits, we round to the place of the third digit from the left. For example:

$$\downarrow$$
14,766   rounds to   14,800

$$\downarrow$$
0.9813   rounds to   0.981

Round to <u>three</u> significant digits.

a) 61,358 _____   c) 701,029 _____

b) 29.4166 _____   d) 0.0098472 _____

a) two   e) five

b) three   f) three

c) two   g) three

d) four   h) four

90. We rounded each measurement below to <u>two</u> significant digits.

         ↓
1,856.75 rounds to 1,900

       ↓
0.06513 rounds to 0.065

Round to <u>two</u> significant digits.

   a) 90,740,000 _____   b) 0.001099 _____

a) 61,400

b) 29.4

c) 701,000

d) 0.00985

---

91. We rounded each measurement below to <u>four</u> significant digits.

       ↓
609,088 rounds to 609,100

       ↓
0.298317 rounds to 0.2983

Round to <u>four</u> significant digits.

   a) 15.00625 _____   b) 5,940.33 _____

a) 91,000,000

b) 0.0011

---

92. We rounded each measurement below to <u>three</u> significant digits. Notice that a final 0 or 0's must be retained.

       ↓
17.034 rounds to 17.0

       ↓
0.049975 rounds to 0.0500

   a) Round 0.06049 to <u>two</u> significant digits. _____

   b) Round 124.963 to <u>four</u> significant digits. _____

   c) Round 1.99518 to <u>three</u> significant digits. _____

   d) Round 0.700035 to <u>four</u> significant digits. _____

a) 15.01   b) 5,94<u>0</u>

---

93. Sometimes we have to drop a final 0 to get the right number of significant digits. For example:

Rounding 0.99753 to <u>two</u> significant digits, we get 1.0 .

   <u>Note</u>: We had to drop a final 0 from 1.00 to get two significant digits.

   a) Round 9.9984 to <u>three</u> significant digits. _____

   b) Round 0.999956 to <u>four</u> significant digits. _____

a) 0.060

b) 125.0

c) 2.00

d) 0.7000

---

a) 10.0   b) 1.000

6-11   RELATIVE ERROR AND ACCURACY

In this section, we will define the relative error and accuracy of measurements. We will show how the number of significant digits in measurements is roughly related to their relative errors and accuracy.

---

94.   The two lengths below were measured to tenths of an inch.

   A steel block.   The reported measurement is 7.1 in.

   A driveway.   The reported measurement is 1,080.6 in.

Though both measurements have the same precision, it is more difficult to measure a long length like 1,080.6 in with only a 0.05 in absolute error than to measure a short length like 7.1 in with a 0.05 in absolute error.   In that sense, the longer measurement is "better".   Relative error and accuracy are concepts used to describe that meaning of the word "better".

---

95.   The relative error of a measurement is a comparison of the absolute error with the measurement itself.   That is:

$$\text{Relative Error} = \frac{\text{Absolute Error}}{\text{Measurement}}$$

Since the absolute error of 5.0 in is 0.05 in,

the relative error of 5.0 in is $\frac{0.05 \text{ in}}{5.0 \text{ in}} = 0.01$.

Since the absolute error of 50.0 in is 0.05 in,

the relative error of 50.0 in is $\frac{0.05 \text{ in}}{50.0 \text{ in}} =$ _____

---

96.   Since relative error is a comparison of two quantities, it is a ratio. Like any ratio, there are no units attached when the ratio is converted to a decimal number.

   a) Find the relative error of 47 cm.   Round to four decimal places.

   Relative Error = _____ = _____

   b) Find the relative error of 8.14 cm.   Round to six decimal places.

   Relative Error = _____ = _____

0.001

---

a) $\frac{0.5 \text{ cm}}{47 \text{ cm}} = 0.0106$

b) $\frac{0.005 \text{ cm}}{8.14 \text{ cm}}$

   $= 0.000614$

97. We can compare two measurements by comparing their relative errors. The one with the smaller relative error is "better" because its absolute error is a smaller part of the measurement itself. For example:

    The relative error of 5.0 in is 0.01 .

    The relative error of 50.0 in is 0.001 .

    Which measurement is "better" because it has a smaller relative error? _____

---

98. Instead of using the word "better" when comparing relative errors, we use the word accurate. We say:

    The measurement with the smaller relative error is more accurate.

    The measurement with the larger relative error is less accurate.

    a) Since the relative error (0.001) of 50.0 in is smaller than the relative error (0.01) of 5.0 in, we say that 50.0 in is _____ accurate than 5.0 in.

    b) Since the relative error (0.0106) of 47 cm is larger than the relative error (0.000614) of 8.14 cm, we say that 47 cm is _____ accurate than 8.14 cm.

50.0 in

---

99. Relative error is related to the number of significant digits in a measurement. The greater the number of significant digits, the smaller the relative error. For example:

Measurement	Significant Digits	Relative Error
9.3 ft	two	0.005376
51.8 ft	three	0.000965
742.6 ft	four	0.000067

Note that 742.6 ft has the greatest number of significant digits and the smallest relative error. It therefore has the greatest accuracy.

   a) Which measurement has the smallest number of significant digits? _____

   b) Which measurement has the least accuracy? _____

a) more

b) less

---

100. Using the fact that the measurement with more significant digits is always more accurate, identify the more accurate measurement in each pair below.

   a) 250.75 cm or 19.22 cm   c) 0.0059 g or 4.05 g

   b) 2.77 ft or 0.1825 ft   d) 0.700 volt or 0.0097 volt

a) 9.3 ft

b) 9.3 ft

---

a) 250.75 cm   c) 4.05 g

b) 0.1825 ft   d) 0.700 volt

101. The two measurements 16.4 cm and 5.28 cm have the same number of significant digits.

   The relative error of 16.4 cm is 0.00305 .

   The relative error of 5.28 cm is 0.00095 .

   When two measurements have the same number of significant digits, we must compare the "sequence" of significant digits in each to determine which is more accurate. The one whose "sequence" of significant digits is larger is more accurate.

   The sequence of significant digits in 16.4 cm is "164".

   The sequence of significant digits in 5.28 cm is "528".

   a) Which sequence of digits is larger? _____

   b) Therefore, which measurement is more accurate? _____

   c) Is that fact confirmed by the relative errors above? _____

102. By comparing the sequences of significant digits, identify the more accurate measurement in each pair.

   a) 508 ft or 2.95 ft   b) 0.309 sec or 0.0382 sec

a) "528"

b) 5.28 cm

c) Yes

103. When two measurements have the same number and sequence of significant digits, they have the same accuracy. To confirm that fact, let's compare the relative errors of 2.5 cm and 0.025 cm.

   a) The relative error of 2.5 cm is $\dfrac{0.05 \text{ cm}}{2.5 \text{ cm}}$ = _____ .

   b) The relative error of 0.025 cm is $\dfrac{0.0005 \text{ cm}}{0.025 \text{ cm}}$ = _____ .

   c) Do the two measurements have the same accuracy? _____

a) 508 ft

b) 0.0382 sec

104. The following statements relate the precision and accuracy of measurements.

   1. Measurements with the same precision do not usually have the same accuracy.

      45.72 ft is more accurate than 3.14 ft.

   2. If two measurements have different precisions, the more precise is not always the more accurate.

      99.8 ℓ is more accurate than 2.54 ℓ .

   3. Two measurements with different precisions can have the same accuracy.

      2.6 m and 0.26 m have the same accuracy.

a) 0.02

b) 0.02

c) Yes

Continued on following page.

104. Continued

Given these measurements: 159.3 g   0.675 g   2.03 g   0.0069 g

    a) The <u>most</u> <u>accurate</u> is _____ .

    b) The <u>least</u> <u>accurate</u> is _____ .

    c) The <u>most</u> <u>precise</u> is _____ .

    d) The <u>least</u> <u>precise</u> is _____ .

---

105. The precision of a measurement is determined by the place of the last significant digit. For example:

    2.47 sec is precise to hundredths.

    0.946 g is precise to thousandths.

The word <u>accurate</u> is commonly misused in making statements about precision. For example, phrases like "accurate to hundredths" or "accurate to thousandths" are used. That use of the word <u>accurate</u> is always incorrect because accurate refers to relative error, not to precision.

Answer "true" or "false" for these.

    a) _____ The measurement 14.5 g is precise to tenths.

    b) _____ The measurement 0.725 sec is accurate to thousandths.

a) 159.3 g

b) 0.0069 g

c) 0.0069 g

d) 159.3 g

---

106. Though the word <u>accurate</u> should not be used to describe the precision of measurements, the phrase "accurate to a certain number of significant digits" is frequently used. For example:

We say: 37.5 ft is accurate to three significant digits.

    a) 0.080 sec is accurate to _____ significant digits.

    b) 6.140 lb is accurate to _____ significant digits.

a) true

b) false

---

a) two   b) four

6-12   OPERATIONS WITH MEASUREMENTS

When doing operations with measurements, we must be careful to report the answer either to a definite precision or accuracy. We will discuss the rules for reporting those answers in this section.

107. When adding or subtracting measurements with the same precision, we use this rule:

> WHEN ADDING (OR SUBTRACTING) MEASUREMENTS WITH THE SAME PRECISION, THE SUM (OR DIFFER-ENCE) SHOULD HAVE THE SAME PRECISION AS THE MEASUREMENTS.

Example 1:   The sum below should be reported to hundredths.

2.46 cm + 4.12 cm = 6.58 cm

Example 2:   The difference below should be reported to tenths. We kept the "0" in the tenths place to report the proper precision.

36.7 g - 15.7 g = 21.0 g

Use the rule for these:

a) 2.65 sec + 7.66 sec + 0.28 sec = _____

b) 4.45 ℓ - 1.05 ℓ = _____

c) 0.344 g + 0.077 g - 0.121 g = _____

108. When adding or subtracting measurements with different precisions, we use this rule:

> WHEN ADDING (OR SUBTRACTING) MEASUREMENTS WITH DIFFERENT PRECISIONS, THE PRECISION OF THE SUM (OR DIFFERENCE) SHOULD BE THE SAME AS THAT OF THE LEAST PRECISE MEASUREMENT.

Two examples of the above rule are given below.

Example 1:   Since 2.6 in is precise to tenths, the sum is rounded to tenths.

2.6 in + 3.78 in = 6.38 in = 6.4 in

Example 2:   Since 0.18 sec is precise to hundredths, the difference is rounded to hundredths. The final 0 must be retained.

0.2829 sec - 0.18 sec = 0.1029 sec = 0.10 sec

Use the rule for these:

a) 18 m - 2.15 m = _____

b) 0.2489 g + 1.24 g + 0.418 g = _____

c) 0.031 sec + 2.95 sec + 0.0153 sec = _____

a) 10.59 sec

b) 3.40 ℓ

c) 0.300 g

109. Two examples involving larger measurements are shown below.

    Example 1: Since the sum is precise to thousands, we underlined the 0 in that place.

        37,000 mi + 153,000 mi = 19<u>0</u>,000 mi

    Example 2: Since 17,600 g is precise to hundreds, we rounded the difference to hundreds.

        17,600 g - 5,942 g = 11,658 g = <u>11,700 g</u>

Do these: a) 547,000 km + 88,700 km = _____

        b) 42,600 lb - 12,600 lb = _____

a) 16 m

b) 1.91 g

c) 3.00 sec

---

110. The following rule is used when multiplying measurements.

> WHEN MULTIPLYING MEASUREMENTS, THE PRODUCT SHOULD CONTAIN THE SAME NUMBER OF SIGNIFICANT DIGITS AS THERE ARE IN THE FACTOR <u>WITH THE LEAST NUMBER OF SIGNIFICANT DIGITS</u>.

    Example 1: Since both factors contain <u>three</u> significant digits, the product should be rounded to <u>three</u> significant digits.

        17.5 ft x 12.5 ft = 218.75 ft^2 = <u>219 ft^2</u>

    Example 2: Since 8.6 cm contains <u>two</u> significant digits, the product should be rounded to two significant digits.

        9.66 cm x 8.6 cm x 17.25 cm = 1,433.061 cm^3

                            = <u>1,400 cm^3</u>

Assuming that the numbers below are measurements, report each product with the correct number of significant digits.

    a) 5.66 x 9.17 = _____   c) 0.200 x 0.30 = _____

    b) 2.629 x 0.404 = _____   d) 48 x 56.47 = _____

a) 636,000 km

b) 30,<u>0</u>00 lb

---

111. Assuming that the numbers below are measurements, report each product with the correct number of significant digits.

    a) 1.79 x 64.8 x 0.925 = _107.2426_

    b) 346 x 7.0 x 15.88 = _38161.36_

    c) 7.8 x 0.925 x 18.0 = _____

a) 51.9   c) 0.060

b) 1.06   d) 2,700

---

a) 107

b) 38,000

c) 130

112. The same rule is used for division of measurements.

> WHEN DIVIDING MEASUREMENTS, THE QUOTIENT SHOULD CONTAIN THE SAME NUMBER OF SIGNIFICANT DIGITS AS THERE ARE IN THE TERM WITH THE LEAST NUMBER OF SIGNIFICANT DIGITS.

Example 1: Since both numbers contain three significant digits, the quotient is rounded to three significant digits.

$$\frac{65.6 \text{ cm}}{2.48 \text{ cm}} = 26.5$$

Example 2: Since 99 sec contains two significant digits, the quotient is rounded to two significant digits.

$$\frac{7.045 \text{ sec}}{99 \text{ sec}} = 0.071$$

Assuming that the numbers below are measurements, report each quotient with the correct number of significant digits.

a) $\frac{0.784}{2.49}$ = _____   c) $\frac{6.20}{9.00}$ = _____

b) $\frac{64.9}{7.6}$ = _____   d) $\frac{483.7}{0.216}$ = _2239_ 2240

113. The square or square root of a measurement should contain the same number of significant digits as the measurement. For example:

$$(2.56 \text{ cm})^2 = 6.55 \text{ cm}^2 \quad (\text{reported \underline{three} significant digits})$$

$$\sqrt{85 \text{ in}^2} = 9.2 \text{ in} \quad (\text{reported \underline{two} significant digits})$$

Assuming that the numbers below are measurements, report each answer with the correct number of significant digits.

a) $(21.8)^2$ = _____   d) $\sqrt{457}$ = _____

b) $(0.095)^2$ = _____   e) $\sqrt{.01980}$ = _____

c) $(899)^2$ = _____   f) $\sqrt{7,900,000}$ = _____

a) 0.315   c) 0.689

b) 8.5   d) 2,240

114. In any combined operation with measurements, the answer should contain as many significant digits as the measurement with the least number of significant digits. For example:

$$\frac{(1.547)(6.03)}{0.917} = 10.2 \quad (\underline{three} \text{ significant digits})$$

Assuming that the numbers below are measurements, report each answer with the correct number of significant digits.

a) $\frac{4.70}{(86.3)(6.2)}$ = _____   c) $\sqrt{\frac{1.480}{6.61}}$ = _____

b) $(3.14)(8.77)^2$ = _____

a) 475

b) 0.0090

c) 808,000

d) 21.4

e) 0.1407

f) 2,800

115. The same rule applies when the combined operation contains an addition or subtraction, as long as it also contains a multiplication, division, squaring, or square root. For example:

$$\frac{16.5 + 71.6}{14.3} = 6.16 \qquad (\underline{three} \text{ significant digits})$$

Assuming that all numbers are measurements, do these.

   a) $\dfrac{17.8 + 21.6}{14.9} = $ _____   b) $\sqrt{(15.9)^2 + (27.1)^2} = $ _____

a) 0.0088

b) 242

c) 0.473

a) 2.64   b) 31.4

SELF-TEST 22 (pages 263-278)

State the precision of each measurement.

  1. 5.030 in _____

  2. 0.000176 sec _____

State the number of significant digits in each measurement.

  3. 0.0902 g _____

  4. 14.800 m _____

Length measurements: 0.372 cm   106.45 cm   9.010 cm   0.0053 cm   81.9 cm

Which measurement has the:

  5. greatest precision? _____

  6. least precision? _____

  7. greatest accuracy? _____

  8. least accuracy? _____

Do Problems 9-18 on a calculator. All numbers are measurements. Report each answer in the correct form.

  9. $4.710 \times 0.0836 = $ _____

 10. $0.4084 + 0.0065 + 0.086 = $ _____

 11. $(6.527)^2 = $ _____

 12. $\dfrac{4.150}{80.47} = $ _____

 13. $\dfrac{26.309}{4.117 \times 0.5740} = $ _____

 14. $53.9 \times 0.0062 \times 8.140 = $ _____

 15. $45.72 - 27.7 = $ _____

 16. $\sqrt{(2.28)^2 + (4.57)^2} = $ _____

 17. $\dfrac{0.0504 \times 89}{73.12} = $ _____

 18. $\dfrac{37.4 + 19.8}{6.51} = $ _____

ANSWERS:

1. thousandths	5. 0.0053 cm	9. 0.394	13. 11.13	17. 0.061
2. millionths	6. 81.9 cm	10. 0.501	14. 2.7	18. 8.79
3. three	7. 106.45 cm	11. 42.60	15. 18.0	
4. five	8. 0.0053 cm	12. 0.05157	16. 5.11	

SUPPLEMENTARY PROBLEMS - CHAPTER 6

Assignment 20

Do these conversions.

    1. 16.5 ft = _____ in   2. 400 oz = _____ lb   3. 3.75 mi = _____ ft

    4. 14 qt = _____ gal   5. 12,320 yd = _____ mi 6. 1.8 min = _____ sec

7. A car weighs 3,180 lb. Convert this weight to tons.

8. A research balloon rose 138,400 ft. Convert this height to miles. Round to a whole number.

Do these conversions.

    9. 6,500 m = _____ km   10. 57 cg = _____ g   11. 400 mm = _____ m

    12. 15 g = _____ hg   13. 0.82 ℓ = _____ mℓ   14. 6.1 kg = _____ dag

    15. 2.5 msec = _____ sec   16. 2.8 hm = _____ dm   17. 8 dam = _____ cm

18. The width of a photographic film is 35 millimeters. Convert this width to centimeters.

19. A radioactive burst lasted 0.00217 second. Convert this time to milliseconds.

Do these conversions.

    20. Round to tenths.   21. Round to thousandths.   22. Round to hundredths.

       10 km = _____ mi   2,690 sec = _____ hr   7.50 lb = _____ kg

    23. Round to hundredths.   24. Round to a whole number.   25. Round to tenths.

       4.193 in = _____ cm   5.8 gal = _____ ℓ   500 g = _____ oz

26. The diameter of a gear is 20.6 cm. Convert this diameter to inches, rounded to hundredths.

27. A marathon race is 26.2 miles. Convert this distance to kilometers, rounded to tenths.

Do these conversions.

    28. 4.7 gal/sec = ____ gal/min   29. 9 m/sec = ____ km/hr   30. 8.4 m/min = ____ cm/sec

31. Convert 230 liters per minute to liters per second. Round to tenths.

32. Convert 100 kilometers per hour to miles per hour. Round to tenths.

33. Convert 900 liters per minute to gallons per second. Round to a whole number.

34. The speed of light is 2.998×10^{10} cm/sec. Convert this speed to kilometers per second.

Assignment 21

Do these temperature conversions.

1. 68°F = _____ °C   2. 1,832°F = _____ °C   3. -15°C = _____ °F

4. 100°C = _____ °F   5. -140°C = _____ °F   6. 23°F = _____ °C

7. 300°K = _____ °C   8. -210°C = _____ °K   9. 122°F = _____ °K

Write the power of ten related to each metric prefix.

10. mega- _____   11. pico- _____   12. tera- _____   13. milli- _____

Write the metric prefix related to each power of ten.

14. 10^{-6} _____   15. 10^3 _____   16. 10^{-9} _____   17. 10^9 _____

Write the metric prefix related to each place-name.

18. hundredths _____   19. hundreds _____   20. tenths _____

Do these metric conversions.

21. .43 Mm = _____ km   22. .055 msec = _____ μsec   23. 98 kg = _____ Mg

24. 549 μm = _____ mm   25. .035 Gm = _____ km   26. 850 Gg = _____ Tg

27. .015 msec = _____ nsec 28. 7,500 psec = _____ μsec   29. .059 Tm = _____ Mm

30. .44 μsec = _____ psec 31. 450 g = _____ kg   32. 450 g = _____ Mg

33. 6,000,000 μsec = ____sec 34. .019 Mm = _____ m   35. 88 sec = _____ μsec

36. Iron melts at 2,795°F. Convert to degrees-Celsius.

37. The duration time of a radar pulse is 0.38 microsecond. Convert this time to nanoseconds.

38. The wavelength of an X-ray is 50 angstroms. An angstrom is a unit of length equal to 1 x 10^{-10} meter. Express this wavelength in picometers.

Assignment 22

State the precision of each measurement.

1. 21.8 liters   2. 0.00843 sec   3. 6.950 grams   4. 118.00 meters   5. 762,000 miles

In each pair, which measurement has the greater precision?

6. 2.730 in and 0.0195 in   7. 382 km and 29.6 km   8. 0.00420 g and 0.0753 g

State the number of significant digits in each measurement.

9. 37.5 in   10. 0.019 sec   11. 4.610 cm   12. 8,250.4 ft   13. 0.00700 g

In each pair, which measurement has the greater accuracy?

14. 344 cm and 8.5 cm   15. 0.280 sec and 0.067 sec   16. 41.93 in and 2.708 in

For the five time measurements below, which measurement has the:

17. greatest accuracy?   18. greatest precision?   19. least accuracy?   20. least precision?

4.817 sec   0.016 sec   62.7 sec   1.5130 sec   349 sec

For the five weight measurements below, which measurement has the:

21. greatest precision?   22. least precision?   23. greatest accuracy?   24. least accuracy?

725.7 g   0.0310 g   54.96 g   18.705 g   9.62 g

Do the remaining problems on a calculator. All numbers are measurements. Report each answer in the correct form.

25. 2.618 + 3.72

26. 0.712 + 0.850 + 0.0564

27. 5.86 + 12.1 + 0.034

28. 12.36 - 7.86

29. 0.0887 + 15.59 - 5.604

30. 0.974 - 0.08863 + 1.5350

31. 6.193 x 8.70

32. 0.3059 x 0.72817

33. 1.38 x 93.16 x 0.029

34. $\dfrac{75.38}{24.26}$

35. $\dfrac{0.0080}{18.7}$

36. $\dfrac{573}{0.165}$

37. $\dfrac{0.3258}{6.79}$

38. $(21.44)^2$

39. $(0.0680)^2$

40. $\sqrt{0.053}$

41. $\sqrt{7,839}$

42. $\dfrac{409.5 \times 0.376}{32.8}$

43. $\dfrac{5.819}{0.474 \times 8.6}$

44. $\dfrac{41.63 \times 0.05290}{0.3874 \times 7.0358}$

45. $\sqrt{(1.85)^2 + (3.26)^2}$

46. $87 \times (0.414)^2$

47. $\dfrac{12.285 + 36.274}{2.142}$

7 Geometry

In this chapter, we will review some common geometric figures and facts. We will solve some problems involving similar triangles and review the common formulas for perimeter, area, volume, and surface area. The basic units for area and volume in both the English System and the Metric System are discussed. A brief discussion of density is included.

7-1 PLANE FIGURES

In this section, we will define some plane figures and state some basic facts related to them.

1. An <u>angle</u> is generated by rotating a half-line about its endpoint in a counterclockwise direction.

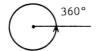
One Revolution

Straight Angle

Right Angle

One complete revolution is an angle of 360 <u>degrees</u>, written 360°.

A <u>straight</u> <u>angle</u> is 180°. It is one half of a revolution.

A <u>right</u> <u>angle</u> is 90°. It is one fourth of a revolution, and is marked with a small square ⌐ .

Continued on following page.

1. Continued

   Angle ABC, written ∠ABC, is a right angle.

   Since lines AB and BC form a right angle, we say that they are <u>perpendicular</u>.

   Since AB and BC intersect at point B, point B is called the <u>vertex</u> of ∠ABC.

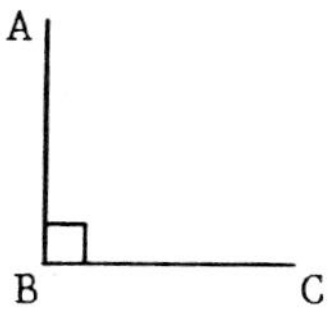

   a) A 180° angle is called a _____ angle.

   b) A 90° angle is called a _____ angle.

   c) Two lines that form a right angle are said to be _____.

---

2. An <u>acute</u> angle is an angle less than 90°. An <u>obtuse</u> angle is an angle between 90° and 180°.

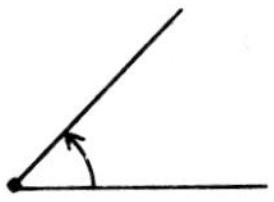

   Acute Angle 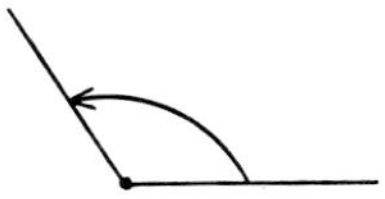   Obtuse Angle

   a) A 50° angle is called an _____ angle.

   b) A 125° angle is called an _____ angle.

   a) straight

   b) right

   c) perpendicular

---

3. <u>Supplementary angles</u> are two angles whose sum is 180°. <u>Complementary angles</u> are two angles whose sum is 90°.

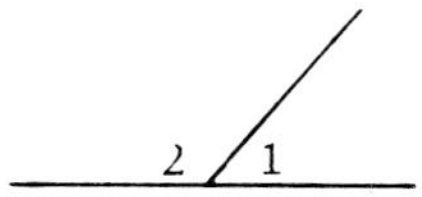

   Supplementary Angles 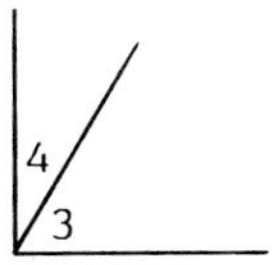   Complementary Angles

   a) At the left, if ∠1 = 50°, ∠2 = _____°.

   b) At the right, if ∠3 = 60°, ∠4 = _____°.

   a) acute

   b) obtuse

---

   a) 130°

   b) 30°

4. Adjacent angles have a common vertex and a common side between
    them. Vertical angles are equal angles "across" the point of inter-
    section of two lines.

    Adjacent Angles   Vertical Angles

    a) ∠1 and ∠2 are adjacent angles. They are also _____
       angles.

    b) ∠3 and ∠5 are vertical angles. Name another pair of vertical
       angles. _____ and _____

---

5. If two lines in a plane never intersect,
    they are parallel. If a line crosses two
    parallel lines, it is called a transversal.
    When a transversal crosses two parallel
    lines, certain pairs of angles are equal.

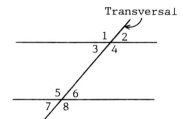

        Corresponding angles are equal.
    That is:

            ∠1 = ∠5
            ∠2 = ∠6
            ∠3 = ∠7
            ∠4 = ∠8

    Alternate interior angles are equal. That is:
            ∠3 = ∠6
            ∠4 = ∠5

    In the figure above, if ∠1 = 130°, give the sizes of the other angles.

    ∠1 = 130°   ∠5 = _____
    ∠2 = _____   ∠6 = _____
    ∠3 = _____   ∠7 = _____
    ∠4 = _____   ∠8 = _____

a) supplementary

b) ∠4 and ∠6

---

6. A polygon is a plane closed figure bounded by straight line segments.
    Polygons are named according to the number of sides they contain.

            A triangle has three sides.
            A quadrilateral has four sides.
            A pentagon has five sides.
            A hexagon has six sides.

∠1 = 130° ∠5 = 130°
∠2 = 50°   ∠6 = 50°
∠3 = 50°   ∠7 = 50°
∠4 = 130° ∠8 = 130°

Continued on following page.

6. Continued

Three special types of triangles are defined below.

Equilateral   Isosceles   Right

In an <u>equilateral</u> <u>triangle</u>, the three sides are equal and the three angles are equal (60° each).

In an <u>isosceles</u> <u>triangle</u>, two sides are equal. The angles opposite the equal sides are equal ($\angle$1 and $\angle$2 are equal).

In a <u>right</u> <u>triangle</u>, one angle is a right angle. The side opposite the right angle is called the <u>hypotenuse</u>. The other two sides are called <u>legs</u>.

This triangle is a right, isosceles triangle.
Sides b and c are equal.

a) The hypotenuse is side _____.

b) The legs are sides _____ and _____.

c) Since $\angle$1 = 45°, $\angle$2 = _____.

	a) d
7. <u>The</u> <u>sum</u> <u>of</u> <u>the</u> <u>three</u> <u>angles</u> <u>of</u> <u>any</u> <u>triangle</u> <u>is</u> 180°. By subtracting the sum of the two known angles from 180°, find the size of the unknown angle in each triangle.	b) b and c
	c) 45°

a)   b)

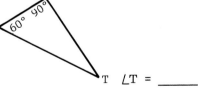

$\angle$A = _____   T $\angle$T = _____

	a) A = 65°
8. The <u>Pythagorean</u> <u>Theorem</u> states the following relationship among the three sides of a right triangle.	b) T = 30°

> IN ANY RIGHT TRIANGLE, THE SQUARE OF THE LENGTH OF THE HYPOTENUSE EQUALS THE SUM OF THE SQUARES OF THE LENGTHS OF THE TWO LEGS.

Continued on following page.

8.   Continued

For this right triangle, the
Pythagorean Theorem says:

$$c^2 = a^2 + b^2$$

We found $\underline{c}$ when a = 3 and b = 4 below.
Find $\underline{c}$ when a = 6 and b = 8.

$c^2 = a^2 + b^2$	$c^2 = a^2 + b^2$
$c^2 = 3^2 + 4^2$	
$c^2 = 9 + 16$	
$c^2 = 25$	
$c = 5$	

---

9.   Three special types of quadrilaterals are defined below.

c = 10

   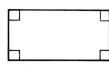

Parallelogram   Rectangle   Square

A parallelogram is a quadrilateral whose opposite sides are both
parallel and equal. Opposite angles of a parallelogram are also
equal.

A rectangle is a parallelogram with four right angles.

A square is a rectangle with four equal sides.

The figure at the right is a parallelogram.

a) How long is side $\underline{a}$? _____

b) How large is $\angle 1$? _____

c) How large is $\angle 2$? _____

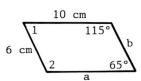

---

10.   Two more special types of quadrilaterals are defined below.

a) 10 cm

b) 65°

c) 115°

   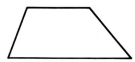

Rhombus   Trapezoid

Continued on following page.

10. Continued

A <u>rhombus</u> is a parallelogram with four equal sides.

A <u>trapezoid</u> is a quadrilateral with only one pair of parallel sides.

The figure at the right is a rhombus.

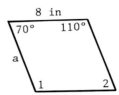

    a) How long is side <u>a</u>? _____

    b) How large is ∠1? _____

    c) How large is ∠2? _____

---

11. The sum of the four angles of any quadrilateral is 360°. Using that fact, find the size of angle D in each quadrilateral. The figure in (b) is a parallelogram.

a)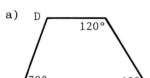
    ∠D = _____

b)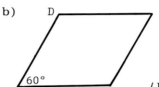
    ∠D = _____

a) 8 in

b) 110°

c) 70°

---

12. A <u>circle</u> is the set of all points at the same distance from a fixed point called the <u>center</u>. A <u>radius</u> is a line from the center to any point on the circle. A <u>diameter</u> is a line through the center that touches two points on the circle. In the circle below:

Point 0 is the <u>center</u>.

0A, 0B, and 0C are <u>radii</u> (pronounced "ray-dee-eye"). All radii of the same circle are equal.

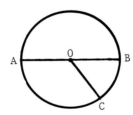

A0B is a <u>diameter</u>. All diameters of the same circle are equal.

A diameter <u>d</u> is twice the radius <u>r</u>. That is: $\boxed{d = 2r}$. Use that formula for these.

    a) Find the diameter of a circle if its radius is 10 cm.

    b) Find the radius of a circle if its diameter is 12 ft.

        d = _____
        r = _____

a) ∠D = 110°

b) ∠D = 120°

a) d = 20 cm

b) r = 6 ft

13.   Five more definitions related to a circle are given below.

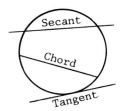 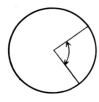

Central Angle   Inscribed Angle

A chord is a line whose endpoints are on the circle. (A diameter is a chord that passes through the center of the circle.)

A secant is a line that intersects the circle at two points.

A tangent is a line that touches the circle at one point.

A central angle is an angle formed by two radii, with its vertex at the center of the circle.

An inscribed angle is an angle formed by two chords, with its vertex on the circle.

7-2   SIMILAR TRIANGLES

In this section, we will define similar triangles and solve some problems involving similar triangles.

14.   Two triangles are similar if their corresponding angles are equal and their corresponding sides are proportional. The two triangles below are similar. Using △ for "triangle" and ~ for "similar to", we write:

$$\triangle ABC \sim \triangle DEF$$

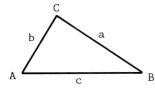 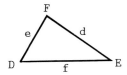

$$\angle A = \angle D, \quad \angle B = \angle E, \quad \angle C = \angle F$$

$$\frac{a}{d} = \frac{b}{e} = \frac{c}{f}$$

In the triangles above:

   a) If $\angle A = 60°$, $\angle D$ = _____   b) If $\angle F = 85°$, $\angle C$ = _____

15. The two right triangles below are similar. Therefore: $\frac{a}{d} = \frac{b}{e} = \frac{c}{f}$.

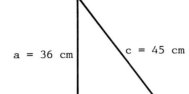

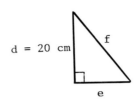

a = 36 cm   c = 45 cm   d = 20 cm   f

b = 27 cm   e

We used a proportion below to find the length of <u>e</u>. Use the other proportion to find the length of <u>f</u>.

$$\frac{a}{d} = \frac{b}{e} \qquad\qquad \frac{a}{d} = \frac{c}{f}$$

$$\frac{36}{20} = \frac{27}{e}$$

$$\cancel{20}e\left(\frac{36}{\cancel{20}}\right) = 20\cancel{e}\left(\frac{27}{\cancel{e}}\right)$$

$$36e = 540$$

$$e = \frac{540}{36}$$

$$e = 15 \text{ cm}$$

a) 60°   b) 85°

f = 25 cm

---

16. In the figure at the right, DE is parallel to BC. Therefore, △ADE ~ △ABC since their corresponding angles are equal. That is:

$$\angle A = \angle A$$
$$\angle 1 = \angle 3$$
$$\angle 2 = \angle 4$$

Since the two triangles are similar:

$$\frac{AD}{AB} = \frac{AE}{AC} = \frac{DE}{BC}$$

If AD = 70 m, AB = 100 m, and DE = 84 m, use a proportion to find the length of BC.

BC = 120 m, from:

$$\frac{70}{100} = \frac{84}{BC}$$

17. In the figure at the right, FG is parallel to DE. Therefore, △CFG ~ △CDE since their corresponding angles are equal. That is:

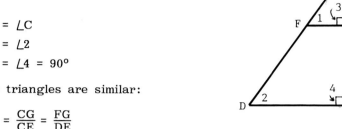

$$\angle C = \angle C$$

$$\angle 1 = \angle 2$$

$$\angle 3 = \angle 4 = 90°$$

Since the two triangles are similar:

$$\frac{CF}{CD} = \frac{CG}{CE} = \frac{FG}{DE}$$

If CF = 20 ft, CG = 16 ft, and CE = 48 ft, use a proportion to find the length of CD.

CD = 60 ft, from: $\frac{20}{CD} = \frac{16}{48}$

7-3   PERIMETER OF PLANE FIGURES

In this section, we will review the formulas for the perimeter of various plane figures.

---

18. In a <u>rectangle</u>:

The longer sides are called the <u>length</u> (L).

The shorter sides are called the <u>width</u> (W).

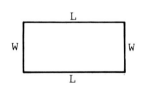

The <u>perimeter</u> (P) of a plane figure is the distance around it. We can find the perimeter of a rectangle by adding the lengths of the four sides (L + W + L + W). Or we can double its length and width and then add. That is:

$$P = 2L + 2W$$

Using the formula, find the perimeter of a rectangle whose length is 5 cm and whose width is 3 cm.

P = 16 cm

19. The perimeter of a <u>square</u> is s + s + s + s. That is:

$$P = 4s$$

Use the formula to find the perimeter of a square whose side is 9.5 ft.

20. The perimeter of a <u>rhombus</u> is also s + s + s + s. That is:

$$P = 4s$$

Find the perimeter of a rhombus whose side is 2.75 m.

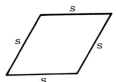

P = 38 ft

21. Find the perimeter of each figure by adding the lengths of the sides.

P = 11 m

a)

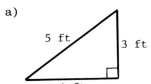

P = _____

b)

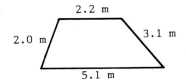

P = _____

22. The distance around a circle is called its <u>circumference</u> (C). By breaking the circle below at A and bending its circumference into a straight line, we can see that its circumference is <u>slightly</u> <u>more</u> <u>then</u> <u>3 times</u> as long as its diameter.

a) P = 12 ft

b) P = 12.4 m

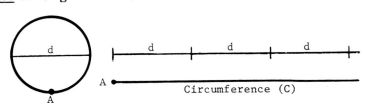

To find the ratio of the circumference of a circle to its diameter, we divide C by <u>d</u>. For any circle, the ratio of C to <u>d</u> is slightly more than 3. The exact value is the irrational number π (pronounced "pie") which equals 3.141592... To confirm that fact, press the $\boxed{\pi}$ key on your calculator. Therefore:

$$\frac{C}{d} = \pi \qquad \text{or} \qquad \frac{C}{d} = 3.141592...$$

Continued on following page.

22. Continued

Rearranging to solve for C, we get the following circumference formula.

$$C = \pi d$$

Use that formula to do these.

a) Find the circumference of a circle whose diameter is 4.28 ft.   Round to tenths.

b) Find the diameter of a circle whose circumference is 14.9 cm.   Round to hundredths.

---

23. Since d = 2r, we can also get a circumference formula in terms of the radius.   That is:

$$C = \pi d$$
$$C = \pi(2r)$$
$$C = 2\pi r$$

Use $C = 2\pi r$ for these.

a) Find the circumference of a circle whose radius is 2.64 cm.   Round to tenths.

b) Find the radius of a circle whose circumference is 23.7 m.   Round to hundredths.

a) C = 13.4 ft

b) d = 4.74 cm

---

24. Some perimeter and circumference formulas are given below.

Perimeter (P) and Circumference (C)	
Rectangle	P = 2L + 2W
Square	P = 4s
Rhombus	P = 4s
Circle	C = πd or C = 2πr

a) C = 16.6 cm

b) r = 3.77 m

7-4   AREA OF PLANE FIGURES

In this section, we will review the formulas for the area of various plane figures.

25. The area of a plane figure is the number of <u>unit squares</u> needed to fill it. Some <u>unit squares</u> are shown below.

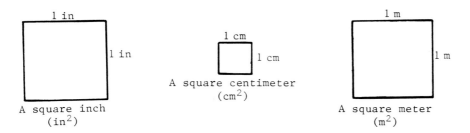

The abbreviations for some area measures are given below.

"10 square centimeters" is written 10 cm^2.

"35 square feet" is written 35 ft^2.

Write the abbreviation for each area measure.

a) 25 square meters = _____   b) 84 square yards = _____

---

26. It takes 6 square centimeters to fill the rectangle at the right. Therefore, its area (A) is 6 cm^2. Notice that we can find its area by multiplying its length (L) and width (W). That is:

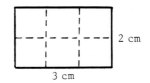

$$6 \text{ cm}^2 = (3 \text{ cm})(2 \text{ cm})$$

Therefore, we get the following formula for the area of a rectangle.

$$A = LW$$

Use the formula for these:

a) Find the area of a rectangle whose length is 3.4 ft and whose width is 2.7 ft. Round to tenths.

b) Find the width of a rectangle whose area is 32.6 m^2 and whose length is 6.86 m. Round to two decimal places.

---

27. Since a square is a rectangle whose sides (s) are equal, the length and width of a square are equal. Therefore, the area of a square is <u>s</u> times <u>s</u>. We get the following formula:

$$A = s^2$$

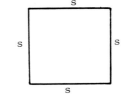

Continued on following page.

---

a) 25 m^2

b) 84 yd^2

---

a) A = 9.2 ft^2

b) W = 4.75 m

27. Continued

Use the formula for these.

a) Find the area of a square
whose side is 18.2 in.
Round to a whole number.

b) Find the side of a square
whose area is 4.8 km².
Round to tenths.

---

28. The symbols " and ' are also used as abbreviations for <u>inches</u> and
<u>feet</u>. For example:

9"   means   9 inches   6'   means   6 feet

Find the area of each figure below.   Round (a) to tenths and (b) to
hundredths.

a)

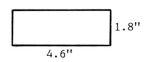

4.6"

b)

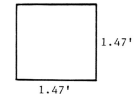

1.47'

A = _____   A = _____

a) A = 331 in²

b) s = 2.2 km

---

29. The <u>height</u> (or <u>altitude</u>) of a parallelogram is the distance between two
opposite sides.   It is found by measuring the length of a perpendicular
between the opposite sides.   The perpendicular can be drawn at any
point.   For example, we drew three heights (h_1, h_2, and h_3) in the
parallelogram below.   The three heights are equal.   The side to which
they are drawn is called their <u>base</u> (b).

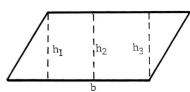

The area of a parallelogram is equal to its base <u>b</u> times its height <u>h</u>.
To show that fact, we can cut off part of the parallelogram and use it
to form a rectangle as we have done below.

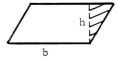 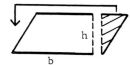

Continued on following page.

a) A = 8.3 in²

b) A = 2.16 ft²

29. Continued

Since the area of the parallelogram equals the area of the rectangle, we use the following formula for the area of a parallelogram.

$$A = bh$$

Using the formula, find the area of each parallelogram. Round to tenths.

a)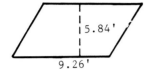

5.84'

9.26'

b)

6.23 m

12.1 m

A = _____

A = _____

---

30. A <u>height</u> of a triangle is the length of a line drawn from the vertex of an angle perpendicular to the opposite side. The side to which the height is drawn is called its <u>base</u>. Since a height can be drawn from any of the three angles in a triangle to the opposite side, there are three <u>height</u>-<u>base</u> pairs in any triangle. For example, we drew the three heights for the triangle below.

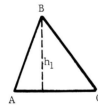 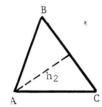

The base for height h_1 is side AC.

   a) The base for height h_2 is side _____.

   b) The base for height h_3 is side _____.

a) A = 54.1 ft²

b) A = 75.4 m²

---

31. The area of a triangle equals one-half of any base times the height drawn to that base. That is:

$$A = \frac{1}{2}bh \qquad or \qquad A = \frac{bh}{2}$$

a) BC

b) AB

Continued on following page.

31. Continued

As you can see from the figure at the right, the formula makes sense because a triangle is simply half a parallelogram. Since the area of the parallelogram is <u>bh</u>, the area of the triangle is half of <u>bh</u>.

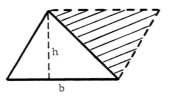

Using the form of the formula at the right above, find the area of these triangles. Round to tenths.

a)

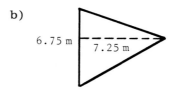

6.5"

4.5"

b)

6.75 m

7.25 m

A = _____

A = _____

---

32. Sometimes we have to extend a side (or base) in order to draw the height to it. For example, we extended side AC below so that a perpendicular could be drawn to it at D. When a side is extended in order to draw a height to it, <u>the original side and not the extended side</u> is the base of that height. Therefore, the area formula for the triangle is:

$$A = \frac{1}{2}(AC)(BD)$$

If AC = 8 cm, AD = 10 cm, and BD = 7 cm, find the area of the triangle.

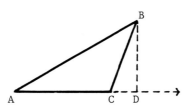

a) A = 14.6 in²

b) A = 24.5 m²

---

33. In the triangle at the right:

a) The height is _____.

b) The base is _____.

c) The area is _____.

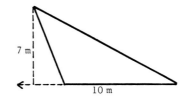

7 m

10 m

A = 28 cm², from:

$$\frac{1}{2}(8)(7)$$

---

34. In the triangle at the right:

a) The height is _____.

b) The base is _____.

c) The area is _____.
Round to thousandths.

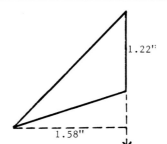

1.22"

1.58"

a) 7 m

b) 10 m

c) 35 m²

35. In a right triangle, each leg is the <u>height</u> to the other leg. That is:

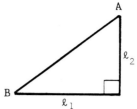

If a height is drawn from angle A to leg ℓ_1, it is identical to leg ℓ_2.

If a height is drawn from angle B to leg ℓ_2, it is identical to leg ℓ_1.

Since the area of a triangle is one half the base times the height, the area of a right triangle is one half the product of the two legs. That is:

$$A = \frac{1}{2}\ell_1\ell_2 \qquad or \qquad A = \frac{\ell_1\ell_2}{2}$$

Using the formula, find the area of each right triangle. Round to tenths.

a)

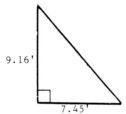

A = _____

b)

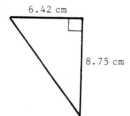

A = _____

a) 1.58 in

b) 1.22 in

c) 0.964 in²

36. The following formula can be used to find the area of a trapezoid.

$$A = \frac{1}{2}h(b_1 + b_2) \qquad or \qquad A = \frac{h(b_1 + b_2)}{2}$$

where: b_1 and b_2 are the parallel sides, and <u>h</u> is the height of the trapezoid.

We can see that the above formula makes sense by analyzing the trapezoid below.

The area of triangle I is: $\frac{1}{2}hb_1$.

The area of triangle II is: $\frac{1}{2}hb_2$.

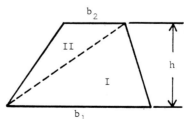

The total area of the trapezoid is the sum of the areas of the two triangles. That is:

$$A = \frac{1}{2}hb_1 + \frac{1}{2}hb_2 = \frac{1}{2}h(b_1 + b_2) \qquad or \qquad \frac{h(b_1 + b_2)}{2}$$

a) A = 34.1 ft²

b) A = 28.1 cm²

Continued on following page.

36. Continued

Using the formula, find the area of each trapezoid below. Round (a) to tenths and (b) to a whole number.

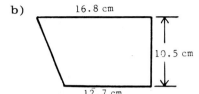

a)

6.5'
4.5'
8.5'

A = _____

b)

16.8 cm
10.5 cm
12.7 cm

A = _____

---

37. The formula below gives the area of a <u>circle</u> in terms of its radius.

$$A = \pi r^2$$

Use the formula for these.

a) Find the area of a circle whose radius is 2.48 cm. Round to tenths.

b) Find the radius of a circle whose area is 6.95 ft². Round to hundredths.

a) 42.3 ft²

b) 155 cm²

---

38. Since $r = \frac{d}{2}$, we can also get an area formula in terms of the diameter. That is:

$$A = \pi r^2$$

$$A = \pi \left(\frac{d}{2}\right)^2$$

$$A = \pi \left(\frac{d^2}{4}\right)$$

$$A = \frac{\pi d^2}{4}$$

Use the formula for these.

a) Find the area of a circle whose diameter is 16.5 in. Round to a whole number.

b) Find the diameter of a circle whose area is 37.3 m². Round to hundredths.

a) A = 19.3 cm²

b) r = 1.49 ft

39. Some area formulas are given below.

Area (A)	
Rectangle	A = LW
Square	A = s²
Parallelogram	A = bh
Triangle	$A = \dfrac{bh}{2}$
Right Triangle	$A = \dfrac{(\text{leg 1})(\text{leg 2})}{2}$
Trapezoid	$A = \dfrac{h(b_1 + b_2)}{2}$
Circle	$A = \pi r^2$ or $A = \dfrac{\pi d^2}{4}$

a) A = 214 in²

b) d = 6.89 m

SELF-TEST 23 (pages 282-300)

The figure has two parallel lines and a transversal. Find those angles.

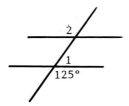

125°

1. ∠1 = _____

2. ∠2 = _____

3. In triangle ABC, DE is parallel to BC. Also, AE = 26 cm, AC = 39 cm, and BC = 54 cm. Find the length of DE.

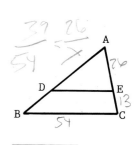

_____

4. Find the side of a square whose area is 184 ft². Round to tenths. _____

A rectangle's length is 17.4 in and its width is 11.8 in.

5. Find its perimeter.

_____

6. Find its area. Round to a whole number.

_____

A circle's radius is 2.75 m.

7. Find its circumference. Round to tenths.

_____

8. Find its area. Round to tenths.

_____

Continued on following page.

(SELF-TEST 23 (pages 282-300) - Continued

9. Find the area of this parallelogram. Round to hundredths.

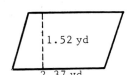

1.52 yd

2.37 yd

_____

10. Find the area of this triangle. Round to hundreds.

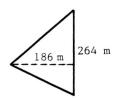

186 m   264 m

_____

11. Find the area of this right triangle. Round to tens.

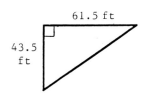

61.5 ft

43.5 ft

_____

12. Find the area of this trapezoid. Round to a whole number.

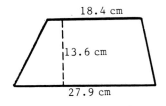

18.4 cm

13.6 cm

27.9 cm

_____

ANSWERS:
1. 55°	5. 58.4 in	9. 3.60 yd²
2. 125°	6. 205 in²	10. 24,600 m²
3. 36 cm	7. 17.3 m	11. 1,340 ft²
4. 13.6 ft	8. 23.8 m²	12. 315 cm²

7-5   COMPOSITE FIGURES

Composite figures are figures that can be divided into more basic geometric figures.   We will find the area of some composite figures in this section.

40.   The figures below are called <u>composite</u> figures.   The dotted lines divide them into more basic figures.

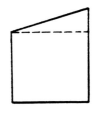

Figure 1

Figure 2

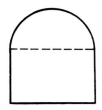

Figure 3

a) Figure 1 is divided into a _____ and a _____.

b) Figure 2 is divided into two _____.

c) Figure 3 is divided into a _____ and a _____.

41. To find the area of the composite figure below, we divide it into two rectangles, find the area of each rectangle, and then add the two areas. Two possible ways of dividing the same figure into rectangles are shown.

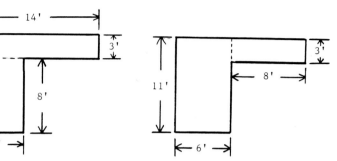

If we compute the total area of the figure at the left, we get:

(3' x 14') + (6' x 8') = 42 ft² + 48 ft² = 90 ft²

a) If we compute the total area of the figure at the right, we get:

(3' x 8') + (6' x 11') = _____ + _____ = _____

b) Did we get the same total area both ways? _____

a) triangle and a square

b) rectangles

c) rectangle and a semi-circle (or half-circle)

---

42. The I-shaped figure at the right can be divided into three rectangles. Do so and compute the area of the total figure.

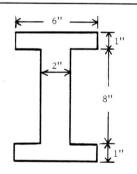

a) 24 ft² + 66 ft²
    = 90 ft²

b) Yes

---

43. The composite figure on the right is divided into a triangle and a rectangle.

a) The area of the triangle is _____.

b) The area of the rectangle is _____.

c) The total area is

_____.

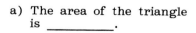
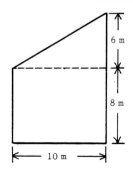

28 in²

44. The figure at the right involves both a <u>square</u> and a <u>circle</u>. To find the area of the shaded part, <u>we must subtract the area of the circle from the area of the square</u>. Find the shaded area by answering the questions below. Round to tenths.

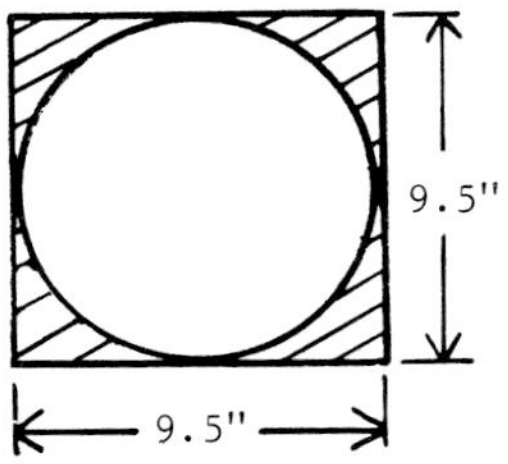

   a) For the square, A = _____.

   b) The diameter of the circle is 9.5" (the same length as the side of the square). Therefore, for the circle, A = _____.

   c) For the shaded part, A = _____ - _____ = _____

   a) 30 m²

   b) 80 m²

   c) 110 m²

45. To find the shaded area at the right, we must subtract the area of the circle from the area of the rectangle. Round to tenths.

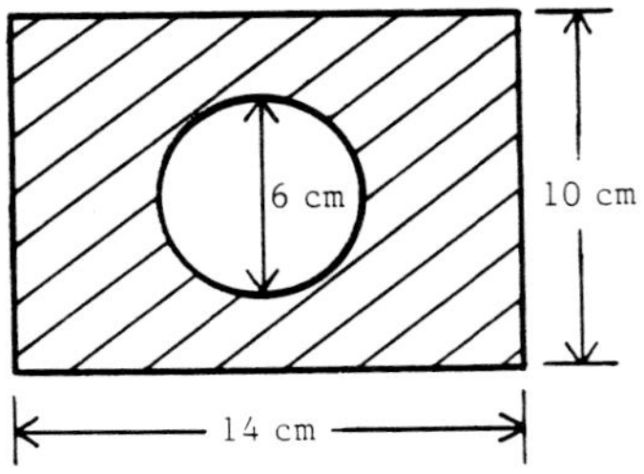

   a) The area of the rectangle is _____.

   b) The area of the circle is _____.

   c) The area of the shaded part is _____.

   a) 90.3 in²

   b) 70.9 in²

   c) 19.4 in², from:
   90.3 in²-70.9 in²

46. At the right, a semi-circular arch with a 10-foot diameter has been cut out of a rectangle. The area of the semi-circle is half the area of a circle with the same diameter. Let's find the area of the shaded part. Round to tenths.

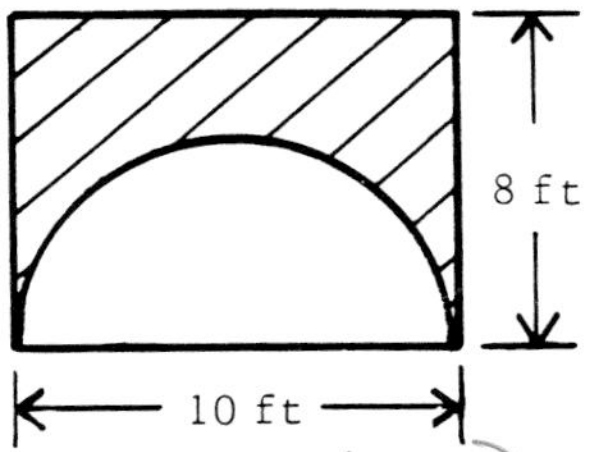

   a) The area of the rectangle is _____.

   b) The area of the semi-circle is _____. $A = \frac{\pi d^2}{4}$ (.5)

   c) The area of the shaded part is _____.

   a) 140 cm²

   b) 28.3 cm²

   c) 111.7 cm²

47. The shaded figure on the right is a cross-section of a large pipe. Two circles are involved. The outer diameter is 17"; the inner diameter is 16".

   To find the area of the cross-section, <u>we subtract the area of the smaller circle from the area of the larger circle</u>. Round to a whole number.

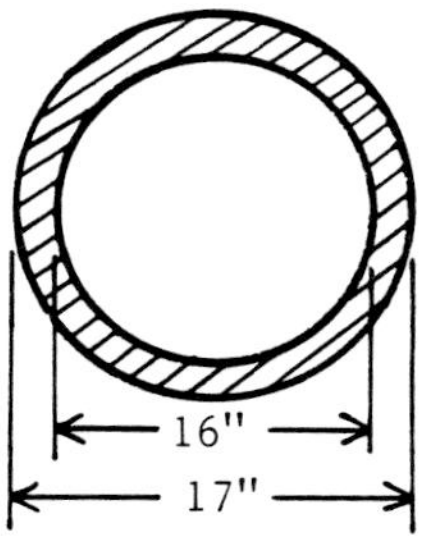

   a) The area of the outer circle is _____.

   b) The area of the inner circle is _____.

   c) The area of the cross-section is _____.

   a) 80 ft²

   b) 39.3 ft²

   c) 40.7 ft²

48. The composite figure at the right is divided into a rectangle and a semi-circle. Find its area. Round to tenths.

a) The area of the rectangle is _____.

b) The area of the semi-circle is _____.

c) The total area is _____.

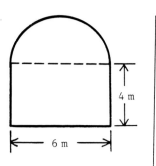

a) 227 in²

b) 201 in²

c) 26 in²

---

a) 24 m²   b) 14.1 m²   c) 38.1 m²

7-6 UNITS OF AREA

In this section, we will discuss the conversion facts for the basic units of area in both the English System and the Metric System.

---

49. Since 12 in = 1 ft, the figure at the right is 1 square foot. We converted 1 ft² to square inches below by substituting 12 in for 1 ft.

$$1 \text{ ft}^2 = (1 \text{ ft})(1 \text{ ft})$$
$$= (12 \text{ in})(12 \text{ in})$$
$$= 144 \text{ in}^2$$

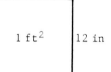

Since 3 ft = 1 yd, the area of the square at the right is 1 square yard. Complete the conversion of 1 yd² to square feet below.

$$1 \text{ yd}^2 = (1 \text{ yd})(1 \text{ yd})$$
$$= (3 \text{ ft})(3 \text{ ft})$$
$$= \underline{\hspace{1cm}} \text{ ft}^2$$

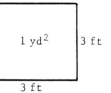

---

50. The basic conversion facts for units of area in the English System are:

9 ft²

1 square foot (ft²)	=	144 square inches (in²)
1 square yard (yd²)	=	9 square feet (ft²)
1 square mile (mi²)	=	640 acres

We used the unity-fraction method to convert 5 mi² to acres below.

$$5 \text{ mi}^2 = 5 \text{ mi}^2 \left(\frac{640 \text{ acres}}{1 \text{ mi}^2} \right) = 5(640 \text{ acres}) = 3,200 \text{ acres}$$

Complete these conversions.

a) 10 ft² = _____ in²   b) 10 yd² = _____ ft²

51. Complete these. Round to hundredths.

    a) 500 in² = _____ ft²   b) 1,000 acres = _____ mi²

a) 1,440 in²

b) 90 ft²

---

52. Since 1,000 m = 1 km, the figure at the right is 1 square kilometer. We converted 1 km² to square meters below by substituting <u>1,000 m</u> for <u>1 km</u>.

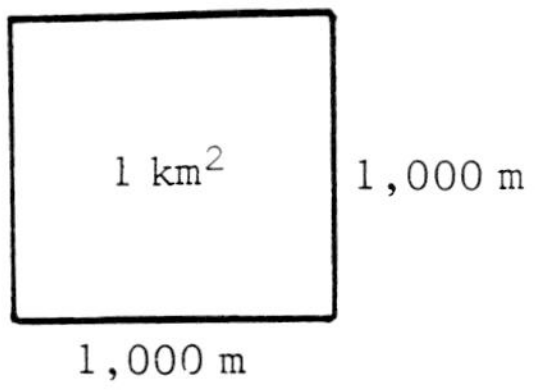

    1 km² = (1 km)(1 km)
          = (1,000 m)(1,000 m)
          = 1,000,000 m²

Since 100 cm = 1 m, the area of the figure at the right is 1 square meter. Complete the conversion of 1 m² to square centimeters below.

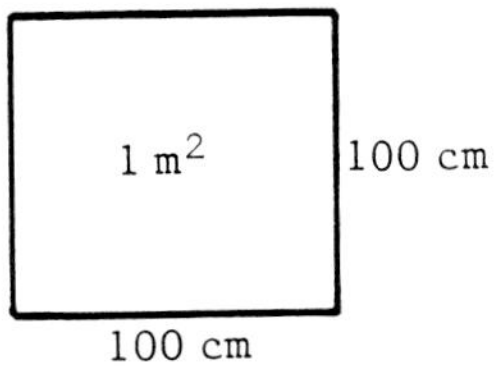

    1 m² = (1 m)(1 m)

       = (100 cm)(100 cm)

       = _____ cm²

a) 3.47 ft²

b) 1.56 mi²

---

53. The basic conversion facts for units of area in the Metric System are:

> 1 square meter (m²) = 10,000 square centimeters (cm²)
>
> 1 square kilometer (km²) = 1,000,000 square meters (m²)
>
> 1 square kilometer (km²) = 100 hectares (ha)

    <u>Note</u>: A <u>hectare</u> (pronounced "hect-air") is similar to an <u>acre</u> in the English System.

Use the unity-fraction method for these:

a) 5m² = _____ cm²   b) 9 km² = _____ hectares

10,000 cm²

---

54. Complete these.

    a) 5,000 cm² = _____ m²   b) 750 hectares = _____ km²

a) 50,000 cm²

b) 900 hectares

---

a) .5 m²

b) 7.5 km²

55. The basic conversion facts relating the English and Metric Systems are:

$$
\begin{aligned}
1 \text{ in}^2 &= 6.452 \text{ cm}^2 \\
1 \text{ yd}^2 &= 0.8361 \text{ m}^2 \\
1 \text{ hectare (ha)} &= 2.471 \text{ acres}
\end{aligned}
$$

Use the unity-fraction method for these. Round (a) to a whole number and (b) to hundredths.

     a) 100 in^2 = _____ cm^2 b) 4 ha = _____ acres

---

56. Complete these. Round (a) to hundredths and (b) to tenths.

     a) 5 m^2 = _____ yd^2 b) 100 acres = _____ ha

a) 645 cm^2

b) 9.88 acres

---

a) 5.98 yd^2 b) 40.5 ha

7-7 VOLUME OF RIGHT PRISMS

In this section, we will review the formulas for the volume of some right prisms, including a rectangular prism, a cube, and a triangular prism.

---

57. A <u>polyhedron</u> is a solid figure bounded by plane polygons. Some polyhedrons are shown below.

       <u>Pyramid</u>   <u>Rectangular Prism</u>   <u>Cube</u>   <u>Triangular Prism</u>

A <u>prism</u> is a polyhedron whose bases are parallel and equal, and whose sides are parallelograms. The three figures at the right above are prisms.

     For a rectangular prism, the bases are rectangles.

     For a cube, the bases are squares.

     For a triangular prism, the bases are _____.

58. The volume of a solid figure is the number of <u>unit</u> <u>cubes</u> needed to fill it. Some <u>unit</u> <u>cubes</u> are shown below.

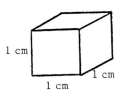
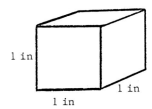

The abbreviations for two volume measures are given below.

125 cubic centimeters is written   125 cm³.

64 cubic yards is written   64 yd³.

Write the abbreviations for each volume measure.

a) 9 cubic inches = _____   b) 1,000 cubic meters = _____

---

59. It takes 24 cubic centimeters to fill the rectangular prism at the right. Therefore, its volume is 24 cm³. Notice that we can find its volume by multiplying its length, width, and height. That is: 24 cm³ = (4 cm)(3 cm)(2 cm).   Therefore, we can use the following formula for the volume of a rectangular prism.

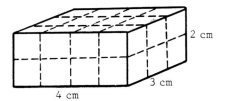

$$V = LWH$$

Use the formula to find the volume of each rectangular prism below. Round to a whole number.

a)
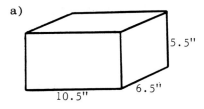

V = _____

b)
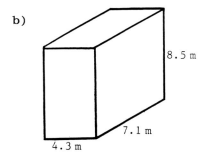

V = _____

triangles

a) 9 in³

b) 1,000 m³

a) 375 in³

b) 260 m³

60. A <u>cube</u> is a special type of rectangular prism in which all sides are squares. Therefore, the volume of a cube equals (s)(s)(s). That is:

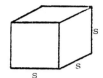

$$V = s^3$$

Using the formula and the $\boxed{y^x}$ key, find the volume of each cube below. Round to a whole number in (a) and to tenths in (b).

a)

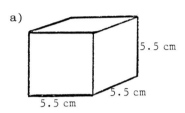

5.5 cm

5.5 cm

5.5 cm

V = _____

b)

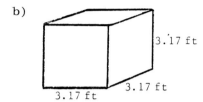

3.17 ft

3.17 ft

3.17 ft

V = _____

61. The general formula for the volume of any right prism is:

Volume = Area of Base x Height

To show that the general formula makes sense, we can use it to derive the specific formulas for the volume of a rectangular prism and a cube.

<u>Rectangular Prism</u> (V = LWH)

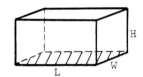

   Area of the rectangular base = LW
   Height = H
   Volume = (LW)(H) or LWH

<u>Cube</u> (V = s^3)

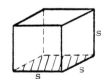

   Area of the square base = (s)(s) or s^2
   Height = s
   Volume = (s^2)(s) = _____

62. When a right triangular prism is <u>vertical</u>, the bases are at the top and bottom and the height is vertical. When it is <u>horizontal</u>, the bases are on the sides and the height is horizontal.

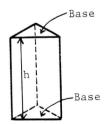

Base

h

Base

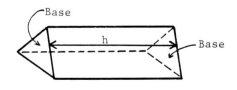

Base

h

Base

Continued on following page.

a) 166 cm³

b) 31.9 ft³

s^3

62.   Continued

The figure at the right is a
right triangular prism. Let's
use the general formula from
the last frame to find its
volume.

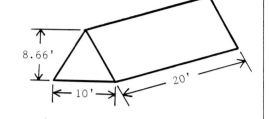

a) The front triangle is a base.
   The area of that triangle is _____.

b) The height of the prism is _____.

c) The volume of the prism is _____.

---

63.   The bases of a right prism can also be composite figures. For example,
the bases of the prism on the left below are L-shaped. The bases of
the prism on the right are I-shaped.

a) 43.3 ft²

b) 20 ft

c) 866 ft³

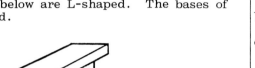

Is the height horizontal or vertical for:

   a) the L-shaped prism? _____

   b) the I-shaped prism? _____

---

64.   The L-shaped figure at the
right is a right prism. Let's
find its volume.

a) vertical

b) horizontal

a) The area of the L-shaped base
   is _____.

b) The height of the prism
   is _____.

c) The volume of the prism
   is _____.

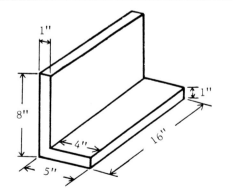

---

a) 12 in²

b) 16 in

c) 192 in³

65. The figure at the right is a rectangular piece of steel with a slot cut out. It is a right prism with a composite figure as its base.

    a) The area of the base is _____.

    b) The height is _____.

    c) The volume is _____.

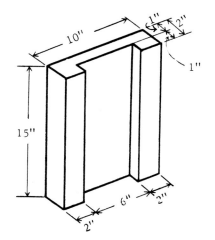

66. The I-beam below is a right prism with a composite figure as its base.

    a) The area of the I-shaped base is _____.

    b) The volume is _____.

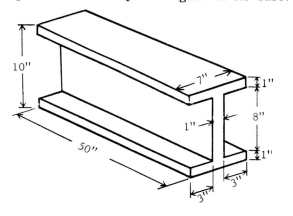

a) 14 in²

b) 15 in

c) 210 in³

a) 22 in²   b) 1,100 in³

7-8 UNITS OF VOLUME

In this section, we will discuss the conversion facts for the basic units of volume in both the English System and the Metric System.

67. Since 12 in = 1 ft, the figure at the right is 1 cubic foot. We converted 1 ft³ to cubic inches below by substituting <u>12 in</u> for <u>1 ft</u>.

$$1 \text{ ft}^3 = (1 \text{ ft})(1 \text{ ft})(1 \text{ ft})$$
$$= (12 \text{ in})(12 \text{ in})(12 \text{ in})$$
$$= 1,728 \text{ in}^3$$

Continued on following page.

67. Continued

Since 3 ft = 1 yd, the figure at the right is 1 cubic yard. We converted 1 yd³ to cubic feet below by substituting <u>3 ft</u> for <u>1 yd</u>.

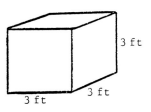

$1 \text{ yd}^3 = (1 \text{ yd})(1 \text{ yd})(1 \text{ yd})$

$= (3 \text{ ft})(3 \text{ ft})(3 \text{ ft})$

$= \underline{\hspace{1cm}} \text{ ft}^3$

---

68. The basic conversion facts for units in the English System are:

1 cubic foot (ft³) = 1,728 cubic inches (in³)
1 cubic yard (yd³) = 27 cubic feet (ft³)

Using the unity-fraction method, complete these conversions.

a) 10 ft³ = _____ in³   b) 100 yd³ = _____ ft³

27 ft³

---

69. Complete these. Round to hundredths.

a) 5,000 in³ = _____ ft³   b) 100 ft³ = _____ yd³

a) 17,280 in³

b) 2,700 ft³

---

70. Since 10 dm = 1 m, the figure at the right is 1 cubic meter. We converted 1 m³ to cubic decimeters below by substituting <u>10 dm</u> for <u>1 m</u>.

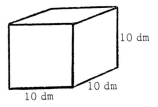

$1 \text{ m}^3 = (1 \text{ m})(1 \text{ m})(1 \text{ m})$

$= (10 \text{ dm})(10 \text{ dm})(10 \text{ dm})$

$= 1,000 \text{ dm}^3$

Since 100 cm = 1 m, the figure at the right is also 1 cubic meter. We converted 1 m³ to cubic centimeters below by substituting <u>100 cm</u> for <u>1 m</u>.

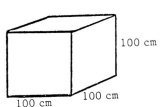

$1 \text{ m}^3 = (1 \text{ m})(1 \text{ m})(1 \text{ m})$

$= (100 \text{ cm})(100 \text{ cm})(100 \text{ cm})$

$= \underline{\hspace{3cm}} \text{ cm}^3$

a) 2.89 ft³

b) 3.70 yd³

1,000,000 cm³

71. Some basic conversion facts for units of volume in the Metric System are:

> 1 cubic meter (m^3) = 1,000 cubic decimeters (dm^3)
>
> 1 cubic meter (m^3) = 1,000,000 cubic centimeters (cm^3)
>
> 1 liter (ℓ) = 1,000 cubic centimeters (cm^3)
>
> 1 milliliter ($m\ell$) = 1 cubic centimeter (cm^3)

Use the unity-fraction method for these.

a) 5 m^3 = _____ dm^3 b) 3 m^3 = _____ cm^3

---

72. Complete these.

a) 2,500 dm^3 = _____ m^3 b) 7,200,000 cm^3 = _____ m^3

a) 5,000 dm^3

b) 3,000,000 cm^3

---

73. Complete these:

a) 5 ℓ = _____ cm^3 b) 75 $m\ell$ = _____ cm^3

a) 2.5 m^3 b) 7.2 m^3

---

74. The basic conversion facts relating volumes in the English and Metric Systems are:

> 1 in^3 = 16.39 cm^3
>
> 1 yd^3 = 0.7646 m^3
>
> 1 gallon (gal) = 231 in^3
>
> 1 liter (ℓ) = 61.02 in^3

Use the unity-fraction method for these.

a) 10 in^3 = _____ cm^3 b) 100 yd^3 = _____ m^3

a) 5,000 cm^3

b) 75 cm^3

---

a) 163.9 cm^3

b) 76.46 m^3

75. Complete these. Round to hundredths.

a) 100 cm³ = _____ in³   b) 1,000 in³ = _____ gal

76. The piston displacement of an automobile engine is expressed in either liters or cubic inches. Complete these.

a) A 5 liter engine is equivalent to a _____ in³ engine. Round to a whole number.

b) A 190 in³ engine is equivalent to a _____ liter engine. Round to tenths.

a) 6.10 in³

b) 4.33 gal

a) 305 in³   b) 3.1 liters

SELF-TEST 24 (pages 300-313)

1. Find the area of this composite figure. Round to a whole number.

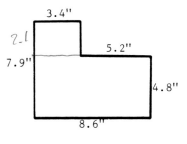

3.4"

2-1

5.2"

7.9"

4.8"

8.6"

_____

2. Find the area of the shaded figure. Round to two decimal places.

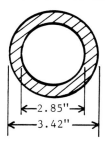

←2.85"→

←—3.42"—→

_____

3. Find the area of this composite figure containing a triangle, rectangle, and semi-circle. Round to a whole number.

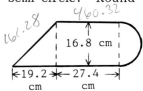

161.28   460.32

16.8 cm

←19.2 ✳ ← 27.4 →
   cm   cm

_____

Do these conversions of area units.

4. Round to tenths.

340 ft² = _____ yd²

5. Round to hundredths.

59.6 cm² = _____ in²

Continued on following page.

SELF-TEST 24 (pages 299-313) - Continued

6. Find the volume of this rectangular prism. Round to thousands.

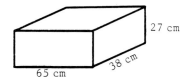

7. Find the volume of this right triangular prism. Round to tenths.

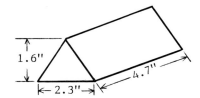

Do these conversions of volume units.

8. 5,000 dm³ = _____ m³

9. Round to hundredths.

4,800 in³ = _____ ft³

10. Round to a whole number.

170 yd³ = _____ m³

ANSWERS: 1. 52 in²   4. 37.8 yd²   7. 8.6 in³   9. 2.78 ft³

2. 2.81 in²   5. 9.24 in²   8. 5 m³   10. 130 m³

3. 732 cm²   6. 67,000 cm³

7-9   VOLUME OF OTHER SOLIDS

In this section, we will review the formulas for the volume of a cylinder, a pyramid, a cone, a frustum of a cone, and a sphere.

77.   The figure shown is a <u>right circular cylinder</u>. To find its volume, we multiply the area of its base by the height.

Area of circular base = πr^2
Height = h
Volume = $(\pi r^2)h$

    or

$V = \pi r^2 h$

Use the formula for these:

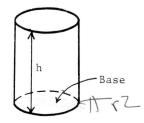

a) Find the volume of a cylinder whose radius is 2 in and whose height is 10 in. Round to a whole number.

b) Find the volume of a cylinder whose diameter is 24 cm and whose height is 25 cm. Round to hundreds.

78. The steel pipe shown involves two right circular cylinders. Its inside diameter is 8.5 cm, its outside diameter is 10 cm, and its height is 16 cm.

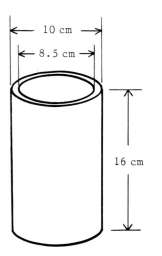

a) V = 126 in³

b) V = 11,300 cm³

To find the volume of the pipe, we can subtract the volume of the inner cylinder from the volume of the outer cylinder.

a) The volume of the outer cylinder rounded to a whole number is _____.

b) The volume of the inner cylinder rounded to a whole number is _____.

c) The volume of the steel pipe is _____.

79. The formula for the volume of a <u>right pyramid</u> is one-third the area of the base (B) times the height (h). That is:

$$V = \frac{1}{3}Bh \qquad \text{or} \qquad V = \frac{Bh}{3}$$

The base of the pyramid at the right is a square. Therefore:

a) The area of the base is _____.

b) The height is _____.

c) The volume is _____.

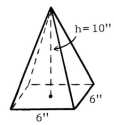

a) 1,257 cm³

b) 908 cm³

c) 350 cm³, from 1,257 - 907

80. The formula for the volume of a <u>right circular cone</u> is also one-third the area of the base (B) times the height (h). That is:

$$V = \frac{1}{3}Bh \qquad \text{or} \qquad V = \frac{Bh}{3}$$

Since the base is a circle whose area is πr^2, we can substitute and get:

$$V = \frac{\pi r^2 h}{3}$$

Using the last formula, find the volume of the cone shown. Round to tens.

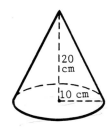

a) 36 in²

b) 10 in

c) 120 in³

81. The figure below is a <u>frustum of a cone</u>. It is the bottom part of a cone with a smaller cone cut off at the top. The formula for the volume of the frustum is:

$$V = \frac{\pi h}{3}(R^2 + Rr + r^2)$$ where <u>h</u> is the height of the frustum.
<u>R</u> is the radius of the large base.
<u>r</u> is the radius of the small base.

Using the formula, find the volume of the frustum on the right. Round to the nearest whole number.

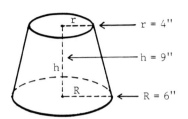

r = 4"

h = 9"

R = 6"

V = _____

V = 2,090 cm³

82. The formula for the volume of a <u>sphere</u> (or ball) is:

$$V = \frac{4}{3}\pi r^3$$ or $$V = \frac{4\pi r^3}{3}$$ where <u>r</u> is the radius.

Using the form of the formula at the right, we found the volume of the sphere. The calculator steps are below. Notice that we have to use the $\boxed{y^x}$ key and the $\boxed{=}$ key to evaluate r^3 before multiplying by 4 and π.

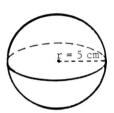

r = 5 cm

Calculator Steps:

5 $\boxed{y^x}$ 3 $\boxed{=}$ $\boxed{×}$ 4 $\boxed{×}$ π $\boxed{÷}$ 3 $\boxed{=}$ 523.59877

Rounding to a whole number, V = _____

V = 716 in³

83. When using the $\boxed{y^x}$ key and then the $\boxed{=}$ key to evaluate r^3 in the formula for the volume of a sphere, some calculators are a little slow in calculating r^3. <u>Be sure to wait until that value appears on the display before multiplying by 4 and</u> π.

In $V = \frac{4\pi r^3}{3}$:   a) when r = 10 in, V = _____ .
Round to tens.

b) when r = 2.5 ft, V = _____ .
Round to tenths.

V = 524 cm³

a) 4,190 in³

b) 65.4 ft³

84. Some volume formulas are summarized below.

VOLUME (V)

Rectangular Prism	$V = LWH$
Cube	$V = s^3$
Right Circular Cylinder	$V = \pi r^2 h$
Right Pyramid	$V = \dfrac{Bh}{3}$
Right Circular Cone	$V = \dfrac{\pi r^2 h}{3}$
Frustum of Cone	$V = \dfrac{\pi h}{3}(R^2 + Rr + r^2)$
Sphere	$V = \dfrac{4\pi r^3}{3}$

7-10 SURFACE AREA

In this section, we will discuss the formulas for the lateral surface area and total surface area of various solid figures.

85. The <u>lateral</u> <u>surface</u> <u>area</u> (S) of a solid is the area other than that of the bases. The <u>total</u> <u>surface</u> <u>area</u> (A) is the lateral area plus the area of the bases.

For the rectangular prism at the right, L = 4", W = 3", and H = 2".

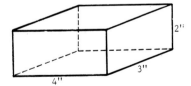

The area of the front and back sides is LH.

The area of the left and right sides is WH.

The area of the bases is LW.

Therefore, the formulas for the lateral area (S) and total area (A) are:

$$S = 2(LH + WH)$$
$$A = 2(LH + WH + LW)$$

Using the formulas, find the lateral area and total area of the rectangular prism.

a) S = _____   b) A = _____

86. For the cube at the right, s = 2.5 m.

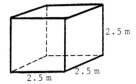

The area is s² for each side and base. Therefore, the formula for the lateral area (S) and total area (A) are:

$$S = 4s^2$$
$$A = 6s^2$$

Using the formulas, find the lateral area and total area of the cube.

a) S = _____   b) A = _____

a) 28 in²

b) 52 in²

---

87. For the right circular cylinder below, r = 3" and h = 12". The lateral area is the area of the circular side. The total area is the lateral area plus the areas of the two circular bases. To show what is meant by the area of the circular side, we cut the cylinder vertically and laid the circular side flat. We got a rectangle whose width is 12" and whose length is the circumference (2πr") of the cylinder.

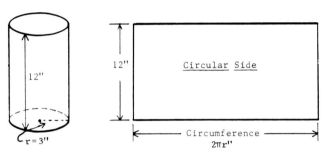

The formulas for the lateral and total area are given below.

$$S = 2\pi rh$$
$$A = 2\pi rh + 2\pi r^2$$

Use the formulas to find S and A. Round each to a whole number.

a) S = _____   b) A = _____

a) S = 25 m²

b) A = 37.5 m²

---

88. For the right pyramid shown, the base is a square whose side (s) is 8'. The slant height (ℓ) of each triangular side is 10'. The area of each of the four triangular sides is $\frac{1}{2} s\ell$. Therefore, the formulas for the lateral area and total area are:

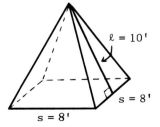

$$S = 4\left(\frac{1}{2} s\ell\right) = 2 s\ell$$
$$A = 2 s\ell + s^2$$

Continued on following page.

a) S = 226 in²

b) A = 283 in²

88. Continued

Use the formulas to find S and A.

a) S = _____   b) A = _____

---

89. The lateral area of a right circular cone is the area of the curved side. The total area is the lateral area plus the area of the circular base. For the cone at the right, the diameter (d) is 8 cm and the slant height (ℓ) is 10 cm.

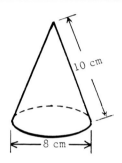

The lateral area is one-half the product of the slant height (ℓ) and the circumference (C). Using the circumference formula based on the radius, we get:

$$S = \frac{1}{2}\ell C$$

$$= \frac{1}{2}\ell(2\pi r)$$

$$= \pi r \ell$$

The total area is the lateral area plus the area of the base. Using the area formula based on the radius, we get:

$$A = \pi r \ell + \pi r^2$$

Use the formulas to find S and A. Round each to a whole number.

a) S = _____   b) A = _____

a) S = 160 ft²

b) A = 224 ft²

---

90. Since a sphere has no base, we can only find its total area. The formula is:

$$A = 4\pi r^2$$

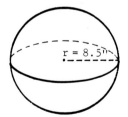

Use the formula to find the total area of the sphere at the right. Round to a whole number.

A = _____

a) S = 126 cm²

b) A = 176 cm²

A = 908 in²

91. The lateral area (S) and total area (A) formulas are summarized below.

LATERAL AREA (S) AND TOTAL AREA (A)	
Rectangular Prism	$S = 2(LH + WH)$ $A = 2(LH + WH + LW)$
Cube	$S = 4s^2$ $A = 6s^2$
Right Circular Cylinder	$S = 2\pi rh$ $A = 2\pi rh + 2\pi r^2$
Right Pyramid (square base)	$S = 2 s\ell$ $A = 2 s\ell + s^2$
Right Circular Cone	$S = \pi r\ell$ $A = \pi r\ell + \pi r^2$
Sphere	$A = 4\pi r^2$

7-11   DENSITY

In this section, we will define the <u>density</u> of materials and liquids and solve some applied problems involving density, volume, and weight.

92. The density (D) of a material is its weight (W) per unit volume (V). The formula is:

$$D = \frac{W}{V}$$

If 100 in³ of steel weighs 28.3 lb, the density of steel is 0.283 lb/in³, since:

$$D = \frac{W}{V} = \frac{28.3 \text{ lb}}{100 \text{ in}^3} = 0.283 \text{ lb/in}^3$$

Find the density of aluminum if 100 cm³ of aluminum weighs 256g.

$$D = \frac{W}{V} = \underline{\hspace{2cm}}$$

---

93. Frequently we know the density and volume of a material and want to find its weight. In that case, we use the formula in the following form:

$$W = DV$$

Since the density of steel is 0.283 lb/in³:

1 in³ of steel weighs 0.283 lb, since W = (0.283)(1).

10 in³ of steel weighs 2.83 lb, since W = (0.283)(10).

How much does 33.4 in³ of steel weigh? $\underline{\hspace{2cm}}$
Round to hundredths.

2.56 g/cm³

94. The density of aluminum is 2.56 g/cm³.

    a) What is the weight of 10 cm³ of aluminum? _____

    b) What is the weight of 1,750 cm³ of aluminum? _____

9.45 lb

---

95. The end of the steel bar below is a square with 2.5" sides. Its length is 40".

    a) What is the volume of the bar? _____

    b) Since the density of steel is 0.283 lb/in³, how much does the steel bar weigh? Round to a whole number. _____

a) 25.6 g

b) 4,480 g

---

96. The length and width of a rectangular aluminum sheet are shown. Its thickness is 0.38 cm.

    100 cm

    230 cm

    a) What is the volume of the sheet? _____

    b) Since the density of aluminum is 2.56 g/cm³, what is the weight of the sheet? Round to hundreds. _____

a) 250 in³

b) 71 lb

---

97. A piece of circular wire is a cylinder. The diameter of 12-gage copper wire is 0.081 in. The density of copper is 0.322 lb/in³.

Rounding to hundredths:

    a) Find the volume in cubic inches of 100 feet of 12-gage copper wire. _____

    b) Find the weight of 100 feet of 12-gage copper wire. _____

a) 8,740 cm³

b) 22,400 g

---

a) 6.18 in³

b) 1.99 lb

98. The density of a liquid is also its weight per unit volume. The density of water is 62.4 lb/ft³.

    a) How much does 1 ft³ of water weigh? _____

    b) How much does 1 yd³ (or 27 ft³) of water weigh? Round to a whole number. _____

99. The density of gasoline is 0.675 g/cm³.

    a) How much does 10 cm³ of gasoline weigh? _____

    b) How much does 1,000 cm³ of gasoline weigh? _____

a) 62.4 lb

b) 1,685 lb

100. Since the density of water is 1 g/cm³:

    a) How much does 1 milliliter of water weigh? _____

    b) How much does 1 liter of water weigh? _____

a) 6.75 g

b) 675 g

101. In many countries, gasoline is sold by the liter. Since the density of gasoline is 0.675 g/cm³:

    a) How much does 1 liter of gasoline weigh in grams? _____

    b) How much does 50 liters of gasoline weigh in kilograms? _____

a) 1 g

b) 1,000 g

a) 675 g   b) 33.75 kg, from: 33,750 g

SELF-TEST 25 (pages 313-322)

A right circular cylinder is shown below.

1. Find its volume. Round to tenths.

2. Find its total area. Round to tenths.

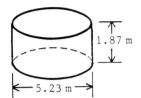

1.87 m

5.23 m

A right circular cone is shown below.

3. Find its volume. Round to tens.

4. Find its total area. Round to tens.

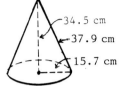

34.5 cm

37.9 cm

15.7 cm

A right pyramid with a square base is shown below.

5. Find its volume. Round to tens.

6. Find its total area. Round to tens.

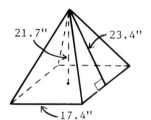

21.7"   23.4"

17.4"

A sphere is shown below.

7. Find its volume. Round to hundreds.

8. Find its surface area. Round to tens.

8.2"

9. Find the weight, in pounds, of a 60" by 80" rectangular steel plate whose thickness is 0.50". Round to tens. The density of steel is 0.283 lb/in^3.

10. Find the weight, in grams, of a 450 cm length of round aluminum wire whose diameter is 0.165 cm. Round to tenths. The density of aluminum is 2.56 g/cm^3.

11. Find the weight, in kilograms, of 64 liters of gasoline. The density of gasoline is 0.675 g/cm^3.

ANSWERS:
1. 40.2 m^3
2. 73.7 m^2
3. 8,910 cm^3
4. 2,640 cm^2
5. 2,190 in^3
6. 1,120 in^2
7. 2,300 in^3
8. 840 in^2
9. 680 lb
10. 24.6 g
11. 43.2 kg

SUPPLEMENTARY PROBLEMS - CHAPTER 7

<u>Assignment 23</u>

The two parallel lines below are cut by a transversal. Find these angles.

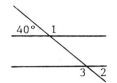

1. $\angle 1$

2. $\angle 2$

3. $\angle 3$

Right triangles ABC and ADE below are similar triangles. AB = 40", AC = 32", BC = 24", and DE = 15". Find these lengths:

4. AD

5. AE

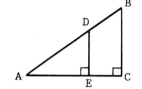

A rectangle's length is 5.62m and its width is 3.17m.

6. Find its area. Round to tenths.

7. Find its perimeter.

The side of a square is 14.9 cm.

8. Find its area. Round to a whole number.

9. Find its perimeter.

10. Find the circumference of a circle whose diameter is 7.94 cm. Round to tenths.

11. Find the circumference of a circle whose radius is 5,280 ft. Round to hundreds.

12. Find the area of this parallelogram. Round to hundredths.

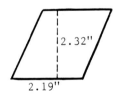

2.32"

2.19"

13. Find the area of this triangle. Round to thousands.

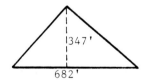

347'

682'

14. Find the area of this trapezoid. Round to hundreds.

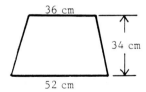

36 cm

34 cm

52 cm

15. Find the area of a circle whose radius is 2.86 m. Round to tenths.

16. Find the area of a circle whose diameter is 0.914". Round to thousandths.

17. Find the area of this triangle. Round to tenths.

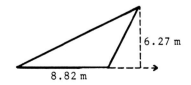

6.27 m

8.82 m

18. Find the area of this right triangle. Round to thousands.

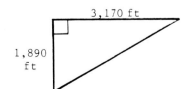

3,170 ft

1,890 ft

19. The area of this square is 1,270 in². Find s, the length of its side. Round to tenths.

s

s

Assignment 24

1. Find the area of this composite rectangular figure.

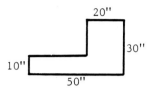

2. Find the area of this composite figure. The lower corner angles are right angles. Round to tenths.

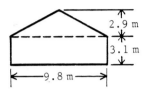

3. Find the area of this composite figure. The two angles at the left are right angles. Round to a whole number.

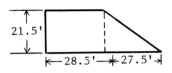

Find the <u>area</u> of each shaded figure. The basic shapes are rectangles and circles.

4. Round to tens.

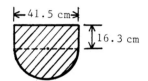

5. Round to tenths.

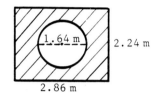

6. Round to hundredths.

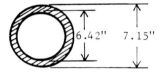

Do these conversions of area units.

7. $648 \ in^2 = $ _____ ft^2

8. $2.93 \ m^2 = $ _____ cm^2

9. $322.6 \ cm^2 = $ _____ in^2

10. Round to hundredths.

    $6 \ m^2 = $ _____ yd^2

11. Round to tenths.

    $180 \ acres = $ _____ hectares

12. Round to a whole number.

    $28.5 \ in^2 = $ _____ cm^2

Find the volume of each rectangular prism.

13. Round to tens.

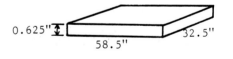

14. Round to tenths.

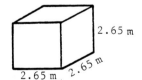

15.

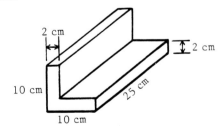

Do these conversions of volume units.

16. $6,048 \ in^3 = $ _____ ft^3

17. $162 \ ft^3 = $ _____ yd^3

18. $3.8 \ m^3 = $ _____ cm^3

19. Round to a whole number.

    $58 \ in^3 = $ _____ cm^3

20. Round to hundredths.

    $7.27 \ m^3 = $ _____ yd^3

21. Round to tens.

    $19.7 \ liters = $ _____ in^3

22. The volume of a cube is $43.7 \ cm^3$. Find the length of a side, rounded to hundredths.

Assignment 25

1. Find the volume of this right circular cylinder. Round to hundreds.

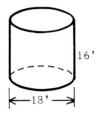

2. Find the volume of this right circular cone. Round to hundredths.

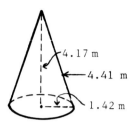

3. Find the volume of this sphere. Round to tens.

4. Find the total surface area of the cylinder in Problem 1. Round to hundreds.

5. Find the total surface area of the cone in Problem 2. Round to tenths.

6. Find the surface area of the sphere in Problem 3. Round to tens.

7. Find the total surface area of a cube whose side is 15 in.

8. Find the total surface area of a rectangular prism whose length is 54 cm, whose width is 35 cm, and whose height is 16 cm.

9. Find the volume of this round bar. Round to a whole number.

10. Find the volume of this pipe. Round to a whole number.

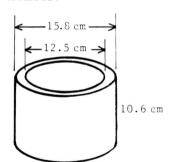

11. Find the volume of this frustum of a cone. Round to a whole number.

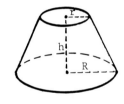

h = 5.4"
R = 3.5"
r = 2.2"

12. Find the weight, in pounds, of a rectangular sheet of aluminum whose length is 36", width is 32", and thickness is 0.138". Round to tenths. The density of aluminum is 0.0924 lb/in³.

13. Find the weight, in grams, of a round copper wire whose diameter is 0.215 cm and whose length is 2,100 centimeters. Round to a whole number. The density of copper is 8.91 g/cm³.

14. Find the weight, in pounds, of a solid steel ball whose diameter is 2.75". Round to hundredths. The density of steel is 0.283 lb/in³.

15. Find the weight, in kilograms, of 25 liters of water. The density of water is 1 g/cm³.

16. Find the weight, in pounds, of 16 gallons of gasoline. Round to a whole number. The density of gasoline is 41.5 lb/ft³. The volume of one gallon is 0.1337 ft³.

8 Algebraic Fractions

In this chapter, we will define algebraic fractions and discuss the basic operations with them. We will also discuss equivalent forms for some patterns of algebraic fractions. A knowledge of the content of this chapter is especially useful for formula rearrangement and formula derivation, topics that are discussed in later chapters.

8-1   MULTIPLICATION

In this section, we will discuss the procedure for multiplications involving algebraic fractions.

1. An algebraic fraction is a fraction that contains one or more variables in its numerator or denominator or both. Some examples are:

$$\frac{x}{a} \qquad \frac{2a}{b} \qquad \frac{p + q}{p} \qquad \frac{c^2d}{m - 1}$$

As we said earlier, in this text we will avoid substitutions in algebraic fractions that do not make sense because they lead to a division by 0. That is:

We would not substitute 0 for $\underline{b}$ in $\frac{2a}{b}$.

We would not substitute "1" for $\underline{m}$ in $\frac{c^2d}{m - 1}$.

2. To multiply algebraic fractions, we multiply their numerators and their denominators. For example:

$$\left(\frac{2}{3}\right)\left(\frac{x}{7}\right) = \frac{(2)(x)}{(3)(7)} = \frac{2x}{21} \qquad\qquad \left(\frac{2a}{b}\right)\left(\frac{5}{4}\right) = \frac{(2a)(5)}{(b)(4)} = \underline{\hspace{2cm}}$$

3. When a multiplication contains more than one variable, <u>we usually write the variables in alphabetical order</u>. For example:

$\dfrac{10a}{4b}$

Instead of 5ta , we write 5at .

Instead of VP^2R , we write P^2RV .

The same convention is used when multiplying fractions. For example:

$$\left(\dfrac{x}{2}\right)\left(\dfrac{y}{3}\right) = \dfrac{xy}{6} \qquad\qquad \left(\dfrac{aF_1}{d}\right)\left(\dfrac{c^2}{bM}\right) = \dfrac{ac^2F_1}{bdM}$$

Complete these.

a) $\left(\dfrac{c}{4}\right)\left(\dfrac{3d}{2}\right) = $ _____   b) $\left(\dfrac{4V_1}{T}\right)\left(\dfrac{P}{bR}\right) = $ _____

---

4. When one factor in a multiplication is an addition or subtraction, we usually write that factor last. For example:

a) $\dfrac{3cd}{8}$ b) $\dfrac{4PV_1}{bRT}$

Instead of $5(p + q)d$, we write $5d(p + q)$.

We also use that convention when writing products. That is:

$$\left(\dfrac{y + 9}{a}\right)\left(\dfrac{x}{t^2}\right) = \dfrac{x(y + 9)}{at^2} \qquad\qquad \left[\dfrac{a}{b(x + y)}\right]\left(\dfrac{c}{f}\right) = $$ _____

---

5. Any non-fractional expression can be written as a fraction whose denominator is "1". For example:

$\dfrac{ac}{bf(x + y)}$

$$x^2 = \dfrac{x^2}{1} \qquad\qquad 3ab = \dfrac{3ab}{1}$$

Using the facts above, we converted each multiplication below to a multiplication of two fractions.

$$\dfrac{b}{5}(x^2) = \dfrac{b}{5}\left(\dfrac{x^2}{1}\right) = \dfrac{bx^2}{5} \qquad\qquad 3ab\left(\dfrac{c}{b}\right) = \left(\dfrac{3ab}{1}\right)\left(\dfrac{c}{d}\right) = \dfrac{3abc}{d}$$

However, we ordinarily use the shorter method below. That is, we multiply the numerator and the non-fractional expression.

$$\dfrac{b}{5}(x^2) = \dfrac{b(x^2)}{5} = \dfrac{bx^2}{5} \qquad\qquad 3ab\left(\dfrac{c}{d}\right) = \dfrac{3ab(c)}{d} = $$ _____

---

6. Complete: a) $x^2\left(\dfrac{3}{b}\right) = $ _____   c) $\dfrac{m}{t}(s^2v) = $ _____

$\dfrac{3abc}{d}$

b) $pq\left(\dfrac{1}{r}\right) = $ _____   d) $\dfrac{1}{2}(bh) = $ _____

---

a) $\dfrac{3x^2}{b}$ c) $\dfrac{ms^2v}{t}$

b) $\dfrac{pq}{r}$ d) $\dfrac{bh}{2}$

7. We used the shorter method for the multiplication below.

$$\left(\frac{1}{a + b}\right)(x + y) = \frac{(1)(x + y)}{a + b} = \frac{x + y}{a + b}$$

Complete these.

a) $\frac{a}{mt}(p - q) = $ _____

b) $\frac{1}{AK(t_2 - t_1)}(LW) = $ _____

---

8. When an algebraic fraction is multiplied by "1", the product is <u>identical</u> to the original fraction.   That is:

$$1\left(\frac{c}{d}\right) = \frac{c}{d} \qquad\qquad \frac{ST}{P}(1) = \frac{ST}{P}$$

Complete these.

a) $1\left(\frac{d^2}{v}\right) = $ _____

b) $1\left(\frac{m + t}{R}\right) = $ _____

c) $\frac{pq}{ad}(1) = $ _____

a) $\frac{a(p - q)}{mt}$

b) $\frac{LW}{AK(t_2 - t_1)}$

a) $\frac{d^2}{v}$ b) $\frac{m + t}{R}$ c) $\frac{pq}{ad}$

8-2   REDUCING TO LOWEST TERMS

In this section, we will discuss the procedure for reducing algebraic fractions to lowest trms.

---

9. When any expression is divided by itself, the quotient is "1".   That is:

$$\frac{x}{x} = 1 \qquad\qquad \frac{p^2}{p^2} = 1 \qquad\qquad \frac{a - b}{a - b} = \text{\_\_\_\_\_}$$

---

10. To reduce an algebraic fraction to lowest terms, we factor out a fraction that equals "1".   For example:

$$\frac{3x}{7x} = \left(\frac{3}{7}\right)\left(\frac{x}{x}\right) = \left(\frac{3}{7}\right)(1) = \frac{3}{7}$$

$$\frac{p^2q}{p^2t} = \left(\frac{p^2}{p^2}\right)\left(\frac{q}{t}\right) = (1)\left(\frac{q}{t}\right) = \frac{q}{t}$$

Reduce these to lowest terms.

a) $\frac{5y^2}{4y^2} = $

b) $\frac{bc}{bd} = $

1

a) $\frac{5}{4}$, from: $\frac{5}{4}(1)$

b) $\frac{c}{d}$, from: $(1)\left(\frac{c}{d}\right)$

11. Following the example, reduce the other fraction to lowest terms.

$$\frac{aSV}{aST} = \left(\frac{a}{a}\right)\left(\frac{S}{S}\right)\left(\frac{V}{T}\right) = (1)(1)\left(\frac{V}{T}\right) = \frac{V}{T}$$

$$\frac{3x^2y}{2x^2y} =$$

---

12. Following the example, reduce the other fraction to lowest terms.

$$\frac{3(x + 5)}{7(x + 5)} = \left(\frac{3}{7}\right)\left(\frac{x + 5}{x + 5}\right) = \left(\frac{3}{7}\right)(1) = \frac{3}{7}$$

$$\frac{2a(b - c)}{3d(b - c)} =$$

$\frac{3}{2}$, from: $\frac{3}{2}(1)(1)$

---

13. If one of the factors is a numerical fraction, it should also be reduced to lowest terms. For example:

$$\frac{8cd}{12cd} = \left(\frac{8}{12}\right)\left(\frac{c}{c}\right)\left(\frac{d}{d}\right) = \left(\frac{2}{3}\right)(1)(1) = \frac{2}{3}$$

Reduce these to lowest terms.

a) $\frac{15x^2}{6x^2} = $ _____   b) $\frac{9PV}{3PV} = $ _____

$\frac{2a}{3d}$, from: $\left(\frac{2a}{3d}\right)(1)$

---

14. Following the example, reduce the other fraction to lowest terms.

$$\frac{7ab}{abd} = \left(\frac{7}{d}\right)\left(\frac{a}{a}\right)\left(\frac{b}{b}\right) = \left(\frac{7}{d}\right)(1)(1) = \frac{7}{d}$$

$$\frac{PQR}{FPR} =$$

a) $\frac{5}{2}$ b) 3

---

15. A fraction can be reduced to lowest terms only if we can factor out a fraction that equals "1". For example, none of the fractions below can be reduced.

$$\frac{H}{T} \qquad\qquad \frac{3mp}{7fq} \qquad\qquad \frac{a(x + y)}{b(x - y)}$$

Reduce to lowest terms if possible.

a) $\frac{cp^2}{p^2t} = $ _____   c) $\frac{m(p + q)}{v(p + q)} = $ _____

b) $\frac{PV}{RT} = $ _____   d) $\frac{20xy}{10xy} = $ _____

$\frac{Q}{F}$

---

a) $\frac{c}{t}$ c) $\frac{m}{v}$

b) not possible   d) 2

16. To reduce the fractions below to lowest terms, we began by substituting $1x$ for $\underline{x}$ and $1m$ for $\underline{m}$.

$$\frac{x}{7x} = \frac{1x}{7x} = \left(\frac{1}{7}\right)\left(\frac{x}{x}\right) = \left(\frac{1}{7}\right)(1) = \frac{1}{7}$$

$$\frac{m}{cm} = \frac{1m}{cm} = \left(\frac{1}{c}\right)\left(\frac{m}{m}\right) = \left(\frac{1}{c}\right)(1) = \frac{1}{c}$$

Reduce these to lowest terms.

a) $\dfrac{t}{9t}$ = _____   b) $\dfrac{S}{ST}$ = _____   c) $\dfrac{a}{3ab}$ = _____

---

17. To reduce the fractions below to lowest terms, we began by substituting $1y$ for $\underline{y}$ and $1a$ for $\underline{a}$.

$$\frac{4y}{y} = \frac{4y}{1y} = \left(\frac{4}{1}\right)\left(\frac{y}{y}\right) = (4)(1) = 4$$

$$\frac{abt}{a} = \frac{abt}{1a} = \left(\frac{a}{a}\right)\left(\frac{bt}{1}\right) = (1)(bt) = bt$$

Reduce these to lowest terms.

a) $\dfrac{7d}{d}$ = _____   b) $\dfrac{bs}{s}$ = _____   c) $\dfrac{8RQ}{R}$ = _____

a) $\dfrac{1}{9}$

b) $\dfrac{1}{T}$

c) $\dfrac{1}{3b}$

---

18. In the reduction below, we began by inserting "1" as a factor in the numerator.

$$\frac{5}{15pq} = \frac{(5)(1)}{15pq} = \left(\frac{5}{15}\right)\left(\frac{1}{pq}\right) = \left(\frac{1}{3}\right)\left(\frac{1}{pq}\right) = \frac{1}{3pq}$$

Following the example, reduce these to lowest terms.

a) $\dfrac{7}{14F_1 T_1}$ = _____   b) $\dfrac{9}{9cm}$ = _____

a) 7

b) b

c) 8Q

---

19. In the reduction below, we also began by inserting "1" as a factor in the numerator.

$$\frac{5c^2}{10ac^2} = \frac{5c^2(1)}{10ac^2} = \left(\frac{5}{10}\right)\left(\frac{c^2}{c^2}\right)\left(\frac{1}{a}\right) = \left(\frac{1}{2}\right)(1)\left(\frac{1}{a}\right) = \frac{1}{2a}$$

Following the example, reduce these to lowest terms.

a) $\dfrac{2m}{8mt}$ = _____   b) $\dfrac{6pq^2}{12hpq^2T}$ = _____

a) $\dfrac{1}{2F_1 T_1}$ b) $\dfrac{1}{cm}$

---

a) $\dfrac{1}{4t}$ b) $\dfrac{1}{2hT}$

39. Two fractions are reciprocals if their product is +1. For example:

    Since $\left(\dfrac{c}{d}\right)\left(\dfrac{d}{c}\right) = \dfrac{cd}{cd} = 1$: the reciprocal of $\dfrac{c}{d}$ is $\dfrac{d}{c}$.

    the reciprocal of $\dfrac{d}{c}$ is $\dfrac{c}{d}$.

    Write the reciprocal of each fraction.

    a) $\dfrac{v}{t}$ _____   b) $\dfrac{ab}{h}$ _____   c) $\dfrac{k}{mp}$ _____   d) $\dfrac{abf}{ct}$ _____

    a) $\dfrac{1}{t}$ c) $\dfrac{1}{cd}$

    b) y d) PV

40. Since $(p + q)\left(\dfrac{1}{p + q}\right) = \dfrac{p + q}{p + q} = 1$:

    a) the reciprocal of $(p + q)$ is _____.

    b) the reciprocal of $\dfrac{1}{p + q}$ is _____.

    a) $\dfrac{t}{v}$ c) $\dfrac{mp}{k}$

    b) $\dfrac{h}{ab}$ d) $\dfrac{ct}{abf}$

41. Since $m(a - c)\left[\dfrac{1}{m(a - c)}\right] = \dfrac{m(a - c)}{m(a - c)} = 1$:

    a) the reciprocal of $m(a - c)$ is _____.

    b) the reciprocal of $\dfrac{1}{m(a - c)}$ is _____.

    a) $\dfrac{1}{p + q}$

    b) $p + q$

42. Since $\left(\dfrac{x - 1}{y}\right)\left(\dfrac{y}{x - 1}\right) = \dfrac{y(x - 1)}{y(x - 1)} = 1$:

    a) the reciprocal of $\dfrac{x - 1}{y}$ is _____.

    b) the reciprocal of $\dfrac{y}{x - 1}$ is _____.

    a) $\dfrac{1}{m(a - c)}$

    b) $m(a - c)$

43. Write the reciprocal of each quantity.

    a) $3y$ _____   d) $x - y$ _____

    b) $\dfrac{1}{4x}$ _____   e) $\dfrac{m}{p + q}$ _____

    c) $\dfrac{1}{a(b + c)}$ _____   f) $\dfrac{h - 1}{t - 7}$ _____

    a) $\dfrac{y}{x - 1}$

    b) $\dfrac{x - 1}{y}$

    a) $\dfrac{1}{3y}$ d) $\dfrac{1}{x - y}$

    b) $4x$ e) $\dfrac{p + q}{m}$

    c) $a(b+c)$ f) $\dfrac{t - 7}{h - 1}$

44. Any division involving an algebraic fraction is written as a <u>complex</u> <u>fraction</u> in algebra. The "major" fraction line separates the numerator and denominator. For example:

$$\frac{x}{3} \div \frac{x}{5} \text{ is written } \frac{\frac{x}{3}}{\frac{x}{5}}$$

To perform divisions like those above, <u>we multiply the numerator by the reciprocal of the denominator</u>. That is:

$$\frac{\frac{x}{3}}{\frac{x}{5}} = \left(\frac{x}{3}\right)\left(\frac{5}{x}\right) = \frac{5}{3}$$

Numerator unchanged ↓

Reciprocal of denominator ↑

a) $\dfrac{\frac{y}{4}}{\frac{2}{3}} = (\quad)(\quad) = $ _____

b) $\dfrac{\frac{pt}{d}}{\frac{a}{bc}} = (\quad)(\quad) = $ _____

---

45. When dividing algebraic fractions, the quotient should always be reduced to lowest terms. We cancel before multiplying to do so. For example:

$$\frac{\frac{hs}{4}}{\frac{ah}{2}} = \left(\frac{\cancel{h}s}{\cancel{4}}\right)\left(\frac{\overset{1}{\cancel{2}}}{a\cancel{h}}\right) = \frac{s}{2a}$$

Complete. Cancel to reduce to lowest terms.

a) $\dfrac{\frac{9t}{5}}{\frac{3t}{10}} = $

b) $\dfrac{\frac{ab}{c}}{\frac{bd}{m}} = $

a) $\left(\dfrac{y}{4}\right)\left(\dfrac{3}{2}\right) = \dfrac{3y}{8}$

b) $\left(\dfrac{pt}{d}\right)\left(\dfrac{bc}{a}\right) = \dfrac{bcpt}{ad}$

---

46. Complete. Cancel when possible.

a) $\dfrac{\frac{b}{c}}{\frac{a}{2t}} = $

b) $\dfrac{\frac{1}{PV}}{\frac{1}{P}} = $

a) $\left(\dfrac{\overset{3}{\cancel{9t}}}{\cancel{5}}\right)\left(\dfrac{\overset{2}{\cancel{10}}}{\cancel{3t}}\right) = 6$

b) $\left(\dfrac{a\cancel{b}}{c}\right)\left(\dfrac{m}{\cancel{b}d}\right) = \dfrac{am}{cd}$

---

47. When writing a division as a complex fraction, the numerator or denominator may be a non-fraction. For example:

$$m \div \frac{a}{b} \text{ is written } \frac{m}{\frac{a}{b}} \qquad \frac{P}{Q} \div S \text{ is written } \frac{\frac{P}{Q}}{S}$$

a) $\dfrac{2bt}{ac}$ b) $\dfrac{1}{V}$

Continued on following page.

47. Continued

To do the preceding divisions, we also multiply the numerator by the reciprocal of the denominator. That is:

$$\frac{m}{\dfrac{a}{b}} = m\left(\frac{b}{a}\right) = \frac{bm}{a} \qquad\qquad \frac{\dfrac{P}{Q}}{S} = \left(\frac{P}{Q}\right)\left(\frac{1}{S}\right) = \frac{P}{QS}$$

Complete each division.

a) $\dfrac{p}{\dfrac{1}{8}} =$ $\underline{\hspace{2cm}}$ $= \underline{\hspace{2cm}}$ b) $\dfrac{\dfrac{5c}{d}}{4a} =$ $\underline{\hspace{2cm}}$ $= \underline{\hspace{2cm}}$

---

48. Complete. Cancel to reduce to lowest terms.

a) $\dfrac{15t}{\dfrac{5}{7}} =$ b) $\dfrac{P}{\dfrac{P}{T}} =$

a) $(p)(8) = 8p$
b) $\left(\dfrac{5c}{d}\right)\left(\dfrac{1}{4a}\right) = \dfrac{5c}{4ad}$

---

49. Complete. Cancel to reduce to lowest terms.

a) $\dfrac{\dfrac{2v}{3}}{6v} =$ b) $\dfrac{\dfrac{7m}{d}}{7d} =$

a) $(\overset{3}{\cancel{15t}})\left(\dfrac{7}{\cancel{5}}\right) = 21t$
b) $\cancel{P}\left(\dfrac{T}{\cancel{P}}\right) = T$

---

50. Complete. Cancel when possible.

a) $\dfrac{t}{\dfrac{3}{4}} =$ b) $\dfrac{\dfrac{2r}{w}}{r} =$

a) $\left(\dfrac{\overset{1}{\cancel{2v}}}{3}\right)\left(\dfrac{1}{\underset{3}{\cancel{6v}}}\right) = \dfrac{1}{9}$
b) $\left(\dfrac{\overset{1}{\cancel{7m}}}{d}\right)\left(\dfrac{1}{\underset{1}{\cancel{7d}}}\right) = \dfrac{m}{d^2}$

---

51. In one fraction below, the numerator is $a - b$.

$$\frac{\dfrac{a-b}{d}}{\dfrac{c}{t}} = \left(\frac{a-b}{d}\right)\left(\frac{t}{c}\right) = \frac{t(a-b)}{cd}$$

Following the example, complete each division.

a) $\dfrac{V}{\dfrac{S-A}{H}} =$ b) $\dfrac{x+y}{\dfrac{a}{b}} =$

a) $\dfrac{4t}{3}$ b) $\dfrac{2}{w}$

---

52. In the division below, we were able to cancel.

$$\frac{\dfrac{c+d}{m}}{\dfrac{c+d}{t}} = \left(\frac{\cancel{c+d}}{m}\right)\left(\frac{t}{\cancel{c+d}}\right) = \frac{t}{m}$$

Following the example, complete each division.

a) $\dfrac{p-q}{\dfrac{p-q}{v}} =$ b) $\dfrac{\dfrac{t^2-1}{h}}{t^2-1} =$

a) $\dfrac{HV}{S-A}$
b) $\dfrac{b(x+y)}{a}$

53. The numerator of each complex fraction below is "1". Following the example, do the other division.

$$\frac{\frac{1}{a}}{b} = (1)\left(\frac{b}{a}\right) = \frac{b}{a} \qquad \qquad \frac{1}{\frac{x + y}{c}} =$$

a) v   b) $\frac{1}{h}$

$\frac{c}{x + y}$

SELF-TEST 26 (pages 326-338)

Multiply.

1. $\left(\frac{3t}{h}\right)\left(\frac{r}{2}\right)$

2. $(x + y)\left(\frac{2s}{3w}\right)$

Reduce each fraction to lowest terms.

3. $\frac{2R^2}{6R^2T}$

4. $\frac{8w(a - b)}{4(a - b)}$

Multiply. Report each product in lowest terms.

5. $\left(\frac{2a}{b}\right)\left(\frac{b}{6d}\right)$

6. $\frac{r}{2t}(2at) =$

7. $\left(\frac{3k}{bh^2}\right)\left(\frac{h^2}{2k}\right) =$

8. $\left(\frac{d}{r + s}\right)\left(\frac{r + s}{dt}\right) =$

Write the reciprocal of: 9. $\frac{1}{4w}$ 10. $\frac{x - 2}{y}$

Divide. Report each quotient in lowest terms.

11. $\frac{\frac{2d}{w}}{\frac{r}{2w}} =$

12. $\frac{3ps}{\frac{s}{p}} =$

13. $\frac{\frac{x - y}{2a}}{\frac{x - y}{4b}} =$

14. $\frac{\frac{t^2 + 1}{at}}{t^2 + 1} =$

ANSWERS:

1. $\frac{3rt}{2h}$

2. $\frac{2s(x + y)}{3w}$

3. $\frac{1}{3T}$

4. $2w$

5. $\frac{a}{3d}$

6. ar

7. $\frac{3}{2b}$

8. $\frac{1}{t}$

9. $4w$

10. $\frac{y}{x - 2}$

11. $\frac{4d}{r}$

12. $3p^2$

13. $\frac{2b}{a}$

14. $\frac{1}{at}$

8-5   ADDITION AND SUBTRACTION WITH LIKE DENOMINATORS

In this section, we will discuss the procedure for adding and subtracting algebraic fractions with <u>like</u> or <u>common</u> denominators.

---

54.   To add or subtract algebraic fractions with like denominators, we add or subtract their numerators and keep the same denominator.   For example:

$$\frac{2x}{5} + \frac{1}{5} = \frac{2x + 1}{5} \qquad\qquad \frac{m}{ab} - \frac{n}{ab} = \frac{m - n}{ab}$$

Complete these.

   a) $\dfrac{b}{y} - \dfrac{3}{y} =$ _____

   b) $\dfrac{5S}{PV} + \dfrac{7R}{PV} =$ _____

---

55.   Notice how we combined like terms in the addition and subtraction below.

$$\frac{6}{t} + \frac{3}{t} = \frac{6 + 3}{t} = \frac{9}{t} \qquad\qquad \frac{4x}{a} - \frac{2x}{a} = \frac{4x - 2x}{a} = \frac{2x}{a}$$

Complete these.

   a) $\dfrac{9}{ab} - \dfrac{5}{ab} =$ _____

   b) $\dfrac{2y}{5} + \dfrac{y}{5} =$ _____

a) $\dfrac{b - 3}{y}$
b) $\dfrac{5S + 7R}{PV}$

---

56.   When possible, we always reduce the sum or difference to lowest terms.   For example:

$$\frac{4}{3x} + \frac{5}{3x} = \frac{4 + 5}{3x} = \frac{\overset{3}{\cancel{9}}}{\underset{1}{\cancel{3}}x} = \frac{3}{x}$$

$$\frac{7}{5a} - \frac{2}{5a} = \frac{7 - 2}{5a} = \frac{\overset{1}{\cancel{5}}}{\underset{1}{\cancel{5}}a} = \frac{1}{a}$$

Complete.   Reduce to lowest terms if possible.

   a) $\dfrac{3y}{10} + \dfrac{3y}{10} =$ _____

   c) $\dfrac{x}{9} + \dfrac{4x}{9} =$ _____

   b) $\dfrac{7}{2t} - \dfrac{3}{2t} =$ _____

   d) $\dfrac{11P}{4m} - \dfrac{7P}{4m} =$ _____

a) $\dfrac{4}{ab}$ b) $\dfrac{3y}{5}$

---

57.   The addition and subtraction below also have like denominators.

$$\frac{x}{x - 2} + \frac{7}{x - 2} = \frac{x + 7}{x - 2} \qquad\qquad \frac{1}{3(a + b)} - \frac{h}{3(a + b)} = \frac{1 - h}{3(a + b)}$$

Complete these.

   a) $\dfrac{S}{2(x - y)} + \dfrac{T}{2(x - y)} =$ _____

   b) $\dfrac{3R}{P + Q} - \dfrac{R}{P + Q} =$ _____

a) $\dfrac{3y}{5}$ c) $\dfrac{5x}{9}$

b) $\dfrac{2}{t}$ d) $\dfrac{P}{m}$

58. Notice how we combined like terms below.

$$\frac{x + 5}{7} + \frac{3}{7} = \frac{(x + 5) + 3}{7} = \frac{x + 8}{7}$$

$$\frac{k + 1}{P - p} - \frac{k}{P - p} = \frac{(k + 1) - k}{P - p} = \frac{1}{P - p}$$

Complete these.   a) $\dfrac{2a + 3b}{v - 7} + \dfrac{2a - 3b}{v - 7} =$ _____

b) $\dfrac{a + 4b}{c} - \dfrac{b}{c} =$ _____

a) $\dfrac{S + T}{2(x - y)}$

b) $\dfrac{2R}{P + Q}$

---

59. To remove the grouping symbols below, we changed the sign of each term. Do the other subtraction.

$$\frac{2x}{x - 3} - \frac{x - 1}{x - 3} = \frac{2x - (x - 1)}{x - 3} = \frac{2x - x + 1}{x - 3} = \frac{x + 1}{x - 3}$$

$$\frac{a}{a + b} - \frac{a - b}{a + b} =$$ _____

a) $\dfrac{4a}{v - 7}$

b) $\dfrac{a + 3b}{c}$

---

60. To add more than two fractions with like denominators, we also add their numerators. For example:

$$\frac{a}{x} + \frac{b}{x} + \frac{c}{x} = \frac{a + b + c}{x} \qquad \frac{s}{7} + \frac{t}{7} + \frac{3}{7} =$$ _____

$\dfrac{b}{a + b}$

---

61. The expression below contains an addition and a subtraction.

$$\frac{x}{7} + \frac{y}{7} - \frac{5}{7} = \frac{x + y - 5}{7}$$

Complete: a) $\dfrac{a}{d} - \dfrac{b}{d} - \dfrac{c}{d} =$ _____

b) $\dfrac{h}{2x} + \dfrac{3}{2x} - \dfrac{h}{2x} =$ _____

$\dfrac{s + t + 3}{7}$

---

a) $\dfrac{a - b - c}{d}$ \qquad b) $\dfrac{3}{2x}$

8-6 EQUIVALENT FORMS

To add or subtract algebraic fractions with unlike denominators, we must convert one or both fractions to an equivalent form. We will discuss the method in this section.

---

62. Below we converted $\frac{x}{2}$ to an equivalent fraction whose denominator is 8. To do so, we multiplied by $\frac{4}{4}$, a fraction that equals "1". We used $\frac{4}{4}$ since $\frac{8}{2} = 4$.

$$\frac{x}{2} = \frac{x}{2}\left(\frac{4}{4}\right) = \frac{4x}{8}$$

To convert an algebraic fraction to an equivalent form, we always multiply by a fraction that equals "1". To find that fraction, we divide the new denominator by the original denominator. That is:

For $\frac{2y}{3} = \frac{(\quad)}{6}$, we multiply by $\frac{2}{2}$ since $\frac{6}{3} = 2$.

For $\frac{m}{5} = \frac{(\quad)}{15}$, we multiply by $\frac{3}{3}$ since $\frac{15}{5} = 3$.

Multiplying by a fraction that equals "1", complete each conversion.

 a) $\frac{3y}{4}\left(\frac{\quad}{\quad}\right) = \frac{(\quad)}{12}$ b) $\frac{ab}{2}\left(\frac{\quad}{\quad}\right) = \frac{(\quad)}{10}$

---

63. Below we converted $\frac{3}{t}$ to an equivalent fraction whose denominator is 5t. To do so, we multiplied by $\frac{5}{5}$ since $\frac{5t}{t} = 5$.

$$\frac{3}{t} = \frac{3}{t}\left(\frac{5}{5}\right) = \frac{15}{5t}$$

To find the fraction to multiply by, we divide the new denominator by the original denominator. For example:

For $\frac{7}{x} = \frac{(\quad)}{2x}$, we multiply by $\frac{2}{2}$ since $\frac{2x}{x} = 2$.

For $\frac{a}{9} = \frac{(\quad)}{9b}$, we multiply by $\frac{b}{b}$ since $\frac{9b}{9} = b$.

Multiplying by a fraction that equals "1", complete each conversion.

 a) $\frac{5}{R}\left(\frac{\quad}{\quad}\right) = \frac{(\quad)}{3R}$ b) $\frac{m}{8}\left(\frac{\quad}{\quad}\right) = \frac{(\quad)}{8d}$

| a) $\frac{3y}{4}\left(\frac{3}{3}\right) = \frac{9y}{12}$ |
| b) $\frac{ab}{2}\left(\frac{5}{5}\right) = \frac{5ab}{10}$ |

---

64. Two more examples of finding the correct fraction to multiply by are given below.

For $\frac{1}{2x} = \frac{(\quad)}{10x}$, we multiply by $\frac{5}{5}$ since $\frac{10x}{2x} = 5$.

For $\frac{a}{5} = \frac{(\quad)}{20t}$, we multiply by $\frac{4t}{4t}$ since $\frac{20t}{5} = 4t$.

Continued on following page.

| a) $\frac{5}{R}\left(\frac{3}{3}\right) = \frac{15}{3R}$ |
| b) $\frac{m}{8}\left(\frac{d}{d}\right) = \frac{dm}{8d}$ |

64. Continued

Multiplying by a fraction that equals "1", complete each conversion.

a) $\dfrac{3a}{4x}\left(\dfrac{\quad}{\quad}\right) = \dfrac{(\quad)}{8x}$ 　　　　b) $\dfrac{1}{3}\left(\dfrac{\quad}{\quad}\right) = \dfrac{(\quad)}{9y}$

---

65. Two more examples of finding the correct fraction to multiply by are given below.

For $\dfrac{a}{b} = \dfrac{(\quad)}{bd}$, we multiply by $\dfrac{d}{d}$ since $\dfrac{bd}{b} = d$.

For $\dfrac{m}{x} = \dfrac{(\quad)}{3xy}$, we multiply by $\dfrac{3y}{3y}$ since $\dfrac{3xy}{x} = 3y$.

Multiplying by a fraction that equals "1", complete each conversion.

a) $\dfrac{1}{m}\left(\dfrac{\quad}{\quad}\right) = \dfrac{(\quad)}{mt}$ 　　　　b) $\dfrac{a}{q}\left(\dfrac{\quad}{\quad}\right) = \dfrac{(\quad)}{pqr}$

a) $\dfrac{3a}{4x}\left(\dfrac{2}{2}\right) = \dfrac{6a}{8x}$

b) $\dfrac{1}{3}\left(\dfrac{3y}{3y}\right) = \dfrac{3y}{9y}$

---

66. Two more examples are given below.

For $\dfrac{4}{x} = \dfrac{(\quad)}{x^2}$, we multiply by $\dfrac{x}{x}$ since $\dfrac{x^2}{x} = x$.

For $\dfrac{1}{2y} = \dfrac{(\quad)}{8y^2}$, we multiply by $\dfrac{4y}{4y}$ since $\dfrac{8y^2}{2y} = 4y$.

Multiplying by a fraction that equals "1", complete each conversion.

a) $\dfrac{b}{y}\left(\dfrac{\quad}{\quad}\right) = \dfrac{(\quad)}{y^2}$ 　　　　b) $\dfrac{1}{4a}\left(\dfrac{\quad}{\quad}\right) = \dfrac{(\quad)}{12a^2}$

a) $\dfrac{1}{m}\left(\dfrac{t}{t}\right) = \dfrac{t}{mt}$

b) $\dfrac{a}{q}\left(\dfrac{pr}{pr}\right) = \dfrac{apr}{pqr}$

---

67. You should be able to make conversions without writing the fraction that equals "1". Do so below.

a) $\dfrac{2x}{7} = \dfrac{(\quad)}{14}$ 　　　　c) $\dfrac{H}{4} = \dfrac{(\quad)}{12T}$

b) $\dfrac{9}{a} = \dfrac{(\quad)}{4a}$ 　　　　d) $\dfrac{3}{2x} = \dfrac{(\quad)}{6x}$

a) $\dfrac{b}{y}\left(\dfrac{y}{y}\right) = \dfrac{by}{y^2}$

b) $\dfrac{1}{4a}\left(\dfrac{3a}{3a}\right) = \dfrac{3a}{12a^2}$

---

68. Do these.

a) $\dfrac{1}{k} = \dfrac{(\quad)}{kt}$ 　　　　c) $\dfrac{b}{m} = \dfrac{(\quad)}{m^2}$

b) $\dfrac{a}{b} = \dfrac{(\quad)}{4bc}$ 　　　　d) $\dfrac{3}{2x} = \dfrac{(\quad)}{6x^2}$

a) $\dfrac{4x}{14}$ 　　　c) $\dfrac{3HT}{12T}$

b) $\dfrac{36}{4a}$ 　　　d) $\dfrac{9}{6x}$

---

a) $\dfrac{t}{kt}$ 　　　b) $\dfrac{4ac}{4bc}$ 　　　c) $\dfrac{bm}{m^2}$ 　　　d) $\dfrac{9x}{6x^2}$

8-7 FINDING LOWEST COMMON DENOMINATORS

To add algebraic fractions with unlike denominators, we use the lowest common denominator. We will discuss a method for finding lowest common denominators in this section.

---

69. We have already discussed a method for finding the lowest common denominator (LCD) when both denominators are numbers.

   If the larger denominator is a multiple of the smaller, the larger denominator is the LCD.

$$\text{For } \frac{x}{8} + \frac{y}{2} \text{ , the LCD is 8.}$$

   If the larger denominator is not a multiple of the smaller, we check multiples of the larger until we find the smallest one that is also a multiple of the smaller.

$$\text{For } \frac{t}{6} + \frac{1}{8} \text{ . the LCD is 24.} \quad \text{For } \frac{m}{5} + \frac{d}{7} \text{ , the LCD is 35.}$$

   Find the LCD for each addition.

   a) $\frac{1}{5} + \frac{y}{20}$ b) $\frac{x}{6} + \frac{3}{4}$ c) $\frac{3x}{4} + \frac{2y}{7}$

   LCD = _____   LCD = _____   LCD = _____

---

70. A <u>prime</u> <u>number</u> is a number that has only itself and "1" as its factors. The first ten prime numbers are given below.

$$2, \ 3, \ 5, \ 7, \ 11, \ 13, \ 17, \ 19, \ 23, \ 29$$

   There is a more general method that can be used to find a lowest common denominator (LCD). In that method, we begin by factoring the denominators into primes. For example:

$$\frac{7}{12} + \frac{11}{30} \qquad \begin{array}{l} 12 = (2)(2)(3) \\ 30 = (2)(3)(5) \end{array}$$

   Then to get the LCD, <u>we use each factor the greatest number of times it appears in any denominator</u>. Therefore, we get the following LCD for the addition above:

$$\text{LCD} = (2)(2)(3)(5) = 60$$

   <u>Note</u>: Since 12 has 2 as a factor twice and 30 has 2 as a factor only once, the LCD has 2 as a factor twice (the greatest number of times it occurs).

   We factored each denominator below. Use the method above to find the LCD.

$$\frac{4}{15} + \frac{7}{18} \qquad \begin{array}{l} 15 = (3)(5) \\ 18 = (2)(3)(3) \end{array} \qquad \text{LCD} = \underline{\hspace{2cm}}$$

---

a) 20

b) 12

c) 28

$(2)(3)(3)(5) = 90$

71. The following principle is used after factoring each denominator.

> To get the LCD, use each factor the greatest number of times it appears in any denominator.

We used the principle above to find the LCD for the three denominators below.

$$\frac{5}{6} + \frac{4}{9} + \frac{17}{21} \qquad \begin{array}{l} 6 = (2)(3) \\ 9 = (3)(3) \\ 21 = (3)(7) \end{array} \qquad LCD = (2)(3)(3)(7) = 126$$

Following the example, find the LCD for this addition.

$$\frac{3}{4} + \frac{9}{10} + \frac{1}{12} \qquad \begin{array}{l} 4 = \underline{\hspace{2cm}} \\ 10 = \underline{\hspace{2cm}} \\ 12 = \underline{\hspace{2cm}} \end{array} \qquad LCD = \underline{\hspace{3cm}}$$

---

72. The same method is used when one or more denominators contains a variable. For example, the LCD below is $\underline{5x}$.

$$\frac{7}{5x} + \frac{1}{x} \qquad \begin{array}{l} 5x = (5)(x) \\ x = (x) \end{array} \qquad LCD = (5)(x) = 5x$$

Find the LCD for each addition.

a) $\frac{3}{p} + \frac{1}{4}$ b) $\frac{a}{y} + \frac{b}{9y}$ c) $\frac{5}{3x} + \frac{1}{x}$

LCD = _____   LCD = _____   LCD = _____

Answer panel (right):
$$\begin{array}{l} 4 = (2)(2) \\ 10 = (2)(5) \\ 12 = (2)(2)(3) \\ LCD = (2)(2)(3)(5) \\ \qquad = 60 \end{array}$$

---

73. We used the same method below. The LCD is $\underline{12a}$, the product of the variable and the smallest common multiple of the two numbers.

$$\frac{7}{12} + \frac{5}{3a} \qquad \begin{array}{l} 12 = (2)(2)(3) \\ 3a = (3)(a) \end{array} \qquad LCD = (2)(2)(3)(a) = 12a$$

Find the LCD for each addition.

a) $\frac{x}{4} + \frac{y}{8d}$ b) $\frac{1}{5t} + \frac{5}{6}$ c) $\frac{3}{10y} + \frac{1}{4}$

LCD = _____   LCD = _____   LCD = _____

Answer panel (right):
a) 4p

b) 9y

c) 3x

---

74. The LCD below is $\underline{10b}$. It is the product of the variable and the smallest common multiple of the coefficients.

$$\frac{a}{5b} + \frac{c}{2b} \qquad \begin{array}{l} 5b = (5)(b) \\ 2b = (2)(b) \end{array} \qquad LCD = (2)(5)(b) = 10b$$

Find the LCD for each addition.

a) $\frac{5}{9x} + \frac{1}{3x}$ b) $\frac{x}{4c} + \frac{y}{3c}$ c) $\frac{t}{8m} + \frac{1}{6m}$

LCD = _____   LCD = _____   LCD = _____

Answer panel (right):
a) 8d

b) 30t

c) 20y

75. The LCD below is 6ab. It is the product of the variables and the smallest common multiple of the coefficients.

$$\frac{5}{3a} + \frac{1}{6b} \qquad \begin{array}{l} 3a = (3)(a) \\ 6b = (2)(3)(b) \end{array} \qquad \text{LCD} = (2)(3)(a)(b) = 6ab$$

Find the LCD for each addition.

a) $\frac{5}{x} + \frac{3}{y}$　　　　　b) $\frac{1}{3P} + \frac{1}{5R}$　　　　　c) $\frac{y}{15m} + \frac{9}{10t}$

　　LCD = _____　　　　LCD = _____　　　　LCD = _____

a) 9x

b) 12c

c) 24m

---

76. The LCD below is <u>xyz</u>. Notice that each variable is used <u>only</u> <u>once</u>.

$$\frac{m}{xy} + \frac{t}{yz} \qquad \begin{array}{l} xy = (x)(y) \\ yz = (y)(z) \end{array} \qquad \text{LCD} = (x)(y)(z) = xyz$$

Find the LCD for each addition.

a) $\frac{1}{ab} + \frac{1}{t}$　　　　　b) $\frac{a}{cd} + \frac{b}{d}$　　　　　c) $\frac{T}{AS} + \frac{R}{ASV}$

　　LCD = _____　　　　LCD = _____　　　　LCD = _____

a) xy

b) 15PR

c) 30mt

---

77. The LCD below is 10PV. It is the product of each variable and the smallest common multiple of the coefficients.

$$\frac{H}{5PV} + \frac{T}{10P} \qquad \begin{array}{l} 5PV = (5)(P)(V) \\ 10P = (2)(5)(P) \end{array} \qquad \text{LCD} = (2)(5)(P)(V) = 10PV$$

Find the LCD for each addition.

a) $\frac{9}{2b} + \frac{8}{3cd}$　　　　　b) $\frac{x}{m} + \frac{y}{3mp}$　　　　　c) $\frac{1}{12RT} + \frac{1}{3R}$

　　LCD = _____　　　　LCD = _____　　　　LCD = _____

a) abt

b) cd

c) ASV

---

78. The LCD below is $6y^2$. Since $3y^2$ has <u>y</u> as a factor twice, the LCD has <u>y</u> as a factor twice.

$$\frac{2}{3y^2} + \frac{5}{6y} \qquad \begin{array}{l} 3y^2 = (3)(y)(y) \\ 6y = (2)(3)(y) \end{array} \qquad \text{LCD} = (2)(3)(y)(y) = 6y^2$$

Find the LCD for each addition.

a) $\frac{1}{2a} + \frac{3}{a^2}$　　　　　　　　b) $\frac{3}{5t} + \frac{1}{4t^2}$

　　LCD = _____　　　　　　LCD = _____

a) 6bcd

b) 3mp

c) 12RT

---

a) $2a^2$　　b) $20t^2$

79. We found the LCD for the addition below.

$$\frac{3}{5xy^2} + \frac{5}{x^2y}$$

$$5xy^2 = (5)(x)(y)(y)$$
$$x^2y = (x)(x)(y)$$

$$LCD = (5)(x)(x)(y)(y) = 5x^2y^2$$

Find the LCD for these.

a) $\frac{1}{3a^2b} + \frac{2}{2ab}$ b) $\frac{3}{4x^2y^2} + \frac{1}{8xy} + \frac{5}{2y^2}$

LCD = _____   LCD = _____

---

80. Find the LCD for each addition.

a) $\frac{R}{V} + \frac{S}{8V}$ b) $\frac{d}{xy} + \frac{f}{a}$ c) $\frac{1}{4m} + \frac{1}{6m}$

LCD = _____   LCD = _____   LCD = _____

a) $6a^2b$ b) $8x^2y^2$

---

81. Find the LCD for each addition.

a) $\frac{v}{8PT} + \frac{m}{2T}$ b) $\frac{a}{x^2} + \frac{b}{x}$ c) $\frac{1}{3a} + \frac{3}{4y} + \frac{5}{2y^2}$

LCD = _____   LCD = _____   LCD = _____

a) $8V$

b) axy

c) $12m$

---

a) $8PT$ b) x^2 c) $12ay^2$

8-8 ADDITION AND SUBTRACTION WITH UNLIKE DENOMINATORS

In this section, we will discuss the procedure for adding and subtracting algebraic fractions with unlike denominators.

82. To add or subtract algebraic fractions with unlike denominators, we use the following steps:

   1. Find the lowest common denominator.
   2. Convert one or both fractions to an equivalent form with the LCD as the denominator.
   3. Then add or subtract in the usual way.

When one of the denominators is the LCD, we only have to make one substitution. Two examples are discussed.

The LCD is 6. We substitute $\frac{3}{6}$ for $\frac{1}{2}$.

$$\frac{x}{6} + \frac{1}{2} = \frac{x}{6} + \frac{3}{6} = \frac{x + 3}{6}$$

Continued on following page.

82. Continued

The LCD is 2y. We substitute $\frac{10}{2y}$ for $\frac{5}{y}$.

$$\frac{5}{y} - \frac{3}{2y} = \frac{10}{2y} - \frac{3}{2y} = \frac{7}{2y}$$

Following the examples, complete these.

a) $\frac{7}{10} + \frac{t}{5} = \frac{7}{10} + \underline{\hspace{1cm}} = \underline{\hspace{2cm}}$

b) $\frac{3}{2} - \frac{5}{8d} = \underline{\hspace{1cm}} - \frac{5}{8d} = \underline{\hspace{2cm}}$

---

83. In each example below, we only have to make one substitution.

The LCD is 12x. We substitute $\frac{3m}{12x}$ for $\frac{m}{4x}$.

$$\frac{1}{12x} + \frac{m}{4x} = \frac{1}{12x} + \frac{3m}{12x} = \frac{1 + 3m}{12x}$$

The LCD is bc. We substitute $\frac{ac}{bc}$ for $\frac{a}{b}$.

$$\frac{a}{b} - \frac{d}{bc} = \frac{ac}{bc} - \frac{d}{bc} = \frac{ac - d}{bc}$$

Following the examples, complete these.

a) $\frac{5}{3x} + \frac{1}{9x} = \underline{\hspace{1cm}} + \frac{1}{9x} = \underline{\hspace{2cm}}$

b) $\frac{R}{PV} - \frac{S}{P} = \frac{R}{PV} - \underline{\hspace{1cm}} = \underline{\hspace{2cm}}$

a) $\frac{7}{10} + \frac{2t}{10} = \frac{7 + 2t}{10}$

b) $\frac{12d}{8d} - \frac{5}{8d} = \frac{12d-5}{8d}$

---

84. In the addition below, we had to reduce the sum to lowest terms.

$$\frac{1}{a} + \frac{12}{2a} = \frac{2}{2a} + \frac{12}{2a} = \frac{\overset{7}{\cancel{14}}}{\underset{1}{\cancel{2a}}} = \frac{7}{a}$$

Complete these. Reduce each answer to lowest terms.

a) $\frac{7a}{8y} + \frac{9a}{24y} = \underline{\hspace{3cm}}$

b) $\frac{1}{3x} - \frac{1}{12x} = \underline{\hspace{3cm}}$

a) $\frac{15}{9x} + \frac{1}{9x} = \frac{16}{9x}$

b) $\frac{R}{PV} - \frac{SV}{PV} = \frac{R-SV}{PV}$

---

a) $\frac{5a}{4y}$, from: $\frac{30a}{24y}$

b) $\frac{1}{4x}$, from: $\frac{3}{12x}$

348 • ALGEBRAIC FRACTIONS

85. When neither denominator is the LCD, we have to substitute for both fractions. Two examples are discussed below.

The LCD is 12. We substitute $\frac{3t}{12}$ for $\frac{t}{4}$ and $\frac{4t}{12}$ for $\frac{t}{3}$.

$$\frac{t}{4} + \frac{t}{3} = \frac{3t}{12} + \frac{4t}{12} = \frac{7t}{12}$$

The LCD is 5x. We substitute $\frac{3x}{5x}$ for $\frac{3}{5}$ and $\frac{5}{5x}$ for $\frac{1}{x}$.

$$\frac{3}{5} - \frac{1}{x} = \frac{3x}{5x} - \frac{5}{5x} = \frac{3x - 5}{5x}$$

Complete these.

a) $\frac{a}{6} + \frac{a}{4} = $ _____ + _____ = _____

b) $\frac{1}{2x} - \frac{5}{3} = $ _____ - _____ = _____

86. Two more examples of the same type are discussed below.

The LCD is cd. We substitute $\frac{ad}{cd}$ for $\frac{a}{c}$ and $\frac{bc}{cd}$ for $\frac{b}{d}$.

$$\frac{a}{c} + \frac{b}{d} = \frac{ad}{cd} + \frac{bc}{cd} = \frac{ad + bc}{cd}$$

The LCD is 2xy. We substitute $\frac{y}{2xy}$ for $\frac{1}{2x}$ and $\frac{2}{2xy}$ for $\frac{1}{xy}$.

$$\frac{1}{2x} - \frac{1}{xy} = \frac{y}{2xy} - \frac{2}{2xy} = \frac{y - 2}{2xy}$$

Complete these.

a) $\frac{2}{p} - \frac{3}{q} = $ _____ - _____ = _____

b) $\frac{t}{my} + \frac{1}{v} = $ _____ + _____ = _____

---

a) $\frac{2a}{12} + \frac{3a}{12} = \frac{5a}{12}$

b) $\frac{3}{6x} - \frac{10x}{6x} = \frac{3 - 10x}{6x}$

a) $\frac{2q}{pq} - \frac{3p}{pq} = \frac{2q - 3p}{pq}$

b) $\frac{tv}{mvy} + \frac{my}{mvy} = \frac{tv + my}{mvy}$

87. Below we have to make one substitution in the top example and two substitutions in the bottom example.

The LCD is x^2. We substitute $\frac{3x}{x^2}$ for $\frac{3}{x}$.

$$\frac{3}{x} - \frac{4}{x^2} = \frac{3x}{x^2} - \frac{4}{x^2} = \frac{3x - 4}{x^2}$$

The LCD is $2a^2$. We substitute $\frac{a}{2a^2}$ for $\frac{1}{2a}$ and $\frac{6}{2a^2}$ for $\frac{3}{a^2}$.

$$\frac{1}{2a} + \frac{3}{a^2} = \frac{a}{2a^2} + \frac{6}{2a^2} = \frac{a + 6}{2a^2}$$

Complete these.

a) $\frac{a}{y^2} + \frac{1}{y} =$ _____ + _____ = _____

b) $\frac{3}{5t} - \frac{1}{4t^2} =$ _____ - _____ = _____

88. When adding or subtracting algebraic fractions, <u>check first to see if one of the denominators is the LCD</u>.

If one of the denominators is the LCD, you only have to substitute for <u>one</u> fraction.

If neither denominator is the LCD, you have to substitute for <u>both</u> fractions.

Complete: a) $\frac{1}{4t} + \frac{1}{t} =$ _____

b) $\frac{a}{3} - \frac{b}{2} =$ _____

c) $\frac{c}{xy} + \frac{d}{x} =$ _____

d) $\frac{3x}{c} - \frac{2y}{d} =$ _____

a) $\frac{a}{y^2} + \frac{y}{y^2} = \frac{a + y}{y^2}$

b) $\frac{12t}{20t^2} - \frac{5}{20t^2} = \frac{12t - 5}{20t^2}$

89. Complete: a) $\frac{3y}{4} - \frac{y}{8} =$ _____

b) $\frac{7}{12x} + \frac{3}{4x} =$ _____

c) $\frac{b}{t^2} - \frac{4}{t} =$ _____

d) $\frac{1}{2a} + \frac{1}{3a^2} =$ _____

a) $\frac{5}{4t}$

b) $\frac{2a - 3b}{6}$

c) $\frac{c + dy}{xy}$

d) $\frac{3dx - 2cy}{cd}$

a) $\frac{5y}{8}$ b) $\frac{4}{3x}$ c) $\frac{b - 4t}{t^2}$ d) $\frac{3a + 2}{6a^2}$

SELF-TEST 27 (pages 339-350)

Add or subtract. Report each sum or difference in lowest terms.

1. $\dfrac{b}{2a} + \dfrac{5b}{2a} =$

2. $\dfrac{8}{3x} - \dfrac{2}{3x} =$

3. $\dfrac{h + 1}{c - w} - \dfrac{h}{c - w} =$

4. $\dfrac{a + t}{p + r} + \dfrac{a - t}{p + r} =$

5. $\dfrac{y}{2} + \dfrac{5y}{6} =$

6. $\dfrac{h}{6} - \dfrac{h}{4} =$

7. $\dfrac{d}{3w} - \dfrac{2}{w} =$

8. $\dfrac{1}{r} + \dfrac{2}{r^2} =$

9. $\dfrac{a}{b} + \dfrac{c}{d} =$ $\quad \dfrac{ad}{bd} + \dfrac{bc}{bd}$ $\quad \dfrac{ad + bc}{bd}$

10. $\dfrac{1}{A} - \dfrac{1}{B} =$ $\quad \dfrac{B}{BA} - \dfrac{A}{BA} \quad \dfrac{B - A}{BA}$

11. $\dfrac{5}{6a} - \dfrac{7}{9a} =$ $\quad \dfrac{15}{18a} - \dfrac{14}{18a}$

12. $\dfrac{2y}{3x} + \dfrac{y}{x^2} =$ $\quad \dfrac{2xx}{3x^2} + \dfrac{3y}{3x^2} \quad \dfrac{2xx + 3y}{3x^2}$

ANSWERS:

1. $\dfrac{3b}{a}$

2. $\dfrac{2}{x}$

3. $\dfrac{1}{c - w}$

4. $\dfrac{2a}{p + r}$

5. $\dfrac{4y}{3}$

6. $-\dfrac{h}{12}$

7. $\dfrac{d - 6}{3w}$

8. $\dfrac{r + 2}{r^2}$

9. $\dfrac{ad + bc}{bd}$

10. $\dfrac{B - A}{AB}$

11. $\dfrac{1}{18a}$

12. $\dfrac{2xy + 3y}{3x^2}$

8-9 ADDING AND SUBTRACTING A FRACTION AND A NON-FRACTION

In this section, we will discuss the procedure for adding and subtracting a fraction and a non-fraction.

90. To convert a non-fraction to a fraction with a specific denominator, we multiply by a fraction that equals "1". The terms of that fraction are determined by the denominator of the desired fraction. For example:

To convert 2 to a fraction whose denominator is 4, we multiply by $\frac{4}{4}$.

$$2\left(\frac{4}{4}\right) = \frac{8}{4}$$

To convert 2t to a fraction whose denominator is $\underline{b}$, we multiply by $\frac{b}{b}$.

$$2t\left(\frac{b}{b}\right) = \frac{2bt}{b}$$

Following the examples, complete these.

a) $5\left(\dfrac{\quad}{\quad}\right) = \dfrac{(\quad)}{x}$ b) $p\left(\dfrac{\quad}{\quad}\right) = \dfrac{(\quad)}{qr}$

91. In such conversions, the numerator can be found <u>by multiplying the non-fraction and the desired denominator</u>. That is:

$$3 = \frac{3y}{y} \text{ , since } (3)(y) = 3y.$$

$$2m = \frac{8m}{4} \text{ , since } (2m)(4) = 8m.$$

Use the shorter method for these.

a) $1 = \dfrac{(\quad)}{9}$ b) $2t = \dfrac{(\quad)}{7}$ c) $7 = \dfrac{(\quad)}{b}$

a) $5\left(\dfrac{x}{x}\right) = \dfrac{5x}{x}$

b) $p\left(\dfrac{qr}{qr}\right) = \dfrac{pqr}{qr}$

92. Complete these.

a) $2 = \dfrac{(\quad)}{x^2}$ b) $5 = \dfrac{(\quad)}{3d}$ c) $4y = \dfrac{(\quad)}{cd}$

a) $\dfrac{9}{9}$ b) $\dfrac{14t}{7}$ c) $\dfrac{7b}{b}$

93. To add or subtract an algebraic fraction and a non-fraction, we convert the non-fraction to a fraction with the same denominator. Two examples are shown below.

$$\frac{x}{3} + 2 = \frac{x}{3} + \frac{6}{3} = \frac{x+6}{3} \qquad 2m - \frac{3}{5} = \frac{10m}{5} - \frac{3}{5} = \frac{10m-3}{5}$$

Following the examples, complete these.

a) $1 - \dfrac{2y}{7} = \underline{\qquad} - \dfrac{2y}{7} = \underline{\qquad}$

b) $\dfrac{1}{2} + b = \dfrac{1}{2} + \underline{\qquad} = \underline{\qquad}$

a) $\dfrac{2x^2}{x^2}$

b) $\dfrac{15d}{3d}$

c) $\dfrac{4cdy}{cd}$

94. Another addition and subtraction are shown below.

$$\frac{8}{x} + \frac{1}{x} = \frac{8x}{x} + \frac{1}{x} = \frac{8x + 1}{x} \qquad \frac{b}{cd} - R = \frac{b}{cd} - \frac{cdR}{cd} = \frac{b - cdR}{cd}$$

Following the examples, complete these.

a) $\dfrac{3}{t} - 1 = \dfrac{3}{t} - \underline{\hspace{1cm}} = \underline{\hspace{2cm}}$

b) $h + \dfrac{P}{V} = \underline{\hspace{1cm}} + \dfrac{P}{V} = \underline{\hspace{2cm}}$

a) $\dfrac{7}{7} - \dfrac{2y}{7} = \dfrac{7 - 2y}{7}$

b) $\dfrac{1}{2} + \dfrac{2b}{2} = \dfrac{1 + 2b}{2}$

---

95. Complete these.

a) $4y + \dfrac{2}{3} = \underline{\hspace{2cm}}$ 　　c) $1 + \dfrac{R}{5} = \underline{\hspace{2cm}}$

b) $\dfrac{1}{4} - d = \underline{\hspace{2cm}}$ 　　d) $4 - \dfrac{5}{x} = \underline{\hspace{2cm}}$

a) $\dfrac{3}{t} - \dfrac{t}{t} = \dfrac{3 - t}{t}$

b) $\dfrac{hV}{V} + \dfrac{P}{V} = \dfrac{hV + P}{V}$

---

96. Complete these.

a) $a + \dfrac{b}{c} = \underline{\hspace{2cm}}$ 　　c) $\dfrac{m}{v^2} + 1 = \underline{\hspace{2cm}}$

b) $\dfrac{c}{de} - 1 = \underline{\hspace{2cm}}$ 　　d) $2T - \dfrac{H}{PV} = \underline{\hspace{2cm}}$

a) $\dfrac{12y + 2}{3}$

b) $\dfrac{1 - 4d}{4}$

c) $\dfrac{5 + R}{5}$

d) $\dfrac{4x - 5}{x}$

a) $\dfrac{ac + b}{c}$ 　　b) $\dfrac{c - de}{de}$ 　　c) $\dfrac{m + v^2}{v^2}$ 　　d) $\dfrac{2PTV - H}{PV}$

8-10 EQUIVALENT FORMS FOR SUMS AND DIFFERENCES

Any fraction whose numerator is an addition or subtraction is a sum or difference of fractions. In this section, we will show how fractions of that type can be written in an equivalent form by breaking up the sum or difference into the original fractions.

---

97. Any fraction whose numerator is a sum or difference is a sum or difference of fractions. It can be broken up into the original fractions. For example:

$$\frac{y + 5}{7} = \frac{y}{7} + \frac{5}{7} \qquad \frac{a - b}{c} = \frac{a}{c} - \frac{b}{c}$$

Break up each sum or difference into the original fractions.

a) $\dfrac{5m + 3t}{8} = \underline{\hspace{1cm}} + \underline{\hspace{1cm}}$ 　　b) $\dfrac{cd - pq}{mt} = \underline{\hspace{1cm}} - \underline{\hspace{1cm}}$

98. When a fraction is broken up, sometimes one or both of the original fractions can be reduced to lowest terms.

In the example below, one fraction can be reduced.

$$\frac{y + 2}{6} = \frac{y}{6} + \frac{2}{6} = \frac{y}{6} + \frac{1}{3}$$

In the example below, both fractions can be reduced.

$$\frac{3a - 4b}{12} = \frac{3a}{12} - \frac{4b}{12} = \frac{a}{4} - \frac{b}{3}$$

Break these up into the two original fractions and then reduce to lowest terms where possible.

a) $\frac{t - 5}{10}$ = _____ - _____ = _____ - _____

b) $\frac{4m + 2t}{8}$ = _____ + _____ = _____ + _____

a) $\frac{5m}{8} + \frac{3t}{8}$

b) $\frac{cd}{mt} - \frac{pq}{mt}$

99. Break these up into the two original fractions and then reduce to lowest terms where possible.

a) $\frac{m + tv}{ct}$ = _____ + _____ = _____ + _____

b) $\frac{ay - cx}{bxy}$ = _____ - _____ = _____ - _____

a) $\frac{t}{10} - \frac{5}{10} = \frac{t}{10} - \frac{1}{2}$

b) $\frac{4m}{8} + \frac{2t}{8} = \frac{m}{2} + \frac{t}{4}$

100. In the example below, one of the fractions reduces to a non-fraction.

$$\frac{y + 7}{7} = \frac{y}{7} + \frac{7}{7} = \frac{y}{7} + 1$$

Write each fraction in an equivalent form with both parts reduced to lowest terms.

a) $\frac{a - 12}{2}$ = _____

b) $\frac{2x + 8y}{4}$ = _____

a) $\frac{m}{ct} + \frac{tv}{ct} = \frac{m}{ct} + \frac{v}{c}$

b) $\frac{ay}{bxy} - \frac{cx}{bxy} = \frac{a}{bx} - \frac{c}{by}$

101. In the example below, one of the fractions also reduces to a non-fraction.

$$\frac{P + QR}{Q} = \frac{P}{Q} + \frac{QR}{Q} = \frac{P}{Q} + R$$

Write each fraction in an equivalent form with both parts reduced to lowest terms.

a) $\frac{bd + m}{d}$ = _____

b) $\frac{LT - AKT}{AKT}$ = _____

a) $\frac{a}{2} - 6$

b) $\frac{x}{2} + 2y$

102. Write each fraction in an equivalent form with both parts reduced to lowest terms.

a) $\frac{x + 1}{x^2}$ = _____

b) $\frac{a + b^2}{b^2}$ = _____

a) $b + \frac{m}{d}$

b) $\frac{L}{AK} - 1$

103. Which of the following is equivalent to $\dfrac{5y + 9}{9}$? _____

    a) $5y + 1$ b) $\dfrac{5y}{9} + 1$ c) $5y$

| a) $\dfrac{1}{x} + \dfrac{1}{x^2}$ |
| b) $\dfrac{a}{b^2} + 1$ |

104. Which of the following is equivalent to $\dfrac{a - bc}{b}$? _____

    a) $a - c$ b) $a - \dfrac{c}{b}$ c) $\dfrac{a}{b} - c$

(b)

105. Which of the following is equivalent to $\dfrac{PQR + T}{QR}$? _____

    a) $P + \dfrac{T}{QR}$ b) $P + T$ c) $\dfrac{P}{QR} + T$

(c)

106. Which of the following is equivalent to $\dfrac{M - STV}{TV}$? _____

    a) $\dfrac{M}{TV} - S$ b) $M - S$ c) $\dfrac{M}{TV} - STV$

(a)

107. Which of the following is equivalent to $\dfrac{2x}{3} + 3$? _____

    a) $2x$ b) $\dfrac{2x + 9}{3}$ c) $\dfrac{5x}{3}$

(a)

108. Which of the following is equivalent to $x - \dfrac{ay}{ct}$? _____

    a) $\dfrac{x - ay}{ct}$ b) $\dfrac{ctx - acty}{ct}$ c) $\dfrac{ctx - ay}{ct}$

(b)

109. Which of the following is equivalent to $\dfrac{33,000H}{\pi dR} + F$? _____

    a) $\dfrac{33,000H + F}{\pi dR}$ b) $\dfrac{33,000\pi dRH + F}{\pi dR}$ c) $\dfrac{33,000H + \pi dRF}{\pi dR}$

(c)

110. Each fraction below is also a sum or difference. Therefore, each can be broken up into the original fractions. That is:

$$\frac{m + 5}{t - 7} = \frac{m}{t - 7} + \frac{5}{t - 7}$$

$$\frac{K - T}{S + V} = \underline{\hspace{2cm}} - \underline{\hspace{2cm}}$$

(c)

111. Following the example, break up the other fraction into the three original fractions.

$$\frac{m + y - 9}{5} = \frac{m}{5} + \frac{y}{5} - \frac{9}{5}$$

    a) $\dfrac{a + b + 4}{4} =$ _____   b) $\dfrac{cV - dS - 1}{c} =$ _____

$\dfrac{K}{S + V} - \dfrac{T}{S + V}$

a) $\dfrac{a}{4} + \dfrac{b}{4} + 1$ b) $V - \dfrac{dS}{c} - \dfrac{1}{c}$

8-11   CONTRASTING TWO PATTERNS

When a fraction contains an addition or subtraction in its numrator, it is the sum or difference of fractions.   However, when a fraction contains an addition or subtraction only in its denominator, it is not a sum or difference of fractions.   We will contrast those two patterns of fractions in this section.

---

112.   When a fraction contains an addition in its numerator, it is the sum of two fractions and can be broken up into the original fractions. For example:

$$\frac{x + 3}{7} = \frac{x}{7} + \frac{3}{7}$$

However, when a fraction contains an addition only in its denominator, it is not the sum of two fractions.   Therefore, it cannot be broken up into two fractions.   That is:

$$\frac{10}{y + 10} \neq \frac{10}{y} + \frac{10}{10}$$

To show that the two expressions above are not equal, we can substitute 2 for y in each and evaluate.

a) $\dfrac{10}{y + 10} = \dfrac{10}{2 + 10} = $ _____   b) $\dfrac{10}{y} + \dfrac{10}{10} = \dfrac{10}{2} + \dfrac{10}{10} = $ _____

---

113.   When a fraction contains a subtraction in its numerator, it is the difference of two fractions and can be broken up into the original fractions.   For example:

$$\frac{w - 5}{w} = \frac{w}{w} - \frac{5}{w}$$

However, when a fraction contains a subtraction only in its denominator, it is not the difference of two fractions.   Therefore, it cannot be broken up into two fractions.   That is:

$$\frac{x}{x - 1} \neq \frac{x}{x} - \frac{x}{1}$$

To show that the two expressions above are not equal, we can substitute 5 for x in each and evaluate.

a) $\dfrac{x}{x - 1} = \dfrac{5}{5 - 1} = $ _____   b) $\dfrac{x}{x} - \dfrac{x}{1} = \dfrac{5}{5} - \dfrac{5}{1} = $ _____

| a) $\dfrac{10}{12} = \dfrac{5}{6}$ |
| b) 5 + 1 = 6 |

---

114.   Which of the following is either a sum or difference of two fractions?   _____

a) $\dfrac{M + 5}{M}$ b) $\dfrac{t}{t + 7}$ c) $\dfrac{ab}{a + b}$ d) $\dfrac{a - b}{ab}$

| a) $\dfrac{5}{4}$ |
| b) 1 - 5 = -4 |

Only (a) and (d)

115. Fractions like those below cannot be written in an equivalent form by breaking them up into two fractions.

$$\frac{P}{P + 6} \qquad\qquad \frac{d}{d - a}$$

If possible, write each of these in an equivalent form by breaking it up into two fractions.

a) $\dfrac{k}{k + 1}$ = _____

b) $\dfrac{k - 1}{k}$ = _____

c) $\dfrac{R + T}{T}$ = _____

d) $\dfrac{T}{R - T}$ = _____

116. A fraction with an addition or subtraction <u>in</u> <u>both</u> <u>its</u> <u>numerator</u> <u>and</u> <u>denominator</u> is the sum or difference of two fractions. Therefore, it can be broken up into the original fractions. That is:

$$\frac{a + b}{c + d} = \frac{a}{c + d} + \frac{b}{c + d}$$

$$\frac{A - CV}{M - T} = \underline{\hspace{4cm}}$$

a) Not possible

b) $\dfrac{k}{k} - \dfrac{1}{k} = 1 - \dfrac{1}{k}$

c) $\dfrac{R}{T} + \dfrac{T}{T} = \dfrac{R}{T} + 1$

d) Not possible

$$\frac{A}{M - T} - \frac{CV}{M - T}$$

8-12   THE DISTRIBUTIVE PRINCIPLE

In this section, we will review the procedures for multiplying and factoring by the distributive principle and extend them to expressions containing more than one variable. Both procedures will be used in the discussion of the equivalent forms of complicated fractions in the next section.

117. Two examples of multiplying by the distributive principle are shown below.

$$3(x + 5) = 3x + 15 \qquad 7(y - 1) = 7y - 7$$

The same procedure is used when the expression contains more than one variable. That is:

$$2(x + y) = 2x + 2y \qquad a(b - c) = ab - ac$$

Following the examples, complete these.

a) 9(y + 7) = _____

b) 4(m - t) = _____

c) 2(R + 3P) = _____

d) M(T - V) = _____

a) 9y + 63

b) 4m - 4t

c) 2R + 6P

d) MT - MV

118. When an expression contains the same factor in each term, we can factor by the distributive principle. For example:

$$2x + 6 = 2(x) + 2(3) = 2(x + 3)$$
$$10y - 5 = 5(2y) - 5(1) = 5(2y - 1)$$

The same procedure is used when the expression contains more than one variable. That is:

$$7a + 7b = 7(a) + 7(b) = 7(a + b)$$
$$6d - 3h = 3(2d) - 3(h) = 3(2d - h)$$

Following the examples, factor each expression.

a) $9p - 9q = $ _____

b) $8x + 4 = $ _____

c) $4d - 20 = $ _____

d) $2h + 8t = $ _____

---

119. In each factoring below, the common factor is a variable.

$$ab + ac = a(b + c)$$
$$cP - dP = P(c - d)$$

Factor out the common factor in these.

a) $mt + mv = $ _____

b) $H_1T_1 - H_1S_2 = $ _____

c) $pq + kq = $ _____

d) $MR - NR = $ _____

a) $9(p - q)$

b) $4(2x + 1)$

c) $4(d - 5)$

d) $2(h + 4t)$

---

120. Two more examples of factoring out a common factor are shown below.

$$bs + st = s(b + t) \qquad CM - AC = C(M - A)$$

Following the examples, factor these.

a) $mt + km = $ _____

b) $C_1D_1 - D_1T_1 = $ _____

a) $m(t + v)$

b) $H_1(T_1 - S_2)$

c) $q(p + k)$

d) $R(M - N)$

---

121. In the factorings below, we substituted $\underline{1S}$ for $\underline{S}$ and $\underline{1b}$ for $\underline{b}$ so that the "1" was not forgotten.

$$ST + S = ST + 1S = S(T + 1)$$
$$b - bc = 1b - bc = b(1 - c)$$

Following the examples, factor these.

a) $MT + T = $ _____

b) $p + pP = $ _____

c) $cd - c = $ _____

d) $R - QR = $ _____

a) $m(t + k)$

b) $D_1(C_1 - T_1)$

---

a) $T(M + 1)$

b) $p(1 + P)$

c) $c(d - 1)$

d) $R(1 - Q)$

122. To factor by the distributive principle, each term must have a common factor. That is:

$bd + cd$ can be factored because <u>d</u> is a common factor.

$HM + TV$ cannot be factored because there is no common factor.

Factor by the distributive principle if possible.

a) $aH - bH =$ _____   c) $KT - SV =$ _____

b) $cp + dq =$ _____   d) $c_1t + c_1v =$ _____

123. Factor by the distributive principle if possible.

a) $ab - c =$ _____   c) $VT - V =$ _____

b) $mM + m =$ _____   d) $RS - Q =$ _____

a) $H(a - b)$
b) Not possible
c) Not possible
d) $c_1(t + v)$

a) Not possible   b) $m(M + 1)$ c) $V(T - 1)$ d) Not possible

---

SELF-TEST 28 (pages 351-358)

Add or subtract. Write each answer as a single fraction.

1. $x + \dfrac{1}{x} =$ $x^2 + 1$

2. $\dfrac{A}{B} - 1 =$ $\dfrac{A}{B} - \dfrac{B}{B}$

3. $3r + \dfrac{2}{5} =$ $\dfrac{15r + 2}{5}$

4. $\dfrac{R}{V} - 2P =$ $\dfrac{R - 2PV}{V}$

5. $1 + \dfrac{1}{F} =$

6. $3x - \dfrac{y}{3w} =$ $\dfrac{9xw}{3w} - \dfrac{y}{3w}$ $\dfrac{9xw - y}{3w}$

Break up each fraction into a sum or difference of two fractions in lowest terms.

7. $\dfrac{5x + 2y}{10x} =$ $\dfrac{5x}{10x} + \dfrac{2y}{10x}$ $\dfrac{1x}{2x} + \dfrac{y}{5x}$

8. $\dfrac{R_1 - R_t}{R_1 R_t} =$

9. $\dfrac{BC + CD}{CD} =$ $\dfrac{BC}{CD} + \dfrac{CD}{CD}$ $\dfrac{B}{D} + \dfrac{2}{D}$

10. $\dfrac{G - H}{G + 1} =$

11. Which of the following are equivalent to $\dfrac{pr - 2rw}{pw}$?

a) $\dfrac{r}{w} - \dfrac{2w}{p}$ b) $\dfrac{r}{w} - \dfrac{2r}{p}$ c) $1 - \dfrac{2w}{p}$ d) $r - \dfrac{2r}{p}$

12. Which of the following can be broken up into two fractions?

a) $\dfrac{b}{d + 2}$ b) $\dfrac{P + 1}{P - 1}$ c) $\dfrac{t - 3a}{r}$ d) $\dfrac{1}{x - y}$

Factor each expression by the distributive principle if possible.

13. $4xy + 12y =$ $4y(x + 3)$ 14. $2hp - dt =$ 15. $G - AG =$ $G(1 - A)$

ANSWERS TO SELF-TEST 28:

1. $\dfrac{x^2 + 1}{x}$

2. $\dfrac{A - B}{B}$

3. $\dfrac{15r + 2}{5}$

4. $\dfrac{R - 2PV}{V}$

5. $\dfrac{F + 1}{F}$

6. $\dfrac{9wx - y}{3w}$

7. $\dfrac{1}{2} + \dfrac{y}{5x}$

8. $\dfrac{1}{R_t} - \dfrac{1}{R_1}$

9. $\dfrac{B}{D} + 1$

10. $\dfrac{G}{G + 1} - \dfrac{H}{G + 1}$

11. Only (b)

12. (b) and (c)

13. $4y(x + 3)$

14. Not possible

15. $G(1 - A)$

8-13   REDUCING PATTERNS TO LOWEST TERMS

In this section, we will show how patterns can be reduced to lowest terms by factoring out a fraction that equals "1". By <u>patterns</u>, we mean fractions whose numerator or denominator or both is an addition or subtraction.

124. To get a fraction equivalent to a given fraction, we multiply it by a fraction that equals "1". For example, we multiplied the fraction below by $\dfrac{a}{a}$. Notice that we multiplied by the distributive principle in the numerator.

$$\frac{p + q}{r} = \left(\frac{a}{a}\right)\left(\frac{p + q}{r}\right) = \frac{a(p + q)}{ar} = \frac{ap + aq}{ar}$$

Following the example, complete these.

   a) $\dfrac{a + b}{c} = \left(\dfrac{3}{3}\right)\left(\dfrac{a + b}{c}\right) = \dfrac{3(a + b)}{3c} = $ _____

   b) $\dfrac{x - 2y}{5} = \left(\dfrac{2}{2}\right)\left(\dfrac{x - 2y}{5}\right) = \dfrac{2(x - 2y)}{10} = $ _____

125. To get a fraction equivalent to the fraction below, we multiplied it by $\dfrac{4}{4}$. Notice that we multiplied by the distributive principle in the denominator.

$$\frac{c}{d - t} = \left(\frac{4}{4}\right)\left(\frac{c}{d - t}\right) = \frac{4c}{4(d - t)} = \frac{4c}{4d - 4t}$$

Following the example, complete these.

   a) $\dfrac{M}{P + Q} = \left(\dfrac{b}{b}\right)\left(\dfrac{M}{P + Q}\right) = \dfrac{bM}{b(P + Q)} = $ _____

   b) $\dfrac{1}{3a - 2b} = \left(\dfrac{3}{3}\right)\left(\dfrac{1}{3a - 2b}\right) = \dfrac{3}{3(3a - 2b)} = $ _____

a) $\dfrac{3a + 3b}{3c}$

b) $\dfrac{2x - 4y}{10}$

a) $\dfrac{bM}{bP + bQ}$

b) $\dfrac{3}{9a - 6b}$

126. To get a fraction equivalent to the fraction below, we multiplied by $\frac{2}{2}$. Notice that we multiplied by the distributive principle in both terms.

$$\frac{x + 3}{y - 2} = \left(\frac{2}{2}\right)\left(\frac{x + 3}{y - 2}\right) = \frac{2(x + 3)}{2(y - 2)} = \frac{2x + 6}{2y - 4}$$

Following the example, complete these.

a) $\frac{a + b}{c - d} = \left(\frac{s}{s}\right)\left(\frac{a + b}{c - d}\right) = \frac{s(a + b)}{s(c - d)} = $ _____

b) $\frac{t - 1}{t + 3} = \left(\frac{4}{4}\right)\left(\frac{t - 1}{t + 3}\right) = \frac{4(t - 1)}{4(t + 3)} = $ _____

127. To get an equivalent fraction below, we multiplied by $\frac{b}{b}$. Notice that each <u>term</u> in the numerator and denominator of the new fraction has <u>b</u> as a common factor.

$$\frac{m}{s - v} = \left(\frac{b}{b}\right)\left(\frac{m}{s - v}\right) = \frac{bm}{bs - bv}$$

Identify the common factor in each term of the numerator and denominator of these.

a) $\frac{mp + mq}{mt}$ _____   b) $\frac{ax}{ay - 3a}$ _____   c) $\frac{at - bt}{ct + dt}$ _____

a) $\frac{as + bs}{cs - ds}$

b) $\frac{4t - 4}{4t + 12}$

128. To get an equivalent fraction below, we multiplied by $\frac{3}{3}$. Notice that each <u>term</u> in the numerator and denominator of the new fraction has 3 as a common factor.

$$\frac{x + 2}{y} = \left(\frac{3}{3}\right)\left(\frac{x + 2}{y}\right) = \frac{3x + 6}{3y}$$

Identify the common factor in each term of the numerator and denominator of these.

a) $\frac{2p}{2q - 2t}$ _____   b) $\frac{5R - 10}{15Q}$ _____   c) $\frac{3y + 6}{3x - 9}$ _____

a) m

b) a

c) t

129. When a pattern has a common variable in each term of its numerator and denominator, we can reduce it to lowest terms by factoring out a fraction that equals "1". For example, we factored out $\frac{B}{B}$ below. Notice that we factored by the distributive principle in the numerator.

$$\frac{BD + BF}{BT} = \frac{B(D + F)}{BT} = \left(\frac{B}{B}\right)\left(\frac{D + F}{T}\right) = (1)\frac{D + F}{T} = \frac{D + F}{T}$$

Following the example, complete these.

a) $\frac{ac}{ap - aq} = \frac{ac}{a(p - q)} = \left(\frac{a}{a}\right)\left(\frac{c}{p - q}\right) = $ _____

b) $\frac{BT + KT}{MT + ST} = \frac{T(B - K)}{T(M + S)} = \left(\frac{T}{T}\right)\left(\frac{B - K}{M + S}\right) = $ _____

a) 2

b) 5

c) 3

130. Reduce to lowest terms by factoring out the common variable.

a) $\dfrac{ax + bx}{cx}$ = _____

b) $\dfrac{cP - cQ}{cR + cV}$ = _____

a) $\dfrac{c}{p - q}$

b) $\dfrac{B - K}{M + S}$

---

131. When a pattern has a common numerical factor in each term of its numerator and denominator, we can also reduce it to lowest terms by factoring out a fraction that equals "1". For example, we factored out $\dfrac{4}{4}$ below. Notice that we factored by the distributive principle in the denominator.

$$\dfrac{4x}{4y - 8} = \dfrac{4x}{4(y - 2)} = \left(\dfrac{4}{4}\right)\left(\dfrac{x}{y - 2}\right) = (1)\left(\dfrac{x}{y - 2}\right) = \dfrac{x}{y - 2}$$

Following the example, complete these.

a) $\dfrac{7m + 7}{7t} = \dfrac{7(m + 1)}{7t} = \left(\dfrac{7}{7}\right)\left(\dfrac{m + 1}{t}\right)$ = _____

b) $\dfrac{2a - 6b}{4c + 2} = \dfrac{2(a - 3b)}{2(2c + 1)} = \left(\dfrac{2}{2}\right)\left(\dfrac{a - 3b}{2c + 1}\right)$ = _____

a) $\dfrac{a + b}{c}$

b) $\dfrac{P - Q}{R + V}$

---

132. Reduce to lowest terms by factoring out the common numerical factor.

a) $\dfrac{5P}{10Q - 15}$ = _____

b) $\dfrac{3x - 6}{3y + 9}$ = _____

a) $\dfrac{m + 1}{t}$

b) $\dfrac{a - 3b}{2c + 1}$

---

133. When the fraction below is reduced to lowest terms, we get a non-fraction.

$$\dfrac{a + ab}{a} = \dfrac{a(1 + b)}{a} = \left(\dfrac{a}{a}\right)(1 + b) = 1 + b$$

Reduce these to lowest terms.

a) $\dfrac{4 + 4x}{4}$ = _____

b) $\dfrac{cd - d}{d}$ = _____

a) $\dfrac{P}{2Q - 3}$

b) $\dfrac{x - 2}{y + 3}$

---

134. To reduce the fraction below, we substituted $\underline{1P}$ for $\underline{P}$ in the numerator. Notice that the numerator of the reduced fraction is "1".

$$\dfrac{P}{P - PQ} = \dfrac{1P}{P(1 - Q)} = \left(\dfrac{P}{P}\right)\left(\dfrac{1}{1 - Q}\right) = \dfrac{1}{1 - Q}$$

Reduce these.

a) $\dfrac{y}{y + ay}$ = _____

b) $\dfrac{m}{dm - m}$ = _____

a) $1 + x$

b) $c - 1$

135. A fraction can be reduced to lowest terms <u>only</u> <u>if</u> <u>it</u> <u>contains</u> <u>a</u> <u>common</u> <u>factor</u> <u>in</u> <u>each</u> <u>term</u> <u>of</u> <u>its</u> <u>numerator</u> <u>and</u> <u>denominator</u>. Therefore, neither fraction below can be reduced.

$$\frac{a}{b-c} \qquad \frac{x+7}{y-3}$$

Reduce to lowest terms if possible.

a) $\dfrac{x-t}{b}$ = _____

c) $\dfrac{2m+4}{2m-8}$ = _____

b) $\dfrac{CP}{AP+BP}$ = _____

d) $\dfrac{5b}{3a-7}$ = _____

a) $\dfrac{1}{1+a}$

b) $\dfrac{1}{d-1}$

136. Though we can factor out <u>t</u> in the numerator below, we cannot reduce to lowest terms because <u>t</u> is not a factor in the denominator.

$$\frac{ct+dt}{am} = \frac{t(c+d)}{am}$$

If possible, reduce to lowest terms.

a) $\dfrac{3B}{6S-3} = \dfrac{3B}{3(2S-1)}$ = _____

b) $\dfrac{5x+5y}{3p-3q} = \dfrac{5(x+y)}{3(p-q)}$ = _____

a) Not possible

b) $\dfrac{C}{A+B}$

c) $\dfrac{m+2}{m-4}$

d) Not possible

a) $\dfrac{B}{2S-1}$ b) Not possible

8-14 CANCELLING TO REDUCE PATTERNS TO LOWEST TERMS

Instead of reducing fractions to lowest terms by factoring out a fraction that equals "1", we can use a shorter <u>cancelling</u> method. We will discuss the cancelling method in this section.

137. Each term in the numerator and denominator of the fraction below contains <u>a</u> as a common factor. Instead of reducing it to lowest terms by factoring out $\dfrac{a}{a}$, we can simply cancel the a's. We get:

$$\frac{\cancel{a}m + \cancel{a}t}{\cancel{a}p} = \frac{m+t}{p}$$

Cancel to reduce these to lowest terms.

a) $\dfrac{CD}{CX-CY}$ = _____

b) $\dfrac{bm-bv}{bs-bt}$ = _____

a) $\dfrac{\cancel{C}D}{\cancel{C}X-\cancel{C}Y} = \dfrac{D}{X-Y}$

b) $\dfrac{\cancel{b}m-\cancel{b}y}{\cancel{b}s-\cancel{b}t} = \dfrac{m-v}{s-t}$

138. Each term in the numerator and denominator of the fraction below contains 2 as a common factor. Instead of reducing it to lowest terms by factoring out $\frac{2}{2}$, we can simply cancel the 2's. We get:

$$\frac{\overset{2}{\cancel{4}x}}{\underset{1}{\cancel{2}}y - \underset{3}{\cancel{6}}} = \frac{2x}{y - 3}$$

Cancel to reduce these to lowest terms.

a) $\dfrac{8a + 8b}{8c} =$ _____   b) $\dfrac{3t - 9}{3s + 6} =$ _____

---

139. Cancel to reduce these to lowest terms.

a) $\dfrac{cS - cT}{cV} =$ _____   c) $\dfrac{FM + KM}{GM - HM} =$ _____

b) $\dfrac{5x}{5y + 10} =$ _____   d) $\dfrac{2d - 4}{4p + 6} =$ _____

a) $\dfrac{\overset{1}{\cancel{8}}a + \overset{1}{\cancel{8}}b}{\underset{1}{\cancel{8}}c} = \dfrac{a + b}{c}$

b) $\dfrac{\overset{1}{\cancel{3}}t - \overset{3}{\cancel{9}}}{\underset{1}{\cancel{3}}s + \underset{2}{\cancel{6}}} = \dfrac{t - 3}{s + 2}$

---

140. After cancelling below, we got a non-fraction.

$$\frac{\overset{1}{\cancel{m}} - \overset{1}{\cancel{m}}t}{\underset{1}{\cancel{m}}} = \frac{1 - t}{1} = 1 - t$$

Cancel to reduce these to lowest terms.

a) $\dfrac{4 + 4x}{4} =$ _____   b) $\dfrac{ay - y}{y} =$ _____

a) $\dfrac{S - T}{V}$ c) $\dfrac{F + K}{G - H}$

b) $\dfrac{x}{y + 2}$ d) $\dfrac{d - 2}{2p + 3}$

---

141. After cancelling below, we got 1 as the numerator of the reduced fraction.

$$\frac{\overset{1}{\cancel{R}}}{B\underset{1}{\cancel{R}} + \underset{1}{\cancel{R}}} = \frac{1}{B + 1}$$

Reduce these to lowest terms.

a) $\dfrac{3}{3t + 3} =$ _____   b) $\dfrac{d}{d - bd} =$ _____

a) $1 + x$

b) $a - 1$

---

142. A pattern can be reduced to lowest terms by cancelling <u>only if</u> <u>each</u> <u>term</u> <u>in</u> <u>its</u> <u>numerator</u> <u>and</u> <u>denominator</u> <u>contains</u> <u>a</u> <u>common</u> <u>factor</u>. Therefore, neither fraction below can be reduced.

$$\frac{P - Q}{R} \qquad\qquad \frac{3x + 7}{2y - 5}$$

Reduce to lowest terms if possible.

a) $\dfrac{d}{dS + dF} =$ _____   c) $\dfrac{5t + 10}{5t - 10} =$ _____

b) $\dfrac{2a - 3}{5y} =$ _____   d) $\dfrac{M - ST}{V} =$ _____

a) $\dfrac{1}{t + 1}$ b) $\dfrac{1}{1 - b}$

143. One of the most common errors with fractions is to cancel the 3's or m's in the fractions below. However, cancelling is not possible because each term of the fraction does not contain a 3 or m as a factor. Therefore, neither fraction can be reduced.

$$\frac{y + 3}{3}$$ (The y term does not have a 3 factor.)

$$\frac{m - t}{m}$$ (The t term does not have an m factor.)

Reduce to lowest terms if possible.

a) $\dfrac{4x + 4}{4}$ = _____

c) $\dfrac{a - b}{a}$ = _____

b) $\dfrac{x + 4}{4}$ = _____

d) $\dfrac{a - ab}{a}$ = _____

a) $\dfrac{1}{S + F}$

b) Not possible

c) $\dfrac{t + 2}{t - 2}$

d) Not possible

---

144. The same type of common error is to cancel the 2's or x's in the fractions below. However, cancelling is again not possible because each term of the fraction does not contain a 2 or x as a factor. Therefore, neither fraction can be reduced.

$$\frac{2}{y + 2}$$ (The y term does not have a 2 factor.)

$$\frac{x}{x - 7}$$ (The 7 term does not have an x factor.)

Reduce to lowest terms if possible.

a) $\dfrac{8}{8t + 8}$ = _____

c) $\dfrac{d}{a - d}$ = _____

b) $\dfrac{8}{8 - t}$ = _____

d) $\dfrac{d}{d + ad}$ = _____

a) x + 1

b) Not possible

c) Not possible

d) 1 - b

---

145. If a pattern is the sum or difference of two fractions and is reducible, it can be written in an equivalent form in two ways:

1) By reducing to lowest terms.

$$\frac{ad + bd}{cd} = \frac{a\!\!\!/d + b\!\!\!/d}{c\!\!\!/d} = \frac{a + b}{c}$$

2) By breaking it up into the two original fractions and reducing each part to lowest terms.

$$\frac{ad + bd}{cd} = \frac{ad}{cd} + \frac{bd}{cd} = \frac{a}{c} + \frac{b}{c}$$

Using the two methods above, write the fraction below in two equivalent forms.

$$\frac{3x - 6y}{9}$$ = _____   or   _____

a) $\dfrac{1}{t + 1}$

b) Not possible

c) Not possible

d) $\dfrac{1}{1 + a}$

$$\frac{x - 2y}{3}$$ or $\dfrac{x}{3} - \dfrac{2y}{3}$

146. When the same two methods are used to write the fraction below in an equivalent form, we get the same non-fractional expression.

$$\frac{5y - 5}{5} = \frac{\overset{1}{\cancel{5}}y - \overset{1}{\cancel{5}}}{\underset{1}{\cancel{5}}} = y - 1$$

$$\frac{5y - 5}{5} = \frac{5y}{5} - \frac{5}{5} = y - 1$$

Using either method above, write each fraction below in an equivalent form.

a) $\dfrac{3 + 3x}{3} =$ _____   b) $\dfrac{PQ - P}{P} =$ _____

147. Though the following fraction cannot be reduced, it can be written in an equivalent form. That is:

$$\frac{C + DF}{C} = \frac{C}{C} + \frac{DF}{C} = 1 + \frac{DF}{C}$$

Write each of these in an equivalent form.

a) $\dfrac{x + 3y}{3} =$ _____   b) $\dfrac{m - 2t}{m} =$ _____

| a) $1 + x$ |
| b) $Q - 1$ |

148. If a pattern is not the sum or difference of two fractions, it can only be written in a simpler equivalent form if it can be reduced. That is:

For $\dfrac{AP}{AC - AF}$, a simpler equivalent form is $\dfrac{P}{C - F}$.

For $\dfrac{H}{1 + CH}$, there is no simpler equivalent form.

Write each of these in a simpler equivalent form if possible.

a) $\dfrac{3t}{6m + 9} =$ _____   c) $\dfrac{mM}{M - m} =$ _____

b) $\dfrac{2x}{1 - 2x} =$ _____   d) $\dfrac{ab}{ac + a} =$ _____

| a) $\dfrac{x}{3} + y$ |
| b) $1 - \dfrac{2t}{m}$ |

149. Which of the following are equivalent to $\dfrac{H + CT}{T}$? _____

a) $H + C$ b) $\dfrac{H + C}{T}$ c) $\dfrac{H}{T} + C$ d) $\dfrac{H}{T} + \dfrac{C}{T}$

| a) $\dfrac{t}{2m + 3}$ |
| b) Not possible |
| c) Not possible |
| d) $\dfrac{b}{c + 1}$ |

150. Which of the following are equivalent to $\dfrac{AKT + KT}{AT}$? _____

a) $K + \dfrac{K}{A}$ b) $K + \dfrac{KT}{A}$ c) $\dfrac{AK + K}{A}$ d) $\dfrac{K + T}{A}$

| Only (c) |

151. Which of the following are equivalent to $\dfrac{KV}{KS - KV}$? _____

     a) $\dfrac{1}{KS}$ b) $\dfrac{V}{S - V}$ c) $\dfrac{1}{S - V}$ d) $\dfrac{V}{S} - 1$

         Both (a) and (c)

152. Which of the following are equivalent to $\dfrac{ab}{ac + bc}$? _____

     a) $\dfrac{b}{c + bc}$ b) $\dfrac{a}{ac + c}$ c) $\dfrac{1}{2c}$ d) $\dfrac{b}{c} + \dfrac{a}{c}$

         Only (b)

153. Though the fraction below cannot be reduced, it can be written as an addition of two fractions. That is:

$$\frac{C_1 + C_T}{C_1 - C_2} = \frac{C_1}{C_1 - C_2} + \frac{C_T}{C_1 - C_2}$$

Which of the following are equivalent to $\dfrac{R_1 - R_2}{R_1 + R_t}$? _____

     a) $\dfrac{1 - R_2}{1 - R_t}$ b) $\dfrac{R_1}{R_1 + R_t} - \dfrac{R_2}{R_1 + R_t}$ c) $\dfrac{1}{R_t} - \dfrac{R_2}{R_1 + R_t}$

         None of them

154. After factoring below, we were able to reduce to lowest terms by cancelling $(c + 1)$.

$$\frac{ac + a}{bc + b} = \frac{a(c + 1)}{b(c + 1)} = \frac{a}{b}$$

Following the example, reduce this one to lowest terms.

$\dfrac{7x + 7}{3x + 3} =$ _____

         Only (b)

         $\dfrac{7}{3}$

8-15 COMPLEX FRACTIONS

In this section, we will discuss the procedure for simplifying complex fractions.

155. Any fraction with a fraction in its numerator or denominator or both is called a <u>complex</u> <u>fraction</u>. Some examples are:

$$\frac{\dfrac{x}{5}}{\dfrac{3}{4}} \qquad\qquad \frac{\dfrac{a + b}{a}}{d} \qquad\qquad \frac{c + d}{\dfrac{p}{q}}$$

Continued on following page.

155. Continued

As we saw earlier, we can simplify complex fractions by performing the division. To do so, we multiply the numerator by the reciprocal of the denominator. That is:

$$\frac{\frac{x}{5}}{\frac{3}{4}} = \left(\frac{x}{5}\right)\left(\frac{4}{3}\right) = \frac{4x}{15}$$

$$\frac{\frac{a+b}{a}}{d} = \left(\frac{a+b}{a}\right)\left(\frac{1}{d}\right) = \frac{a+b}{ad}$$

$$\frac{c+d}{\frac{p}{q}} = (\qquad)(\qquad) = \underline{\qquad}$$

156. Following the example, do the other divisions.

$$\frac{\frac{c}{d}}{\frac{k+r}{t}} = \left(\frac{c}{d}\right)\left(\frac{t}{k+r}\right) = \frac{ct}{d(k+r)}$$

a) $$\frac{\frac{a+b}{c}}{\frac{p}{q}} = (\qquad)(\qquad) = \underline{\qquad}$$

b) $$\frac{\frac{d+f}{a}}{\frac{d-f}{b}} = (\qquad)(\qquad) = \underline{\qquad}$$

157. Following the example, do the other division.

$$\frac{\frac{b}{m}}{1-h} = \frac{b}{m}\left(\frac{1}{1-h}\right) = \frac{b}{m(1-h)}$$

$$\frac{\frac{a-b}{a}}{c-d} = (\qquad)(\qquad) = \underline{\qquad}$$

158. In the division below, we cancelled to reduce to lowest terms. Do the other division.

$$\frac{\frac{a+b}{c}}{\frac{a-b}{c}} = \left(\frac{a+b}{c}\right)\left(\frac{c}{a-b}\right) = \frac{\cancel{c}(a+b)}{\cancel{c}(a-b)} = \frac{a+b}{a-b}$$

$$\frac{\frac{t-1}{v}}{\frac{t-1}{h}} =$$

155 answer:
$$(c+d)\left(\frac{q}{p}\right) = \frac{q(c+d)}{p}$$

156 answers:
a) $$\left(\frac{a+b}{c}\right)\left(\frac{q}{p}\right) = \frac{q(a+b)}{cp}$$

b) $$\left(\frac{d+f}{a}\right)\left(\frac{b}{d-f}\right) = \frac{b(d+f)}{a(d-f)}$$

157 answer:
$$\left(\frac{a-b}{a}\right)\left(\frac{1}{c-d}\right) = \frac{a-b}{a(c-d)}$$

159. In the division below, we also cancelled to reduce to lowest terms. Do the other division.

$$\frac{c-d}{\dfrac{c-d}{t}} = (c-d)\left(\frac{t}{c-d}\right) = \frac{t(c-d)}{c-d} = t$$

$$\frac{x^2-1}{\dfrac{x^2-1}{y}} =$$

$\dfrac{h}{v}$, from: $\dfrac{h(t-1)}{v(t-1)}$

---

160. The complex fraction below contains an addition in its numerator. To simplify it, we performed that addition and then divided. Simplify the other fraction.

$$\frac{1+\dfrac{a}{b}}{t} = \frac{\dfrac{b}{b}+\dfrac{a}{b}}{t} = \frac{\dfrac{b+a}{b}}{t} = \left(\frac{b+a}{b}\right)\left(\frac{1}{t}\right) = \frac{b+a}{bt}$$

$$\frac{1-\dfrac{c}{d}}{m} =$$

$\dfrac{1}{y}$, from: $\dfrac{x^2-1}{y(x^2-1)}$

---

161. To simplify the complex fraction below, we performed the subtraction in the denominator and then divided.

$$\frac{R}{1R} - \frac{Q}{1R} \quad \frac{P}{1-\dfrac{Q}{R}} = \frac{P}{\dfrac{R-Q}{R}} = P\left(\frac{R}{R-Q}\right) = \frac{PR}{R-Q}$$

$$\frac{h}{m+\dfrac{h}{t}} =$$

$\dfrac{d-c}{dm}$

---

162. To simplify the complex fraction below, we performed the additions in the numerator and denominator and then divided. Simplify the other fraction.

$$\frac{1+\dfrac{1}{c}}{1+\dfrac{1}{d}} = \frac{\dfrac{c+1}{c}}{\dfrac{d+1}{d}} = \left(\frac{c+1}{c}\right)\left(\frac{d}{d+1}\right) = \frac{d(c+1)}{c(d+1)}$$

$$\frac{h-\dfrac{p}{q}}{h-\dfrac{v}{t}} =$$

$\dfrac{ht}{mt+h}$

$\dfrac{t(hq-p)}{q(ht-v)}$

163. In the simplification below, we were able to cancel to reduce the answer to lowest terms. Simplify the other fraction.

$$\frac{x - \dfrac{2x}{b}}{3x} = \frac{\dfrac{bx - 2x}{b}}{3x} = \left(\frac{bx - 2x}{b}\right)\left(\frac{1}{3x}\right) = \frac{b\cancel{x} - 2\cancel{x}}{3b\cancel{x}} = \frac{b - 2}{3b}$$

$$\frac{y + \dfrac{5y}{c}}{7y} =$$

---

164. Simplify each fraction.

a) $\dfrac{p}{p - \dfrac{p}{q}} =$

b) $\dfrac{t - \dfrac{1}{t}}{t + \dfrac{1}{t}} =$

$\dfrac{c + 5}{7c}$

---

165. Following the example, simplify the other fraction.

$$\frac{\dfrac{x + y}{x}}{\dfrac{1}{x} + \dfrac{1}{y}} = \frac{\dfrac{x + y}{x}}{\left(\dfrac{1}{x}\right)\left(\dfrac{y}{y}\right) + \left(\dfrac{1}{y}\right)\left(\dfrac{x}{x}\right)} = \frac{\dfrac{x + y}{x}}{\dfrac{y + x}{xy}} = \left(\frac{\cancel{x + y}}{\cancel{x}}\right)\left(\frac{\cancel{x}y}{\cancel{y + x}}\right) = y$$

$$\frac{\dfrac{1}{b} - \dfrac{1}{a}}{\dfrac{a - b}{b}}$$

a) $\dfrac{q}{q - 1}$

b) $\dfrac{t^2 - 1}{t^2 + 1}$

$\dfrac{1}{a}$, from: $\left(\dfrac{a - b}{ab}\right)\left(\dfrac{b}{a - b}\right)$

SELF-TEST 29 (pages 359-370)

Reduce each fraction to lowest terms.

1. $\dfrac{4t}{2r - 6w} =$

2. $\dfrac{GP - PV}{KP} =$

3. $\dfrac{8K - 12}{16T} =$

4. $\dfrac{cm - mw}{dm + mr} =$

5. $\dfrac{x}{x - xy} =$

6. $\dfrac{GH + 2H}{4HR - H} =$

7. Which of the following are equivalent to $\dfrac{hk + kw}{bk}$?

     a) $\dfrac{h}{b} + kw$ b) $\dfrac{h}{b} + \dfrac{w}{b}$ c) $hk + \dfrac{w}{b}$ d) $\dfrac{h + w}{b}$

Simplify each complex fraction. Report each answer in lowest terms.

8. $\dfrac{\frac{a - d}{4b}}{\frac{3}{2b}} =$

9. $\dfrac{x}{1 + \frac{1}{x}} =$

10. $\dfrac{1 - \frac{t}{r}}{\frac{t}{r}} =$

11. $\dfrac{\frac{p}{v} - p}{p} =$

12. $\dfrac{1 + \frac{1}{y}}{1 - \frac{1}{y}} =$

13. $\dfrac{\frac{1}{A} - A}{\frac{1}{A} + A} =$

ANSWERS:

1. $\dfrac{2t}{r - 3w}$

2. $\dfrac{G - V}{K}$

3. $\dfrac{2K - 3}{4T}$

4. $\dfrac{c - w}{d + r}$

5. $\dfrac{1}{1 - y}$

6. $\dfrac{G + 2}{4R - 1}$

7. (b) and (d)

8. $\dfrac{a - d}{6}$

9. $\dfrac{x^2}{x + 1}$

10. $\dfrac{r - t}{t}$

11. $\dfrac{1 - v}{v}$

12. $\dfrac{y + 1}{y - 1}$

13. $\dfrac{1 - A^2}{1 + A^2}$

SUPPLEMENTARY PROBLEMS - CHAPTER 8

Assignment 26

Reduce each fraction to lowest terms.

1. $\dfrac{x^2y}{x^2y}$

2. $\dfrac{ab}{2bd}$

3. $\dfrac{2G(R-1)}{6P(R-1)}$

4. $\dfrac{6tw}{8tw}$

5. $\dfrac{3r}{3cr}$

6. $\dfrac{4PT}{P}$

7. $\dfrac{d^2t}{bd^2}$

8. $\dfrac{2ks}{10krs}$

Multiply. Report each product in lowest terms.

9. $\left(\dfrac{1}{x}\right)\left(\dfrac{h}{y}\right)$

10. $\left(\dfrac{a}{2}\right)\left(\dfrac{4}{ab}\right)$

11. $\dfrac{1}{k}(p^2t)$

12. $\left(\dfrac{p}{r^2}\right)\left(\dfrac{ar^2}{p}\right)$

13. $\left(\dfrac{L+R}{G}\right)\left(\dfrac{E}{L+R}\right)$

14. $\left(\dfrac{s}{m}\right)\left(\dfrac{m}{st}\right)$

15. $xy\left(\dfrac{2t}{y}\right)$

16. $\left(\dfrac{v}{k-1}\right)\left(\dfrac{1}{v}\right)$

Write the reciprocal of: 17. $\dfrac{1}{h+rw}$ 18. $\dfrac{x-y}{v}$

Divide. Report each quotient in lowest terms.

19. $\dfrac{\frac{a}{d}}{\frac{a}{b}}$

20. $\dfrac{\frac{3y}{4x}}{\frac{9y}{12}}$

21. $\dfrac{\frac{2R}{R}}{P}$

22. $\dfrac{\frac{1}{t}}{ks}$

23. $\dfrac{\frac{4m}{r}}{8p}$

24. $\dfrac{\frac{F}{T}}{1-A}$

25. $\dfrac{1}{\frac{1}{c+k}}$

26. $\dfrac{2t^2}{\frac{mt^2}{2d}}$

27. $\dfrac{\frac{x}{y+1}}{2x}$

28. $\dfrac{\frac{h+2}{3}}{\frac{h+2}{2}}$

29. $\dfrac{\frac{1}{h}}{\frac{1}{2h}}$

30. $\dfrac{\frac{r}{dv}}{\frac{r}{dv}}$

Assignment 27

Add or subtract. Report each sum or difference in lowest terms.

1. $\dfrac{3w}{8}+\dfrac{w}{8}$

2. $\dfrac{v}{r}+\dfrac{w}{r}$

3. $\dfrac{ab}{2}-\dfrac{5ab}{2}$

4. $\dfrac{3t}{w}-\dfrac{2}{w}$

5. $\dfrac{gh}{m}-\dfrac{d}{m}$

6. $\dfrac{9}{4p}+\dfrac{7}{4p}$

7. $\dfrac{a}{rt}-\dfrac{a}{rt}$

8. $\dfrac{4R}{3}+\dfrac{2R}{3}$

9. $\dfrac{F}{E-2}-\dfrac{H}{E-2}$

10. $\dfrac{h}{x+y}-\dfrac{h-1}{x+y}$

11. $\dfrac{c}{h}+\dfrac{d}{h}+\dfrac{p}{h}$

12. $\dfrac{1}{F}+\dfrac{1}{F}+\dfrac{1}{F}$

13. $\dfrac{P+1}{T}-\dfrac{P}{T}$

14. $\dfrac{d+2}{a+b}+\dfrac{1}{a+b}$

15. $\dfrac{P}{2}-\dfrac{P}{3}$

16. $\dfrac{r}{t}+\dfrac{1}{4}$

17. $\dfrac{3r}{8}+\dfrac{r}{2}$

18. $\dfrac{2}{G}-\dfrac{1}{3G}$

19. $\dfrac{d}{x}+\dfrac{r}{5x}$

20. $\dfrac{5a}{6t}-\dfrac{a}{2t}$

21. $\dfrac{m}{3}-\dfrac{1}{2}$

22. $\dfrac{d}{8}+\dfrac{r}{6}$

23. $\dfrac{k}{r}-\dfrac{p}{v}$

24. $\dfrac{1}{Q}+\dfrac{1}{P}$

25. $\dfrac{h}{rw} + \dfrac{s}{r}$ 26. $\dfrac{5}{6} - \dfrac{1}{2m}$ 27. $\dfrac{w}{a} + \dfrac{t}{2b}$ 28. $\dfrac{1}{pv} - \dfrac{1}{cp}$

29. $\dfrac{2}{x^2} + \dfrac{1}{2x}$ 30. $\dfrac{1}{d} - \dfrac{1}{a}$ 31. $\dfrac{r}{s} + \dfrac{t}{w}$ 32. $\dfrac{2v}{pw} - \dfrac{b}{w^2}$

Assignment 28

Add or subtract. Write each answer as a single fraction.

1. $\dfrac{E}{2} - 1$ 2. $1 + \dfrac{w}{r}$ 3. $\dfrac{2}{3} + P$ 4. $5 - \dfrac{4}{B}$

5. $4 + \dfrac{1}{t}$ 6. $m - \dfrac{w}{k}$ 7. $\dfrac{r}{av} + 2d$ 8. $\dfrac{1}{x} - 1$

9. $a - \dfrac{1}{r}$ 10. $\dfrac{h}{3} + 2$ 11. $1 - \dfrac{B}{A}$ 12. $\dfrac{m}{t} + w$

Break up each fraction into a sum or difference of two fractions in lowest terms.

13. $\dfrac{r + s}{w}$ 14. $\dfrac{4r - 3}{6r}$ 15. $\dfrac{AB - C}{AC}$ 16. $\dfrac{ct - br}{rt}$

17. $\dfrac{F + H}{FH}$ 18. $\dfrac{x + 1}{y + 1}$ 19. $\dfrac{a + 8}{4}$ 20. $\dfrac{1 - r}{r}$

21. $\dfrac{12a - 8d}{6ad}$ 22. $\dfrac{st + r}{t}$ 23. $\dfrac{4PV + 24P}{12P}$ 24. $\dfrac{3adh - 5d}{dh}$

State whether each of the following is "True" or "False".

25. $\dfrac{R - W}{W} = \dfrac{R}{W} - 1$ 26. $\dfrac{H}{H + K} = 1 + \dfrac{H}{K}$ 27. $\dfrac{6}{4x - 8} = \dfrac{3}{2x} - \dfrac{3}{4}$

Factor each expression by the distributive principle if possible.

28. $18at + 12t$ 29. $F - FP$ 30. $cd + dh$ 31. $2rs - 3tw$

32. $4A_1 - 4A_2$ 33. $RV + V$ 34. $ac - bd$ 35. $pw - 2p$

Assignment 29

Reduce each fraction to lowest terms.

1. $\dfrac{6x + 12v}{9w}$ 2. $\dfrac{rt}{pr - 2r}$ 3. $\dfrac{dm}{mw + m}$

4. $\dfrac{15s}{5s - 10v}$ 5. $\dfrac{bp + ab}{4b}$ 6. $\dfrac{V - PV}{RV}$

7. $\dfrac{xy + x}{x - xy}$ 8. $\dfrac{2k - 4}{4r - 2}$ 9. $\dfrac{6t + 3w}{3p + 9v}$

10. $\dfrac{F}{AF + F}$ 11. $\dfrac{2}{4y - 6}$ 12. $\dfrac{bd - d}{d}$

State whether each of the following is "True" or "False".

13. $\dfrac{P}{P - KP} = \dfrac{1}{1 - K}$ 14. $\dfrac{cr + ct}{ct} = \dfrac{r}{t} + 1$ 15. $\dfrac{2x}{2x + 2} = 1 + x$

16. Which of the following are equivalent to $\dfrac{bp - pt}{dp}$?

    a) $\dfrac{b}{d} - \dfrac{t}{d}$ b) $bp - \dfrac{t}{d}$ c) $\dfrac{p(b - t)}{d}$ d) $\dfrac{b}{d} - t$ e) $\dfrac{b - t}{d}$

Simplify each complex fraction.

17. $\dfrac{\dfrac{km}{m}}{r - a}$ 18. $\dfrac{d}{\dfrac{t + w}{b}}$ 19. $\dfrac{\dfrac{x - y}{p}}{2v}$ 20. $\dfrac{\dfrac{F}{D + H}}{F}$

21. $\dfrac{1 + \dfrac{1}{x}}{2y}$ 22. $\dfrac{at}{\dfrac{a}{r} - a}$ 23. $\dfrac{P}{P + \dfrac{2P}{R}}$ 24. $\dfrac{1 - \dfrac{1}{w}}{b}$

25. $\dfrac{x + y}{\dfrac{t}{3}}$ 26. $\dfrac{\dfrac{d}{c - k}}{\dfrac{d}{p - t}}$ 27. $\dfrac{1 + \dfrac{1}{R}}{1 - \dfrac{1}{R}}$ 28. $\dfrac{\dfrac{a}{w} - 1}{\dfrac{b}{t} + 1}$

29. $\dfrac{\dfrac{V}{R}}{1 - \dfrac{V}{R}}$ 30. $\dfrac{x + \dfrac{1}{x}}{x - \dfrac{1}{x}}$ 31. $\dfrac{\dfrac{1}{A} + \dfrac{1}{B}}{1 + \dfrac{B}{A}}$ 32. $\dfrac{1 - \dfrac{b}{a}}{\dfrac{a}{2} - \dfrac{b}{2}}$

9 Formula Rearrangement

Formulas are frequently rearranged to solve for a variable. The rearrangements are made to make evaluations easier, to emphasize a different meaning or relationship, or to eliminate a variable or variables from a system of formulas so that a new formula can be derived. We will discuss formula rearrangement in this chapter. We will show that the same algebraic principles used to solve equations are used to rearrange formulas.

9-1   FORMULA REARRANGEMENT

In this section, we will give examples of the three major purposes of formula rearrangement.

1.   Formula rearrangement is used to make evaluations easier. For example, in the formula below, v_1 is <u>not-solved-for</u>. Therefore, to find the value of v_1 when $a = 10$, $v_2 = 60$, and $t = 2$, we must solve an equation. We get:

$$a = \frac{v_2 - v_1}{t}$$

$$10 = \frac{60 - v_1}{2}$$

$$2(10) = 2\left(\frac{60 - v_1}{2}\right)$$

$$20 = 60 - v_1$$

$$-40 = -v_1$$

$$v_1 = \frac{-40}{-1} = 40$$

Continued on following page.

374

1. Continued

However, we can also rearrange the preceding formula to solve for v_1. Doing so, we get:

$$v_1 = v_2 - at$$

Using the rearranged formula in which v_1 is <u>solved for</u>, we can do the same evaluation without solving an equation. We get:

$$v_1 = v_2 - at = 60 - (10)(2) = 60 - 20 = 40$$

We got $v_1 = 40$ as the solution with both methods. Which method is easier, the one with or without equation-solving? _____

2. Formula rearrangement is used to emphasize a different meaning or relationship. For example, in the formula below, V (voltage) is solved-for in terms of I (current) and R (resistance). As you can see, V varies jointly as I and R.

$$V = IR$$

However, we can rearrange the formula to solve for I in terms of V and R. We get:

$$I = \frac{V}{R}$$

From the rearranged formula, we can see these facts:

    a) I varies directly as _____.

    b) I varies inversely as _____.

3. Formula rearrangement is also used to eliminate a variable from a system of formulas so that a new formula can be derived. The two formulas below form a system because each contains the variable <u>r</u>.

$$d = 2r$$
$$A = \pi r^2$$

By rearranging the top formula to solve for <u>r</u> and then substituting that expression in the bottom formula, we can eliminate <u>r</u> and derive a formula for A in terms of <u>d</u>.

    Solving for <u>r</u> in the top formula, we get:

$$r = \frac{d}{2}$$

Substituting $\frac{d}{2}$ for <u>r</u> in the bottom formula and simplifying, we get:

$$A = \pi r^2$$
$$A = \pi\left(\frac{d}{2}\right)^2$$
$$A = \pi\left(\frac{d^2}{4}\right)$$
$$A = \frac{\pi d^2}{4}$$

By the substitution, did we get a formula for A in terms of <u>d</u>? _____

---

the one without equation-solving

a) V

b) R

Yes

9-2 TERMS AND COEFFICIENTS

In this section, we will identify <u>terms</u> in formulas and define the <u>coefficient</u> of a variable in a non-fractional term.

---

4. In a formula, terms are separated by addition or subtraction symbols. <u>Any</u> <u>single</u> <u>variable</u> <u>or</u> <u>squared</u> <u>variable</u> is a term. For example, we have drawn boxes around each term in the formulas below.

$$\boxed{F_t} = \boxed{F_1} + \boxed{F_2} + \boxed{F_3} \qquad\qquad \boxed{a^2} = \boxed{c^2} - \boxed{b^2}$$

Draw a box around each term in these formulas.

  a) $I_C = I_E - I_B$ b) $A = s^2$ c) $R = C + M$

---

5. <u>Any</u> <u>multiplication</u> <u>of</u> <u>two</u> <u>or</u> <u>more</u> <u>factors</u> <u>is</u> <u>one</u> <u>term</u>. For example, we have drawn boxes around each term in the formulas below.

$$\boxed{E} = \boxed{\tfrac{1}{2}mv^2} \qquad\qquad \boxed{I_A} = \boxed{I_C} + \boxed{md^2}$$

Draw a box around each term in these formulas.

  a) $T_1 V_2 = T_2 V_1$ b) $E = IR_1 + IR_2$

a) $\boxed{I_C} = \boxed{I_E} - \boxed{I_B}$

b) $\boxed{A} = \boxed{s^2}$

c) $\boxed{R} = \boxed{C} + \boxed{M}$

---

6. <u>A</u> <u>multiplication</u> <u>is</u> <u>one</u> <u>term</u> <u>even</u> <u>if</u> <u>one</u> <u>of</u> <u>the</u> <u>factors</u> <u>is</u> <u>an</u> <u>addition</u> <u>or</u> <u>subtraction</u>. For example, we have drawn boxes around each term in the formulas below.

$$\boxed{W_s} = \boxed{P(V_1 - V_2)} \qquad\qquad \boxed{A} = \boxed{\tfrac{1}{2}h(b_1 + b_2)}$$

Draw a box around each term in these formulas.

  a) $C = \dfrac{5}{9}(F - 32)$ b) $H = ms(t_2 - t_1)$

a) $\boxed{T_1 V_2} = \boxed{T_2 V_1}$

b) $\boxed{E} = \boxed{IR_1} + \boxed{IR_2}$

---

7. <u>Any</u> <u>fraction</u> <u>is</u> <u>one</u> <u>term</u>. For example, we have drawn boxes around each term in the formulas below.

$$\boxed{P} = \boxed{\dfrac{V^2}{R}} \qquad\qquad \boxed{H} = \boxed{\dfrac{\pi R(F_1 - F_2)}{33,000}}$$

Draw a box around each term in these formulas.

  a) $\dfrac{P_1}{P_2} = \dfrac{V_2}{V_1}$ b) $m = \dfrac{y_2 - y_1}{x_2 - x_1}$

a) $\boxed{C} = \boxed{\dfrac{5}{9}(F - 32)}$

b) $\boxed{H} = \boxed{ms(t_2 - t_1)}$

---

8. Draw a box around each term.

  a) $y = mx + b$ c) $M = \dfrac{WLX}{2} - \dfrac{WX^2}{2}$

  b) $A = \tfrac{1}{2}bh$ d) $D = P(Q + R)$

a) $\boxed{\dfrac{P_1}{P_2}} = \boxed{\dfrac{V_2}{V_1}}$

b) $\boxed{m} = \boxed{\dfrac{y_2 - y_1}{x_2 - x_1}}$

9.   Draw a box around each term.

        a) $C_t C_2 = C_1 C_2 - C_1 C_t$ c) $v_o{}^2 = v_f{}^2 = 2gs$

        b) $v_{av} = \dfrac{v_o + v_f}{2}$ d) $P = p_1 + w(h - h_1)$

a) $\boxed{y} = \boxed{mx} + \boxed{b}$

b) $\boxed{A} = \boxed{\tfrac{1}{2}bh}$

c) $\boxed{M} = \boxed{\dfrac{WLX}{2}} - \boxed{\dfrac{WX^2}{2}}$

d) $\boxed{D} = \boxed{P(Q + R)}$

---

10.   In any two-factor term containing a number and a variable, the <u>num</u>-<u>ber</u> is the coefficient of the variable. That is:

        In 1.5t, the coefficient of <u>t</u> is 1.5 .

In any two-factor term containing two variables, the coefficient of each factor is <u>the</u> <u>other</u> <u>factor</u>. That is:

        In LW, the coefficient of W is L.

        In $V_1 T_2$, the coefficient of V_1 is _____.

a) $\boxed{C_t C_2} = \boxed{C_1 C_2} - \boxed{C_1 C_t}$

b) $\boxed{v_{av}} = \dfrac{v_o + v_f}{2}$

c) $\boxed{v_o{}^2} = \boxed{v_f{}^2} - \boxed{2gs}$

d) $\boxed{P} = \boxed{p_1} + \boxed{w(h-h_1)}$

---

11.   In any term with more than two factors, the coefficient of each factor is <u>the</u> <u>other</u> factors. That is:

        In LWH, the coefficient of H is LW.

        In $.24I^2 Rt$, the coefficient of I^2 is _____.

T_2

---

12.   In any term, the coefficient of any one factor is <u>the</u> <u>other</u> <u>factor</u> <u>or</u> <u>factors</u>. That is:

        In P(Q + R): the coefficient of (Q + R) is P.

                the coefficient of P is (Q + R).

      In $ms(t_2 - t_1)$: a) the coefficient of $(t_2 - t_1)$ is _____.

                  b) the coefficient of <u>s</u> is _____.

                  c) the coefficient of <u>m</u> is _____.

.24Rt

---

13.   a) In $.24I^2Rt$, the coefficient of R is _____.

      b) In W(R - r), the coefficient of W is _____.

      c) In $AKT(t_2 - t_1)$, the coefficient of T is _____.

a) ms

b) $m(t_2 - t_1)$

c) $s(t_2 - t_1)$

---

a) $.24I^2t$

b) $(R - r)$

c) $AK(t_2 - t_1)$

9-3   THE MULTIPLICATION AXIOM

When a formula has only one non-fractional term on each side, we can use the multiplication axiom to solve for any variable that is a factor.   We will discuss the method in this section.   A shortcut for the multiplication axiom is shown.

---

14.   We used the multiplication axiom to solve for <u>h</u> below.   That is, we multiplied both sides by $\frac{1}{.433}$, <u>the</u> <u>reciprocal</u> <u>of</u> <u>the</u> <u>coefficient</u> <u>of</u> <u>h</u>. Solve for <u>t</u> in the other formula by multiplying both sides by $\frac{1}{10}$.

$$P = .433h \qquad\qquad v = 10t$$

$$\frac{1}{.433}(P) = \frac{1}{.433}(.433h)$$

$$\frac{P}{.433} = \quad 1h$$

$$h = \frac{P}{.433}$$

---

15.   To solve for R below, we multiplied both sides by $\frac{1}{I^2}$, <u>the</u> <u>reciprocal</u> <u>of</u> <u>the</u> <u>coefficient</u> <u>of</u> <u>R</u>.   Solve for V_1 in the other formula by multiplying both sides by $\frac{1}{P_1}$.

$$P = I^2 R \qquad\qquad P_1 V_1 = P_2 V_2$$

$$\frac{1}{I^2}(P) = \frac{1}{I^2}(I^2 R)$$

$$\frac{P}{I^2} = 1R$$

$$R = \frac{P}{I^2}$$

$t = \frac{v}{10}$, from:

$$\frac{1}{10}(v) = \frac{1}{10}(10t)$$

$$\frac{v}{10} = \quad 1t$$

---

16.   To solve for <u>s</u> below, we multiplied both sides by $\frac{1}{2a}$, <u>the</u> <u>reciprocal</u> <u>of</u> <u>the</u> <u>coefficient</u> <u>of</u> <u>s</u>.   Solve for L in the other formula by multiplying both sides by $\frac{1}{WH}$.

$$v^2 = 2as \qquad\qquad V = LWH$$

$$\frac{1}{2a}(v^2) = \frac{1}{2a}(2as)$$

$$\frac{v^2}{2a} = \quad 1s$$

$$s = \frac{v^2}{2a}$$

$V_1 = \frac{P_2 V_2}{P_1}$, from:

$$\frac{1}{P_1}(P_1 V_1) = \frac{1}{P_1}(P_2 V_2)$$

$$1V_1 = \frac{P_2 V_2}{P_1}$$

17. To solve for P below, we multiplied both sides by $\frac{1}{Q + R}$, the reciprocal of the coefficient of P. Solve for b in the other formula by multiplying both sides by $\frac{1}{c - d}$.

$$D = P(Q + R) \qquad\qquad a = b(c - d)$$

$$\left(\frac{1}{Q + R}\right)(D) = P(Q + R)\left(\frac{1}{Q + R}\right)$$

$$\frac{D}{Q + R} = P \qquad\qquad 1$$

$$P = \frac{D}{Q + R}$$

$L = \frac{V}{WH}$, from:

$$\frac{1}{WH}(V) = LWH\left(\frac{1}{WH}\right)$$

$$\frac{V}{WH} = L \cdot 1$$

---

18. Two solutions from the preceding frames are shown below.

If V = LWH,   If D = P(Q + R)

$$L = \frac{V}{WH} \qquad\qquad P = \frac{D}{Q + R}$$

As you can see from the above examples, there is a shortcut for the multiplication axiom. That is, to solve for each variable, we can simply divide the other side of the formula by the coefficient of that variable. For example:

In V = IR : The coefficient of I is R.

Therefore, $I = \frac{V}{R}$

a) In $2Pt^2 = rw$: The coefficient of P is $2t^2$.

Therefore, P = _____

b) In $H = ms(t_2 - t_1)$: The coefficient of s is $m(t_2 - t_1)$.

Therefore, s = _____

$b = \frac{a}{c - d}$, from:

$$\frac{1}{c-d}(a) = b(c-d)\left(\frac{1}{c-d}\right)$$

$$\frac{a}{c - d} = b \cdot 1$$

---

19. When rearranging a formula, a capital letter should not be changed to a small letter and a small letter should not be changed to a capital letter. That is:

Don't change A to a.
Don't change e to E.

Using the shortcut of the multiplication axiom, solve for the indicated variables in this frame and the next.

a) Solve for F.   b) Solve for R.

$$d^2Fr = m_1m_2 \qquad\qquad H = .24I^2Rt$$

F = _____   R = _____

---

a) $P = \frac{rw}{2t^2}$

b) $s = \frac{H}{m(t_2 - t_1)}$

---

a) $F = \frac{m_1m_2}{d^2r}$

b) $R = \frac{H}{.24I^2t}$

20. a) Solve for $\underline{b}$.

$$2ab = c(d - f)$$

$$b = \underline{\hspace{2cm}}$$

b) Solve for $\underline{m}$.

$$H = ms(t_2 - t_1)$$

$$m = \underline{\hspace{2cm}}$$

a) $b = \dfrac{c(d - f)}{2a}$ b) $m = \dfrac{H}{s(t_2 - t_1)}$

9-4   FRACTIONAL FORMULAS

In this section, we will discuss the method for rearranging formulas that contain one or two fractions. The formulas are limited to those with only one term on each side.

21. To rearrange a formula containing one fraction, we begin by clearing the fraction. To do so, we use the multiplication axiom, <u>multiplying</u> <u>both</u> <u>sides</u> <u>by</u> <u>the</u> <u>denominator</u> <u>of</u> <u>the</u> <u>fraction</u>. For example, to solve for B below, we multiplied both sides by N. Solve for $\underline{a}$ in the other formula.

$$D = \frac{B}{N}$$

$$N(D) = \cancel{N}\left(\frac{B}{\cancel{N}}\right)$$

$$DN = B$$

$$B = DN$$

$$h = \frac{a}{s^2}$$

22. To solve for $\underline{s}$ below, we cleared the fraction and then used the short-cut of the multiplication axiom. Complete the other rearrangement.

Solve for $\underline{s}$.

$$P = \frac{Fs}{t}$$

$$t(P) = \cancel{t}\left(\frac{Fs}{\cancel{t}}\right)$$

$$Pt = Fs$$

$$s = \frac{Pt}{F}$$

Solve for m_1.

$$F = \frac{m_1 m_2}{rd^2}$$

$a = hs^2$, from:

$$s^2(h) = \cancel{s^2}\left(\frac{a}{\cancel{s^2}}\right)$$

$$hs^2 = a$$

23. Following the example, complete the other rearrangement.

Solve for R.

$$I = \frac{V}{R}$$

$$R(I) = \cancel{R}\left(\frac{V}{\cancel{R}}\right)$$

$$IR = V$$

$$R = \frac{V}{I}$$

Solve for P_1.

$$V_1 = \frac{P_2 V_2}{P_1}$$

$m_1 = \dfrac{Frd^2}{m^2}$, from:

$$rd^2(F) = \cancel{rd^2}\left(\frac{m_1 m_2}{\cancel{rd^2}}\right)$$

$$Frd^2 = m_1 m_2$$

24. Following the example, complete the other rearrangement.

Solve for $\underline{t}$.

$$a = \frac{v_2 - v_1}{t}$$

$$t(a) = \not{t}\left(\frac{v_2 - v_1}{\not{t}}\right)$$

$$at = v_2 - v_1$$

$$t = \frac{v_2 - v_1}{a}$$

Solve for $\underline{r}$.

$$m = \frac{c - d}{2r}$$

$$P_1 = \frac{P_2 V_2}{V_1} \text{ , from:}$$

$$P_1(V_1) = \not{P_1}\left(\frac{P_2 V_2}{\not{P_1}}\right)$$

$$P_1 V_1 = P_2 V_2$$

---

25. Following the example, complete the other rearrangement.

Solve for A.

$$H = \frac{AKT(t_2 - t_1)}{L}$$

$$L(H) = \not{L}\left[\frac{AKT(t_2 - t_1)}{\not{L}}\right]$$

$$HL = AKT(t_2 - t_1)$$

$$A = \frac{HL}{KT(t_2 - t_1)}$$

Solve for R.

$$H = \frac{\pi dR(F_1 - F_2)}{33,000}$$

$$r = \frac{c - d}{2m} \text{ , from:}$$

$$2r(m) = 2\not{r}\left(\frac{c - d}{2\not{r}}\right)$$

$$2mr = c - d$$

---

26. a) Solve for $\underline{t}$.

$$H = \frac{M[(V_1)^2 - (V_2)^2]}{1100gt}$$

b) Solve for C.

$$X_c = \frac{1}{2\pi fC}$$

$$R = \frac{33,000H}{\pi d(F_1 - F_2)}$$

---

27. a) Solve for L.

$$H = \frac{AKT(t_2 - t_1)}{L}$$

b) Solve for F.

$$Ft = \frac{mv}{g}$$

a) $t = \dfrac{M[(V_1)^2-(V_2)^2]}{1100gH}$

b) $C = \dfrac{1}{2\pi fX_c}$

---

a) $L = \dfrac{AKT(t_2 - t_1)}{H}$ b) $F = \dfrac{mv}{gt}$

28. In the formula at the left below, $t \times 10^8$ means: t times 10^8. Therefore, we can rewrite the formula as we have done at the right below.

$$E = \frac{N\phi}{t \times 10^8} \qquad \text{or} \qquad E = \frac{N\phi}{t(10^8)}$$

a) Solve for ϕ.

b) Solve for $\underline{t}$.

---

29. Since $\frac{1}{2}bh = \frac{1}{2}(bh)$, we can write the formula below in two equivalent forms.

$$A = \frac{1}{2}bh \qquad \text{or} \qquad A = \frac{bh}{2}$$

It is easier to clear the fraction in the form at the right above. Therefore, before clearing the fraction in formulas of that type, we should write them in that form. That is:

$$s = \frac{1}{2}at^2 \quad \text{should be written} \quad s = \frac{at^2}{2}$$

$$A = \frac{1}{2}h(b_1 + b_2) \quad \text{should be written} \quad A = \underline{\qquad\qquad}$$

a) $\phi = \dfrac{Et(10^8)}{N}$

b) $t = \dfrac{N\phi}{E(10^8)}$

---

30. Following the example, complete the other rearrangement.

Solve for $\underline{h}$.

$$A = \frac{1}{2}bh$$

$$A = \frac{bh}{2}$$

$$2(A) = 2\left(\frac{bh}{2}\right)$$

$$2A = bh$$

$$h = \frac{2A}{b}$$

Solve for $\underline{a}$.

$$s = \frac{1}{2}at^2$$

$$\frac{h(b_1 + b_2)}{2}$$

$$a = \frac{2s}{t^2}$$

31. Following the example, complete the other rearrangement.

    Solve for <u>h</u>.   Solve for v^2.

    $$A = \frac{1}{2}h(b_1 + b_2)$$ $$E = \frac{1}{2}(m_1 + m_2)v^2$$

    $$A = \frac{h(b_1 + b_2)}{2}$$

    $$2(A) = 2\left[\frac{h(b_1 + b_2)}{2}\right]$$

    $$2A = h(b_1 + b_2)$$

    $$h = \frac{2A}{b_1 + b_2}$$

---

32. To rearrange a formula containing two fractions, we also begin by clearing the fractions. To do so, <u>we</u> <u>multiply</u> <u>both</u> <u>sides</u> <u>by</u> <u>both</u> <u>denominators</u> <u>at</u> <u>the</u> <u>same</u> <u>time</u>. For example, to solve for P_1 below, we multiplied both sides by P_2V_1. Solve for T_1 in the other formula.

    $$\frac{P_1}{P_2} = \frac{V_2}{V_1}$$ $$\frac{V_1}{V_2} = \frac{T_1}{T_2}$$

    $$P_2V_1\left(\frac{P_1}{P_2}\right) = P_2V_1\left(\frac{V_2}{V_1}\right)$$

    $$V_1P_1 = P_2V_2$$

    $$P_1 = \frac{P_2V_2}{V_1}$$

    $$v^2 = \frac{2E}{m_1 + m_2}$$

---

33. We solved for I_2 below. To clear the fractions, we multiplied both sides by $I_2(d_1)^2$. Solve for T_2 in the other formula.

    $$\frac{I_1}{I_2} = \frac{(d_2)^2}{(d_1)^2}$$ $$\frac{P_1V_1}{T_1} = \frac{P_2V_2}{T_2}$$

    $$I_2(d_1)^2\left(\frac{I_1}{I_2}\right) = I_2(d_1)^2\left[\frac{(d_2)^2}{(d_1)^2}\right]$$

    $$I_1(d_1)^2 = I_2(d_2)^2$$

    $$I_2 = \frac{I_1(d_1)^2}{(d_2)^2}$$

    $$T_1 = \frac{T_2V_1}{V_2} \text{ , from:}$$

    $$V_2T_2\left(\frac{V_1}{V_2}\right) = V_2T_2\left(\frac{T_1}{T_2}\right)$$

    $$T_2V_1 = V_2T_1$$

---

34. a) Solve for I_1.   b) Solve for V_1.

    $$\frac{I_1}{I_2} = \frac{(d_2)^2}{(d_1)^2}$$ $$\frac{P_1}{P_2} = \frac{V_2}{V_1}$$

    $$T_2 = \frac{P_2T_1V_2}{P_1V_1}$$

    a) $$I_1 = \frac{I_2(d_2)^2}{(d_1)^2}$$

    b) $$V_1 = \frac{P_2V_2}{P_1}$$

SELF-TEST 30 (pages 374-384)

1. Solve for t.

$$d = kt$$

2. Solve for h.

$$r = h(w - v)$$

3. Solve for F_1.

$$d_1 F_1 = d_2 F_2$$

4. Solve for H.

$$L = \frac{V}{HW}$$

5. Solve for m.

$$E = \frac{1}{2}mv^2$$

6. Solve for f.

$$X_c = \frac{1}{2\pi fC}$$

7. Solve for R.

$$W = \frac{BR(P_2 - P_1)}{M}$$

8. Solve for s_1.

$$\frac{t_1}{t_2} = \frac{s_1}{s_2}$$

9. Solve for A_1.

$$\frac{A_2}{A_1} = \frac{(d_1)^2}{(d_2)^2}$$

ANSWERS:

1. $t = \dfrac{d}{k}$

2. $h = \dfrac{r}{w - v}$

3. $F_1 = \dfrac{d_2 F_2}{d_1}$

4. $H = \dfrac{V}{LW}$

5. $m = \dfrac{2E}{v^2}$

6. $f = \dfrac{1}{2\pi CX_c}$

7. $R = \dfrac{MW}{B(P_2 - P_1)}$

8. $s_1 = \dfrac{s_2 t_1}{t_2}$

9. $A_1 = \dfrac{A_2(d_2)^2}{(d_1)^2}$

9-5   THE SQUARE ROOT PRINCIPLE

In this section, we will show how the square root principle can be used to solve for a variable that is squared in a formula.

---

35.   The principal square root of the square of a number is the number itself.   For example.

$$\sqrt{7^2} = \sqrt{49} = 7 \qquad \sqrt{3^2} = \sqrt{9} = 3$$

Similarly, the principal square root of the square of a letter is the letter itself.   That is:

$$\sqrt{x^2} = x \qquad \sqrt{y^2} = y \qquad \text{a) } \sqrt{s^2} = \underline{\qquad} \qquad \text{b) } \sqrt{P^2} = \underline{\qquad}$$

---

36.   Equations like $x^2 = 16$ have two roots, one positive and one negative. However, as we saw earlier, the negative root does not ordinarily make sense in a formula evaluation.   For example:

If $s^2 = 100$ when s stands for <u>distance</u>, -10 does not ordinarily make sense as a value for distance.

When rearranging formulas to solve for a variable that is squared, negative square roots also do not ordinarily make sense.   Therefore, to solve for s in the formula below, <u>we simply take the principal (or positive) square root of each side</u>.   That is:

$$s^2 = A$$
$$\sqrt{s^2} = \sqrt{A}$$
$$s = \sqrt{A}$$

In the formula, s stands for the length of the side of a square. Would the solution $s = -\sqrt{A}$ make sense?   _____

a) s   b) P

---

37.   To solve for V below, we took the principal square root of each side. Complete the other solutions.

If $V^2 = PR$ a) If $v^2 = 2as$ b) If $s^2 = \dfrac{a}{h}$

$\sqrt{V^2} = \sqrt{PR}$ $\sqrt{v^2} = \sqrt{2as}$ $\sqrt{s^2} = \sqrt{\dfrac{a}{h}}$

$V = \sqrt{PR}$ $v = \underline{\qquad}$ $s = \underline{\qquad}$

No, because a length is not ordinarily negative.

---

a) $v = \sqrt{2as}$

b) $s = \sqrt{\dfrac{a}{h}}$

38. When solving for a variable that is squared, <u>be sure to take the square root of all of the other side</u>. For example:

If $\quad a^2 = \dfrac{b + c}{d}$,

$\qquad a = \sqrt{\dfrac{b + c}{d}}$

If $\quad c^2 = a^2 + b^2$,

$\qquad c = \sqrt{a^2 + b^2}$

a) If $\quad d^2 = \dfrac{I_A - I_C}{m}$

$\qquad d = \underline{\hspace{2cm}}$

b) If $\quad v_o^2 = v_f^2 - 2gs$,

$\qquad v_o = \underline{\hspace{2cm}}$

---

39. To solve for I below, we isolated I^2 first and then took the square root of each side. Complete the other rearrangements.

Solve for I.

$P = I^2 R$

$I^2 = \dfrac{P}{R}$

$I = \sqrt{\dfrac{P}{R}}$

a) Solve for <u>w</u>.

$a = w^2 r$

b) Solve for <u>d</u>.

$A = .7854d^2$

a) $d = \sqrt{\dfrac{I_A - I_C}{m}}$

b) $v_o = \sqrt{v_f^2 - 2gs}$

---

40. To isolate v^2 below, we began by clearing the fraction. Complete the other rearrangement.

Solve for <u>v</u>.

$F = \dfrac{mv^2}{r}$

$r(F) = \cancel{r}\left(\dfrac{mv^2}{\cancel{r}}\right)$

$Fr = mv^2$

$v^2 = \dfrac{Fr}{m}$

$v = \sqrt{\dfrac{Fr}{m}}$

Solve for <u>d</u>.

$P = \dfrac{pL}{d^2}$

a) $w = \sqrt{\dfrac{a}{r}}$

b) $d = \sqrt{\dfrac{A}{.7854}}$

---

41. a) Solve for D.

$H = \dfrac{D^2 N}{2.5}$

b) Solve for <u>d</u>.

$F = \dfrac{m_1 m_2}{rd^2}$

c) Solve for <u>t</u>.

$P = \dfrac{rw}{2t^2}$

$d = \sqrt{\dfrac{pL}{P}}$

42. To solve for $\underline{t}$ below, we began by writing $\frac{1}{2}at^2$ as $\frac{at^2}{2}$. Complete the other rearrangement.

   Solve for $\underline{t}$.

   $$s = \frac{1}{2}at^2$$

   $$s = \frac{at^2}{2}$$

   $$2s = at^2$$

   $$t^2 = \frac{2s}{a}$$

   $$t = \sqrt{\frac{2s}{a}}$$

   Solve for $\underline{v}$.

   $$E = \frac{1}{2}mv^2$$

   a) $D = \sqrt{\dfrac{2.5H}{N}}$

   b) $d = \sqrt{\dfrac{m_1 m_2}{Fr}}$

   c) $t = \sqrt{\dfrac{rw}{2P}}$

43. To solve for $\underline{r}$ below, we began by clearing both fractions. Solve for $\underline{t}$ in the other formula.

   $$\frac{E}{d} = \frac{2r^2}{a}$$

   $$ad\left(\frac{E}{d}\right) = ad\left(\frac{2r^2}{a}\right)$$

   $$aE = 2dr^2$$

   $$r^2 = \frac{aE}{2d}$$

   $$r = \sqrt{\frac{aE}{2d}}$$

   $$\frac{s}{t^2} = \frac{m}{h}$$

   $v = \sqrt{\dfrac{2E}{m}}$

44. Though neither H is squared in the formula below, we get an H^2 when the fractions are cleared. Complete the other rearrangement.

   Solve for H.

   $$\frac{M}{H} = \frac{H}{T}$$

   $$HT\left(\frac{M}{H}\right) = HT\left(\frac{H}{T}\right)$$

   $$MT = H^2$$

   $$H = \sqrt{MT}$$

   Solve for $\underline{p}$.

   $$\frac{cd}{p} = \frac{pq}{a}$$

   $t = \sqrt{\dfrac{hs}{m}}$

   $p = \sqrt{\dfrac{acd}{q}}$

9-6 ADDITIVE INVERSES AND ADDITION-SUBTRACTION CONVERSIONS

In this section, we will extend the concept of <u>additive inverses</u> and addition-subtraction conversions to literal expressions.

---

45. <u>Two literal terms are additive inverses if their sum is 0.</u>

   Since $4P + (-4P) = 0$, $4P$ and $-4P$ are additive inverses.

   Since $2ab + (-2ab) = 0$, $2ab$ and $-2ab$ are additive inverses.

   Write the additive inverse of each term.

   a) $10t$ _____   b) $-3s$ _____   c) $7TV$ _____   d) $-5pq$ _____

---

46. To show that the following terms are additive inverses, we wrote "1" and -1 explicitly.

   Since $m^2 + (-m^2) = 1m^2 + (-1m^2) = 0$, m^2 and $-m^2$ are additive inverses.

   Since $AB + (-AB) = 1AB + (-1AB) = 0$, AB and $-AB$ are additive inverses.

   Write the additive inverse of each term.

   a) t^2 _____   b) $-I_A$ _____   c) F_1r_1 _____   d) $-pq$ _____

| a) $-10t$ |
| b) $3s$ |
| c) $-7TV$ |
| d) $5pq$ |

---

47. To convert a subtraction to an addition, <u>we add the additive inverse of the second term.</u>

   Change - to +.

   Additive inverse of <u>b</u>.

   $$a - b = a + (-b)$$

   Convert each subtraction to an addition.

   a) $F_1r_1 - F_2r_2 =$ _____   b) $c^2 - d^2 =$ _____

| a) $-t^2$ |
| b) I_A |
| c) $-F_1r_1$ |
| d) pq |

---

48. Similarly, any addition in which the second term is negative can be converted to a subtraction. To do so, <u>we subtract the additive inverse of the second term.</u>

   Change + to -.

   Additive inverse of $-cd$.

   $$ab + (-cd) = ab - cd$$

   Convert each addition to a subtraction.

   a) $v + (-t) =$ _____   c) $F_1r_1 + (-F_2r_2) =$ _____

   b) $V_2 + (-V_1) =$ _____   d) $c^2 + (-a^2) =$ _____

| a) $F_1r_1 + (-F_2r_2)$ |
| b) $c^2 + (-d^2)$ |

49. In the addition below, both the second and third terms are negative. We converted both to subtraction form.

$$D + (-S) + (-T) = D - S - T$$

Convert these additions to subtraction form.

a) $p + (-q) + (-r) = $ _____

b) $R_t + (-R_1) + (-R_2) = $ _____

a) v - t

b) $V_2 - V_1$

c) $F_1 r_1 - F_2 r_2$

d) $c^2 - a^2$

---

a) p - q - r   b) $R_t - R_1 - R_2$

9-7 THE ADDITION AXIOM

When a formula contains more than one term on one side, we use the addition axiom to solve for a variable on that side. We will discuss the method in this section.

---

50. To solve for A below, we added L to both sides. Solve for P in the other formula by adding D to both sides.

$$C = A - L \qquad\qquad N = P - D$$

$$L + C = A \underbrace{- L + L}$$
$$\qquad\qquad\quad\downarrow$$
$$L + C = A + \quad 0$$

$$A = L + C$$

---

51. Using the same steps, solve for the indicated variable in each formula.

a) Solve for K.   b) Solve for V.

$$C = K - 273 \qquad\qquad v = V - Ir$$

P = D + N

---

52. To solve for C below, we added -M to both sides. To solve for V below, we added -100 to both sides.

$$R = C + M \qquad\qquad F = V + 100$$

$$R + (-M) = C + \underbrace{M + (-M)} \qquad F + (-100) = V + \underbrace{100 + (-100)}$$
$$\qquad\qquad\qquad\qquad\downarrow \qquad\qquad\qquad\qquad\qquad\qquad\qquad\downarrow$$
$$R + (-M) = C + \quad 0 \qquad\qquad F + (-100) = V + \quad 0$$

$$C = R + (-M) \qquad\qquad\qquad V = F + (-100)$$

a) K = C + 273

b) V = v + Ir

Continued on following page.

52. Continued

In each solution on the preceding page, the right side is an addition in which <u>the second term is negative</u>. Ordinarily, additions of that type are converted to subtraction form. That is:

Instead of $C = R + (-M)$, we write $C = R - M$.

Instead of $V = F + (-100)$, we write _____ .

---

53. Solve for the indicated variable. Write each solution in subtraction form.

a) Solve for E.

$$H = E + PV$$

b) Solve for R.

$$D = R + S + T$$

$V = F - 100$

---

54. To solve for P below, we used the addition axiom to isolate PV. Then we used the multiplication axiom. Solve for <u>m</u> in the other formula.

$$H = E + PV \qquad\qquad I_a = I_c + md^2$$

$$H + (-E) = (-E) + E + PV$$

$$H - E = PV$$

$$P = \frac{H - E}{V}$$

a) $E = H - PV$

b) $R = D - S - T$

---

55. After clearing the fraction below, we used the addition axiom to solve for v_2. Solve for F_i in the other formula.

$$a = \frac{v_2 - v_1}{t} \qquad\qquad w = \frac{F_o + F_i}{t}$$

$$at = v_2 - v_1$$

$$at + v_1 = v_2 - v_1 + v_1$$

$$v_2 = at + v_1$$

$m = \dfrac{I_a - I_c}{d^2}$

---

56. To solve for v_f below, we used the addition axiom to isolate v_f^2 first. Solve for c in the other formula.

$$v_o^2 = v_f^2 - 2gs \qquad\qquad a^2 = c^2 - b^2$$

$$v_o^2 + 2gs = v_f^2 - 2gs + 2gs$$

$$v_o^2 + 2gs = v_f^2$$

$$v_f = \sqrt{v_o^2 + 2gs}$$

$F_i = wt - F_o$

$c = \sqrt{a^2 + b^2}$

57. To solve for X below, we isolated X^2 first. Solve for b in the other formula.

$$Z^2 = X^2 + R^2 \qquad\qquad c^2 = a^2 + b^2$$

$$Z^2 + (-R^2) = X^2 + R^2 + (-R^2)$$

$$Z^2 - R^2 = X^2$$

$$X = \sqrt{Z^2 - R^2}$$

---

58. The steps needed to solve for F_1 in the formula below are described.

$$F_1 r_1 + F_2 r_2 + F_3 r_3 = 0$$

1) Add $-F_2 r_2$ and $-F_3 r_3$ to both sides and get:

$$F_1 r_1 = -F_2 r_2 + (-F_3 r_3)$$

2) Change the addition to subtraction form.

$$F_1 r_1 = -F_2 r_2 - F_3 r_3$$

3) Use the multiplication axiom.

$$F_1 = \frac{-F_2 r_2 - F_3 r_3}{r_1}$$

Solve for r_2 in the same formula.

$$F_1 r_1 + F_2 r_2 + F_3 r_3 = 0$$

$b = \sqrt{c^2 - a^2}$

$$r_2 = \frac{-F_1 r_1 - F_3 r_3}{F_2}$$

9-8   THE INVERSE PRINCIPLE

In this section, we will define the inverse principle and show how it is used in formula rearrangement.

---

59. We can get the <u>additive inverse of a subtraction</u> by simply interchanging the two terms.   For example:

The additive inverse of <u>(5 - 3)</u> is <u>(3 - 5)</u>, since:

$$(5 - 3) + (3 - 5) = 2 + (-2) = 0$$

Continued on following page.

59. Continued

The additive inverse of (2R - 1) is (1 - 2R), since:

$$(2R - 1) + (1 - 2R) = 2R - 1 + 1 - 2R = 0$$

The additive inverse of (a - b) is (b - a), since:

$$(a - b) + (b - a) = a - b + b - a = 0$$

Write the additive inverse of each subtraction.

a) $4y - 1$ _____

b) $p - q$ _____

c) $a^2 - c^2$ _____

d) $v_f^2 - 2gs$ _____

---

60. If we replace each side of an equation with its additive inverse, the new equation is equivalent to the original one. For example:

$$\left.\begin{array}{c} -5x = 15 \\ \text{and} \\ 5x = -15 \end{array}\right\} \begin{array}{l} \text{are equivalent because} \\ \text{the solution of each is -3.} \end{array}$$

Replacing each side of an equation by its additive inverse is called the inverse principle for equations. That principle was used to solve for y below. y is not-solved-for in the top equation because it has a - in front of it.

$$-y = 10 - 2x$$
$$y = 2x - 10$$

The same principle was used to solve for D and v_1 below.

$$-D = N - L \qquad\qquad -v_1 = at - v_2$$
$$D = L - N \qquad\qquad v_1 = v_2 - at$$

Use the inverse principle to solve for t in each formula below.

a) $-t = a - b$ 　　　　　　b) $-t = cd - pq$

　　$t =$ _____ 　　　　$t =$ _____

a) $1 - 4y$

b) $q - p$

c) $c^2 - a^2$

d) $2gs - v_f^2$

---

61. Use the inverse principle for these.

a) Solve for I. 　　　　b) Solve for a.

　　$-I = E_1 - E_2$ 　　　　$-a = bF_1 - cF_2$

a) $t = b - a$

b) $t = pq - cd$

---

62. In solving for L below, we had to use the inverse principle for the final step. Solve for D in the other formula.

$$C = A - L \qquad\qquad N = P - D$$
$$C + (-A) = (-A) + A - L$$
$$C - A = -L$$
$$L = A - C$$

a) $I = E_2 - E_1$

b) $a = cF_2 - bF_1$

63. In solving for T_C below, we used the inverse principle for the final step. Solve for R_1 in the other formula.

$$T_K - T_C = 273 \qquad\qquad R_t - R_1 = R_2$$

$$(-T_K) + T_K - T_C = 273 + (-T_K)$$

$$-T_C = 273 - T_K$$

$$T_C = T_K - 273$$

$D = P - N$

64. We solved for v_1 in the formula below. Use the same steps to solve for d in the other formula.

$$a = \frac{v_2 - v_1}{t} \qquad\qquad m = \frac{b - d}{r}$$

$$at = v_2 - v_1$$

$$at + (-v_2) = (-v_2) + v_2 - v_1$$

$$at - v_2 = -v_1$$

$$v_1 = v_2 - at$$

$R_1 = R_t - R_2$

65. We solved for I below. Use the same steps to solve for s in the other formula.

$$e = E - Ir \qquad\qquad v_o{}^2 = v_f{}^2 - 2gs$$

$$e + (-E) = (-E) + E - Ir$$

$$e - E = -Ir$$

$$Ir = E - e$$

$$I = \frac{E - e}{r}$$

$d = b - mr$

66. We solved for b below. Use the same steps to solve for q in the other formula.

$$a^2 = c^2 - b^2 \qquad\qquad p^2 - q^2 = r^2$$

$$a^2 + (-c^2) = (-c^2) + c^2 - b^2$$

$$a^2 - c^2 = -b^2$$

$$b^2 = c^2 - a^2$$

$$b = \sqrt{c^2 - a^2}$$

$s = \dfrac{v_f{}^2 - v_o{}^2}{2g}$

$q = \sqrt{p^2 - r^2}$

SELF-TEST 31 (pages 385-394)

1. Solve for $\underline{r}$.

$$A = \pi r^2$$

2. Solve for $\underline{v}$.

$$s = \frac{v^2}{g}$$

3. Solve for $\underline{d}$.

$$F = \frac{kQ_1Q_2}{d^2}$$

4. Solve for $\underline{t}$.

$$\frac{2b}{at^2} = \frac{c}{m}$$

5. Solve for K.

$$\frac{G}{K} = \frac{K}{T}$$

6. Solve for d_3.

$$d_1w_1 + d_2w_2 + d_3w_3 = 0$$

7. Solve for y_2.

$$m = \frac{y_2 - y_1}{x}$$

8. Solve for $\underline{i}$.

$$p = P - ei$$

9. Solve for N.

$$F^2 = R^2 - N^2$$

ANSWERS:

1. $r = \sqrt{\dfrac{A}{\pi}}$

2. $v = \sqrt{gs}$

3. $d = \sqrt{\dfrac{kQ_1Q_2}{F}}$

4. $t = \sqrt{\dfrac{2bm}{ac}}$

5. $K = \sqrt{GT}$

6. $d_3 = \dfrac{-d_1w_1 - d_2w_2}{w_3}$

7. $y_2 = y_1 + mx$

8. $i = \dfrac{P - p}{e}$

9. $N = \sqrt{R^2 - F^2}$

9-9   MULTIPLYING BY THE DISTRIBUTIVE PRINCIPLE

In this section, we will discuss formula rearrangements that involve multiplying by the distributive principle.

---

67.   When literal terms are involved, we multiply by the distributive principle in the usual way. That is:

$$P(Q + R) = PQ + PR$$

$$C(T_1 - T_2) = CT_1 - CT_2$$

In $D = P(Q + R)$, the variable Q appears within the parentheses. To solve for Q, we must get it out of the parentheses. To do so, we can begin by multiplying by the distributive principle as we have done below. Use the same method to solve for $\underline{t}$ in the other formula.

$$D = P(Q + R) \qquad\qquad m = b(d + t)$$

$$D = PQ + PR$$

$$D + (-PR) = PQ + PR + (-PR)$$

$$D - PR = PQ$$

$$Q = \frac{D - PR}{P}$$

---

68.   In the formula below, T_1 is within the parentheses. Therefore, to solve for it, we began by multiplying by the distributive principle. Solve for L in the other formula.

$$Q = C(T_1 - T_2) \qquad\qquad M = P(L - X)$$

$$Q = CT_1 - CT_2$$

$$Q + CT_2 = CT_1 - CT_2 + CT_2$$

$$Q + CT_2 = CT_1$$

$$T_1 = \frac{Q + CT_2}{C}$$

$$t = \frac{m - bd}{b}$$

---

69.   To solve for V_2 below, we began by multiplying by the distributive principle. <u>Notice that we had to use the inverse principle.</u> Solve for t_1 in the other formula.

$$W_s = P(V_1 - V_2) \qquad\qquad H = ms(t_2 - t_1)$$

$$W_s = PV_1 - PV_2$$

$$W_s + (-PV_1) = (-PV_1) + PV_1 - PV_2$$

$$W_s - PV_1 = -PV_2$$

$$PV_1 - W_s = PV_2$$

$$V_2 = \frac{PV_1 - W_s}{P}$$

$$L = \frac{M + PX}{P}$$

70. To solve for either b_1 or b_2 below, we have to get them out of the parentheses. Notice the steps.

$$A = \frac{1}{2}h(b_1 + b_2)$$

$$A = \frac{h(b_1 + b_2)}{2}$$

$$2A = h(b_1 + b_2)$$

$$2A = hb_1 + hb_2$$

Beginning with the last formula, solve for both b_1 and b_2.

a) Solve for b_1.   b) Solve for b_2.

$$t_1 = \frac{mst_2 - H}{ms}$$

---

71. To solve for R below, we began by multiplying by the distributive principle. Solve for <u>d</u> in the other formula.

$$A = \pi(R^2 - r^2) \qquad\qquad m = (d^2 - 1)t$$

$$A = \pi R^2 - \pi r^2$$

$$A + \pi r^2 = \pi R^2 - \pi r^2 + \pi r^2$$

$$A + \pi r^2 = \pi R^2$$

$$R^2 = \frac{A + \pi r^2}{\pi}$$

$$R = \sqrt{\frac{A + \pi r^2}{\pi}}$$

---

a) $b_1 = \dfrac{2A - hb_2}{h}$

b) $b_2 = \dfrac{2A - hb_1}{h}$

---

72. To solve for C, we also began by multiplying by the distributive principle. <u>Notice</u> <u>that</u> <u>we</u> <u>had</u> <u>to</u> <u>use</u> <u>the</u> <u>inverse</u> <u>principle.</u> Solve for <u>r</u> in the other formula.

$$H = (1 - C^2)h \qquad\qquad A = \pi(R^2 - r^2)$$

$$H = h - C^2h$$

$$H + (-h) = (-h) + h - C^2h$$

$$H - h = -C^2h$$

$$h - H = C^2h$$

$$C^2 = \frac{h - H}{h}$$

$$C = \sqrt{\frac{h - H}{h}}$$

$$d = \sqrt{\frac{m + t}{t}}$$

$$r = \sqrt{\frac{\pi R^2 - A}{\pi}}$$

73. To solve for t_2 below, we began by clearing the fraction. Solve for t in the other formula.

$$H = \frac{AKT(t_2 - t_1)}{L} \qquad\qquad p = \frac{k(t^2 - b^2)}{ar}$$

$$HL = AKT(t_2 - t_1)$$

$$HL = AKTt_2 - AKTt_1$$

$$HL + AKTt_1 = AKTt_2$$

$$t_2 = \frac{HL + AKTt_1}{AKT}$$

---

74. To solve for V_2 below, we also began by clearing the fraction. <u>Notice that we had to use the inverse principle.</u> Solve for F_2 in the other formula.

$$H = \frac{M(V_1^2 - V_2^2)}{1,100gt} \qquad\qquad H = \frac{\pi dR(F_1 - F_2)}{33,000}$$

$$1,100gtH = M(V_1^2 - V_2^2)$$

$$1,100gtH = MV_1^2 - MV_2^2$$

$$1,100gtH - MV_1^2 = -MV_2^2$$

$$MV_1^2 - 1,100gtH = MV_2^2$$

$$V_2^2 = \frac{MV_1^2 - 1,100gtH}{M}$$

$$V_2 = \sqrt{\frac{MV_1^2 - 1,100gtH}{M}}$$

$$t = \sqrt{\frac{apr + b^2k}{k}}$$

$$F_2 = \frac{\pi dRF_1 - 33,000H}{\pi dR}$$

9-10 ANOTHER METHOD AND EQUIVALENT FORMS OF SOLUTIONS

There is another method that can be used for the rearrangements in the last section. The other method leads to a solution that looks different but is really equivalent. We will discuss the other method in this section and show that the two forms of the solution are equivalent.

75. To solve for L at the left below, we began by multiplying by the distributive principle. At the right below, we used another method to solve for L. In the second method, we began by isolating the subtraction.

$$M = P(L - X) \qquad\qquad M = P(L - X)$$

$$M = PL - PX \qquad\qquad \frac{M}{P} = L - X$$

$$M + PX = PL - PX + PX$$

$$\frac{M}{P} + X = L - X + X$$

$$M + PX = PL$$

$$L = \frac{M + PX}{P} \qquad\qquad L = \frac{M}{P} + X$$

Though the two solutions look different, they are equivalent. We have shown that fact below.

To show that the left solution is equivalent to the right solution, we can break up the sum of fractions and reduce.

$$L = \frac{M + PX}{P} = \frac{M}{P} + \frac{\cancel{P}X}{\cancel{P}} = \frac{M}{P} + X$$

To show that the right solution is equivalent to the left solution, we can add the fraction and non-fraction.

$$L = \frac{M}{P} + X = \frac{M}{P} + \frac{PX}{P} = \underline{\qquad\qquad}$$

76. To solve for $\underline{r}$ below, we began by multiplying by the distributive principle. At the right below, we used another method in which we began by isolating the subtraction.

$$\frac{M + PX}{P}$$

$$A = \pi(R^2 - r^2) \qquad\qquad A = \pi(R^2 - r^2)$$

$$A = \pi R^2 - \pi r^2 \qquad\qquad \frac{A}{\pi} = R^2 - r^2$$

$$A - \pi R^2 = -\pi r^2$$

$$\frac{A}{\pi} - R^2 = -r^2$$

$$\pi R^2 - A = \pi r^2$$

$$R^2 - \frac{A}{\pi} = r^2$$

$$r^2 = \frac{\pi R^2 - A}{\pi}$$

$$r = \sqrt{\frac{\pi R^2 - A}{\pi}} \qquad\qquad r = \sqrt{R^2 - \frac{A}{\pi}}$$

Continued on following page.

76. Continued

Though the two solutions look different, they are also equivalent. We have shown that fact below.

a) To show that the left radicand is equivalent to the right radicand, we can break up the sum of fractions and reduce.

$$r = \sqrt{\frac{\pi R^2 - A}{\pi}} = \sqrt{\frac{\pi R^2}{\pi} - \frac{A}{\pi}} = \underline{\hspace{3cm}}$$

b) To show that the right radicand is equivalent to the left radicand, we can add the non-fraction and fraction.

$$r = \sqrt{R^2 - \frac{A}{\pi}} = \sqrt{\frac{\pi R^2}{\pi} - \frac{A}{\pi}} = \underline{\hspace{3cm}}$$

---

77. To solve for t_2 by the second method below, we cleared the fraction and then isolated the subtraction. Use the same method to solve for F_1 in the other formula.

$$H = \frac{AKT(t_2 - t_1)}{L}$$

$$HL = AKT(t_2 - t_1)$$

$$\frac{HL}{AKT} = t_2 - t_1$$

$$t_2 = \frac{HL}{AKT} + t_1$$

$$H = \frac{\pi dR(F_1 - F_2)}{33,000}$$

a) $\sqrt{R^2 - \dfrac{A}{\pi}}$

b) $\sqrt{\dfrac{\pi R^2 - A}{\pi}}$

---

78. To solve for V_2 by the second method below, we isolated the subtraction after clearing the fraction. Notice how we used the inverse principle. Use the same method to solve for b in the other formula.

$$H = \frac{M(V_1^2 - V_2^2)}{1,100gt}$$

$$1,100gtH = M(V_1^2 - V_2^2)$$

$$\frac{1,100gtH}{M} = V_1^2 - V_2^2$$

$$\frac{1,100gtH}{M} - V_1^2 = -V_2^2$$

$$V_1^2 - \frac{1,100gtH}{M} = V_2^2$$

$$V_2 = \sqrt{V_1^2 - \frac{1,100gtH}{M}}$$

$$p = \frac{k(t^2 - b^2)}{ar}$$

$$F_1 = \frac{33,000H}{\pi dR} + F_2$$

$$b = \sqrt{t^2 - \frac{apr}{k}}$$

79. Two methods are possible for the rearrangements below. Use the one that you prefer.

    a) Solve for V_1.

$$W_s = P(V_1 - V_2)$$

    b) Solve for t_1.

$$H = \frac{AKT(t_2 - t_1)}{L}$$

a) $V_1 = \dfrac{W_s}{P} + V_2$

   or

$V_1 = \dfrac{W_s + PV_2}{P}$

---

80. Use either method for the rearrangements below.

    a) Solve for $\underline{m}$.

$$b = c(d^2 - m^2)$$

    b) Solve for G.

$$V = \frac{R(G^2 - H^2)}{100}$$

b) $t_1 = t_2 - \dfrac{HL}{AKT}$

   or

$t_1 = \dfrac{AKTt_2 - HL}{AKT}$

---

81. We solved for R_1 below. Solve for R_3 in the same formula.

$$E = IR_1 + IR_2 + IR_3$$

$$IR_1 = E - IR_2 - IR_3$$

$$R_1 = \frac{E - IR_2 - IR_3}{I}$$

    or

$$R_1 = \frac{E}{I} - R_2 - R_3$$

$$E = IR_1 + IR_2 + IR_3$$

$$R_3 = \frac{E}{I} - R_2 - R_1$$

a) $m = \sqrt{d^2 - \dfrac{b}{c}}$

   or

$m = \sqrt{\dfrac{cd^2 - b}{c}}$

b) $G = \sqrt{\dfrac{100V}{R} + H^2}$

   or

$G = \sqrt{\dfrac{100V + RH^2}{R}}$

$$R_3 = \frac{E - IR_1 - IR_2}{I}$$

   or

$$R_3 = \frac{E}{I} - R_1 - R_2$$

9-11   FACTORING BY THE DISTRIBUTIVE PRINCIPLE

To solve for a variable that appears in two terms in a formula, we must factor by the distributive principle. We will discuss rearrangements of that type in this section.

---

82.   In a solution for a variable, <u>the same variable cannot appear in the expression on the other side</u>. For example, the following is not a solution for H because H also appears on the right side.

$$H = \frac{A + BH}{C}$$

Which of the following is a solution for <u>v</u>? _____

a) $v = \dfrac{p + qr}{t}$ 　　　　　 b) $v = \dfrac{p + qv}{t}$

---

83.   In am = b + cm , <u>m</u> appears in one term on each side. We tried to solve for <u>m</u> below by isolating <u>am</u> alone and <u>cm</u> alone. Neither expression is a solution for <u>m</u> because <u>m</u> also appears on the right side.

Only (a). In (b), <u>v</u> also appears on the right side.

　　Isolating "am" alone. 　　　 Isolating "cm" alone.

　　　am = b + cm 　　　　　 cm = am - b

　　　$m = \dfrac{b + cm}{a}$ 　　　　　 $m = \dfrac{am - b}{c}$

To solve for <u>m</u> in am = b + cm , we must get both <u>m</u> terms on one side, <u>factor by the distributive principle</u>, and then complete the solution as we have done below.

　　　　　am = b + cm
　　　　am - cm = b
　　　　m(a - c) = b
　　　　　　$m = \dfrac{b}{a - c}$

Before rearranging formulas like those above, let's review the procedure for factoring by the distributive principle. Two examples are shown.

$$bd + bt = b(d + t)$$

$$AV_1 - CV_1 = V_1(A - C)$$

Following the examples, factor these.

a) mx + my = _____ 　　　 c) $HR_1 + HR_2$ = _____

b) aR - bR = _____ 　　　 d) $c_1T_1 - c_2T_1$ = _____

84. To solve for $\underline{d}$ below, we factored by the distributive principle and then used the multiplication axiom. Solve for I in the other formula.

$$m = dp + dq$$

$$V = IR_1 + IR_2$$

$$m = d(p + q)$$

$$d = \frac{m}{p + q}$$

a) $m(x + y)$

b) $R(a - b)$

c) $H(R_1 + R_2)$

d) $T_1(c_1 - c_2)$

---

85. To solve for S below, we isolated both S terms on one side before factoring. Solve for P_2 in the other formula.

$$MS = R - QS$$

$$P_1 P_2 = I - RP_2$$

$$MS + QS = R$$

$$S(M + Q) = R$$

$$S = \frac{R}{M + Q}$$

$$I = \frac{V}{R_1 + R_2}$$

---

86. To solve for $\underline{t}$ below, we isolated both $\underline{t}$ terms on one side before factoring. Solve for T in the other formula.

$$bt + cm = dt$$

$$RV + ST = QT$$

$$cm = dt - bt$$

$$cm = t(d - b)$$

$$t = \frac{cm}{d - b}$$

$$P_2 = \frac{I}{P_1 + R}$$

---

87. To solve for $\underline{c}$ below, we began by clearing the fractions. To do so, we multiplied both sides by $\underline{abc}$. We then factored by the distributive principle. Solve for $\underline{f}$ in the other formula.

$$\frac{1}{a} + \frac{1}{b} = \frac{1}{c}$$

$$\frac{1}{D} + \frac{1}{d} = \frac{1}{f}$$

$$abc\left(\frac{1}{a} + \frac{1}{b}\right) = abc\left(\frac{1}{c}\right)$$

$$abc\left(\frac{1}{a}\right) + abc\left(\frac{1}{b}\right) = abc\left(\frac{1}{c}\right)$$

$$bc + ac = ab$$

$$c(b + a) = ab$$

$$c = \frac{ab}{b + a}$$

$$T = \frac{RV}{Q - S}$$

$$f = \frac{Dd}{D + d}$$

88. To solve for R_t below, we also began by clearing the fractions. To do so, we multiplied both sides by $R_t R_1 R_2$. Solve for C_1 in the other formula.

$$\frac{1}{R_t} = \frac{1}{R_1} + \frac{1}{R_2}$$

$$\frac{1}{C_1} = \frac{1}{C_t} - \frac{1}{C_2}$$

$$R_t R_1 R_2 \left(\frac{1}{R_t}\right) = R_t R_1 R_2 \left(\frac{1}{R_1} + \frac{1}{R_2}\right)$$

$$\cancel{R_t} R_1 R_2 \left(\frac{1}{\cancel{R_t}}\right) = R_t \cancel{R_1} R_2 \left(\frac{1}{\cancel{R_1}}\right) + R_t R_1 \cancel{R_2} \left(\frac{1}{\cancel{R_2}}\right)$$

$$R_1 R_2 = R_t R_2 + R_t R_1$$

$$R_1 R_2 = R_t (R_2 + R_1)$$

$$R_t = \frac{R_1 R_2}{R_2 + R_1}$$

---

89. To solve for W below, we cleared the fractions by multiplying both sides by 2. Solve for A in the other formula.

$$M = \frac{WLX}{2} - \frac{WX^2}{2}$$

$$T = \frac{AFG}{3} - \frac{AG^2}{3}$$

$$2(M) = 2\left(\frac{WLX}{2} - \frac{WX^2}{2}\right)$$

$$2M = 2\left(\frac{WLX}{\cancel{2}}\right) - 2\left(\frac{WX^2}{\cancel{2}}\right)$$

$$2M = WLX - WX^2$$

$$2M = W(LX - X^2)$$

$$W = \frac{2M}{LX - X^2}$$

$$C_1 = \frac{C_t C_2}{C_2 - C_t}$$

---

90. To solve for $\underline{d}$ below, we had to isolate the $\underline{d}$ terms before factoring. Solve for C_t in the other formula.

$$\frac{1}{D} + \frac{1}{d} = \frac{1}{f}$$

$$\frac{1}{C_1} = \frac{1}{C_t} - \frac{1}{C_2}$$

$$Ddf\left(\frac{1}{D} + \frac{1}{d}\right) = Ddf\left(\frac{1}{f}\right)$$

$$\cancel{D}df\left(\frac{1}{\cancel{D}}\right) + D\cancel{d}f\left(\frac{1}{\cancel{d}}\right) = Dd\cancel{f}\left(\frac{1}{\cancel{f}}\right)$$

$$df + Df = Dd$$

$$Df = Dd - df$$

$$Df = d(D - f)$$

$$d = \frac{Df}{D - f}$$

$$A = \frac{3T}{FG - G^2}$$

91. We solved for T_R below. Notice how we factored $T - ET$ on the right side of the solution. Solve for R in the other formula.

$$E = \frac{T - T_R}{T}$$

$$V = \frac{S - R}{S}$$

$$ET = T - T_R$$

$$ET - T = -T_R$$

$$T - ET = T_R$$

$$T_R = T(1 - E)$$

$$C_t = \frac{C_1 C_2}{C_1 + C_2}$$

---

92. To solve for M below, we had to isolate the M terms after multiplying by the distributive principle. Solve for β in the other formula.

$$\mu = \frac{Mm}{M + m}$$

$$\alpha = \frac{\beta}{\beta + 1}$$

$$\mu(M + m) = (M + m)\left(\frac{Mm}{M + m}\right)$$

$$\mu M + \mu m = Mm$$

$$\mu m = Mm - \mu M$$

$$\mu m = M(m - \mu)$$

$$M = \frac{\mu m}{m - \mu}$$

$R = S(1 - V)$,

from:

$$R = S - SV$$

---

93. Using the same steps, complete these.

a) Solve for $\underline{m}$.

$$\mu = \frac{Mm}{M + m}$$

b) Solve for A.

$$A_f = \frac{A}{1 - BA}$$

$$\beta = \frac{\alpha}{1 - \alpha}$$

---

94. To solve for $\underline{a}$ below, we factored by the distributive principle. Solve for I in the other formula.

$$m = ab + ac + ad$$

$$V = IR_1 + IR_2 + IR_3$$

$$m = a(b + c + d)$$

$$a = \frac{m}{b + c + d}$$

a) $m = \dfrac{\mu M}{M - \mu}$

b) $A = \dfrac{A_f}{1 + A_f B}$

95. Two methods of solving for v in the formula below are shown. At the left, we did not clear the fraction first. At the right, we did clear the fraction first.

$$I = \frac{V}{R_1 + R_2 + R_3}$$

$$P = \frac{mv^2}{r} - mg$$

$$P + mg = \frac{mv^2}{r}$$

$$r(P + mg) = mv^2$$

$$v^2 = \frac{r(P + mg)}{m}$$

$$v = \sqrt{\frac{r(P + mg)}{m}}$$

$$P = \frac{mv^2}{r} - mg$$

$$r(P) = \not{r}\left(\frac{mv^2}{\not{r}}\right) - r(mg)$$

$$Pr = mv^2 - mgr$$

$$Pr + mgr = mv^2$$

$$v^2 = \frac{Pr + mgr}{m}$$

$$v = \sqrt{\frac{r(P + mg)}{m}}$$

Using either method above, do these.

a) Solve for a.

$$a^2 + 1 = \frac{b}{c}$$

b) Solve for d.

$$T = \frac{bd^2}{p} - S$$

a) $a = \sqrt{\dfrac{b}{c} - 1}$ or $a = \sqrt{\dfrac{b - c}{c}}$

b) $d = \sqrt{\dfrac{p(T + S)}{b}}$

<u>SELF-TEST 32</u> (<u>pages 395-406</u>)

1. Solve for t_1.

$$d = r(t_1 + t_2)$$

2. Solve for <u>p</u>.

$$H = \frac{1}{2}k(P^2 - p^2)$$

3. Solve for R.

$$V = \frac{AH(R - r)}{M}$$

4. Solve for <u>i</u>. $e_1 i = p - e_2 i$

5. Solve for F. $\frac{1}{F} = \frac{1}{P} + \frac{1}{Q}$

6. Solve for h_2. $\frac{1}{h_1} + \frac{1}{h_2} = \frac{1}{H}$

7. Solve for <u>d</u>. $v = \frac{ad}{a - d}$

<u>ANSWERS:</u>

1. $t_1 = \dfrac{d - rt_2}{r}$

2. $p = \sqrt{\dfrac{kP^2 - 2H}{k}}$

   or $p = \sqrt{P^2 - \dfrac{2H}{k}}$

3. $R = \dfrac{MV + AHr}{AH}$

   or $R = \dfrac{MV}{AH} + r$

4. $i = \dfrac{p}{e_1 + e_2}$

5. $F = \dfrac{PQ}{P + Q}$

6. $h_2 = \dfrac{h_1 H}{h_1 - H}$

7. $d = \dfrac{av}{a + v}$

9-12   A PREFERRED FORM FOR FRACTIONAL SOLUTIONS

When rearranging formulas, fractional solutions are usually written so that they contain as few negative signs as possible. We will discuss that preferred form for solutions in this section.

96.   Two fractions are equivalent if their numerators are inverses and their denominators are inverses. That is:

$\frac{6}{3}$ and $\frac{-6}{-3}$ are equivalent, because both equal +2.

Two fractions with subtractions as denominators are also equivalent if their numerators are inverses and their denominators are inverses. That is:

$\frac{10}{6-4}$ and $\frac{-10}{4-6}$ are equivalent, because both equal +5.

Write each of these in an equivalent form by replacing each term with its inverse.

a) $\frac{12}{7-4}$ = $\frac{-12}{4-7}$ c) $\frac{P_1}{R_1 - R_2}$ = $\frac{-P_1}{R_2 - R_1}$

b) $\frac{-15}{3-8}$ = $\frac{15}{8-3}$ d) $\frac{-ab}{c-d}$ = $\frac{ab}{d-c}$

97.   Depending on the method used, you can get either of the following equivalent solutions when solving dm = cm - b for m.

$$m = \frac{b}{c-d} \qquad\qquad m = \frac{-b}{d-c}$$

The solution on the left is preferred because it contains fewer negative signs. If you get the solution on the right, you can convert it to the preferred form by replacing each term with its inverse. That is:

$$m = \frac{-b}{d-c} = \frac{b}{c-d}$$

Write each solution in the preferred form.

a) $P = \frac{-S}{V-T}$ = _____ .   b) $R = \frac{-wr}{2F-w}$ = _____

a) $\frac{-12}{4-7}$

b) $\frac{15}{8-3}$

c) $\frac{-P_1}{R_2 - R_1}$

d) $\frac{ab}{d-c}$

a) $P = \frac{S}{T-V}$

b) $R = \frac{wr}{w-2F}$

98. At the left below, we used the inverse principle when solving for T. At the right below, we did not use the inverse principle when solving for T.

$$E = \frac{T - T_R}{T}$$

$$ET = T - T_R$$

$$ET - T = -T_R$$

$$T - ET = T_R$$

$$T(1 - E) = T_R$$

$$T = \frac{T_R}{1 - E}$$

$$E = \frac{T - T_R}{T}$$

$$ET = T - T_R$$

$$ET - T = -T_R$$

$$T(E - 1) = -T_R$$

$$T = \frac{-T_R}{E - 1}$$

The solution at the left is preferred. If you get the solution at the right, you can convert it to the preferred form by replacing each term with its inverse. That is:

$$T = \frac{-T_R}{E - 1} = \underline{\hspace{3cm}}$$

99. Do these. Write each solution in the preferred form.

   a) Solve for q.

   $$\frac{1}{p} - \frac{1}{q} = \frac{1}{r}$$

   b) Solve for C_2.

   $$\frac{1}{C_1} = \frac{1}{C_t} - \frac{1}{C_2}$$

$$T = \frac{T_R}{1 - E}$$

a) $q = \dfrac{pr}{r - p}$

b) $C_2 = \dfrac{C_1 C_t}{C_1 - C_t}$

100. In the formula below, $\underline{b}$ appears on both sides. To solve for $\underline{b}$, we multiplied by the distributive principle and then isolated the $\underline{b}$ terms. Notice how we wrote the solution in the preferred form. Solve for R.

$$ab = c(b - t) \qquad\qquad 2RF = w(R - r)$$

$$ab = bc - ct$$

$$ab - bc = -ct$$

$$b(a - c) = -ct$$

$$b = \frac{-ct}{a - c}$$

$$b = \frac{ct}{c - a}$$

---

101. Two fractions with subtractions as numerators are also equivalent $\underline{if}$ $\underline{their}$ $\underline{numerators}$ $\underline{are}$ $\underline{inverses}$ $\underline{and}$ $\underline{their}$ $\underline{denominators}$ $\underline{are}$ $\underline{inverses}$. That is:

$$\frac{8 - 2}{3} \text{ and } \frac{2 - 8}{-3} \text{ are equivalent because both equal } +2.$$

Write each of these in an equivalent form by replacing each term with its inverse.

a) $\dfrac{c - d}{-b} =$ _____   b) $\dfrac{R - V}{-ST} =$ _____

$$R = \frac{wr}{w - 2F}$$

---

102. Depending on the method used, you can get either of the following $\underline{equivalent}$ solutions when solving $Q = R - MP$ for P.

$$P = \frac{R - Q}{M} \qquad\qquad P = \frac{Q - R}{-M}$$

The solution on the left is preferred because it contains fewer negative signs. If you get the solution on the right, you can convert it to the preferred form $\underline{by}$ $\underline{replacing}$ $\underline{each}$ $\underline{term}$ $\underline{with}$ $\underline{its}$ $\underline{inverse}$. That is:

$$P = \frac{Q - R}{-M} =$$ _____

a) $\dfrac{d - c}{b}$ b) $\dfrac{V - R}{ST}$

---

103. We solved for $\underline{t}$ below. Notice how we wrote the solution in the preferred form. Solve for B in the other formula.

$$a = \frac{d}{1 - dt} \qquad\qquad A_f = \frac{A}{1 - AB}$$

$$a(1 - dt) = d$$

$$a - adt = d$$

$$-adt = d - a$$

$$t = \frac{d - a}{-ad}$$

$$t = \frac{a - d}{ad}$$

$$P = \frac{R - Q}{M}$$

$$B = \frac{A_f - A}{A_f A}$$

9-13 DISTRIBUTIVE-PRINCIPLE TERMS

In this section, we will show how the addition axiom is used in formula rearrangements when one of the terms is an instance of the distributive principle. We will begin by discussing the additive inverses of instances of the distributive principle.

104. Since 2(3 + 4) = 14, the additive inverse of 2(3 + 4) must equal -14. To get the inverse of 2(3 + 4), we can replace 2 with its inverse, -2. That is: The additive inverse of 2(3 + 4) is (-2)(3 + 4). (-2)(3 + 4) is the inverse of 2(3 + 4) because (-2)(3 + 4) = ___-14___ .	

105. By replacing 4 with its inverse, we can get the additive inverse of 4(y + 5). That is:
 	
    The additive inverse of 4(y + 5) is (-4)(y + 5).
 	
To show that 4(y + 5) and (-4)(y + 5) are inverses, we added them below to show that their sum is 0.
 	
    4(y + 5) + (-4)(y + 5) = 4y + 20 + (-4y) + (-20)
 	
        = [4y + (-4y)] + [20 + (-20)]
        =   0   +   0   = 0
 	
Write the inverse of each expression.
 	
    a) 7(x + 3) __-7(x+3)__   b) R(S + T) __-R(S+T)__
	-14

106. By replacing 2 with its inverse, we can get the additive inverse of 2(x - 5). That is:
 	
    The additive inverse of 2(x - 5) is (-2)(x - 5).
 	
To show that 2(x - 5) and (-2)(x - 5) are inverses, we added them below to show that their sum is 0.
 	
    2(x - 5) + (-2)(x - 5) = 2x - 10 + (-2x) + 10
 	
        = [2x + (-2x)] [- 10 + 10]
        =   0   +   0   = 0
 	
Write the inverse of each expression.
 	
    a) 8(m - 1) __(-8)(m-1)__   b) a(b - c) __-a(b-c)__
	a) (-7)(x + 3)
	
	b) (-R)(S + T)

	a) (-8)(m - 1) b) (-a)(b - c)

107. To convert each subtraction below to addition, we added the inverse of the instance of the distributive principle.

$$7 - 6(y + 3) = 7 + (-6)(y + 3)$$
$$V - M(T - S) = V + (-M)(T - S)$$

Convert each subtraction to addition.

a) $5 - 2(x - 5) =$ _$5 + (-2)(x - 5)$_   b) $a - b(c + d) =$ _$a + (-b)(c + d)$_

---

108. To convert each addition below to subtraction, we subtracted the inverse of the instance of the distributive principle.

$$4 + (-3)(t + 7) = 4 - 3(t + 7)$$
$$H + (-d)(b + c) = H - d(b + c)$$

Convert each addition to subtraction.

a) $8 + (-9)(y - 1) =$ _$8 - 9(y - 1)$_   b) $d + (-R)(P + Q) =$ _$d - R(P + Q)$_

a) $5 + (-2)(x - 5)$

b) $a + (-b)(c + d)$

---

109. To solve for m below, we add the inverse of $a(c - d)$ to both sides. We get:

$$V = m + a(c - d)$$

$m(-a)(c-d)$

$$V + (-a)(c - d) = m + \underline{a(c - d) + (-a)(c - d)}$$
$$V + (-a)(c - d) = m \quad + \quad 0$$
$$V + (-a)(c - d) = m$$

$V = a(c-d)$

To complete the solution, we convert the addition on the left side to subtraction form. Do so.

$$m = \underline{V - a(c - d)}$$

a) $8 - 9(y - 1)$

b) $d - R(P + Q)$

---

110. Using the same steps, complete these.

a) Solve for p_1.   b) Solve for T.

$$P = p_1 + w(h - h_1) \qquad T + k(a + b) = S$$

$P - w(h - h_1)$

m = V - a(c - d)

---

a) $p_1 = P - w(h - h_1)$

b) $T = S - k(a + b)$

111. To solve for D below, we used the addition axiom to isolate CD. Solve for I_b in the other formula.

$$B = CD + (v + 1)F \qquad I_c = \beta I_b + (\beta + 1)I_{co}$$

$$B - (v + 1)F = CD$$

$$-(v + 1)F$$

$$D = \frac{B - (v + 1)F}{C}$$

---

112. To solve for <u>a</u> in the formula below, we began by using the addition axiom to isolate the instance of the distributive principle. Solve for <u>w</u> in the other formula.

$$m = t + a(b + c) \qquad P = p_1 + w(h - h_1)$$

$$m - t = a(b + c)$$

$$a = \frac{m - t}{b + c}$$

$$I_b = \frac{I_c - (\beta + 1)I_{co}}{\beta}$$

---

113. To solve for R in the formula below, we isolated the instance of the distributive principle and then multiplied to get R out of the parentheses. Solve for <u>h</u> in the other formula.

$$b = v + m(R - 1) \qquad P = p_1 + w(h - h_1)$$

$$b - v = m(R - 1)$$

$$b - v = mR - m$$

$$b - v + m = mR$$

$$R = \frac{b - v + m}{m}$$

or

$$R = \frac{b - v}{m} + 1$$

$$w = \frac{P - p_1}{h - h_1}$$

$$h = \frac{P - p_1 + wh_1}{w}$$

or

$$h = \frac{P - p_1}{w} + h_1$$

114. a) Solve for I_{co}.

$$I_c = \beta I_b + (\beta + 1)I_{co}$$

b) Solve for p.

$$R = ab + m(p + q)$$

a) $I_{co} = \dfrac{I_c - \beta I_b}{\beta + 1}$

b) $p = \dfrac{R - ab - mq}{m}$

or

$p = \dfrac{R - ab}{m} - q$

115. To solve for v below, we began by multiplying by the distributive principle to get v out of the parentheses. Then we isolated the v terms. Solve for β in the other formula.

$$R = vt + (v + 1)d$$

$$R = vt + vd + d$$

$$R - d = vt + vd$$

$$R - d = v(t + d)$$

$$v = \dfrac{R - d}{t + d}$$

$$I_c = \beta I_b + (\beta + 1)I_{co}$$

$\beta = \dfrac{I_c - I_{co}}{I_b + I_{co}}$

116. At the left below, we used the inverse principle when solving for h_1. At the right below, we did not use the inverse principle.

$$P = p_1 + w(h - h_1)$$
$$P - p_1 = w(h - h_1)$$
$$P - p_1 = wh - wh_1$$
$$P - p_1 - wh = -wh_1$$
$$p_1 + wh - P = wh_1$$
$$h_1 = \dfrac{p_1 + wh - P}{w}$$

$$P = p_1 + w(h - h_1)$$
$$P - p_1 = w(h - h_1)$$
$$P - p_1 = wh - wh_1$$
$$P - p_1 - wh = -wh_1$$
$$h_1 = \dfrac{P - p_1 - wh}{-w}$$

The form of the solution at the left is preferred because it contains fewer negative signs. If you get the solution on the right, you can convert it to the preferred form by replacing each term of the fraction with its inverse. That is:

$$h_1 = \dfrac{P - p_1 - wh}{-w} = \underline{\hspace{2cm}}$$

$h_1 = \dfrac{p_1 + wh - P}{w}$

9-14 DISTRIBUTIVE-PRINCIPLE DENOMINATORS

In this section, we will rearrange formulas containing one fraction whose denominator is an instance of the distributive principle.

---

117. To solve for $\underline{a}$ below, we cleared the fraction by multiplying both sides by $t(p - r)$. We then used the addition axiom to isolate $\underline{a}$. Notice <u>how</u> <u>we</u> <u>were</u> <u>able</u> <u>to</u> <u>factor</u> <u>the</u> <u>solution</u>. Solve for L_2 in the other formula.

$$m = \frac{a - t}{t(p - r)} \qquad\qquad \alpha = \frac{L_2 - L_1}{L_1(t_2 - t_1)}$$

$$mt(p - r) = a - t$$

$$a = mt(p - r) + t$$

or

$$a = t[m(p - r) + 1]$$

$L_2 = \alpha L_1(t_2 - t_1) + L_1$
or
$L_2 = L_1[\alpha(t_2 - t_1) + 1]$

---

118. We solved for $\underline{b}$ below. Solve for T in the other formula.

$$H = \frac{b(t_2 - t_1)}{c(d_2 - d_1)} \qquad\qquad B = \frac{T(V_1 - V_2)}{V_1(P_2 - P_1)}$$

$$cH(d_2 - d_1) = b(t_2 - t_1)$$

$$b = \frac{cH(d_2 - d_1)}{t_2 - t_1}$$

$T = \dfrac{BV_1(P_2 - P_1)}{V_1 - V_2}$

---

119. We solved for $\underline{k}$ below. Solve for V_1 in the other formula.

$$t = \frac{a(b - c)}{k(p - q)} \qquad\qquad B = \frac{T(R_1 - R_2)}{V_1(P_2 - P_1)}$$

$$kt(p - q) = a(b - c)$$

$$k = \frac{a(b - c)}{t(p - q)}$$

$V_1 = \dfrac{T(R_1 - R_2)}{B(P_2 - P_1)}$

120. To solve for $\underline{r}$ below, we had to multiply by the distributive principle on the left side to get $\underline{r}$ out of the parentheses. Notice how we were able to write the solution in an equivalent form. Solve for P_2 in the other formula.

$$b = \frac{m(d_1 - d_2)}{d_1(r - s)} \qquad\qquad B = \frac{T(V_1 - V_2)}{V_1(P_2 - P_1)}$$

$$bd_1(r - s) = m(d_1 - d_2)$$

$$bd_1r - bd_1s = m(d_1 - d_2)$$

$$bd_1r = m(d_1 - d_2) + bd_1s$$

$$r = \frac{m(d_1 - d_2) + bd_1s}{bd_1}$$

or

$$r = \frac{m(d_1 - d_2)}{bd_1} + s$$

---

121. To solve for b_2 below, we also had to multiply by the distributive principle on the left side to get b_2 out of the parentheses. Notice how we were able to write the solution in an equivalent form. Solve for t_2 in the other formula.

$$R = \frac{P - Q}{Q(b_2 - b_1)} \qquad\qquad \alpha = \frac{L_2 - L_1}{L_1(t_2 - t_1)}$$

$$QR(b_2 - b_1) = P - Q$$

$$QRb_2 - QRb_1 = P - Q$$

$$QRb_2 = P - Q + QRb_1$$

$$b_2 = \frac{P - Q + QRb_1}{QR}$$

or

$$b_2 = \frac{P - Q}{QR} + b_1$$

$$P_2 = \frac{T(V_1 - V_2) + BV_1P_1}{BV_1}$$

or

$$P_2 = \frac{T(V_1 - V_2)}{BV_1} + P_1$$

$$t_2 = \frac{L_2 - L_1 + \alpha L_1 t_1}{\alpha L_1}$$

or

$$t_2 = \frac{L_2 - L_1}{\alpha L_1} + t_1$$

122. Notice how we used the inverse principle in solving for t_1 below. Complete the solution.

$$\alpha = \frac{L_2 - L_1}{L_1(t_2 - t_1)}$$

$$\alpha L_1(t_2 - t_1) = L_2 - L_1$$

$$\alpha L_1 t_2 - \alpha L_1 t_1 = L_2 - L_1$$

$$-\alpha L_1 t_1 = L_2 - L_1 - \alpha L_1 t_2$$

$$\alpha L_1 t_1 = L_1 - L_2 + \alpha L_1 t_2$$

$$t_1 = \underline{\hspace{3cm}}$$

---

123. Some of the steps needed to solve for L_1 below are shown.

$$\alpha = \frac{L_2 - L_1}{L_1(t_2 - t_1)}$$

$$\alpha L_1(t_2 - t_1) = L_2 - L_1$$

$$\alpha L_1 t_2 - \alpha L_1 t_1 = L_2 - L_1$$

As you can see, L_1 appears in <u>three</u> terms. Isolate those three terms on the left side, factor by the distributive principle, and complete the solution.

$$t_1 = \frac{L_1 - L_2 + \alpha L_1 t_2}{\alpha L_1}$$

or

$$t_1 = \frac{L_1 - L_2}{\alpha L_1} + t_2$$

$$\alpha L_1 t_2 - \alpha L_1 t_1 + L_1 = L_2$$

$$L_1(\alpha t_2 - \alpha t_1 + 1) = L_2$$

$$L_1 = \frac{L_2}{\alpha t_2 - \alpha t_1 + 1}$$

$$\text{or } L_1 = \frac{L_2}{\alpha(t_2 - t_1) + 1}$$

SELF–TEST 33 (pages 407-417)

1. Solve for H. $P = \dfrac{H - F}{2H}$

2. Solve for R_2. $\dfrac{1}{R_1} = \dfrac{1}{R_t} - \dfrac{1}{R_2}$

3. Solve for P. $T = \dfrac{V}{1 - PV}$

4. Solve for s. $w = s + k(d - b)$

5. Solve for r. $p = rt + a(r - 1)$

$\dfrac{a + p}{t + a} \quad \dfrac{rt + area}{r(t + a)}$

6. Solve for V_2. $F = \dfrac{P_1 + P_2}{C(V_1 + V_2)}$

ANSWERS:

1. $H = \dfrac{F}{1 - 2P}$

2. $R_2 = \dfrac{R_1 R_t}{R_1 - R_t}$

3. $P = \dfrac{T - V}{TV}$

4. $s = w - k(d - b)$

5. $r = \dfrac{a + p}{a + t}$

6. $V_2 = \dfrac{P_1 + P_2 - CFV_1}{CF}$

or $V_2 = \dfrac{P_1 + P_2}{CF} - V_1$

SUPPLEMENTARY PROBLEMS - CHAPTER 9

Assignment 30

1. Solve for s.

$$P = 4s$$

2. Solve for L.

$$A = LW$$

3. Solve for r.

$$C = 2\pi r$$

4. Solve for R.

$$E^2 = PR$$

5. Solve for h.

$$V = \pi r^2 h$$

6. Solve for v.

$$d = v(t_2 - t_1)$$

7. Solve for M.

$$F(p - a) = 2Mm$$

8. Solve for V_2.

$$T_1 V_2 = T_2 V_1$$

9. Solve for W.

$$F = \frac{W}{d}$$

10. Solve for f.

$$t = \frac{1}{f}$$

11. Solve for c.

$$T = \frac{Q}{cm}$$

12. Solve for G.

$$W = \frac{GKN}{R^2}$$

13. Solve for R.

$$I = \frac{E + e}{R}$$

14. Solve for C.

$$s = \frac{1}{2}Cgt^2$$

15. Solve for h.

$$p = \frac{bh(1 - a)}{2d}$$

16. Solve for L.

$$\frac{AL}{H} = K(b_1 + b_2)$$

17. Solve for A_2.

$$\frac{P_1}{P_2} = \frac{A_2}{A_1}$$

18. Solve for R_2.

$$\frac{(E_1)^2}{(E_2)^2} = \frac{R_1}{R_2}$$

19. Solve for d.

$$\frac{bd}{h} = \frac{pt}{w}$$

20. Solve for a.

$$\frac{r}{a} = \frac{m}{r}$$

Assignment 31

1. Solve for H.

$$H^2 = D + T$$

2. Solve for c.

$$E = mc^2$$

3. Solve for v.

$$a = \frac{v^2}{2s}$$

4. Solve for t.

$$2k = \frac{d}{t^2}$$

5. Solve for h.

$$M = \frac{1}{2}bh^2$$

6. Solve for P.

$$\frac{P}{N} = \frac{G}{P}$$

7. Solve for r.

$$\frac{2c}{r} = \frac{ar}{w}$$

8. Solve for V.

$$\frac{F}{W} = \frac{V^2}{gR}$$

9. Solve for F_t.

$$F_1 = F_t - F_2$$

10. Solve for W.

$$P = 2L + 2W$$

11. Solve for Z.

$$X^2 = Z^2 - R^2$$

12. Solve for E.

$$i = \frac{E - e}{r}$$

13. Solve for s_2.

$$s_1 f_1 + s_2 f_2 = 0$$

14. Solve for h.

$$r^2 - h^2 = v^2$$

15. Solve for P_1.

$$P_2 - P_1 = EI$$

16. Solve for k.

$$w = ah - ks$$

17. Solve for d_2.

$$v = \frac{d_1 - d_2}{t}$$

18. Solve for r.

$$hN = R^2 - r^2$$

19. Solve for G_1.

$$W = \frac{G_1 + G_2}{P}$$

20. Solve for c.

$$d = \frac{1 - cm}{2a}$$

33. Four lines are drawn at the right.
Write the equation of each line.

A: _y = 1_

B: _x = -1_

C: _x = 3_

D: _y = -4_

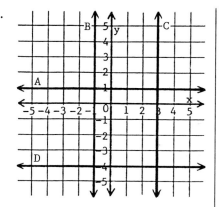

---

34. a) The graph of $x = 0$ is the _____ (x-axis, y-axis).

b) The graph of $y = 0$ is the _____ (x-axis, y-axis).

A: $y = 1$	C: $x = 3$
B: $x = -1$	D: $y = -4$

a) y-axis	b) x-axis

SELF-TEST 34 (pages 420-434)

1. Which of these are linear equations?

a) $y = x^2 - 2$ b) $y = x + 4$ c) $3x - 2y = 6$ d) $y = 5x^2$

2. Which of these ordered pairs are solutions of $2x - 3y = 12$?

a) (0,4)   b) (-3,2)   c) (6,0)   d) (0,0)   e) (-6,-8)

Complete each solution for $y = -3x + 2$.

3. (-2, 8)   4. ($\frac{2}{3}$,0)

In what quadrant does each point lie?

5. (10,-6)   6. (-3,7)

7. The point (0,0) is called the _____ .

For $4x - 3y = 24$, write the coordinates of the:

8. x-intercept. _6, 0_

9. y-intercept. _0, -8_

Graph each equation at the right.

10. $5x - 2y = 10$ 11. $y = -\frac{1}{2}x$

2y = 5x - 10

y = $\frac{5}{2}$·x - 5

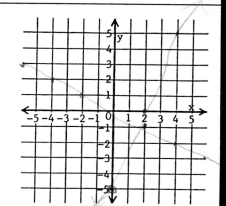

Continued on following page.

SELF-TEST 34 (pages 420-434) - Continued

Write the equation of:

12. line A. _x = 3_

13. line B. _Y = -4_

14. the x-axis. _Y = 0_

15. the y-axis. _x = 0_

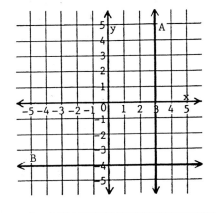

ANSWERS:

1. b,c
2. c,e
3. (-2,8)
4. $\left(\frac{2}{3}, 0\right)$

5. Quadrant 4
6. Quadrant 2
7. origin
8. (6,0)
9. (0,-8)

10-11.

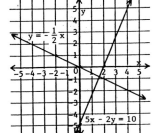

12. x = 3
13. y = -4
14. y = 0
15. x = 0

10-4 SLOPE OF A LINE

The slope of a graphed line is a measure of the steepness of the rise or fall of the line from left to right. We will discuss the slope of a line and the two-point formula for slope in this section.

35. Points P (1,1) and Q (4,5) are plotted on the line at the right. Δx and Δy are the changes in $\underline{x}$ and $\underline{y}$ from P to Q. The symbol Δ (pronounced delta) is used as an abbreviation for the phrase change in. That is:

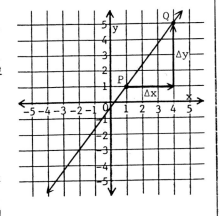

Δx means change in x.

Δy means change in y.

The slope of a line is a ratio of the change in $\underline{y}$ to the change in $\underline{x}$ from one point to another point on the line. That is:

$$\text{Slope} = \frac{\Delta y}{\Delta x} = \frac{\text{increase or decrease in y}}{\text{increase in x}}$$

35. Continued

Let's use the changes from P to Q to compute the slope of the line on the graph.

    Since <u>x</u> increases from 1 to 4, Δx = 3.

    Since <u>y</u> increases from 1 to 5, Δy = 4.

    Therefore, the slope = $\frac{\Delta y}{\Delta x}$ = _____

---

36. Let's use the changes from S (-4,2) to T (3,-2) to compute the slope of the line at the right.

    Since <u>x</u> increases from -4 to 3, Δx = 7.

    Since <u>y</u> decreases from 2 to -2, Δy = -4.

    Therefore, the slope is:

$$\frac{\Delta y}{\Delta x} = \underline{\hspace{1cm}}$$

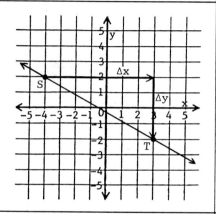

$\frac{4}{3}$

---

37. Slope is a ratio or fraction. When computing a slope, <u>the ratio should always be reduced to lowest terms</u>.

    a) If Δx = 8 and Δy = -6, the slope is _____.

    b) If Δx = 4 and Δy = 12, the slope is _____.

$\frac{-4}{7}$ or $-\frac{4}{7}$

---

38. No matter which pair of points we choose to compute the slope of a line, we always get the same value for the slope. As an example, we graphed the changes from A to B and from C to D on the line at the right.

    a) For A and B, Δx = 2 and Δy = 1. Therefore, the slope = _____.

    b) For C and D, Δx = 4 and Δy = 2. Therefore, the slope = _____.

    c) Did we get the same value for the slope with each pair? _____

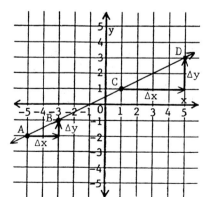

a) $-\frac{3}{4}$

b) 3

---

a) $\frac{1}{2}$

b) $\frac{1}{2}$ $\left(\text{from } \frac{2}{4}\right)$

c) Yes

39. The <u>sign</u> of the slope tells us whether a line rises or falls from left to right.

   If its slope is <u>positive</u>, the line <u>rises</u>.

   If its slope is <u>negative</u>, the line <u>falls</u>.

On the graph at the right, we have drawn four lines and labeled them #1, #2, #3, and #4.

   a) Which lines have a positive slope? _____

   b) Which lines have a negative slope? _____

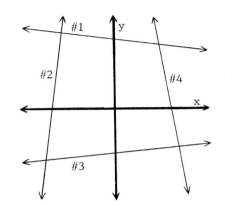

40. The <u>absolute value</u> of the slope tells us <u>how steep the rise or fall of the line is</u>.

We graphed three lines with positive slopes at the right. A is the steepest, then B, then C.

   Line A has a slope of 3.

   Line B has a slope of "1". (It forms a 45° angle with the x-axis.)

   Line C has a slope of $\frac{1}{3}$.

Does the steepest line A have the slope with the largest absolute value? _____

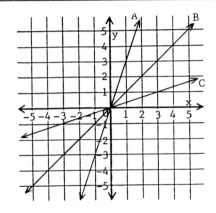

a) lines #2 and #3

b) lines #1 and #4

41. We graphed three lines with negative slopes at the right. A is the steepest, then B, then C.

   Line A has a slope of -4.

   Line B has a slope of -1. (It forms a 45° angle with the x-axis.)

   Line C has a slope of $-\frac{1}{4}$.

Does the steepest line A have the slope with the largest absolute value? _____

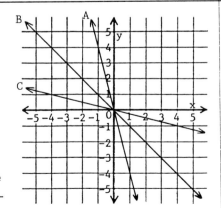

Yes

Yes

42. On the line at the right, $P_1(x_1,y_1)$ and $P_2(x_2,y_2)$ represent any two points. We can use their co-ordinates to find Δy and Δx. That is:

$$\Delta y = y_2 - y_1$$

$$\Delta x = x_2 - x_1$$

Using the letter $\underline{m}$ for slope, we can write the following formula for the slope of a line.

$$m = \frac{\Delta y}{\Delta x} = \frac{y_2 - y_1}{x_2 - x_1}$$

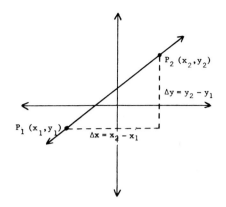

The line at the right passes through the points (2,-1) and (4,3). From the graph, you can see this fact:

$$m = \frac{\Delta y}{\Delta x} = \frac{4}{2} = 2$$

Let's use the two-point formula above to compute the slope. We can use either (2,-1) or (4,3) for (x_2,y_2).

Using (4,3) as (x_2,y_2), we get: $m = \frac{3 - (-1)}{4 - 2} = \frac{4}{2} = 2$

Using (2,-1) as (x_2,y_2), we get: $m = \frac{(-1) - 3}{2 - 4} = \frac{-4}{-2} = 2$

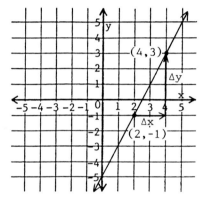

Did we get the correct value for the slope using both methods? _____

---

43. Let's use the same formula to find the slope of the line through (-4,0) and (0,2).

Note: When using the formula, it is helpful to sketch the two points as we have done to avoid gross errors.

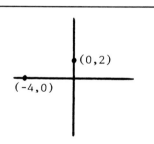

Using either (-4,0) or (0,2) as (x_2,y_2):

$$m = \frac{y_2 - y_1}{x_2 - x_1} = \underline{\hspace{2in}}$$

Yes

$m = \frac{1}{2}$, from:

$$\frac{2 - 0}{0 - (-4)} = \frac{2}{4}$$

or

$$\frac{0 - 2}{-4 - 0} = \frac{-2}{-4}$$

44. Use the formula to find the slope of the line through each pair of points below. (Sketch the points first.)

a) (5,-7) and (7,1)   b) (1,0) and (-5,4)

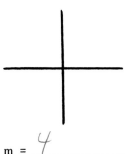 

m = 4   m = -3

---

45. Find the slope of the line through each pair of points. (Sketch the points first.)

a) (-8,0) and (0,20)   b) (17,-11) and (-12,18)

a) m = 4

b) m = $-\frac{2}{3}$

 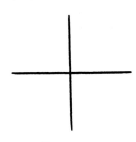 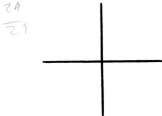

m = 5   m = -1

---

46. We know that the line at the right passes through (0,0) and (2,10). Let's use the two-point formula to find its slope.

$$m = \frac{y_2 - y_1}{x_2 - x_1} = \frac{10 - 0}{2 - 0} = \frac{10}{2} = 5$$

or

$$m = \frac{y_2 - y_1}{x_2 - x_1} = \frac{0 - 10}{0 - 2} = \underline{\hspace{1cm}}$$

a) m = $\frac{5}{2}$

b) m = -1

$$\frac{-10}{-2} = 5$$

47. Use the two-point formula to find the slope of a line passing through the origin and each point below. (Make a sketch.)

    a) (-10,8)   b) (-6,-7)

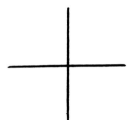   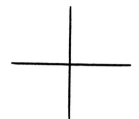

        m = _____   m = _____

---

48. The equation of the horizontal line at the right is y = 2. Since a horizontal line does not rise or fall, it seems that its slope should be 0. Let's use points C (1,2) and D (4,2) to confirm that fact.

$$m = \frac{y_2 - y_1}{x_2 - x_1} = \frac{2 - 2}{4 - 1} = \underline{\hspace{1.5cm}}$$

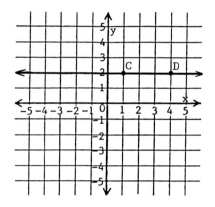

a) $m = -\frac{4}{5}$

b) $m = \frac{7}{6}$

---

49. The equation of the vertical line at the right is x = 3. Let's use P (3,2) and Q (3,4) to examine the slope of the line.

$$m = \frac{y_2 - y_1}{x_2 - x_1} = \frac{4 - 2}{3 - 3} = \frac{2}{0}$$

But division by 0 is undefined. Therefore, <u>the slope of the line is undefined</u>.

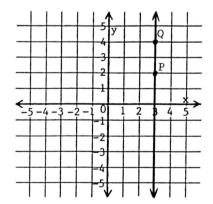

$\frac{0}{3} = 0$

50. In the last two frames, we saw these facts:

    1) The slope of any <u>horizontal</u> line is <u>0</u>.

    2) The slope of any <u>vertical</u> line is <u>undefined</u>.

What is the slope of each line below?

    a) y = 7   m = _____   d) x = 20   m = _____

    b) x = -6   m = _____   e) y-axis   m = _____

    c) y = -30   m = _____   f) x-axis   m = _____

a) 0	c) 0	e) undefined
b) undefined	d) undefined	f) 0

10-5   SLOPE-INTERCEPT FORM

In this section, we will discuss the slope-intercept form of linear equations. We will show how linear equations in other forms can be rearranged to slope-intercept form. We will also show how the y-intercept and the slope can be used to graph a linear equation.

51. The following form of a linear equation is called <u>slope-intercept</u> form.

$$y = mx + b$$ where: <u>m</u> is the <u>slope</u> of the line.

                                        <u>b</u> is the <u>y-intercept</u> of the line.

As an example, we graphed   y = 2x - 3   below. The coordinates of points A and B are given.

1. In the equation, m = 2. To show that 2 is the slope of the line, we can use the changes in <u>y</u> and <u>x</u> from A to B.

$$m = \frac{\Delta y}{\Delta x} = \frac{4}{2} = 2$$

2. In the equation, b = -3. A is the y-intercept. Its coordinates are (0,-3). -3 is the y-coordinate of A. Sometimes we say "-3 is the the y-intercept".

3. Therefore, for   y = 2x - 3   the slope is 2 and the y-intercept is (0,-3).

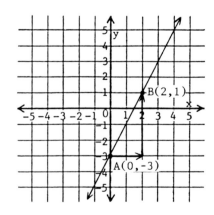

52. Following the examples, write the slope and y-intercept of the linear equations with the following slope-intercept forms.

Slope-intercept form	Slope	y-intercept
y = x - 5	1	(0,-5)
y = -3x + 1	-3	(0,1)
a) $y = \frac{3}{4}x + 8$	_____	_____
b) $y = -\frac{1}{2}x - \frac{5}{4}$	_____	_____

---

53. Following the example, write the slope-intercept form of the linear equations with the following slopes and y-intercepts.

Slope	y-intercept	Slope-intercept form
4	(0,-1)	y = 4x - 1
a) $-\frac{5}{2}$	(0,4)	_____
b) -1	$\left(0,-\frac{1}{2}\right)$	_____

a) $\frac{3}{4}$ (0,8)

b) $-\frac{1}{2}$ $\left(0,-\frac{5}{4}\right)$

---

54. To put the equation below in slope-intercept form, we solved for **y**. Notice how we wrote the x-term first on the right side. Put the other equation in slope-intercept form.

$$y - 2x = 7 \qquad\qquad x + y = 10$$

$$y - 2x + 2x = 2x + 7$$

$$y = 2x + 7$$

a) $y = -\frac{5}{2}x + 4$

b) $y = -x - \frac{1}{2}$

---

55. To get the additive inverse of an addition or subtraction, we can replace each term with its additive inverse.

The additive inverse of -9x + 5 is 9x - 5.

The additive inverse of x - 1 is -x + 1.

We used the inverse principle to put each equation below in slope-intercept form. That is, we replaced each side with its inverse.

$$-y = -2x + 3 \qquad\qquad -y = x - 9$$
$$y = 2x - 3 \qquad\qquad y = -x + 9$$

Notice how we used the principle above in the last step to put each equation below in slope-intercept form.

$$3x - y = 10 \qquad\qquad x = 5 - y$$
$$-3x + 3x - y = -3x + 10 \qquad\qquad x + (-5) = -5 + 5 - y$$
$$-y = -3x + 10 \qquad\qquad x - 5 = -y$$
$$y = 3x - 10 \qquad\qquad y = -x + 5$$

y = -x + 10

Continued on following page.

55. Continued

Put these in slope-intercept form.

a) $x - y = 4$ b) $8 - y = 7x$

---

56. To put the equation below in slope-intercept form, we divided $x + 6$ by 3. Put the other equation in slope-intercept form.

$$3y = x + 6 \qquad\qquad 2y = -6x + 1$$

$$y = \frac{x + 6}{3}$$

$$y = \frac{x}{3} + \frac{6}{3}$$

$$y = \frac{1}{3}x + 2$$

a) $y = x - 4$

b) $y = -7x + 8$

---

57. Following the example, put the other equation in slope-intercept form.

$$4x + 3y = 12 \qquad\qquad x + 5y = 3$$

$$3y = -4x + 12$$

$$y = \frac{-4x + 12}{3}$$

$$y = -\frac{4}{3}x + 4$$

$y = -3x + \frac{1}{2}$

---

58. To get the additive inverse of $7 - x$, we can replace each term with its inverse and get $-7 + x$ or $x - 7$. Notice how we used the inverse principle below to get $4y = x - 7$. Put the other equation in slope-intercept form.

$$x - 4y = 7 \qquad\qquad 5x - 3y = 15$$

$$-4y = 7 - x$$

$$4y = x - 7$$

$$y = \frac{x - 7}{4}$$

$$y = \frac{1}{4}x - \frac{7}{4}$$

$y = -\frac{1}{5}x + \frac{3}{5}$

$y = \frac{5}{3}x - 5$

59. After putting an equation in slope-intercept form, we can easily identify its slope and y-intercept. For example:

Since $2x + y = 5$ is equivalent to $y = -2x + 5$:

Its slope is -2. Its y-intercept is $(0,5)$.

Since $5x - 2y = 6$ is equivalent to $y = \frac{5}{2}x - 3$:

a) Its slope is _____. b) Its y-intercept is _____.

---

60. The general slope-intercept form of linear equations is $y = mx + b$. However, the y-intercept of all lines through the origin is $(0,0)$. Since $b = 0$ for all lines through the origin, their slope-intercept form is:

$$y = mx$$

If we know the slope of a line through the origin, we can write its equation. That is:

If $m = \frac{3}{4}$, $y = \frac{3}{4}x$ If $m = -\frac{5}{2}$, $y = -\frac{5}{2}x$

We can find the slope of $y - 5x = 0$ by putting it in slope-intercept form. We get: $y = 5x$. Therefore, $m = 5$. Find the slope of each line below by putting it in slope-intercept form.

a) $4x + 3y = 0$ b) $x - y = 0$

m = _____   m = _____

a) $\frac{5}{2}$

b) $(0,-3)$

a) $m = -\frac{4}{3}$ b) $m = 1$

10-6 GRAPHING NON-LINEAR EQUATIONS

A non-linear equation is an equation whose graph is a curved line rather than a straight line. We will graph some non-linear equations in this section.

---

61. The graph of a non-linear equation is a curved line. The sketches of the graphs of two non-linear equations are shown below. Notice the different shapes of the two curves.

$y = x^2$

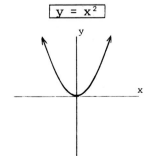

$xy = 12$

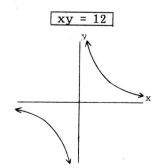

62. We graphed both $y = x^2$ and $y = -x^2$ below. To do so, we made up a solution-table, plotted the points, and drew a smooth curve through the plotted points.

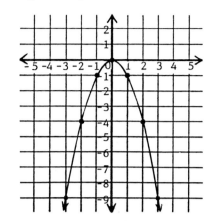

$y = x^2$

x	y
3	9
2	4
1	1
0	0
-1	1
-2	4
-3	9

$y = -x^2$

x	y
3	-9
2	-4
1	-1
0	0
-1	-1
-2	-4
-3	-9

The graph of each equation is called a <u>parabola</u>. For $y = x^2$, the parabola opens upward. For $y = -x^2$, the parabola opens downward.

63. When graphing a parabola, plot enough points so that the outline of the parabola is clear. Then draw a smooth curve through the plotted points. Use the solution-table below to graph $y = 2x^2$.

$y = 2x^2$

x	y
5	50
3	18
2	8
1	2
0	0
-1	2
-2	8
-3	18
-5	50

64. Use the solution-table below to graph $y = x^2 - 2x$.

$y = x^2 - 2x$

x	y
4	8
3	3
2	0
1	-1
0	0
-1	3
-2	8

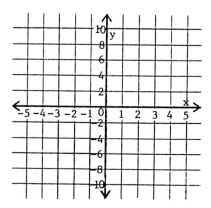

<u>Answer to Frame 63</u>:

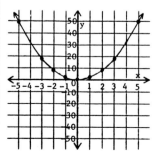

65. Use the solution-table below to graph $y = -2x^2 + 4x + 1$.

$y = -2x^2 + 4x + 1$

x	y
-1	-5
0	1
1	3
2	1
3	-5
$\frac{1}{2}$	$2\frac{1}{2}$
$1\frac{1}{2}$	$2\frac{1}{2}$

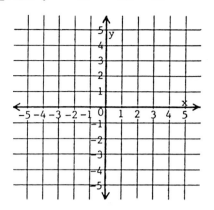

Answer to Frame 64:

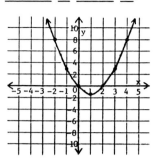

66. Graph $y = x^2 + 1$ below. Make up your own solution-table.

$y = x^2 + 1$

x	y
3	10
2	5
1	2
0	1
-1	2
-2	5

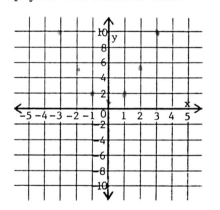

Answer to Frame 65:

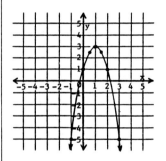

67. Graph $y = 4x - x^2$ below. Make up your own solution-table.

$y = 4x - x^2$

x	y

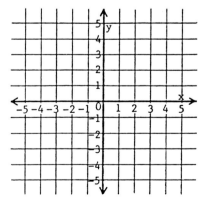

Answer to Frame 66:

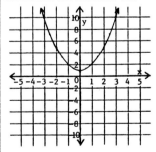

Answer to Frame 67:

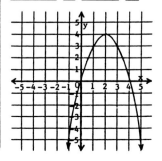

68. We used the solution-table below to graph $xy = 24$. The graph is called a <u>hyperbola</u>. Notice these points about it:

   1. When <u>x</u> = 0, $y = \frac{24}{0}$ which is an impossible or undefined operation. Therefore, there is no solution for $x = 0$ and no corresponding point on the graph.

   2. The graph has two parts, one in Quadrant 1 and one in Quadrant 3. The two parts are not connected.

   3. Though the curves $xy = 24$ or $y = \frac{24}{x}$ approach the axes, they do not touch the axes.

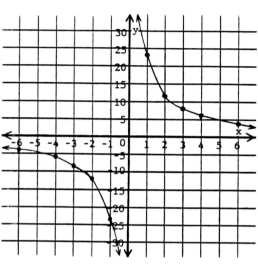

x	y		x	y
0	--		0	--
1	24		-1	-24
2	12		-2	-12
3	8		-3	-8
4	6		-4	-6
6	4		-6	-4

Complete the solution-table for $xy = 48$ and then graph the equation. Its graph is also a <u>hyperbola</u>.

$xy = 48$ or $y = \frac{48}{x}$

x	y		x	y
0			0	
2			-2	
3			-3	
4			-4	
6			-6	
8			-8	
12			-12	

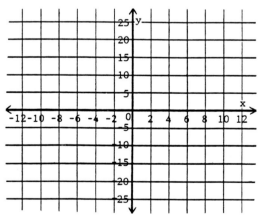

Answer to Frame 68:

x	y		x	y
0	--		0	--
2	24		-2	-24
3	16		-3	-16
4	12		-4	-12
6	8		-6	-8
8	6		-8	-6
12	4		-12	-4

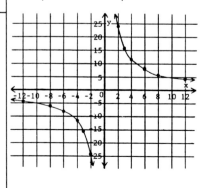

SELF–TEST 35 (pages 434-447)

1. Using Δx and Δy, write the formula for slope. Slope = _____

Find the slope of each line.

2. Line A

   m = _____

3. Line B

   m = _____

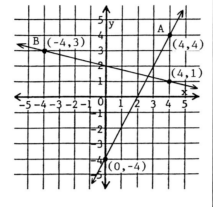

Find the slope of the line through:

4. (-3,1) and (2,-4)

   m = _____

5. (0,0 and (-4,-6)

   m = _____

Given: x + y = 5

6. Put the equation in slope-intercept form.

7. The slope is _____.

8. The y-intercept is _____.

Given: 2x - 5y = 10

9. Put the equation in slope-intercept form.

10. The slope is _____.

11. The y-intercept is _____.

---

12. Find the slope and y-intercept of 2x - y = 0.

---

13. Graph this non-linear equation.

   $y = x^2 - 3$

x	y

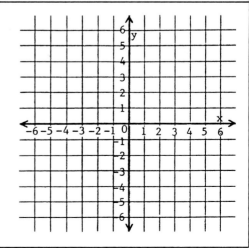

Answers to Self-Test 35:

1. $\frac{\Delta y}{\Delta x}$

2. $m = 2$

3. $m = -\frac{1}{4}$

4. $m = -1$

5. $m = \frac{3}{2}$

6. $y = -x + 5$

7. $m = -1$

8. $(0,5)$

9. $y = \frac{2}{5}x - 2$

10. $m = \frac{2}{5}$

11. $(0,-2)$

12. $m = 2$
    $(0,0)$

13.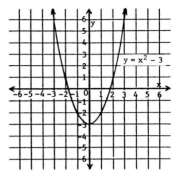

10-7   GRAPHS OF FORMULAS

In this section, we will discuss the graphs of formulas and show how they can be used to solve applied problems.

69.   The formula below shows the relationship between degrees-Celsius (C) and degrees-Fahrenheit (F). We used three points to graph the formula. Notice that we plotted C on the horizontal axis and F on the vertical axis. This choice is arbitrary.

$$F = \frac{9}{5}C + 32$$

C	F
30°	86°
0°	32°
-40°	-40°

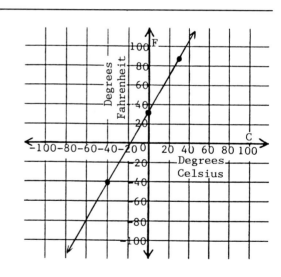

The formula above is in slope-intercept form. Therefore:

    a) The slope of the line is _____.

    b) The coordinates of the F-intercept are _____.

a) $\frac{9}{5}$

b) $(0,32)$

70. Though temperature has negative values, most quantities (like distances, time, weight, pressure, and so on) do not have negative values. Therefore, <u>only quadrant 1 is used when graphing most formulas</u>. An example is discussed below.

The formula below shows the relationship between distance traveled (d) and time (t) for an object traveling at a constant velocity (or speed) of 50 miles per hour. We used three points to graph the formula.

d = 50t

t	d
0	0
2	100
5	250

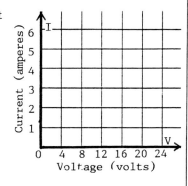

The formula is in slope-intercept form for a line through the origin. Therefore:

a) The slope of the line is _____.

b) The coordinates of the d-intercept are _____.

71. The formula below is related to the concept of <u>load line</u> in transistor electronics. The two variables are voltage V and current I.

V + 8I = 24

Use the intercept method to graph the formula.

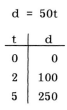

a) 50

b) (0,0)

72. The graph below shows the relationship between distance traveled and time at an average speed of 100 kilometers per hour.

We can use the graph to solve this problem.

How far will a train travel in 3 hours at an average speed of 100 kilometers per hour?

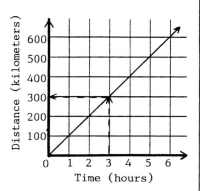

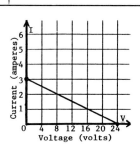

Continued on following page.

72.   Continued

To solve the problem, we use these steps:

1) Draw an arrow from 3 on the <u>time</u> axis to the graphed line.

2) Then draw an arrow from that point to the <u>distance</u> axis.

3) Since the horizontal arrow points to 300, we know that the train will travel 300 kilometers in 3 hours.

Use the graph for these.

a) How far would the train travel in 2 hours? _____

b) How far would the train travel in 5 hours? _____

---

73.   The graph below shows the relationship between distance traveled and time at an average speed of 50 miles per hour.

We can use the graph to solve this problem.

How long would it take a car to drive 150 miles at an average speed of 50 miles per hour?

To solve the problem, we use these steps:

1) Draw an arrow from 150 on the <u>distance</u> axis to the graphed line.

2) Then draw an arrow from that point to the <u>time axis</u>.

3) Since the vertical arrow points to 3, we know that it would take the car 3 hours to drive 150 miles at that speed.

Use the graph to answer these.

a) How long would it take the car to drive 100 miles? _____

b) How long would it take the car to drive 250 miles? _____

a) 200 kilometers

b) 500 kilometers

---

74.   The graph below shows how far a spring will stretch when various amounts of force are applied to it. Use it to complete these.

a) If a force of 150 grams is applied, the spring will stretch approximately _____ centimeters.

b) If a force of 75 grams is applied, the spring will stretch approximately _____ centimeters.

c) To stretch the spring 4 centimeters, we need a force of _____ grams.

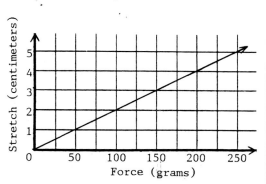

a) 2 hours

b) 5 hours

75. The graph below shows how much current we get in an electric circuit with constant resistance when various voltages are applied. Use it to complete these.

   a) If a voltage of 40 volts is applied, we get a current of _____ amperes.

   b) To get a current of 6 amperes, we must apply a voltage of _____ volts.

   c) To get a current of 13 amperes, we must apply a voltage of _____ volts.

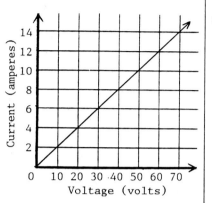

a) 3 centimeters

b) 1.5 centimeters

c) 200 grams

76. The graph below shows the distance (in meters) a dropped object will fall from rest in a given period of time (in seconds). The graph is half a parabola.

   By drawing arrows, we used the graph to complete these.

   In 4 seconds, an object would fall 80 meters.

   To fall 20 meters, it would take 2 seconds.

   Use the graph to get approximate answers for these.

   a) How far would an object fall in 3 seconds? _____ meters

   b) How long would it take an object to fall 130 meters?

   _____ seconds

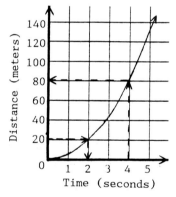

a) 8 amperes

b) 30 volts

c) 65 volts

77. The graph below shows the relationship between the pressure and volume of a gas at a constant temperature. The graph is half a hyperbola.

   Use the graph to get approximate answers for these.

   a) When the volume is 100 milliliters, the pressure is _____ millimeters of mercury.

   b) When the volume is 50 milliliters, the pressure is _____ millimeters of mercury.

   c) When the pressure is 20 millimeters of mercury the volume is _____ milliliters.

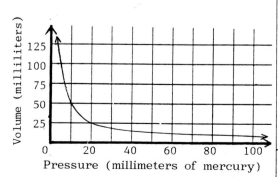

a) 45 meters

b) 5 seconds

a) 5
b) 10
c) 25

10-8   FUNCTIONS

In this section, we will define functions and discuss functional notation. The definition of functions is extended to formulas.

---

78.   A <u>relation</u> is a set of ordered pairs. All of the following two-variable equations are relations including an infinite number of ordered pairs.

$$y = 5x$$
$$y = x^2 + 2x + 3$$
$$y = \pm\sqrt{x}$$
$$y = \sqrt{x - 4}$$
$$y = \frac{7}{x}$$

In each equation above, <u>y</u> is solved for. <u>y</u> is called the <u>dependent variable</u> because its value "depends" on the value substituted for <u>x</u>. <u>x</u> is called the <u>independent variable</u>. That is:

In $y = 3x - 1$: a) <u>y</u> is called the _____ variable.

b) <u>x</u> is called the _____ variable.

---

79.   A <u>function</u> is a relation in which, for each value of the independent variable, there is only one value of the dependent variable.

$y = x^2 + 2x + 3$ <u>is</u> <u>a</u> <u>function</u> because for each value of <u>x</u>, there is only one value of <u>y</u>. That is:

If $x = 3$, $y = 18$.

If $x = 0$, $y = 3$.

If $x = -1$, $y = 2$.

$y = \pm\sqrt{x}$ <u>is</u> <u>not</u> <u>a</u> <u>function</u> because for each positive value of <u>x</u>, there are two values of <u>y</u>. That is:

If $x = 4$, $y = 2$ and -2.

If $x = 9$, $y =$ _____ and _____.

a) dependent

b) independent

---

80.   Since a relation is a function <u>only</u> if there is just one value of <u>y</u> for each value of <u>x</u>, we can use a <u>vertical line</u> test to determine if a graph represents a function. To show that fact, the two graphs below are discussed on the following page.

3 and -3

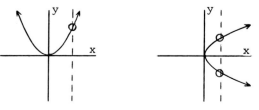

Continued on following page.

80. Continued

By mentally sliding the dotted vertical line along the horizontal axis, we can see these facts.

At the left, since the dotted line never intersects the graph more than once, there is only one y-value for each x-value. Therefore, the graph is a function.

At the right, since the dotted line usually intersects the graph twice, there is usually more than one y-value for each x-value. Therefore, the graph is not a function.

Which of the following represent functions? _____

a) 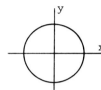 b) 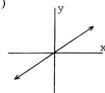 c)

---

81. Which of the following represent functions? _____ | Only (b)

a) 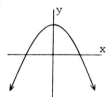 b) 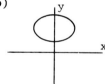 c)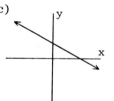

---

82. The functions below are not defined for one value of x because it leads to a division by 0 which is impossible. | Both (a) and (c)

$y = \dfrac{6}{x}$ is not defined for x = 0, since $y = \dfrac{6}{0}$.

$y = \dfrac{5}{x - 3}$ is not defined for x = 3, since $y = \dfrac{5}{3 - 3} = \dfrac{5}{0}$.

Each function below is not defined for one value of x. Name that x-value for each.

a) $y = \dfrac{9}{2x}$ b) $y = \dfrac{5}{x^2}$ c) $y = \dfrac{11}{x - 7}$

x = _____   x = _____   x = _____

---

a) x = 0

b) x = 0

c) x = 7

83. The functions below are not defined for various values of $\underline{x}$ because those values lead to square roots of negative numbers, and negative numbers do not have real number square roots.

$\quad\quad y = \sqrt{x}\quad\quad$ is not defined for negative values of $\underline{x}$ ($x < 0$).

$\quad\quad y = \sqrt{x - 5}$ is not defined for values of $\underline{x}$ less than 5 ($x < 5$).

Each function below is not defined for various values of $\underline{x}$. Name those x-values for each.

$\quad$ a) $y = \sqrt{2x}$ _____

$\quad$ b) $y = \sqrt{x - 1}$ _____

$\quad$ c) $y = \sqrt{x - 8}$ _____

---

84. For a function, the set of first coordinates (or $\underline{x}$) is called the <u>domain</u>; the set of second coordinates (or $\underline{y}$) is called the <u>range</u>.

For some functions, the domain includes all real numbers. Two functions of that type are:

$\quad\quad y = 10x \quad\quad\quad\quad\quad\quad y = x^2 - 7x + 2$

For other functions, the domain is limited because the function is not defined for certain values of x. Two functions of that type are:

$\quad\quad y = \dfrac{7}{x - 6} \quad\quad$ (The domain is all real numbers except $x = 6$.)

$\quad\quad y = \sqrt{x - 9} \quad\quad$ (The domain is all real numbers except values of $\underline{x}$ less than 9.)

Identify the domain for each function below.

$\quad$ a) $y = \dfrac{3}{5x}$ $\quad\quad\quad$ b) $y = 2x^2 - 1$ $\quad\quad\quad$ c) $y = \sqrt{x - 10}$

a) negative values of $\underline{x}$ ($x < 0$)

b) values of $\underline{x}$ less than "1" ($x < 1$)

c) values of $\underline{x}$ less than 8 ($x < 8$)

---

85. In a function, we say that $\underline{y}$ is a function of $\underline{x}$. For example, $\underline{y}$ is a function of $\underline{x}$ in each equation below.

$\quad\quad y = 3x \quad\quad\quad y = x^2 - 1 \quad\quad\quad y = 2x^2 + x - 3$

For the phrase "function of $\underline{x}$", we use the symbol f(x), which is read "$\underline{f}$ of $\underline{x}$". The symbol f(x) <u>does</u> <u>not</u> <u>mean</u> "multiply $\underline{f}$ and $\underline{x}$". Rather it is a general symbol for any function of $\underline{x}$. Therefore, we can use it for each function above. We write:

$\quad\quad f(x) = 3x$

$\quad\quad f(x) = x^2 - 1$

$\quad\quad f(x) = 2x^2 + x - 3$

As you can see, f(x) is another way of stating $\underline{y}$. Therefore:

Instead of $y = 10x$, we can write $f(x) = 10x$.

Instead of $y = x^2 - 5x - 9$, we can write _____.

a) All real numbers except $x = 0$.

b) All real numbers.

c) All real numbers except values of $\underline{x}$ less than 10.

$f(x) = x^2 - 5x - 9$

86. We can use functional notation to evaluate a function for a particular value of <u>x</u>. For example, we used it below to evaluate $f(x) = 2x + 1$ for $x = 4$ and $x = -3$. To do so, we substituted 4 and -3 for <u>x</u>.

$$f(x) = 2x + 1 \qquad\qquad f(x) = 2x + 1$$
$$f(4) = 2(4) + 1 \qquad\qquad f(-3) = 2(-3) + 1$$
$$f(4) = 9 \qquad\qquad\qquad f(-3) = -5$$

The expressions $f(4)$ and $f(-3)$ are read "<u>f</u> of 4" and "<u>f</u> of -3".

$f(4)$ means: Find the value of the function when $x = 4$.

$f(-3)$ means: Find the value of the function when $x = -3$.

Given the same function $f(x) = 2x + 1$, find the following.

   a) $f(2)$ b) $f(0)$ c) $f(-2)$

---

87. Given the function $f(x) = 5x^2$, find these.

   a) $f(3)$ b) $f(0)$ c) $f(-4)$

a) $f(2) = 5$
b) $f(0) = 1$
c) $f(-2) = -3$

---

88. a) Find $f(-5)$ for this function.   b) Find $f(29)$ for this function.

$$f(x) = \frac{10}{x} \qquad\qquad\qquad f(x) = \sqrt{x - 4}$$

a) $f(3) = 45$
b) $f(0) = 0$
c) $f(-4) = 80$

---

89. Other letters besides <u>f</u> can be used to represent functions. For example, we can use $\bar{g}(x)$, $h(x)$, $A(x)$, $P(x)$, and so on.

   a) If $g(x) = x^2 - x$,   b) If $A(x) = x^2 + 5x - 10$,
      find $g(-4)$.   find $A(0)$.

a) $f(-5) = -2$
b) $f(29) = 5$

---

90. Notice the steps used for the evaluation below.

   If $f(x) = 2x - 1$, find $f(4) - f(-2)$.

$$f(x) = 2x - 1$$
$$f(4) = 2(4) - 1 = 8 - 1 = 7$$
$$f(-2) = 2(-2) - 1 = -4 - 1 = -5$$

Therefore, $f(4) - f(-2) = 7 - (-5) = 7 + 5 = 12$

Continued on following page.

a) $g(-4) = 20$
b) $A(0) = -10$

90.   Continued

For the same function f(x) = 2x - 1, do these.

a) Find f(3) + f(-5)   b) Find $\dfrac{f(10) - f(6)}{4}$

---

91.   We can also evaluate functions for algebraic expressions other than
numbers. For example, we evaluated f(x) = 3x + 5 below for
x = 2a to get f(2a) and x = a + 1 to get f(a + 1). To do so,
we substituted 2a and a + 1 for x.

f(x) = 3x + 5	f(x) = 3x + 5
f(2a) = 3(2a) + 5	f(a + 1) = 3(a + 1) + 5
f(2a) = 6a + 5	f(a + 1) = 3a + 3 + 5
	f(a + 1) = 3a + 8

Given the same function f(x) = 3x + 5, find these.

a) f(5a)   b) f(a - 4)   c) f(2a + 7)

a) -6   b) 2

---

92.   We evaluated f(x) = 2x - 7 below for x = 5b to get f(5b) and
x = 3b - 1 to get f(3b - 1).

f(x) = 2x - 7	f(x) = 2x - 7
f(5b) = 2(5b) - 7	f(3b - 1) = 2(3b - 1) - 7
f(5b) = 10b - 7	f(3b - 1) = 6b - 2 - 7
	f(3b - 1) = 6b - 9

Given the same function f(x) = 2x - 7, find these.

a) f(8b)   b) f(a + 4)   c) f(4b - 3)

a) f(5a) = 15a + 5

b) f(a - 4) = 3a - 7

c) f(2a+7) = 6a + 26

93. For $y = 2x + 1$, $f(x)$ is equivalent to $\underline{y}$. Therefore, when graphing the function, we sometimes use $f(x)$ instead of $\underline{y}$ on the vertical axis as we have done below.

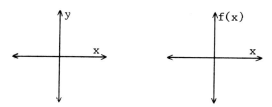

Using $f(x)$ on the vertical axis, we graphed the linear function $y = 2x + 1$ below. Graph the linear function $y = -x - 2$ on the same graph.

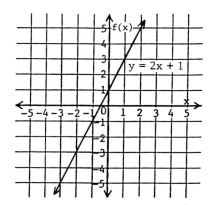

a) $f(8b) = 16b - 7$

b) $f(a + 4) = 2a + 1$

c) $f(4b-3) = 8b - 13$

94. Formulas can also be functions. Three examples are given below.

$d = 50t$ Distance traveled $\underline{d}$ is a function of time traveled $\underline{t}$.

$C = \pi d$ The circumference C of a circle is a function of the diameter $\underline{d}$ of the circle.

$A = s^2$ The area A of a square is a function of the side $\underline{s}$ of the square.

In a formula, letters other than $\underline{x}$ and $\underline{y}$ are the dependent and independent variables. That is:

In $d = 50t$: $\underline{d}$ is the dependent variable.
               $\underline{t}$ is the independent variable.

In $A = s^2$:   a) the dependent variable is _____.

               b) the independent variable is _____.

<u>Answer to Frame 93:</u>

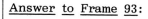

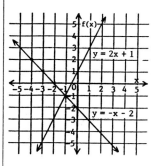

a) A

b) s

95. When a formula is a function, it can be stated in functional notation. That is:

For d = 50t,   d = f(t).

For C = πd,   C = f(d).

For A = s²,   A = f(s).

Usually we use the letter of the dependent variable instead of <u>f</u> for the functions above. That is:

Instead of f(t) = 50t, we write d(t) = 50t.

Instead of f(d) = πd, we write C(d) = πd.

Instead of f(s) = s², we write _____.

---

96. We found d(2) and d(10) for d(t) = 50t below.

d(t) = 50t	d(t) = 50t
d(2) = 50(2)	d(10) = 50(10)
d(2) = 100	d(10) = 500

a) Find C(4) for C(d) = πd.   b) Find A(5) for A(s) = s².

> A(s) = s²

---

97. When a function is a formula, the domain is not usually defined for negative values because they usually do not make sense. For example:

d = 50t   is not defined for negative values of <u>t</u> because negative periods of time do not make sense.

Therefore, when graphing a formula, usually only quadrant 1 is used. As an example, we graphed the linear formula d = 50t below.

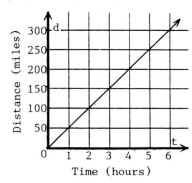

Given that d(t) = 50t, use the graph to complete these.

a) d(1) = _____   c) d(4) = _____

b) d(2) = _____   d) d(6) = _____

> a) C(4) = 4π
>
> b) A(5) = 25

---

a) 50   b) 100   c) 200   d) 300

SELF-TEST 36 (pages 448-459)

1. For the electrical power formula EI = 12 , complete
   the solution-table and construct the non-linear graph.

EI = 12

E	I
0	~
1	12
2	6
4	3

E	I
6	2
8	
16	
24	

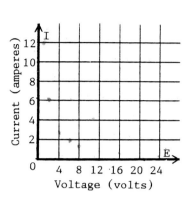

For a ball projected vertically upward, the graph shows the relation between the ball's height and the elapsed time.

2. After 2 seconds, the height is _____ meters.

3. The greatest height is _____ meters.

4. The height is 50 meters for _____ second and _____ seconds.

5. The ball returns to the ground after _____ seconds.

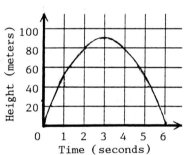

6. In y = 7x - 5 , the independent variable is _____.

7. In $s = 16t^2$, the dependent variable is _____.

8. The function $y = \dfrac{5}{x - 2}$ is not defined for one value of x. It is x = _____.

9. For the function $y = x^2 - 4x$, the set of x-values is called the _____ (domain/range).

10. If f(x) = 2x - 5,
    f(2) = _____.

11. If $g(x) = 3x^2$,
    g(-1) = _____.

12. If f(x) = 1 - x,
    f(-2) - f(0) = _____.

13. If A(x) = 5x
    A(2b) = _____.

14. If P(x) = 4x + 3
    P(a + 1) = _____.

15. If $g(s) = 4s - s^2$,
    g(4) = _____.

ANSWERS: 1.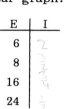

2. 80 meters

3. 90 meters

4. 1 sec and 5 sec

5. 6 sec

6. x

7. s

8. x = 2

9. domain

10. -1

11. 3

12. 2

13. 10b

14. 4a + 7

15. 0

SUPPLEMENTARY PROBLEMS - CHAPTER 10

<u>Assignment 34</u>

1. Which of these are linear equations?

    a) $y = 2x$ b) $y = x - x^2$ c) $5x - y = 8$ d) $y = -3$ e) $x^2 + y^2 = 9$

2. Which of these ordered pairs are solutions of $x - 2y = 6$?

    a) $(4,-1)$ b) $(10,-2)$ c) $(0,3)$ d) $(-2,-4)$ e) $(6,0)$ f) $(-8,-1)$

Complete each solution for $y = 4 - 2x$.   Complete each solution for $3x + 2y = 8$.

    3. (,6)   4. (3,)   5. (,0)   6. (-2,)   7. (,1)   8. (0,)

State the number of the quadrant in which each point lies.

    9. $(-7,2)$ 10. $(12,8)$ 11. $(-6,-1)$ 12. $(9,-5)$ 13. $(-4,20)$

For $2x - 5y = 20$, find the coordinates of the: For $y = 4x + 8$, find the coordinates of the:

    14. x-intercept   15. y-intercept   16. x-intercept   17. y-intercept

Graph each equation. Use intercepts.

    18. $y = x + 2$ 19. $x + 3y = 3$ 20. $3x - 2y = 6$

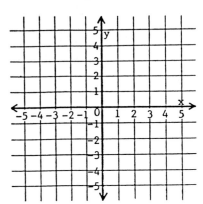 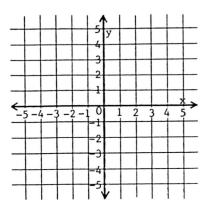 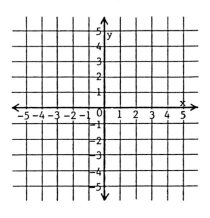

Write the equation of each line.

    21. Line A _____

    22. Line B _____

    23. Line C _____

    24. Line D _____

    25. The x-axis _____

    26. The y-axis _____

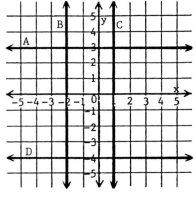

Assignment 35

Find the slope of each line graphed below.

1. Line A

2. Line B

3. Line C

4. Line D

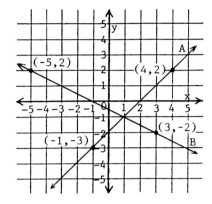

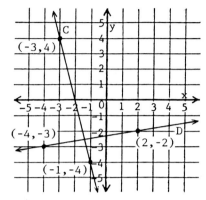

Five lines have these slopes: Line A: 4   Line B: -1   Line C: -5   Line D: $\frac{1}{2}$ Line E: -3

5. Which lines rise from left to right?

6. Which line has the steepest rise?

7. Which lines fall from left to right?

8. Which line has the steepest fall?

Find the slope of the line through each pair of points.

9. (2,3) and (-1,-3)

10. (6,8) and (10,2)

11. (-4,4) and (0,8)

12. (-4,2) and (2,-4)

13. (-3,-4) and (5,-2)

14. (0,6) and (10,0)

15. (-20,10) and (30,35)

16. (0,0) and (-6,-12)

17. (0,0) and (2,-8)

What is the slope of: 18. Any horizontal line?   19. Any vertical line?

For each equation, identify the slope and the coordinates of the y-intercept.

20. $y = 5x + 2$

21. $y = x - 3$

22. $y = -\frac{2}{5}x + \frac{3}{2}$

23. $y = \frac{1}{2}x$

Put each equation in slope-intercept form.

24. $y = 6 - 4x$

25. $y - 3x = 2$

26. $x - y = 5$

27. $4y = x - 8$

28. $3x + 6y = 10$

29. $x - 5y = 0$

Graph each non-linear equation.

30. $y = x^2 - 4$

31. $y = -2x^2 + 3$

32. $y = x^2 - 2x - 3$

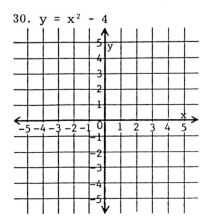

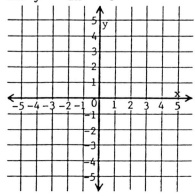

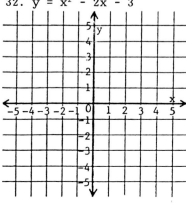

Assignment 36

Complete the solution-table and graph each formula.

1. $w = 10p^2$

p	w
0	
1	
2	
3	
4	
5	

2. $E + 2I = 20$

E	I
0	
4	
8	
12	
16	
20	

The graph at the right shows how much current flows in an electric circuit of constant resistance when various voltages are applied.

How much current flows for these applied voltages?

3. 20 volts _____   4. 55 volts _____

How much voltage must be applied to get these currents?

5. 12 amperes _____   6. 7 amperes _____

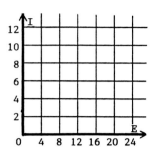

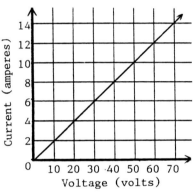

The graph below shows the relation between the length and width of a rectangle of constant area.

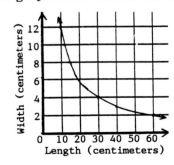

7. If the length is 40 cm, the width is approximately _____ cm.

8. If the length is 15 cm, the width is approximately _____ cm.

9. If the width is 10 cm, the length is approximately _____ cm.

10. If the width is 5 cm, the length is approximately _____ cm.

11. The constant area is approximately _____ cm^2.

12. In $y = x^2 + 2x - 3$, the dependent variable is _____.

13. In $F = \frac{9}{5}C + 32$, the independent variable is _____.

14. The function $y = \frac{24}{x}$ is not defined for one value of x. It is x = _____.

15. For the function $y = 2x - x^2$, the set of y-values is called the _____ (domain/range).

16. For which of the following functions does the domain consist of <u>all</u> real numbers?

    a) $y = 3x + 8$ b) $y = \sqrt{3x}$ c) $y = x^2$ d) $y = 5 - 2x^2$ e) $y = \frac{15}{x - 6}$

Evaluate the following functions.

17. If $f(x) = 2x - 7$, $f(0) = $ _____ .

18. If $f(x) = 2x^2$, $f(-1) = $ _____ .

19. If $f(x) = 1 - 2x$, $f(1) - f(-2) = $ _____ .

20. If $g(x) = 3x + 1$, $g(2a) = $ _____ .

21. If $P(x) = \dfrac{12}{x - 2}$, $P(-6) = $ _____ .

22. If $h(x) = \dfrac{10}{x}$, $h(t + 2) = $ _____ .

23. If $f(x) = 4x$, $f(t) - f(t + 1) = $ _____ .

24. If $p(x) = 2 - 3x$, $p(r) - p(r - 1) = $ _____ .

25. If $A(d) = d - d^2$, $A(5) = $ _____ .

26. If $s(t) = \sqrt{12t}$, $s(3) = $ _____ .

Identify the domain of each function.   27. $y = \dfrac{4}{x - 9}$ 28. $y = \sqrt{2x}$

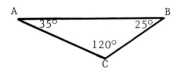

11 Right Triangles

In this chapter, we will show how the angle-sum principle, the Pythagorean Theorem, and the three basic trigonometric ratios (sine, cosine, and tangent) can be used to find unknown angles and sides in right triangles.

11-1 THE ANGLE-SUM PRINCIPLE FOR TRIANGLES

In any triangle, the sum of the three angles is 180°. We will discuss that principle in this section.

1.   There are three angles in any triangle. The angles are usually labeled with capital letters.

   Any angle between 0° and 90° is called an acute angle.
   Any angle with exactly 90° is called a right angle.
   Any angle between 90° and 180° is called an obtuse angle.

In triangle ABC:

   Angle A is an acute angle.

   a) Angle B is an _____ angle.

   b) Angle C is an _____ angle.

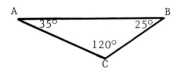

2.   In triangle MPR:

   a) There are two acute angles, angle _____ and angle _____.

   b) The obtuse angle is angle _____.

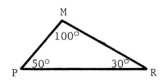

a) acute

b) obtuse

3. The <u>angle-sum</u> <u>principle</u> for triangles is this:

| THE SUM OF THE THREE ANGLES OF ANY TRIANGLE IS 180° |

By subtracting the total number of degrees of the two known angles from 180°, find the number of degrees of the unknown angle in each triangle below.

a)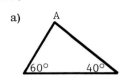

A = _____

b)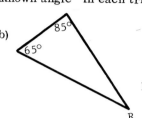

R = _____

a) Angles P and R

b) Angle M

---

4. The angle-sum principle also applies to right triangles (those that contain a right angle). Therefore, in right triangle CDE:

a) The sum of the angles is _____.

b) Angle D = _____.

c) Angle E = _____.

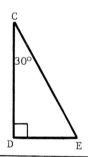

a) A = 80°

b) R = 30°

---

5. In a right triangle, the right angle contains 90°. Since the <u>sum</u> of the other two angles must be 90°, <u>the</u> <u>other</u> <u>two</u> <u>angles</u> <u>must</u> <u>be</u> <u>acute</u> <u>angles</u>.

In right triangle TBC:

a) Angles T and B are both _____ angles.

b) The sum of angles T and B must be _____°.

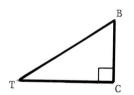

a) 180°

b) 90°

c) 60°

---

6. Since <u>the</u> <u>sum</u> <u>of</u> <u>the</u> <u>two</u> <u>acute</u> <u>angles</u> <u>in</u> <u>a</u> <u>right</u> <u>triangle</u> <u>is</u> <u>90°</u>, it is easy to find the size of one acute angle if we know the size of the other. To do so, we simply subtract the known angle from 90°. Find the unknown acute angle in each right triangle.

a)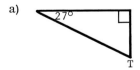

T = _____

b)

R = _____

a) acute

b) 90°

---

a) T = 63°

b) R = 45°

11-2 LABELING ANGLES AND SIDES IN TRIANGLES

In this section, we will discuss the conventional way to label the angles and sides in any triangle.

---

7. We labeled the angles and sides of the triangle at the right in the conventional way. Notice these points:

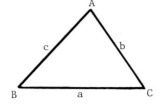

1) Each <u>angle</u> is labeled with a <u>capital</u> <u>letter</u>.

2) Each <u>side</u> is labeled with the <u>small</u> <u>letter</u> corresponding to the capital letter of the angle opposite it. That is:

The side opposite angle A is labeled "a".
The side opposite angle B is labeled "b".
The side opposite angle C is labeled "c".

Sometimes we use the two capital letters at each end of the side to represent the side. For example:

Instead of "a", we use BC (or CB).

a) Instead of "b", we use _____.

b) Instead of "c", we use _____.

---

8. When the sides of a triangle are not labeled with small letters, we must use two capital letters to represent the sides. For example, in the triangle below.

The side opposite angle M is TV (or VT).

a) The side opposite angle T is _____.

b) The side opposite angle V is _____.

a) AC (or CA)

b) AB (or BA)

---

9. In any triangle:

THE <u>LONGEST</u> <u>SIDE</u> IS OPPOSITE THE <u>LARGEST</u> <u>ANGLE</u>.

THE <u>SHORTEST</u> <u>SIDE</u> IS OPPOSITE THE <u>SMALLEST</u> <u>ANGLE</u>.

In triangle DFH:

a) Angle F is the <u>largest</u> angle. Therefore, _____ is the <u>longest</u> side.

b) Angle H is the <u>smallest</u> angle. Therefore, _____ is the <u>shortest</u> side.

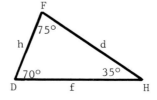

a) MV (or VM)

b) MT (or TM)

---

a) f   b) h

10. In the triangle at the right:

   a) The longest side is _____.

   b) The shortest side _____.

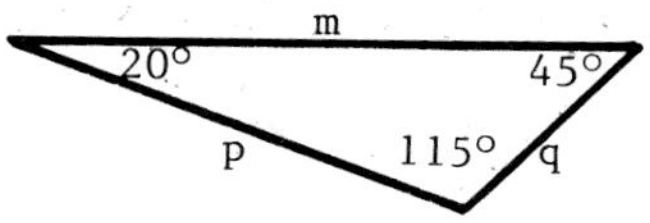

---

11. In a right triangle, the side opposite the right angle is called the "hypotenuse". The sides opposite the two acute angles are simply called "legs". For example:

   In right triangle DFH:

   "f" is the hypotenuse, and "d" and "h" are the legs.

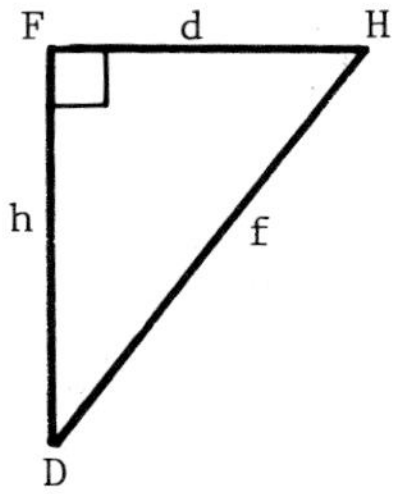

   a) The hypotenuse "f" can be labeled in two other ways: _____ or _____.

   b) Leg "d" can be labeled in two other ways: _____ or _____

a) m   b) q

---

12. Ordinarily we use small letters to represent the hypotenuse and legs of a right triangle. Use small letters to complete the questions below.

   In right triangle FET:

   a) The hypotenuse is _____.

   b) The legs are _____ and _____.

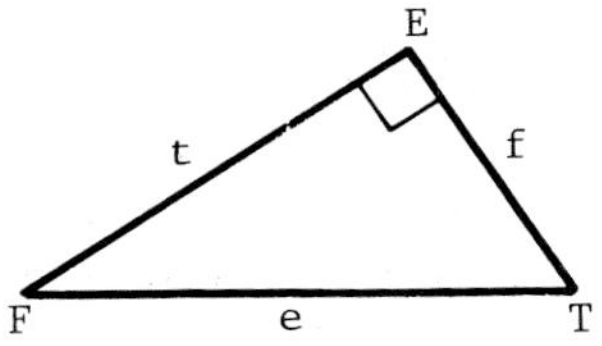

a) DH or HD

b) FH or HF

---

13. Since the right angle is the largest angle in any right triangle, the hypotenuse is the longest side of any right triangle.

   In right triangle DMF:

   a) The longest side is _____.

   b) The shortest side is _____.

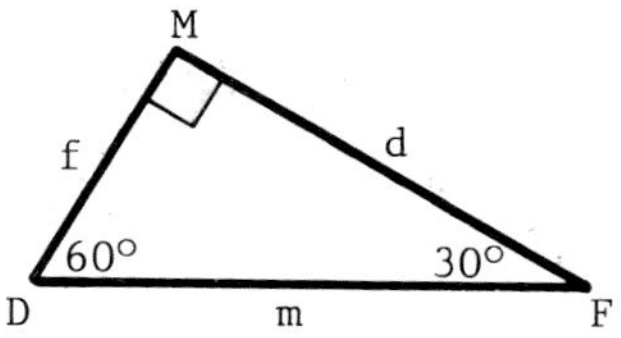

a) e

b) t and f

---

a) m (the hypotenuse)   b) f

11-3   THE PYTHAGOREAN THEOREM

In this section, we will discuss the Pythagorean Theorem and show how it can be used to find the length of an unknown side in a right triangle.

14. The Pythagorean Theorem states the following relationship among the three sides of a right triangle:

> IN ANY RIGHT TRIANGLE, THE SQUARE OF THE LENGTH OF THE HYPOTENUSE IS EQUAL TO THE SUM OF THE SQUARES OF THE LENGTHS OF THE TWO LEGS.

For right triangle ABC, the Pythagorean Theorem says:

$$c^2 = a^2 + b^2$$

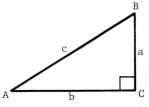

Using small letters, state the Pythagorean Theorem for each right triangle below.

a)

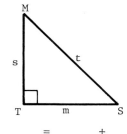

b)

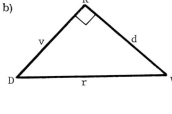

_____ = _____ + _____   _____ = _____ + _____

a) $t^2 = s^2 + m^2$

b) $r^2 = v^2 + d^2$

15. The lengths of the two legs of the right triangle below are given. We can use the Pythagorean Theorem to find the length of the hypotenuse. The steps are:

1) Write the Pythagorean Theorem.

$$m^2 = d^2 + p^2$$

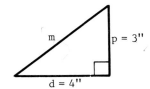

2) Substitute the known values and simplify.

$$m^2 = (4")^2 + (3")^2$$
$$= 16 \text{ in}^2 + 9 \text{ in}^2$$
$$= 25 \text{ in}^2$$

3) Find "m" by taking the square root of the right side.

$$m = \sqrt{25 \text{ in}^2} = \underline{\hspace{1cm}}$$

m = 5 in

16. We can use the same steps to find the length of the hypotenuse below.

$$h^2 = t^2 + b^2$$
$$= (7m)^2 + (4 \text{ m})^2$$
$$= 49 \text{ m}^2 + 16 \text{ m}^2$$
$$= 65 \text{ m}^2$$

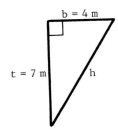

Use a calculator to find "h" by taking the square root of the right side. Round to tenths.

$$h = \sqrt{65 \text{ m}^2} = \underline{\hspace{1cm}}$$

17. When using the Pythagorean Theorem to find the hypotenuse of a right triangle, all of the calculations can be done in one process on a calculator. To prepare to do so, we solve for the hypotenuse by taking the square root of the other side before substituting. That is:

$$\text{If } c^2 = a^2 + b^2, \quad c = \sqrt{a^2 + b^2}$$

Using the method above, we set up the solution below so that "t" can be found in one calculator process.

$$t = k^2 + d^2$$
$$t = \sqrt{k^2 + d^2}$$
$$t = \sqrt{(5.68')^2 + (9.75')^2}$$

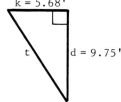
k = 5.68'
d = 9.75'
t

The calculator steps for the solution are shown. Notice that we added the squares of 5.68 and 9.75, pressed $\boxed{=}$ to complete that addition, and then pressed $\boxed{\sqrt{x}}$.

Calculator Steps: 5.68 $\boxed{x^2}$ $\boxed{+}$ 9.75 $\boxed{x^2}$ $\boxed{=}$ $\boxed{\sqrt{x}}$ 11.283834

Rounding to tenths, we get: t = _____

h = 8.1 m

---

18. In right triangle TMB, the length of the two legs and the size of two angles are given.

a) Using the angle-sum principle, find the size of angle B. _____

b) Using the Pythagorean Theorem, find the length of the hypotenuse "m". Round to tenths.

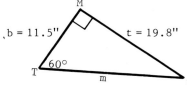
M
b = 11.5"
t = 19.8"
60°
T   m   B

m = _____

t = 11.3 ft

---

19. The distance between the opposite corners of a rectangle or square is called the "diagonal" of the rectangle or square. The diagonal of each figure is the hypotenuse of a right triangle. Use the Pythagorean Theorem to find the diagonal of the rectangle and square below. Round to tenths.

a)
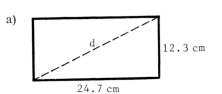
d
12.3 cm
24.7 cm

d = _____

b)
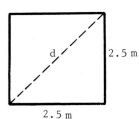
d
2.5 m
2.5 m

d = _____

a) Angle B = 30°

b) m = 22.9 in

20. The lengths of the hypotenuse and one leg are given for the triangle below. We can use the Pythagorean Theorem to find the length of the unknown leg "t". The steps are:

a) d = 27.6 cm

b) d = 3.5 m

    1) Write the Pythagorean Theorem.

       $d^2 = t^2 + k^2$

    2) Solve for "t^2".

       $t^2 = d^2 - k^2$

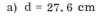

    3) Substitute and simplify.

       $t^2 = (10')^2 - (6')^2$

         $= 100 \text{ ft}^2 - 36 \text{ ft}^2$

         $= 64 \text{ ft}^2$

    4) Find "t" by taking the square root of the right side.

       $t = \sqrt{64 \text{ ft}^2} = $ _____

---

21. We can use the same steps to find the length of leg "v" below.

    $h^2 = v^2 + b^2$

    $v^2 = h^2 - b^2$

    $v^2 = (9 \text{ cm})^2 - (7 \text{ cm})^2$

    $v^2 = 81 \text{ cm}^2 - 49 \text{ cm}^2$

    $v^2 = 32 \text{ cm}^2$

8 ft

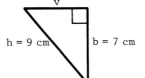

Find "v" by taking the square root of the right side. Round to tenths.

    $v = \sqrt{32 \text{ cm}^2} = $ _____

---

22. When using the Pythagorean Theorem to find an unknown leg in a right triangle, we can also do all of the calculations in one process on a calculator. To prepare to do so, we solve for the leg before substituting. As an example, let's solve for leg "y" below.

v = 5.7 cm

    $z^2 = y^2 + x^2$

    $y^2 = z^2 - x^2$

    $y = \sqrt{z^2 - x^2}$

    $y = \sqrt{(18 \text{ m})^2 - (12 \text{ m})^2}$

The calculator steps for the solution are shown. Notice again that we pressed ☐= before pressing ☐$\sqrt{x}$.

    Calculator Steps: 18 ☐x^2 ☐- 12 ☐x^2 ☐= ☐$\sqrt{x}$ 13.416408

Rounding to tenths, we get: y = _____

23. We set up the calculator solution for leg "a" below. Complete the solution. Round to tenths.

$$r^2 = m^2 + a^2$$
$$a^2 = r^2 - m^2$$
$$a = \sqrt{r^2 - m^2}$$
$$a = \sqrt{(27.5')^2 - (21.9')^2}$$
$$a = \underline{\hspace{1.5cm}}$$

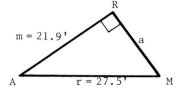

y = 13.4 m

---

24. In right triangle RQS, the length of the hypotenuse and one leg and the size of two angles are given.

a) Using the angle-sum principle, find the size of angle S. _____

b) Using the Pythagorean Theorem, find the length of leg "s". Round to tenths.

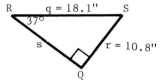

s = _____

a = 16.6 ft

---

25. Use the Pythagorean Theorem to find the width of the rectangle at the right. Round to tenths.

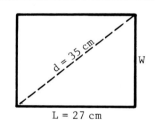

W = _____

a) Angle S = 53°

b) s = 14.5 in

---

26. There are two holes in the metal plate at the right. Use the Pythagorean Theorem to find the distance between the holes, measured center-to-center. Round to tenths.

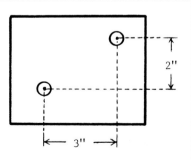

W = 22.3 cm

3.6 in

27. In the rectangular prism below, BD is the diagonal of the prism and BC is the diagonal of the base of the prism. To find BD, we must find BC first. The Pythagorean Theorem can be used to find both diagonals.

a) ABC is a right triangle. Find BC, the diagonal of the base.

BC = _____

b) BCD is a right triangle. Find BD, the diagonal of the prism. Round to tenths.

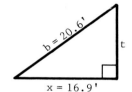

BD = _____

a) BC = 50 m

b) BD = 53.9 m

28. We can find the <u>area</u> of this right triangle in two steps.

a) First find the length of leg "t". Round to tenths.

t = _____

b) Then use the area formula below, rounding to tenths.

$$A = \frac{(\text{leg 1})(\text{leg 2})}{2}$$

a) t = 11.8 ft   b) A = 99.7 ft²

11-4   ISOSCELES AND EQUILATERAL TRIANGLES

In this section, we will discuss <u>isosceles</u> and <u>equilateral</u> triangles and solve problems involving triangles of those types.

29. Triangles with two equal sides are called <u>isosceles</u> triangles. In an isosceles triangle, <u>the angles opposite the equal sides are equal</u>.

The triangle at the right is an isosceles triangle because sides "c" and "d" are equal.

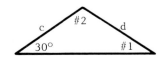

a) Since the angles opposite "c" and "d" are equal, angle #1 = _____ .

b) Therefore, angle #2 = _____ .

30. In the isosceles triangle on the right, sides "t" and "p" are equal.

Angles #1 and #2 must be equal because they are opposite the equal sides. How many degrees are there in each of these two equal angles? _____

a) 30°   b) 120°

---

31. The triangle at the right is an <u>isosceles</u> <u>right</u> <u>triangle</u> since it contains a right angle, and sides "c" and "k" are equal.

    a) Angle #1 = _____ °

    b) Angle #2 = _____ °

50°

---

32. Triangles with <u>three</u> <u>equal</u> <u>sides</u> are called <u>equilateral</u> <u>triangles</u>. In an equilateral triangle, <u>all</u> <u>three</u> <u>angles</u> <u>are</u> <u>equal</u>.

The triangle at the right is an equilateral triangle because all three sides are equal. How many degrees are there in each of the three equal angles? _____

a) 45°

b) 45°

---

33. When a height is drawn to the unequal side in an <u>isosceles</u> triangle, it bisects that side. That is, it cuts that side into two equal parts. We can use that fact to find the height and the area of the triangle when all three sides are known.

Triangle ABC is an isosceles triangle. AD is a height drawn to the unequal side BC.

    a) Since AD bisects BC, how long are BD and DC? _____

    b) Using the Pythagorean Theorem, find the length of AD to the nearest tenth.

$$AD = \underline{\hspace{2cm}}$$

    c) The area of the triangle is $\dfrac{(BC)(AD)}{2}$ = _____

60°

---

a) 10 in

b) 6.6 in

c) 66 in²

34. When a height is drawn to any side in an <u>equilateral</u> triangle, it bisects that side. We can use that fact to find the height and the area of the triangle when the length of the sides is known.

Triangle DEF is an equilateral triangle. FG is a height drawn to DE.

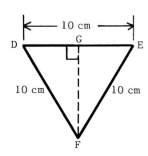

   a) Since FG bisects DE, how long are DG and GE? _____

   b) Using the Pythagorean Theorem, find the length of FG to the nearest tenth.

FG = _____

   c) The area of the triangle is $\dfrac{(DE)(FG)}{2}$ = _____

---

35. In the rectangular figure at the right, the shaded part is an isosceles triangle. Find the area of the triangle.

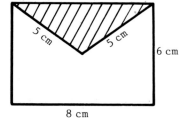

A = _____

a) 5 cm

b) 8.7 cm

c) 43.5 cm^2

---

36. In the figure at the right, a V-shaped cut has been made in the top of a rectangle. The cut is an isosceles triangle whose base is 3" and whose height is 2".

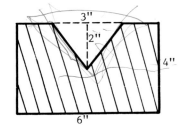

To find the area of the shaded figure, <u>we</u> <u>must</u> <u>subtract</u> <u>the</u> <u>area</u> <u>of</u> <u>the</u> <u>triangle</u> <u>from</u> <u>the</u> <u>area</u> <u>of</u> <u>the</u> <u>rectangle</u>.

   a) The area of the rectangle is _____.

   b) The area of the triangle is _____.

   c) The area of the shaded figure is _____.

A = 12 cm^2

a) 24 in^2

b) 3 in^2

c) 21 in^2

SELF–TEST 37   (pages 464-475)

1. Angle P = _____

2. Which side is <u>shortest</u>? _____

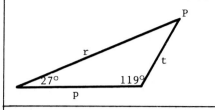

3. Angle F = _____

4. Which side is <u>longest</u>? _____

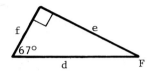

5. Find side "c". Round to tenths.

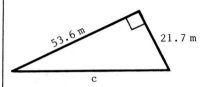

c = _____

6. Find side "p". Round to hundredths.

7. Find the area. Round to hundredths.

p = _____

A = _____

8. FHG is an <u>isosceles</u> triangle.
    Find angle G and angle H.

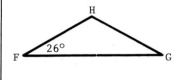

G = _____

H = _____

9. ABC is an <u>isosceles</u> <u>right</u> <u>angle</u>.
    Find angle A and angle B.

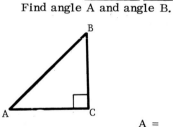

A = _____

B = _____

10. PQR is an <u>equilateral</u> <u>triangle</u>.
     Find angle P, angle Q, and
     angle R.

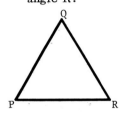

P = _____

Q = _____

R = _____

11. Find "d", the diagonal of the rectangle.
     Round to the nearest whole number.

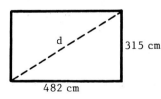

d = _____

12. Find the area of the shaded figure. Round to the
     nearest whole number.

13.266
132.664
24'   24'
14'
40'   560

A = _____

ANSWERS:
1. 34°
2. t
3. 23°
4. d

5. c = 57.8 m
6. p = 1.12 in
7. A = 1.61 in^2
8. G = 26°
    H = 128°

9. A = 45°
    B = 45°

10. P = 60°
     Q = 60°
     R = 60°

11. d = 576 cm
12. A = 825 ft^2

11-5   THE TANGENT RATIO

When the Pythagorean Theorem or angle-sum principle cannot be used to find an unknown side or unknown angle in a right triangle, we can usually use one of the three basic trigonometric ratios to do so. In this section, we will define the <u>tangent</u> ratio and show how it can be used to find an unknown side or angle in a right triangle.

37.   In right triangle ABC, "c" is the hypotenuse, and "a" and "b" are the two legs.

    "a" is the <u>side opposite</u> angle A.

    "b" is the <u>side adjacent</u> to angle A.

<u>Note</u>: The word "adjacent" means "next to". Though both "c" and "b" are "next to" angle A, only "b" is the "side adjacent" because "c" is the hypotenuse.

In the same triangle: a) the <u>side opposite</u> angle B is _____.

                      b) the <u>side adjacent</u> to angle B is _____.

---

38.   In right triangle PRT:

   a) The hypotenuse is _____.

   b) The side opposite angle T is _____.

   c) The side adjacent to angle T is _____.

   d) The side opposite angle R is _____.

   e) The side adjacent to angle R is _____.

a) b   b) a

---

39.   In right triangle HSV:

   a) The hypotenuse is _____.

   b) The side opposite angle V is _____.

   c) The side adjacent to angle V is _____.

   d) The side opposite angle H is _____.

   e) The side adjacent to angle H is _____.

a) p   d) r
b) t   e) t
c) r

---

40.   The <u>tangent ratio</u> is one of the three basic trigonometric ratios. In a right triangle, the "<u>tangent of an acute angle</u>" is a comparison of the "<u>length of the side opposite</u>" to the "<u>length of the side adjacent</u>" to the angle. That is:

$$\text{THE TANGENT OF AN ANGLE} = \frac{\text{SIDE OPPOSITE}}{\text{SIDE ADJACENT}}$$

a) s   d) h
b) v   e) v
c) h

Continued on following page.

40. Continued

In right triangle ABC:

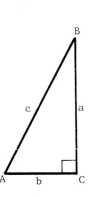

The side opposite angle A is "a".

The side adjacent to angle A is "b".

Therefore, the <u>tangent</u> <u>of</u> <u>angle</u> <u>A</u> is $\frac{a}{b}$.

The side opposite angle B is "b".

The side adjacent to angle B is "a".

Therefore, the tangent of angle B is _____.

---

41. In right triangle MPT:

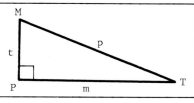

a) The tangent of angle T is _____

b) The tangent of angle M is _____ .

$\frac{b}{a}$

---

42. Using capital letters to label the sides, complete these for right triangle CDF.

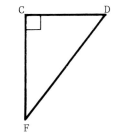

a) The tangent of angle D is

_____ .

b) The tangent of angle F is

_____ .

a) $\frac{t}{m}$ b) $\frac{m}{t}$

---

43. In right triangle ADV:

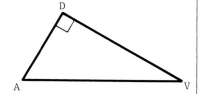

a) The tangent of angle A is

_____ .

b) The tangent of angle V is

_____ .

a) $\frac{CF}{CD}$ b) $\frac{CD}{CF}$

---

44. The tangent of any specific angle has the same numerical value in right triangles of any size. For example, in the diagram below, there are three right triangles: ABG, ACF, and ADE. Each right triangle contains the common angle A which is a 37° angle.

a) $\frac{DV}{AD}$ b) $\frac{AD}{DV}$

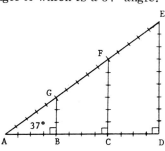

Continued on following page.

44. Continued

To show that the tangent of a 37° angle is the same in right triangles of any size, we computed the tangent of angle A in each of the three right triangles by counting units to approximate the lengths of the sides. As you can see, the tangent of 37° is approximately $\frac{3}{4}$ in each triangle.

In triangle ABG, the tangent of A is $\frac{BG}{AB} = \frac{3}{4}$

In triangle ACF, the tangent of A is $\frac{CF}{AC} = \frac{6}{8}$ or $\frac{3}{4}$

In triangle ADE, the tangent of A is $\frac{DE}{AD} = \frac{9}{12}$ or $\frac{3}{4}$

Tangents are usually expressed as decimal numbers. Therefore, instead of saying that the tangent of 37° is approximately $\frac{3}{4}$, we would say that the tangent of 37° is approximately _____ .

---

45. Instead of "the tangent of a 37° angle", the abbreviation "tan 37°" is used. Therefore:

tan 15° means: the tangent of a 15° angle

tan 74° means: _____

0.75

---

46. A partial table of tangents is shown at the right. Each tangent is rounded to four decimal places. Notice this fact:

As the size of the angle increases from 0° to 89°, the size of the tangent <u>increases</u> from 0.0000 to 57.2900 .

Angle A	tan A
0°	0.0000
10°	0.1763
20°	0.3640
30°	0.5774
40°	0.8391
50°	1.1918
60°	1.7321
70°	2.7475
80°	5.6713
89°	57.2900

Using the table, complete these:

a) The tangent of any 70° angle is _____ .

b) 0.1763 is the tangent of any _____ angle.

the tangent of a 74° angle

---

47. A calculator can be used to find the tangent of any angle. To do so, we simply enter the angle and press ⌐tan⌐ .

> Note: Some calculators are designed to give tangents of angles measured in degrees, radians, and even grads. <u>Be sure that your calculator is set to give the tangents of angles measured in degrees.</u>

a) 2.7475

b) 10°

Continued on following page.

47. Continued

Following the steps below, find tan 15°, tan 47°, and tan 86°.

Calculator Steps: 15 | tan | .26794919

47 | tan | 1.0723687

86 | tan | 14.300666

Tangents are usually rounded <u>to four</u> decimal <u>places</u>. Therefore:

a) tan 15° = _____   b) tan 47° = _____   c) tan 86° = _____

---

48. Use a calculator for these. Round each tangent <u>to four</u> decimal <u>places</u>.

a) tan 34° = _____   b) tan 78° = _____

a) 0.2679

b) 1.0724

c) 14.3007

---

49. A calculator can also be <u>used to find the size of an angle</u> whose tangent is known. Either | 2nd | | tan | , | INV | | tan | , or | tan⁻¹ | is used. Following the steps below, find the angle whose tangent is 0.3173.

Calculator Steps: 0.3173 | 2nd | | tan | 17.604233

or | INV | | tan |

or | tan⁻¹ |

The angle is usually rounded <u>to the</u> nearest <u>whole number</u>. Therefore:

The angle whose tangent is 0.3173 is _____.

a) 0.6745

b) 4.7046

---

50. Complete these. Round each angle to the nearest whole number.

a) If tan F = 0.7541,   b) If tan A = 2.3555,

F = _____   A = _____

18°

---

51. We cannot use the Pythagorean Theorem to find side MR in the right triangle below because the length of only one side is known. However, we can use tan P to do so. The steps are:

$$\tan P = \frac{MR}{PR}$$

$$\tan 25° = \frac{MR}{52.5}$$

$$MR = (52.5)(\tan 25°)$$

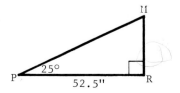

a) 37°   b) 67°

Continued on following page.

51.   Continued

To evaluate (52.5)(tan 25°) on a calculator, follow the steps below. Wait until tan 25° appears on the display before pressing $\boxed{=}$.

Calculator Steps: 52.5 $\boxed{\text{x}}$ 25 $\boxed{\text{tan}}$ $\boxed{=}$ 24.481152

Rounding to the nearest tenth of an inch, MR = _____

24.5 in

---

52.   We cannot use the Pythagorean Theorem to find side CE in the right triangle below because only one side is known. However, we can use tan C to do so.   The steps are:

$$\tan C = \frac{DE}{CE}$$

$$\tan 55° = \frac{125}{CE}$$

$$(\tan 55°)(CE) = 125$$

$$CE = \frac{125}{\tan 55°}$$

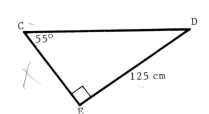

To complete the evaluation on a calculator, follow the steps below. Wait until tan 55° appears on the display before pressing $\boxed{=}$.

Calculator Steps: 125 $\boxed{÷}$ 55 $\boxed{\text{tan}}$ $\boxed{=}$ 87.525942

Rounding to the nearest tenth of a centimeter, CE = _____

87.5 cm

---

53.   We cannot use the angle-sum principle to find angle R in the right triangle at the right because angle T is not known. However, we can use tan R to do so.   The steps are:

$$\tan R = \frac{ST}{RS}$$

$$\tan R = \frac{19.8}{34.3}$$

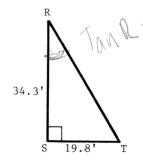

To find R, follow the steps below. Notice that we pressed $\boxed{=}$ to complete the division before pressing $\boxed{\text{2nd}}\ \boxed{\text{tan}}$, $\boxed{\text{INV}}\ \boxed{\text{tan}}$, or $\boxed{\text{tan}^{-1}}$.

Calculator Steps: 19.8 $\boxed{÷}$ 34.3 $\boxed{=}$ $\boxed{\text{2nd}}$ $\boxed{\text{tan}}$ 29.996098

                        or $\boxed{\text{INV}}$ $\boxed{\text{tan}}$

                        or $\boxed{\text{tan}^{-1}}$

Rounding to the nearest whole number degree, R = _____

30°

11-6   THE SINE RATIO

The <u>sine</u> <u>ratio</u> is also one of the three basic trigonometric ratios. In this section, we will define the sine ratio and show how it can be used to find an unknown side or angle in a right triangle.

---

54.   In a right triangle, the "<u>sine</u> <u>of</u> <u>an</u> <u>acute</u> <u>angle</u>" is a comparison of the "<u>length</u> <u>of</u> <u>the</u> <u>side</u> <u>opposite</u>" the angle to the "<u>length</u> <u>of</u> <u>the</u> <u>hypotenuse</u>". That is:

> THE SINE OF AN ANGLE $= \dfrac{\text{SIDE OPPOSITE}}{\text{HYPOTENUSE}}$

    <u>Note</u>: The word "sine" is pronounced "sign". It is not pronounced "sin".

In right triangle ABC:

    The side opposite angle A is "a".

    The hypotenuse is "c".

    Therefore, the <u>sine</u> <u>of</u> <u>angle</u> <u>A</u> is $\dfrac{a}{c}$.

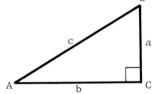

What is the sine of angle B? _____

---

55.   In right triangle XYT:

    a) The sine of angle T is _____.

    b) The sine of angle Y is _____.

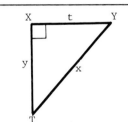

$\dfrac{b}{c}$

---

56.   Using capital letters to label the sides, complete these for right triangle PQS.

    a) The sine of angle P is _____.

    b) The sine of angle S is _____.

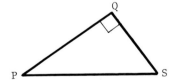

a) $\dfrac{t}{x}$

b) $\dfrac{y}{x}$

---

57.   To show that the sine of a specific angle has the same numerical value in right triangles of any size, we will again use a 37° angle as an example. Let's count units to find the sine of angle A in each of the three triangles at the right.

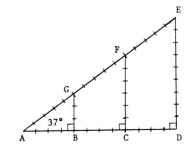

a) $\dfrac{QS}{PS}$

b) $\dfrac{PQ}{PS}$

Continued on following page.

57. Continued

In triangle ABG, the sine of A is $\dfrac{BG}{AG} = \dfrac{3}{5}$

In triangle ACF, the sine of A is $\dfrac{CF}{AF} = \dfrac{6}{10}$ or $\dfrac{3}{5}$

In triangle ADE, the sine of A is $\dfrac{DE}{AE} = \dfrac{9}{15}$ or $\dfrac{3}{5}$

Sines are usually expressed as decimal numbers. Therefore, instead of saying that the sine of 37° is approximately $\dfrac{3}{5}$, we would say that the sine of 37° is approximately _____ .

---

58. Instead of "the sine of a 37° angle", the abbreviation "sin 37°" is used. Though the abbreviation is "sin", it is still pronounced "sign".

sin 15° means: the sine of a 15° angle

sin 48° means: _____

0.6

---

59. A partial table of sines is shown at the right. Each sine is rounded to four decimal places. Notice this fact:

Angle A	sin A
0°	0.0000
10°	0.1736
20°	0.3420
30°	0.5000
40°	0.6428
50°	0.7660
60°	0.8660
70°	0.9397
80°	0.9848
90°	1.0000

As the size of the angle increases from 0° to 90°, the size of the sine <u>increases</u> from 0.0000 to 1.0000.

Using the table, complete these:

a) The sine of any 50° angle is _____ .

b) 0.9848 is the sine of any _____ angle.

the sine of a 48° angle

---

60. To find the sine of an angle on a calculator, we enter the angle and press $\boxed{\text{sin}}$. Use a calculator for these. Round to four decimal places.

a) sin 7° = _____   b) sin 41° = _____   c) sin 83° = _____

a) 0.7660

b) 80°

---

61. To find an angle whose sine is known on a calculator, we enter the sine and then press either $\boxed{\text{2nd}}\ \boxed{\text{sin}}$ or $\boxed{\text{INV}}\ \boxed{\text{sin}}$ or $\boxed{\text{sin}^{-1}}$. Use a calculator for these. Round to the nearest whole-number degree.

a) If sin D = 0.2666,   b) If sin G = 0.8511,

D = _____   G = _____

a) 0.1219

b) 0.6561

c) 0.9925

---

a) 15°   b) 58°

62. We cannot use the Pythagorean Theorem to find side ST in the right triangle below because only one side is known. However, we can use sin Q to do so. The steps are:

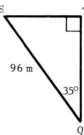

$$\sin Q = \frac{ST}{SQ}$$

$$\sin 35° = \frac{ST}{96}$$

$$ST = (96)(\sin 35°)$$

To evaluate (96)(sin 35°) on a calculator, follow the steps below. Wait until sin 35° appears on the display before pressing [=].

Calculator Steps: 96 [x]   35 [sin] [=]   55.063338

Rounding to the nearest tenth of a meter, ST = _____.

55.1 m

63. We cannot use the Pythagorean Theorem to find hypotenuse CF in the right triangle below because only one side is known. However, we can use sin F to do so. The steps are:

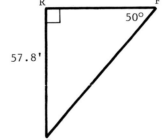

$$\sin F = \frac{CR}{CF}$$

$$\sin 50° = \frac{57.8}{CF}$$

$$(\sin 50°)(CF) = 57.8$$

$$CF = \frac{57.8}{\sin 50°}$$

To complete the evaluation on a calculator, follow the steps below. Wait until sin 50° appears on the display before pressing [=].

Calculator Steps: 57.8 [÷]   50 [sin] [=]   75.452541

Rounding to the nearest tenth of a foot, CF = _____

75.5 ft

64. We cannot use the angle-sum principle to find angle D below because angle R is not known. However, we can use sin D to do so. The steps are:

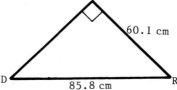

$$\sin D = \frac{PR}{DR}$$

$$\sin D = \frac{60.1}{85.8}$$

Continued on following page.

64.   Continued

To find D on a calculator, following the steps below. Notice that we pressed ⬜=⬜ to complete the division before pressing ⬜2nd⬜ ⬜sin⬜ , ⬜INV⬜ ⬜sin⬜ , or ⬜sin⁻¹⬜

Calculator Steps: 60.1 ⬜÷⬜ 85.8 ⬜=⬜ ⬜2nd⬜ ⬜sin⬜   44.464419

or ⬜INV⬜ ⬜sin⬜

or ⬜sin⁻¹⬜

Rounding to the nearest whole-number degree, D = _____

44°

11-7   THE COSINE RATIO

The cosine ratio is the last of the three basic trigonometric ratios. In this section, we will define the cosine ratio and show how it can be used to find an unknown side or angle in a right triangle.

65.   In a right triangle, the "cosine of an acute angle" is a comparison of the "length of the side adjacent" to the angle to the "length of the hypotenuse". That is:

$$\text{THE COSINE OF AN ANGLE} = \frac{\text{SIDE ADJACENT}}{\text{HYPOTENUSE}}$$

Note: The word "cosine" is pronounced "co-sign".

In right triangle CDE:

The side adjacent to angle C is "d".

The hypotenuse is "e".

Therefore, the cosine of angle C is $\frac{d}{e}$

What is the cosine of angle D?   _____

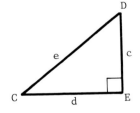

$\frac{c}{e}$

66.   The abbreviation for "cosine" is "cos". Therefore, in right triangle MFT:

a) cos M = _____

b) cos T = _____

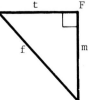

a) $\frac{t}{f}$ b) $\frac{m}{f}$

67. Using capital letters to label
the sides, complete these.

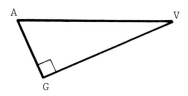

a) cos A = _____

b) cos V = _____

a) $\dfrac{AG}{AV}$ b) $\dfrac{GV}{AV}$

68. To show that the cosine of a
specific angle has the same
numerical value in right
triangles of any size, we will
again use a 37° angle as the
example. Let's count units
to find the cosine of angle A
in each of the three triangles
at the right.

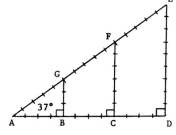

In triangle ABG, $\cos A = \dfrac{AB}{AG} = \dfrac{4}{5}$

In triangle ACF, $\cos A = \dfrac{AC}{AF} = \dfrac{8}{10}$ or $\dfrac{4}{5}$

In triangle ADE, $\cos A = \dfrac{AD}{AE} = \dfrac{12}{15}$ or $\dfrac{4}{5}$

Expressed as a decimal number, cos 37° in each of the three triangles
is approximately _____ .

69. A partial table of cosines is shown
at the right. Each cosine is rounded
to four decimal places. Notice this
fact:

   As the size of the angle increases
   from 0° to 90°, the size of the
   cosine <u>decreases</u> from 1.0000
   to 0.0000.

Using the table, complete these:

a) cos 50° = _____

b) If cos D = 0.1736, D = _____

Angle A	cos A
0°	1.0000
10°	0.9848
20°	0.9397
30°	0.8660
40°	0.7660
50°	0.6428
60°	0.5000
70°	0.3420
80°	0.1736
90°	0.0000

0.8

70. To find the cosine of an angle on a calculator, we enter the angle and
press ⃞cos . Use a calculator for these. Round to four decimal places.

a) cos 9° = _____   b) cos 27° = _____   c) cos 81° = _____

a) 0.6428

b) 80°

a) 0.9877

b) 0.8910

c) 0.1564

71. To find an angle whose cosine is known on a calculator, we enter the cosine and then press either $\boxed{\text{2nd}}$ $\boxed{\text{cos}}$ or $\boxed{\text{INV}}$ $\boxed{\text{cos}}$ or $\boxed{\text{cos}^{-1}}$. Use a calculator for these. Round to the nearest whole-number degree.

   a) If cos H = 0.7561,　　　　b) If cos V = 0.2099,

   　　　H = _____　　　　　　　　V = _____

---

72. We cannot use the Pythagorean Theorem to find side AC in the right triangle below because only one side is known. However, we can use cos A to do so. The steps are:

$$\cos A = \frac{AC}{AB}$$

$$\cos 53° = \frac{AC}{60.7}$$

$$AC = (60.7)(\cos 53°)$$

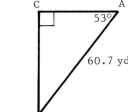

To evaluate (60.7)(cos 53°) on a calculator, follow the steps below. Wait until cos 53° appears before pressing $\boxed{=}$.

　　Calculator Steps: 60.7 $\boxed{\text{x}}$ 53 $\boxed{\text{cos}}$ $\boxed{=}$ 　36.530172

Rounding to the nearest tenth of a yard, AC = _____

a) 41°　　　b) 78°

---

73. We cannot use the Pythagorean Theorem to find hypotenuse FM in the right triangle below because only one side is known. However, we can use cos F to do so. The steps are:

$$\cos F = \frac{FG}{FM}$$

$$\cos 33° = \frac{256}{FM}$$

$$(\cos 33°)(FM) = 256$$

$$FM = \frac{256}{\cos 33°}$$

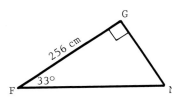

To complete the evaluation on a calculator, follow the steps below. Wait until cos 33° appears before pressing $\boxed{=}$.

　　Calculator Steps: 256 $\boxed{\div}$ 33 $\boxed{\text{cos}}$ $\boxed{=}$ 　305.245

Rounding to the nearest centimeter, FM = _____

36.5 yd

305 cm

74. We cannot use the angle-sum principle to find angle D below because angle S is not known. However, we can use cos D to do so. The steps are:

$$\cos D = \frac{DH}{DS}$$

$$\cos D = \frac{14.7}{18.7}$$

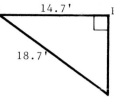

Find D on a calculator. Be sure to press ⎡=⎤ before pressing ⎡2nd⎤ ⎡cos⎤ or ⎡INV⎤ ⎡cos⎤ or ⎡cos⁻¹⎤.

To the nearest whole-number degree, D = _____

38°

11-8 CONTRASTING THE BASIC TRIGONOMETRIC RATIOS

In this section, we will give some exercises contrasting the definitions of the three basic trigonometric ratios.

75. The definitions of the three basic trigonometric ratios for angle A in the right triangle are:

$$\tan A = \frac{\text{side opposite angle A}}{\text{side adjacent to angle A}}$$

$$\sin A = \frac{\text{side opposite angle A}}{\text{hypotenuse}}$$

$$\cos A = \frac{\text{side adjacent to angle A}}{\text{hypotenuse}}$$

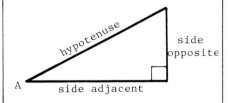

When writing the trig ratios for an angle in a right triangle, it is helpful to locate the hypotenuse first. For example, in triangle BFT:

a) The hypotenuse is _____.

b) sin B = _____

c) cos B = _____

d) tan B = _____

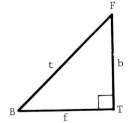

a) t c) $\frac{f}{t}$

b) $\frac{b}{t}$ d) $\frac{b}{f}$

76. Using capital letters for the sides, complete these:

a) cos P = _____

b) tan S = _____

c) sin P = _____

d) cos S = _____

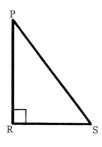

---

77. Using capital letters for the sides, complete these:

a) tan H = _____

b) tan V = _____

c) sin H = _____

d) sin V = _____

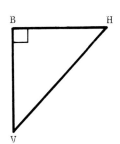

a) $\dfrac{PR}{PS}$ c) $\dfrac{RS}{PS}$

b) $\dfrac{PR}{RS}$ d) $\dfrac{RS}{PS}$

---

78. Complete:

a) tan T = _____

b) cos F = _____

c) sin T = _____

d) tan F = _____

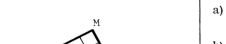

a) $\dfrac{BV}{BH}$ c) $\dfrac{BV}{HV}$

b) $\dfrac{BH}{BV}$ d) $\dfrac{BH}{HV}$

---

79. Using small letters for the sides, complete these:

a) sin D = _____

b) tan C = _____

c) sin C = _____

d) cos D = _____

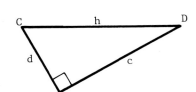

a) $\dfrac{FM}{MT}$ c) $\dfrac{FM}{FT}$

b) $\dfrac{FM}{FT}$ d) $\dfrac{MT}{FM}$

---

a) $\dfrac{d}{h}$ c) $\dfrac{c}{h}$

b) $\dfrac{c}{d}$ d) $\dfrac{c}{h}$

80. If a ratio does not involve the hypotenuse, it is a "tangent" ratio. For example, in right triangle FGH:

   a) The ratio $\dfrac{FG}{GH}$ is the tangent of angle _____.

   b) The ratio $\dfrac{GH}{FG}$ is the tangent of angle _____.

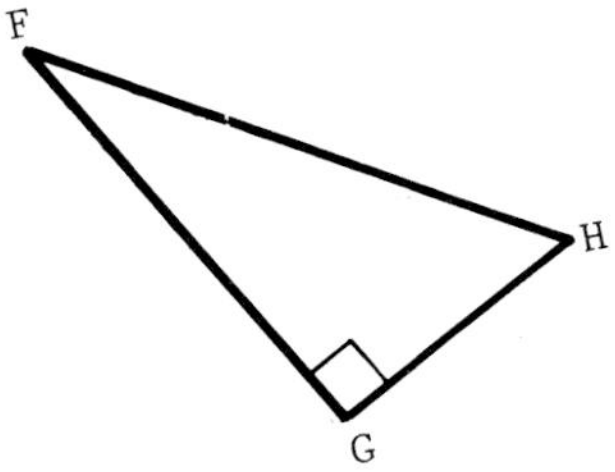

---

81. If a ratio involves the hypotenuse, it is the sine of one angle and the cosine of the other. For example, in the triangle on the right:

   $\dfrac{m}{d}$ is the sine of angle M and the cosine of angle T.

   $\dfrac{t}{d}$ is the sine of angle ____ and the cosine of angle ____.

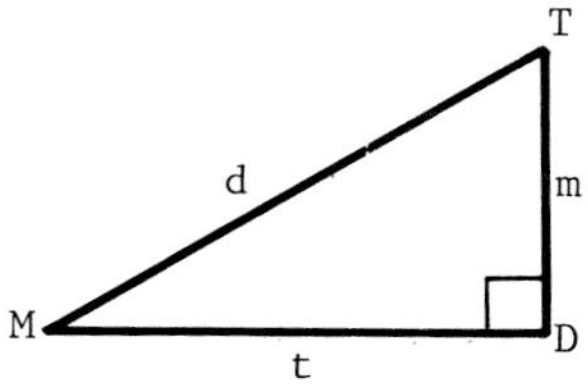

   a) H   b) F

---

82. In this right triangle:

   $\dfrac{MN}{NP}$ is tan P

   a) $\dfrac{NP}{MN}$ is _____

   b) $\dfrac{MN}{MP}$ is either _____ or _____

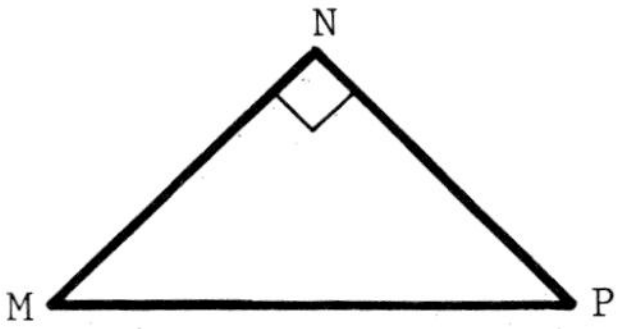

   sine of angle T, cosine of angle M

---

83. In this right triangle:

   a) $\dfrac{a}{b}$ = _____

   b) $\dfrac{b}{a}$ = _____

   c) $\dfrac{a}{c}$ = _____ or _____

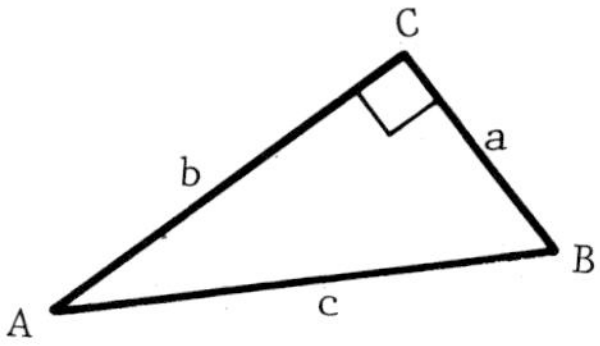

   a) tan M

   b) sin P or cos M

---

84. In this right triangle:

   a) $\dfrac{m}{t}$ = _____ or _____

   b) $\dfrac{f}{t}$ = _____ or _____

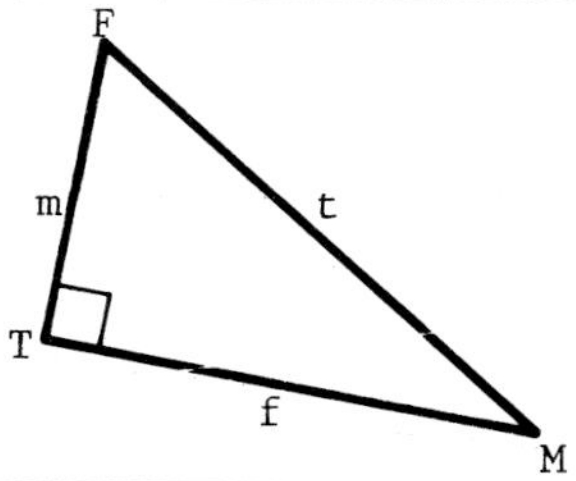

   a) tan A

   b) tan B

   c) sin A or cos B

---

a) sin M or cos F   b) sin F or cos M

SELF-TEST 38   (pages 476-490)

Find the numerical value of each trigonometric ratio.   Round to four decimal places.

1. cos 35° = _____

2. tan 23° = _____

3. sin 66° = _____

Find each angle.   Round to the nearest whole-number degree.

4. If sin B = 0.1614,

B = _____

5. If cos F = 0.2597,

F = _____

6. If tan R = 1.3318,

R = _____

Evaluate each of the following.   Round as directed.

7. Round to tenths.

$$\frac{27.8}{\cos 19°} = \underline{\qquad}$$

8. Round to hundreds.

$(42,900)(\tan 61°) = \underline{\qquad}$

9. Round to hundredths.

$$\frac{1.53}{\sin 40°} = \underline{\qquad}$$

Find each angle.   Round to the nearest whole-number degree.

10. $\tan A = \frac{5,730}{1,980}$

A = _____

11. $\sin P = \frac{2.19}{7.55}$

P = _____

12. $\cos H = \frac{66.7}{78.4}$

H = _____

13. In right triangle PRT, define the following trigonometric ratios.   Use capital letters for labeling the sides.

a) sin T = _____

c) tan P = _____

b) tan T = _____

d) cos P = _____

14. In right triangle DEH, define the following trigonometric ratios.   Use small letters for labeling the sides.

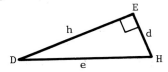

a) cos H = _____

c) sin D = _____

b) sin H = _____

d) tan H = _____

ANSWERS:
1. 0.8192
2. 0.4245
3. 0.9135
4. B = 9°
5. F = 75°
6. R = 53°

7. 29.4
8. 77,400
9. 2.38
10. A = 71°
11. P = 17°
12. H = 32°

13. a) $\frac{PR}{PT}$

b) $\frac{PR}{RT}$

c) $\frac{RT}{PR}$

d) $\frac{PR}{PT}$

14. a) $\frac{d}{e}$

b) $\frac{h}{e}$

c) $\frac{d}{e}$

d) $\frac{h}{d}$

11-9   FINDING UNKNOWN SIDES IN RIGHT TRIANGLES

Either the Pythagorean Theorem or the trig ratios are used to find unknown sides in right triangles.   We will discuss the use of both methods in this section.

85.   When only two sides of a right triangle are known, we use the Pythagorean Theorem to find the third side.

In right triangle ABC, AB and BC are known.   Let's use the Pythagorean Theorem to find AC.

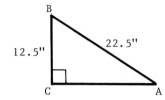

$$(AB)^2 = (BC)^2 + (AC)^2$$
$$(22.5)^2 = (12.5)^2 + (AC)^2$$
$$(AC)^2 = (22.5)^2 - (12.5)^2$$
$$AC = \sqrt{(22.5)^2 - (12.5)^2}$$

Use a calculator to complete the solution.   Round to the nearest tenth of an inch.   AC = _____

---

86.   When only one side and one acute angle of a right triangle are known, we use a trig ratio to find an unknown side.

18.7 in

In right triangle CDF, side CD and angle C are known.   We can use the <u>sine ratio</u> to find DF.

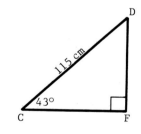

$$\sin C = \frac{DF}{CD}$$
$$\sin 43° = \frac{DF}{115}$$
$$DF = (115)(\sin 43°)$$

Use a calculator to complete the solution.   Round to the nearest tenth of a centimeter.   DF = _____

---

87.   To find PS in this right triangle, we can use the <u>cosine ratio</u>.

78.4 cm

$$\cos P = \frac{PR}{PS}$$
$$\cos 75° = \frac{400}{PS}$$
$$(\cos 75°)(PS) = 400$$
$$PS = \frac{400}{\cos 75°}$$

Use a calculator to complete the solution.   Round to the nearest yard.
PS = _____

88. Before using a trig ratio to find an unknown side in a right triangle, we must decide whether to use the sine, cosine, or tangent of the given angle. The following strategy can be used.

Find "c" in right triangle CHM.

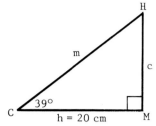

1) Identify the side you want to find ("c") and the known side ("h") in terms of the known angle (C).

"c" is the side opposite angle C.

"h" is the side adjacent to angle C.

2) Identify the ratio that includes both "side opposite" and "side adjacent".

The ratio is the tangent of angle C.

3) Use that ratio to set up an equation.

$$\tan C = \frac{c}{h} \quad \text{or} \quad \tan 39° = \frac{c}{20}$$

Use a calculator to complete the solution. Round to the nearest tenth of a centimeter.   c = _____

1,545 yd

89. Let's use the same steps to solve for DM in this right triangle.

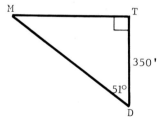

1) Identify the side you want to find (DM) and the known side (DT) in terms of the known angle (D).

DM is the hypotenuse.

DT is the side adjacent to angle D.

2) Identify the ratio that includes both "side adjacent" and "hypotenuse".

The ratio is the cosine of angle D.

3) Use that ratio to set up an equation.

$$\cos D = \frac{DT}{DM} \quad \text{or} \quad \cos 51° = \frac{350}{DM}$$

Use a calculator to complete the solution. Round to the nearest foot.
DM = _____

16.2 cm, from:
$$c = (20)(\tan 39°)$$

556 ft, from:
$$DM = \frac{350}{\cos 51°}$$

90.   We want to find GF in the triangle
      at the right.   The known angle is G.

         GF is the side adjacent to angle G.

         FH is the side opposite angle G.

      a) Should we use sin 29°, cos 29°, or
          tan 29° to solve for GF? _____

      b) Complete the solution.   Round to the nearest
          tenth of a meter.

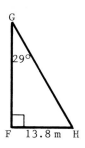

                        GF = _____

---

91.   We want to find "d" in the
      triangle at the right.   The
      known angle is B.

         "d" is the hypotenuse.

         "b" is the side opposite angle B.

      a) Should we use sin 24°, cos 24°, or
          tan 24° to solve for "d"? _____

      b) Complete the solution.   Round to the
          nearest tenth of an inch.

                        d = _____

a) tan 29°

b) 24.9 m, from:

    $\tan 29° = \dfrac{13.8}{GF}$

---

92.   We want to find "f" in the
      triangle at the right.   The
      known angle is S.

         "f" is the side adjacent to angle S.

         "v" is the hypotenuse.

      a) Should we use sin 63°, cos 63° or
          tan 63° to solve for "f"? _____

      b) Complete the solution.   Round to the
          nearest centimeter.

                        f = _____

a) sin 24°

b) 19.1 in, from:

    $\sin 24° = \dfrac{7.75}{d}$

---

93.   Let's solve for "p" in this triangle.

      a) Should we use sin 65°, cos 65°, or
          tan 65°? _____

      b) Complete the solution.   Round to
          the nearest foot.

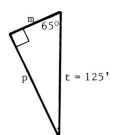

                        p = _____

a) cos 63°

b) 113 cm, from:

    $\cos 63° = \dfrac{f}{250}$

94. Let's solve for TV in this triangle.

   a) Should we use sin 70°, cos 70°, or tan 70°? _____

   b) Complete the solution. Round to the nearest tenth of a meter.

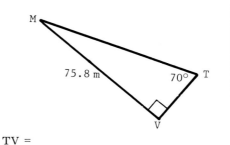

TV = _____

a) sin 65°

b) 113 ft, from:

$$\sin 65° = \frac{p}{125}$$

---

95. Let's solve for AD in this triangle.

   a) Should we use sin 25°, cos 25°, or tan 25°? _____

   b) Complete the solution. Round to the nearest mile.

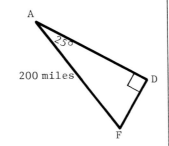

AD = _____

a) tan 70°

b) 27.6 m, from:

$$\tan 70° = \frac{75.8}{TV}$$

---

96. If <u>only</u> <u>one side</u> and <u>an acute angle of a right triangle are known</u>, we must use one of the trig ratios to find an unknown side.

   If <u>only two sides of a right triangle are known</u>, we should use the Pythagorean Theorem to find the third side.

   In which triangles below would we use the Pythagorean Theorem in order to find MP? _____

   a) 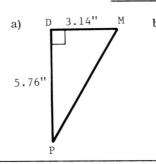   b)   c)

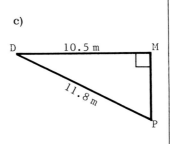

a) cos 25°

b) 181 miles, from:

$$\cos 25° = \frac{AD}{200}$$

---

97. In which triangles below would we have to use a trigonometric ratio to find CD? _____

   a)   b)   c)

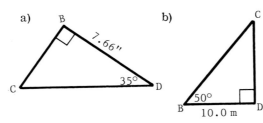

In (a) and (c)

In (a) and (b)

98. If we know <u>two</u> <u>sides</u> and <u>one</u> <u>acute</u> <u>angle</u> of a <u>right</u> <u>triangle</u>, we can use either of two trig ratios or the Pythagorean Theorem to find the third side. Here is an example. Round each answer to the nearest tenth of an inch.

a) Use cos 33° to find "t".

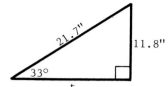

t = _____

b) Use tan 33° to find "t".

t = _____

c) Use the Pythagorean Theorem to find "t".

t = _____

a) t = 18.2 in, from:	b) t = 18.2 in, from:	c) t = 18.2 in, from:
$\cos 33° = \dfrac{t}{21.7}$	$\tan 33° = \dfrac{11.8}{t}$	$t = \sqrt{(21.7)^2 - (11.8)^2}$

11-10 FINDING UNKNOWN ANGLES IN RIGHT TRIANGLES

Either the angle-sum principle or the trig ratios are used to find unknown angles in right triangles. We will discuss the use of both methods in this section.

99. If one acute angle in a right triangle is known, we can subtract it from 90° to find the other acute angle. Find angle A in each right triangle below.

a)

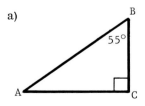

A = _____

b)

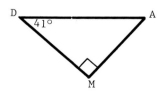

A = _____

a) 35°

b) 49°

100. When neither acute angle in a right triangle is known, we can use a trig ratio to find an acute angle <u>if two sides of the triangle are known</u>. An example is shown.

To find angle A at the right, we can use the <u>sine</u> ratio.

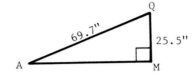

$$\sin A = \frac{MQ}{AQ}$$

$$\sin A = \frac{25.5}{69.7}$$

Use a calculator to complete the solution. Be sure to press $\boxed{=}$ to complete the division before pressing $\boxed{2nd}$ $\boxed{\sin}$ or $\boxed{INV}$ $\boxed{\sin}$ or $\boxed{\sin^{-1}}$. Round to the nearest degree.   A = _____

---

101. To find angle P at the right, we can use the <u>cosine</u> ratio.

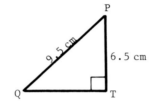

$$\cos P = \frac{PT}{PQ}$$

$$\cos P = \frac{6.5}{9.5}$$

Use a calculator to complete the solution. Press $\boxed{2nd}$ $\boxed{\cos}$ or $\boxed{INV}$ $\boxed{\cos}$ or $\boxed{\cos^{-1}}$ after dividing. Round to the nearest degree.   P = _____

21°

---

102. To decide which trig ratio to use to find an unknown angle in a right triangle, we identify the known sides in terms of the desired angle. An example is shown.

To find angle M at the right we identify PV and MP in terms of angle M.

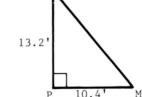

PV is the <u>side opposite</u> angle M.

MP is the <u>side adjacent</u> to angle M.

Therefore, we can use tan M to find angle M.

$$\tan M = \frac{PV}{MP}$$

$$\tan M = \frac{13.2}{10.4}$$

Use $\boxed{2nd}$ $\boxed{\tan}$ or $\boxed{INV}$ $\boxed{\tan}$ or $\boxed{\tan^{-1}}$ to complete the solution. Round to the nearest degree.   M = _____

47°

52°

103. To find angle T at the right, we identify TV and DT in terms of angle T.

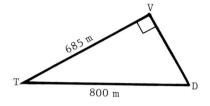

TV is the <u>side adjacent</u> to angle T.

DT is the <u>hypotenuse</u>.

a) Should we use sin T, cos T, or tan T to find angle T? _____

b) To the nearest degree, angle T = _____

---

104. We want to find angle M in this right triangle.

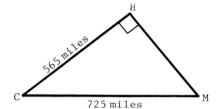

a) Should we use sin M, cos M, or tan M? _____

b) To the nearest degree, angle M = _____

a) cos T

b) 31°, from:

$$\cos T = \frac{685}{800}$$

---

105. We want to find angle C in this right triangle.

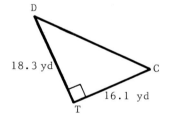

a) Should we use sin C, cos C, or tan C? _____

b) To the nearest degree, angle C = _____

a) sin M

b) 51°, from:

$$\sin M = \frac{565}{725}$$

---

106. After finding one acute angle in a right triangle, we can find the other acute angle by subtracting from 90°.

In right triangle PDM:

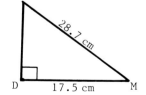

a) To the nearest degree, angle P = _____

b) To the nearest degree, angle M = _____

a) tan C

b) 49°, from:

$$\tan C = \frac{18.3}{16.1}$$

---

a) 38°, from:

$$\sin P = \frac{17.5}{28.7}$$

b) 52°, from:
90° – 38°

107.   In right triangle FGR:

a) Angle F = _____

b) To the nearest tenth of a foot,
    FG = _____

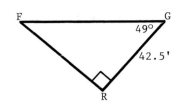

108.   In right triangle ABC:

a) Angle A = _____

b) To the nearest tenth of a centimeter,
    AB = _____

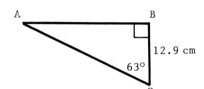

a) 41°

b) 64.8 ft

a) 27°   b) 25.3 cm

11-11 APPLIED PROBLEMS

In this section, we will discuss some applied problems that involve solving a right triangle.

109.   To measure the height of a tall building, a surveyor set his transit
       (angle-measuring instrument) 100 feet horizontally from the base of
       the building, measured the angle of elevation to the top of the building,
       and found it to be 72°.

       Find "h", the height of the building.
       Round to the nearest foot.

       h = _____

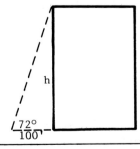

110.   A right triangle called an "impedance"
       triangle is used in analyzing alternating
       current circuits. Angle A, shown in the
       diagram, is the "phase angle" of the circuit.

       In a particular circuit, X = 2,630 and Z = 6,820.
       Find angle A to the nearest degree.

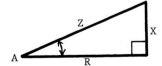

       A = _____

h = 308 ft, from:
$$\tan 72° = \frac{h}{100}$$

111. The end-view of the roof of a building is an isosceles triangle, as shown at the right.

Find angle A, the angle of slope of the roof to the nearest degree.

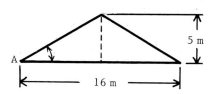

A = _____

A = 23°, from:

$$\sin A = \frac{2,630}{6,820}$$

---

112. In this metal bracket, find D, the distance between the two holes. Round to the nearest hundredth of an inch.

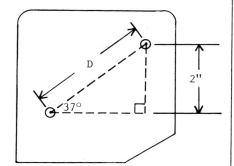

D = _____

A = 32°, from:

$$\tan A = \frac{5}{8}$$

---

113. The metal shape at the right is called a "template". Figure ABCD is a rectangle. We want to find side BE to the nearest tenth of a centimeter. (Note: AE can be found by subtracting BC from DE.)

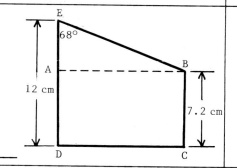

BE = _____

D = 3.32 in, from:

$$\sin 37° = \frac{2}{D}$$

---

114. Here is a cross-sectional view of a metal shaft with a tapered end.

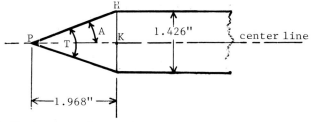

Find T, the taper angle, to the nearest degree.   T = _____

Note: Angle A is half of angle T. Angle A is an acute angle in right triangle PHK.

T = _____

BE = 12.8 cm, from:

$$\cos 68° = \frac{4.8}{BE}$$

T = 40°, since A = 20°, from: $\tan A = \dfrac{0.713}{1.968}$

SELF–TEST 39   (pages 491–500)

1. Find sides "h" and "v". Round to the nearest tenth of a centimeter.

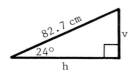

h = _____
v = _____

2. Find angles F and H. Round to the nearest degree.

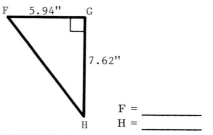

F = _____
H = _____

3. Find angle T. Round to the nearest degree.

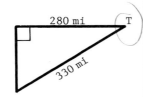

T = _____

4. Find side GP. Round to hundredths.

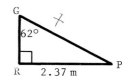

GP = _____

5. Find side "w". Round to tenths.

w = _____

6. Find angle A. Round to the nearest degree.

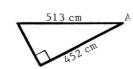

A = _____

7. Find "a" in isosceles triangle DEF. Round to hundredths.

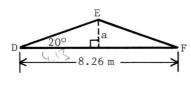

a = _____

8. Find angle P in parallelogram PQRS. Round to the nearest degree.

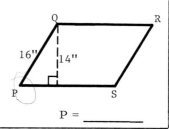

P = _____

ANSWERS:
1. h = 75.6 cm
   v = 33.6 cm
2. F = 52°
   H = 38°
3. T = 32°
4. GP = 2.68 m
5. w = 42.7 ft
6. A = 28°
7. a = 1.50 m
8. P = 61°

SUPPLEMENTARY PROBLEMS - CHAPTER 11

Assignment 37

Use the angle-sum principle to find each unknown angle.

1.
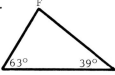

2.

3.
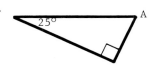

Use the Pythagorean Theorem to find each unknown dimension. Round as directed.

4. Round to hundreds.
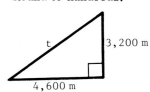

5. Round to tenths.
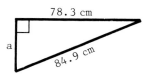

6. Round to hundredths.
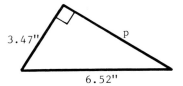

7. Round to tenths.
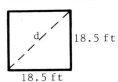

8. Round to the nearest whole number.
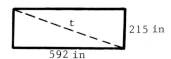

9. Round to thousandths.
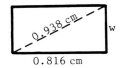

Each triangle is an isosceles triangle. Find each unknown angle.

10.
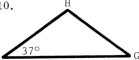

11.
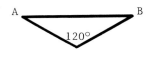

12.
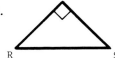

Find the area of each figure. Round as directed.

13. Round to hundreds.
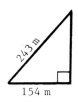

14. Round to the nearest whole number.

15. Round to tenths.
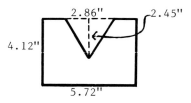

Assignment 38

Find the numerical value of each trig ratio. Round to four decimal places.

1. $\tan 12°$	2. $\sin 87°$	3. $\cos 71°$	4. $\cos 0°$
5. $\sin 30°$	6. $\cos 30°$	7. $\tan 65°$	8. $\sin 1°$

Find each angle. Round to the nearest degree.

9. cos A = 0.9715   10. tan P = 3.2168   11. sin H = 0.9999   12. cos Q = 0.3014

13. tan T = 0.1172   14. cos F = 0.1382   15. tan A = 1.0000   16. sin B = 0.2105

Evaluate each of the following. Round to tenths.

17. (31.8)(cos 24°)   18. $\dfrac{136}{\tan 85°}$ 19. $\dfrac{51.9}{\sin 42°}$ 20. (71.3)(tan 30°)

Evaluate each of the following. Round to the nearest whole number.

21. $\dfrac{429}{\tan 51°}$ 22. (1,136)(sin 14°)   23. (250)(cos 64°)   24. $\dfrac{98.4}{\cos 79°}$

Find each angle. Round to the nearest degree.

25. cos A = $\dfrac{38.4}{41.7}$ 26. sin G = $\dfrac{7,850}{9,160}$ 27. tan R = $\dfrac{5.62}{1.93}$ 28. sin E = $\dfrac{0.138}{0.897}$

Using sides "r", "s", and "t" in right triangle RST, define the following trig ratios.

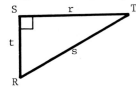

29. sin R   30. cos R   31. tan R

32. sin T   33. cos T   34. tan T

Using sides "a", "b", and "c" in right triangle ABC, define the following trig ratios.

35. cos B   36. sin A   37. tan B

38. tan A   39. cos A   40. sin B

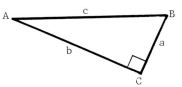

Assignment 39

In problems 1-3, round each answer to hundredths.

1. Find "h".   2. Find "w".   3. Find "p".

   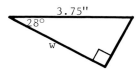   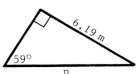

In problems 4-6, round each answer to the nearest whole number.

4. Find "d".   5. Find "t".   6. Find "b".

      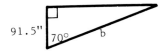

In problems 7-9, round each angle to the nearest degree.

7. Find angles F and G.

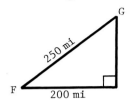

8. Find angles A and B.

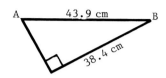

9. Find angles R and S.

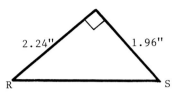

10. Find "h" and "v" in the diagram below. Round to hundredths.

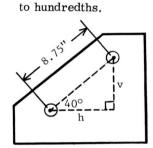

11. The roof diagram below is an isosceles triangle. Find angle A. Round to the nearest degree.

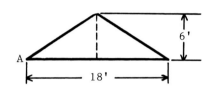

12. In the figure below, the bottom is rectangular. Find "w". Round to tenths.

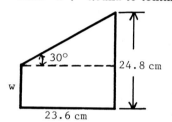

12 Oblique Triangles

In this chapter, we will show how the Law of Sines and the Law of Cosines can be used to find unknown sides and angles in oblique triangles. The strategies used to solve oblique triangles are emphasized. Problems involving obtuse angles are delayed until all types of problems involving acute angles are examined.

12-1 OBLIQUE TRIANGLES

In this section, we will define oblique triangles. We will show that the angle-sum principle applies to oblique triangles, but that the Pythagorean Theorem and the three trigonometric ratios <u>do</u> <u>not</u> <u>apply</u> to oblique triangles.

1. Any triangle which does not contain a right angle (90°) is called an <u>oblique</u> triangle. (The word "oblique" is pronounced "ō-bleek".) Both triangles below are oblique triangles.	a) obtuse b) acute

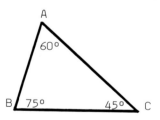 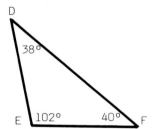

In triangle ABC, <u>all three angles</u> are acute angles. Triangles of that type are called "<u>ACUTE</u> oblique" triangles.

Continued on following page.

ɪ.   Continued

In triangle DEF, one angle is an obtuse angle. Triangles of that type are called "OBTUSE oblique" triangles.

   a) If a triangle contains angles of 20°, 60°, and 100°, it is an
       _____ (acute/obtuse) oblique triangle.

   b) If a triangle contains angles of 44°, 58°, and 78°, it is an
       _____ (acute/obtuse) oblique triangle.

---

2   The angle-sum principle applies to oblique triangles. That is, the sum of the three angles is 180°.

Find the unknown angle in each oblique triangle.

a)

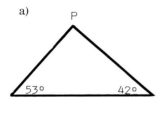

P = _____

b)

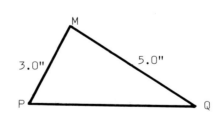

Q = _____

a) obtuse

b) acute

---

3   The Pythagorean Theorem applies only to right triangles. It does not apply to oblique triangles.

Triangle DEF is a right triangle. Triangle MPQ is an oblique triangle.

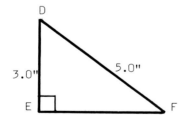

   a) Can we use the Pythagorean Theorem to find the length of EF
       in triangle DEF? _____

   b) Can we use the Pythagorean Theorem to find the length of PQ
       in triangle MPQ? _____

a) P = 85°

b) Q = 25°

a) Yes

b) No

4.   The three basic trigonometric ratios (sine, cosine, and tangent) are comparisons of the sides of <u>right</u> triangles. They <u>are</u> <u>not</u> comparisons of the sides of <u>oblique</u> triangles.

Triangle ABC is a <u>right</u> triangle. Triangle FGH is an <u>oblique</u> triangle.

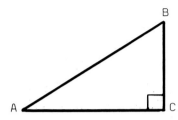
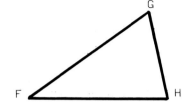

In triangle ABC, sin A is a comparison of two sides of the triangle. That is:

$$\sin A = \frac{BC}{AB}$$

In triangle FGH, is sin F a comparison of two sides of the triangle?

_____

---

5.   Triangle CDE is a <u>right</u> triangle. Triangle PQR is an <u>oblique</u> triangle.

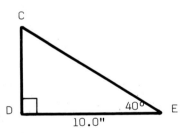
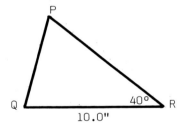

In triangle CDE, we can use the tangent ratio to find the length of CD. We get:

$$\tan E = \frac{CD}{DE}$$

$$\tan 40° = \frac{CD}{10.0''} \quad \text{and} \quad CD = 8.39''$$

Can we use the tangent ratio to find the length of PQ in triangle PQR?

_____

No

No

12-2   RIGHT-TRIANGLE METHODS AND OBLIQUE TRIANGLES

"Solving triangles" means finding the lengths of unknown sides and the sizes of unknown angles. If an oblique triangle is divided into two right triangles, right-triangle methods can be used for the solution. We will discuss that method in this section.

---

6. If a line drawn from the vertex of an angle of a triangle meets the opposite side at right angles, the line is called an "altitude" of the triangle. In the triangles below, DF and ST are altitudes.

Any oblique triangle can be divided into two right triangles by drawing an altitude from one angle to the opposite side.

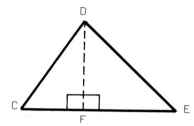 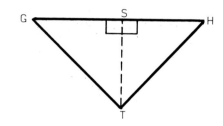

The altitude DF divides oblique triangle CDE into the two right triangles CDF and DEF.

The altitude ST divides oblique triangle GHT into the two right triangles _____ and _____ .

---

7. When an altitude divides an oblique triangle into two right triangles, the Pythagorean Theorem applies to each right triangle.

GST and HST

Altitude AD divides triangle ABC into the two right triangles ABD and ACD. By using the Pythagorean Theorem twice to find the lengths of BD and CD, we can then add to find the length of BC.

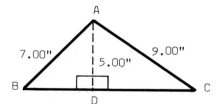

Rounding to hundredths:

a) The length of BD is _____ inches.

b) The length of CD is _____ inches.

c) The length of BC is _____ inches.

---

a) 4.90

b) 7.48

c) 12.38

8.   When an altitude divides an oblique triangle into two right triangles, the three basic trigonometric ratios apply to each right triangle.

Altitude RT divides triangle PQR into right triangles PRT and QRT. The length of RT is given.

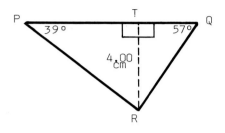

a) In **right** triangle PRT, we can use sin P to find the length of PR. Round to hundredths.

Since $\sin P = \dfrac{RT}{PR}$, $\sin 39° = \dfrac{4.00}{PR}$ and PR = _____ .

b) In right triangle QRT, we can use sin Q to find the length of QR. Round to hundredths.

Since $\sin Q = \dfrac{RT}{QR}$, $\sin 57° = \dfrac{4.00}{QR}$ and QR = _____ .

---

9.   Altitude DT divides oblique triangle MST into two right triangles. We can use sin M and sin S to find the sizes of angles M and S to the nearest degree.

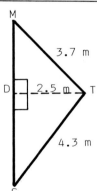

a) Since $\sin M = \dfrac{DT}{MT} = \dfrac{2.5}{3.7} = 0.6757$,

angle M = _____

b) Since $\sin S = \dfrac{DT}{ST} = \dfrac{2.5}{4.3} = 0.5814$,

angle S = _____

a) 6.36 cm

b) 4.77 cm

---

10.   When drawing an altitude to divide an oblique triangle into two right triangles, the length of the altitude is usually not known. Therefore, we usually need a two-step process to find an unknown side or angle. The first step is finding the length of the altitude. An example is given below.

In oblique triangle CDE, the length of altitude EF is not known. Therefore, we need a two-step process to find the length of DE.

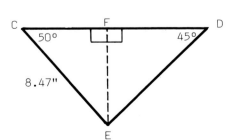

a) 43°

b) 36°

Continued on following page.

10.   Continued

      a) Find the length of EF. Round to hundredths.

                                                  EF = _____

      b) Find the length of DE. Round to hundredths.

                                                  DE = _____

11.   A two-step process is needed to find the size of angle T at the right.

      a) Find altitude MP.
          Round to tenths.

                      MP = _____

      b) Find angle T. Round to the nearest whole-number degree.

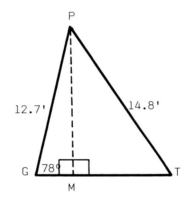

                  Angle T = _____

a) 6.49"

b) 9.18"

a) 12.4 feet   b) 57°

12-3   THE LAW OF SINES

Instead of solving oblique triangles by right-triangle methods, we use one of two direct methods: the Law of Sines and the Law of Cosines. We will discuss the Law of Sines in this section.

12.   The Law of Sines involves the ratios of the sides of a triangle to the sines of the angles opposite them. It says that the three ratios of that type for any triangle are equal.

For oblique triangle ABC, the Law of Sines says this:

$$\frac{\text{side opposite angle A}}{\text{sine of A}} = \frac{\text{side opposite angle B}}{\text{sine of B}} = \frac{\text{side opposite angle C}}{\text{sine of C}}$$

$$\text{or}$$

$$\frac{a}{\sin A} = \frac{b}{\sin B} = \frac{c}{\sin C}$$

Continued on following page.

12.   Continued

Since the Law of Sines says that all three ratios are equal, we can write the following three proportions:

$$\frac{a}{\sin A} = \frac{b}{\sin B} \qquad \frac{a}{\sin A} = \frac{c}{\sin C} \qquad \frac{b}{\sin B} = \frac{c}{\sin C}$$

13.   The numerical values of the three angles and three sides of triangle PQR are shown on the figure. The values reported for the sides involve some rounding. We will use this triangle to confirm the Law of Sines which says this:

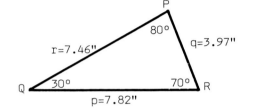

$$\frac{p}{\sin P} = \frac{q}{\sin Q} = \frac{r}{\sin R}$$

Use your calculator to evaluate each ratio. Round to hundredths.

a) $\dfrac{p}{\sin P} = \dfrac{7.82}{\sin 80°} =$ _____

b) $\dfrac{q}{\sin Q} = \dfrac{3.97}{\sin 30°} =$ _____

c) $\dfrac{r}{\sin R} = \dfrac{7.46}{\sin 70°} =$ _____

d) Are the three ratios equal? _____

14.   According to the Law of Sines for this triangle:

a) $\dfrac{c}{\sin C} = \dfrac{d}{\boxed{}}$

b) $\dfrac{c}{\sin C} = \dfrac{\boxed{}}{\sin E}$

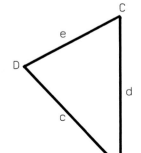

a) 7.94

b) 7.94

c) 7.94

d) Yes

a) $\dfrac{d}{\boxed{\sin D}}$

b) $\dfrac{\boxed{e}}{\sin E}$

15. According to the Law of Sines for this triangle:

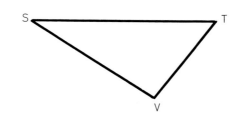

   a) $\dfrac{SV}{\sin T} = \dfrac{ST}{\boxed{}}$

   b) $\dfrac{SV}{\sin T} = \dfrac{\boxed{}}{\sin S}$

---

6. When writing the Law of Sines for the triangle at the right, we have written the ratios with the <u>sides</u> as the <u>numerators</u>. That is:

$$\boxed{\dfrac{a}{\sin A} = \dfrac{b}{\sin B} = \dfrac{c}{\sin C}}$$

However, we can also write the Law of Sines with the "sines" as the numerators. We get:

$$\boxed{\dfrac{\sin A}{a} = \dfrac{\sin B}{b} = \dfrac{\sin C}{c}}$$

Write the Law of Sines in two different ways for triangle DFM.

$\dfrac{}{} = \dfrac{}{} = \dfrac{}{}$

$\dfrac{}{} = \dfrac{}{} = \dfrac{}{}$

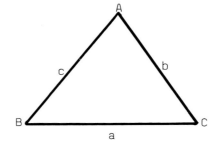

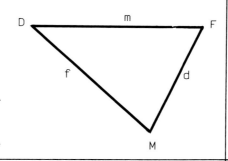

a) $\dfrac{ST}{\boxed{\sin V}}$

b) $\dfrac{\boxed{VT}}{\sin S}$

---

17. According to the Law of Sines for this triangle:

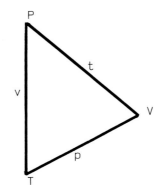

   a) $\dfrac{\sin P}{p} = \dfrac{\boxed{}}{t}$

   b) $\dfrac{\sin P}{p} = \dfrac{\sin V}{\boxed{}}$

$\dfrac{d}{\sin D} = \dfrac{f}{\sin F} = \dfrac{m}{\sin M}$

$\dfrac{\sin D}{d} = \dfrac{\sin F}{f} = \dfrac{\sin M}{m}$

a) $\dfrac{\boxed{\sin T}}{t}$ b) $\dfrac{\sin V}{\boxed{v}}$

18. When writing a proportion based on the Law of Sines, we must be consistent. That is, the numerators of both ratios must either be "sides" or "sines".

Using triangle BDF, complete these with one of the two possible ratios. Be consistent.

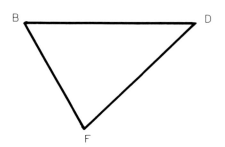

a) $\dfrac{BD}{\sin F}$ = _____

b) $\dfrac{\sin D}{BF}$ = _____

a) $\dfrac{BF}{\sin D}$ or $\dfrac{DF}{\sin B}$

b) $\dfrac{\sin B}{DF}$ or $\dfrac{\sin F}{BD}$

19. In this frame we will prove the Law of Sines. We will not expect you to memorize the proof or reproduce it. The proof is given simply to show that a proof is possible.

Using the triangle at the right, we will prove the following proportion.

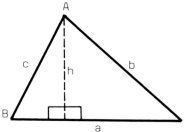

$$\frac{c}{\sin C} = \frac{b}{\sin B}$$

The steps are:

1. Draw an altitude from angle A to side BC.

2. Define sin B and sin C.

$$\sin B = \frac{h}{c} \qquad \sin C = \frac{h}{b}$$

3. Solve for "h" in each equation and equate the solutions.

$$h = c(\sin B) \quad \text{and} \quad h = b(\sin C)$$
$$c(\sin B) = b(\sin C)$$

4. Multiple both sides by $\dfrac{1}{\sin B}$ and $\dfrac{1}{\sin C}$.

$$\left(\frac{1}{\sin B}\right)\left(\frac{1}{\sin C}\right)c(\sin B) = \left(\frac{1}{\sin B}\right)\left(\frac{1}{\sin C}\right)b(\sin C)$$

$$\left(\frac{\sin B}{\sin B}\right)\left(\frac{1}{\sin C}\right)c = \left(\frac{\sin C}{\sin C}\right)\left(\frac{1}{\sin B}\right)b$$

$$\frac{c}{\sin C} = \frac{b}{\sin B}$$

By drawing altitudes from angles B and C, we can prove the following two proportions in a similar manner.

$$\frac{a}{\sin A} = \frac{b}{\sin B} \quad \text{and} \quad \frac{a}{\sin A} = \frac{c}{\sin C}$$

Continued on following page.

19.   Continued

We can also write the Law of Sines with the "sines" in the numerator. We get:

$$\frac{\sin A}{a} = \frac{\sin B}{b} = \frac{\sin C}{c}$$

12-4   USING THE LAW OF SINES TO FIND SIDES

In this section, we will use the Law of Sines to find unknown sides in oblique triangles. To simplify the solutions, proportions involving obtuse angles will be delayed until a later section.

20.   Let's use the Law of Sines to find the length of MP. According to the Law of Sines:

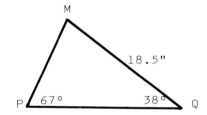

$$\frac{MP}{\sin Q} = \frac{MQ}{\sin P}$$

$$\frac{MP}{\sin 38°} = \frac{18.5"}{\sin 67°}$$

$$MP = \frac{(18.5")(\sin 38°)}{\sin 67°}$$

To complete the solution on a calculator, follow these steps.

Calculator Steps: 18.5 $\boxed{\text{x}}$ 38 $\boxed{\sin}$ $\boxed{÷}$ 67 $\boxed{\sin}$ $\boxed{=}$ 12.373359

Rounding to tenths,   MP = _____

21.   When using the Law of Sines <u>to find an unknown side</u> of a triangle, we set up a proportion like the one below. Notice that we put the unknown side <u>in the numerator on the left side</u>.

$$\frac{RS}{\sin 52°} = \frac{17.9"}{\sin 71°}$$

We can solve for RS by multiplying both sides by sin 52°. We get:

$$\sin 52° \left(\frac{RS}{\sin 52°}\right) = \left(\frac{17.9"}{\sin 71°}\right)\sin 52°$$

$$RS = \frac{(17.9")(\sin 52°)}{\sin 71°}$$

Having isolated RS, we can find RS in one process on a calculator by following the steps in the last frame. Do so. Round to tenths.

RS = _____

12.4"

22.   Let's use the same process to find MT below.

$$\frac{MT}{\sin 49°} = \frac{6.34 \text{ cm}}{\sin 56°}$$

a) Isolate MT by multiplying both sides by sin 49°.

MT = _____

b) Use a calculator to find MT. Round to hundredths.

MT = _____

14.9"

---

23.   The complete Law of Sines for triangle DFH is:

$$\frac{d}{\sin D} = \frac{f}{\sin F} = \frac{h}{\sin H}$$

or $\frac{d}{\sin 70°} = \frac{5.00m}{\sin 50°} = \frac{h}{\sin 60°}$

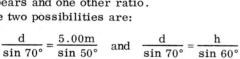

If we want to solve for "d" we must use the ratio in which "d" appears and one other ratio. The two possibilities are:

$$\frac{d}{\sin 70°} = \frac{5.00m}{\sin 50°} \quad \text{and} \quad \frac{d}{\sin 70°} = \frac{h}{\sin 60°}$$

A proportion can be solved only if it contains <u>only</u> <u>one</u> unknown.

The proportion on the left contains <u>one</u> unknown, "d".

The proportion on the right contains <u>two</u> unknowns, "d" and "h".

Which proportion (the one on the left or the one on the right) must we use to solve for "d"? _____

a) MT = $\frac{(6.34 \text{ cm})(\sin 49°)}{\sin 56°}$

b) MT = 5.77 cm

---

24.   To solve for "s" at the right, we must set up a proportion in which "s" is the only unknown. Let's do so.

a) Which ratio contains side "s"?

_____

b) Which other ratio should we use?

_____

c) The proportion is:

_____ = _____

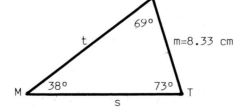

The one on the left.

25. Sometimes we have to use the angle-sum principle before using the Law of Sines.

    a) How large is angle R?

    _____

Let's set up the proportion needed to find CD at the right.

    b) The proportion is:

    _____ = _____

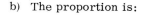

a) $\dfrac{s}{\sin S}$ or $\dfrac{s}{\sin 69°}$

b) $\dfrac{m}{\sin M}$ or $\dfrac{8.33 \text{ cm}}{\sin 38°}$

(since we know both "m" and "sin M")

c) $\dfrac{s}{\sin 69°} = \dfrac{8.33 \text{ cm}}{\sin 38°}$

---

26. Let's set up the proportion needed to solve for FV at the right.

    a) How large is angle F? _____

    b) The proportion is:

    _____ = _____

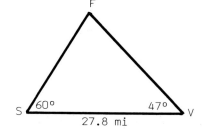

a) 81°

b) $\dfrac{CD}{\sin 81°} = \dfrac{16.5'}{\sin 51°}$

---

27. a) Set up the proportion needed to solve for BC.

    _____ = _____

    b) Set up the proportion needed to solve for CF.

    _____ = _____

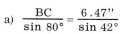

a) 73°

b) $\dfrac{FV}{\sin 60°} = \dfrac{27.8 \text{ mi}}{\sin 73°}$

---

28. After using the Law of Sines to calculate an angle or a side, you can check the sensibleness of your answer by applying a relationship involving the sizes of the sides and angles of a triangle. The relationship is:

a) $\dfrac{BC}{\sin 80°} = \dfrac{6.47''}{\sin 42°}$ b) $\dfrac{CF}{\sin 58°} = \dfrac{6.47''}{\sin 42°}$

The <u>longest</u> side is opposite the <u>largest</u> angle.

The <u>shortest</u> side is opposite the <u>smallest</u> angle.

We can verify this relationship in triangle FGH.

    a) The largest angle is _____.

    b) The longest side is _____.

    c) The smallest angle is _____.   d) The shortest side is _____.

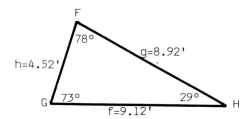

a) F = 78°   b) f = 9.12'   c) H = 29°   d) h = 4.52'

29.　In triangle ABC, which side would be larger?

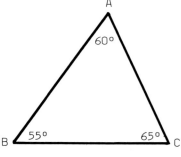

　　a)　AB or AC _____

　　b)　AB or BC _____

　　c)　AC or BC _____

30.　Let's solve for side FR in triangle FRT.

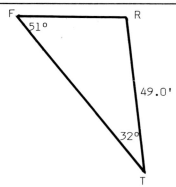

　　a)　Is FR longer or shorter than 49.0? _____

　　b)　Set up the proportion needed to solve for FR.

　　　　_____ = _____

　　c)　Complete the solution. Round to tenths. FR = _____

a)　AB

b)　AB

c)　BC

31.　Let's solve for "d" in triangle CDM.

　　a)　How large is angle D? _____

　　b)　Is "d" longer or shorter than 25.9"? _____

　　c)　Set up the proportion needed to solve for "d".

　　　　_____ = _____

　　d)　Complete the solution. Round to tenths. d = _____

a)　Shorter, because angle T is smaller than angle F.

b)　$\dfrac{FR}{\sin 32°} = \dfrac{49.0'}{\sin 51°}$

c)　FR = 33.4'

---

a)　D = 47°　　b)　a little longer　　c)　$\dfrac{d}{\sin 47°} = \dfrac{25.9 \text{ cm}}{\sin 45°}$　　d)　d = 26.8 cm

12-5   USING THE LAW OF SINES TO FIND ANGLES

In this section, we will use the Law of Sines to find unknown angles in oblique triangles. To simplify the solutions, proportions involving obtuse angles will be delayed until a later section.

---

32.   Let's use the Law of Sines to find the size of angle T. According to the Law of Sines:

$$\frac{\sin T}{11.2'} = \frac{\sin 64°}{18.5'}$$

$$\sin T = \frac{(11.2')(\sin 64°)}{18.5'}$$

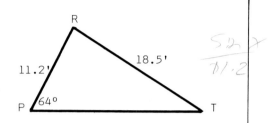

To complete the solution on a calculator, use these steps:

Calculator Steps: 11.2 $\boxed{\text{x}}$ 64 $\boxed{\text{sin}}$ $\boxed{÷}$ 18.5 $\boxed{=}$ $\boxed{\text{2nd}}$ $\boxed{\text{sin}}$ 32.965557

or $\boxed{\text{INV}}$ $\boxed{\text{sin}}$

or $\boxed{\text{sin}^{-1}}$

Rounded to the nearest whole-number degree,   T = _____

---

33.   When using the Law of Sines <u>to find an unknown</u> angle of a triangle, we set up a proportion <u>with the</u> "sines" <u>in the numerator and the unknown sine on the left side</u>.

Which proportion below would we use to solve for angle B in this triangle? _____

a) $\dfrac{11.0''}{\sin B} = \dfrac{7.0''}{\sin 38°}$

b) $\dfrac{\sin B}{11.0''} = \dfrac{\sin 38°}{7.0''}$

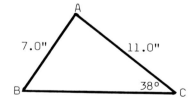

33°

b

34. Set up the proportion needed to solve for angle D in each triangle below.

a)

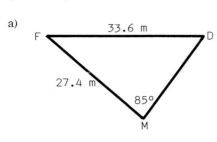

b)

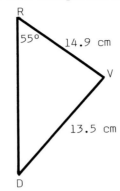

_____ = _____   _____ = _____

---

35. When solving for an angle in a triangle, remember these relations:

    1. The <u>largest</u> angle is opposite the <u>longest</u> side.
    2. The <u>smallest</u> angle is opposite the <u>shortest</u> side.

In triangle FHT:

    a) The largest angle is _____ .

    b) The smallest angle is _____ .

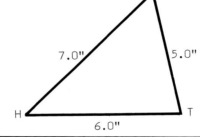

a) $\dfrac{\sin D}{27.4m} = \dfrac{\sin 85°}{33.6m}$

b) $\dfrac{\sin D}{14.9 \text{ cm}} = \dfrac{\sin 55°}{13.5 \text{ cm}}$

---

36. In triangle BKV, angle B contains 52°.

    a) Is angle K larger or smaller than 52° ? _____

    b) Is angle V larger or smaller than 52°? _____

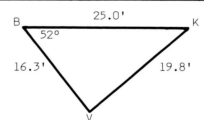

a) T

b) H

---

37. Let's use the Law of Sines to find angle B.

    a) Is angle B larger or smaller than 47°? _____

    b) Set up the proportion needed to solve for angle B.

_____ = _____

    c) Complete the solution. Round to the nearest whole-number degree.

B = _____

a) smaller, because BV is shorter than KV.

b) larger, because BK is longer than KV.

38. Let's solve for angle F in this triangle.

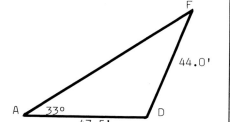

a) Is angle F larger or smaller than 33°? _____

b) Set up the proportion needed to solve for angle F.

_____ = _____

c) Complete the solution. Round to the nearest whole-number degree. F = _____

a) larger, since MV is longer than BV

b) $\dfrac{\sin B}{14.0 \text{ km}} = \dfrac{\sin 47°}{12.5 \text{ km}}$

c) 55°

a) larger, since AD is longer than DF

b) $\dfrac{\sin F}{47.5'} = \dfrac{\sin 33°}{44.0'}$

c) 36°

SELF-TEST 40   (pages 504-520)

1. Find the unknown angle in each triangle.

a)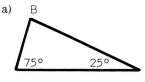

B = _____

b)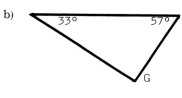

G = _____

c)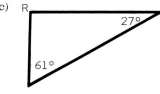

R = _____

2. Identify each triangle in Problem 1 as "right", "acute oblique", or "obtuse oblique".

a) _____ b) _____ c) _____

3. For triangle PRT at the right, use the Law of Sines to complete these proportions:

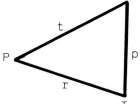

a)

b)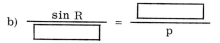

SELF-TEST 4 Continued on following page

SELF-TEST 40 (pages 504-520) - Continued

4. a) Using the Law of Sines, set up the proportion for finding side DE.

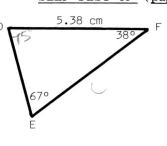

b) Find side DE. Round to hundredths.

DE = _____

5. a) Using the Law of Sines, set up the proportion for finding angle N.

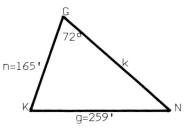

b) Find angle N. Round to a whole number.

N = _____

6. Refer to triangle DEF in Problem 4.

a) Which angle (D, E, or F) is largest? _____

b) Which side (DE, DF, or EF) is smallest?

_____

c) Which side is larger, DF or EF? _____

7. Refer to triangle GKN in Problem 5.

a) What is the size of angle K? K = _____

b) Which angle is smaller, G or K? _____

c) Which side (g, k, or n is largest?

_____

ANSWERS:

1. a) B = 80°
   b) G = 90°
   c) R = 92°

2. a) acute oblique
   b) right
   c) obtuse oblique

3. a) $\dfrac{t}{\sin T} = \dfrac{p}{\sin P}$

   b) $\dfrac{\sin R}{r} = \dfrac{\sin P}{p}$

4. a) $\dfrac{DE}{\sin 38°} = \dfrac{5.38 \text{ cm}}{\sin 67°}$

   b) DE = 3.60 cm

5. a) $\dfrac{\sin N}{165'} = \dfrac{\sin 72°}{259'}$

   b) N = 37°

6. a) angle D
   b) side DE
   c) side EF

7. a) K = 71°
   b) angle K
   c) side g

12-6   LIMITATIONS OF THE LAW OF SINES

In this section, we will show some more complex solutions involving the Law of Sines. Then we will discuss some cases in which the Law of Sines cannot be used.

39. To use the Law of Sines to find a side or angle, we must be able to set up a proportion in which that side or the sine of that angle is the only unknown. We have seen two types of solutions.

  1. Those in which the proportion can be set up directly.

   To solve for angle D in this triangle, we can directly set up a proportion in which sin D is the only unknown. We get:

   $$\frac{\sin D}{10.2m} = \frac{\sin 55°}{8.5m}$$

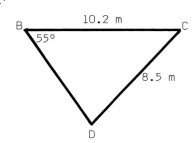

Continued on following page.

39.   Continued

2. Those in which the proportion can be set up only after the angle-sum principle is used.

To solve for SV in this triangle, we must use the angle-sum principle to find angle V. Then we can set up a proportion in which SV is the only unknown. We get:

$$\frac{SV}{\sin 38°} = \frac{12.0''}{\sin 72°}$$

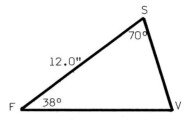

40.   Some solutions involving the Law of Sines are more complex than those in the last frame. Examples are given in this frame and the next frame.

To solve for angle P in this triangle, we cannot use the Law of Sines directly because side "p" is unknown. We would get either of these two proportions:

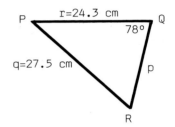

$$\frac{\sin P}{p} = \frac{\sin 78°}{27.5 \text{ cm}}$$

or $\dfrac{\sin P}{p} = \dfrac{\sin R}{24.3 \text{ cm}}$

However, we can use the Law of Sines to find angle P in two steps. That is:

1. Use the Law of Sines to find angle R.
2. Then use the angle-sum principle to find angle P.

a) Set up the proportion needed to find angle R.

_____ = _____

b) Angle R contains _____°. (Round to a whole-number.)

c) Angle P contains _____°.

41.   To find HS in triangle HST, we cannot use the Law of Sines directly because angle T is unknown. However, we can find HS in three steps. They are:

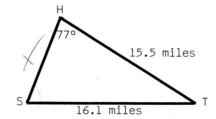

1. a) $\dfrac{\sin R}{24.3 \text{ cm}} = \dfrac{\sin 78°}{27.5 \text{ cm}}$

    b) 60°

    c) 42°

a) Use the Law of Sines to find angle S.

Angle S = _____ (Round to a whole number.)

Continued on following page.

41. Continued

    b) Use the angle-sum principle to find angle T.

Angle T = _____

    c) Now use the Law of Sines to find HS. Round to tenths.

HS = _____ miles

---

42. The Law of Sines cannot be used for all solutions of oblique triangles. Examples are given in this frame and the next frame.

In triangle ABD, we are given two sides (AB and AD) and the angle between them called the "included" angle. The three ratios are:

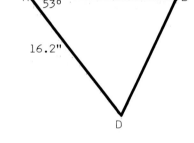

$$\frac{AB}{\sin D} = \frac{AD}{\sin B} = \frac{BD}{\sin A}$$

or

$$\frac{12.5''}{\sin D} = \frac{16.2''}{\sin B} = \frac{BD}{\sin 53°}$$

Can we set up a proportion to solve for side BD or angles B and D? _____

a) 70°

b) 33°

c) 9.0 miles

---

43. In triangle DFK, we are given the lengths of the three sides. Here are the three ratios:

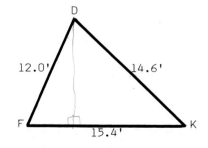

$$\frac{DK}{\sin F} = \frac{DF}{\sin K} = \frac{FK}{\sin D}$$

or

$$\frac{14.6'}{\sin F} = \frac{12.0'}{\sin K} = \frac{15.4'}{\sin D}$$

Can we set up a proportion to solve for any of the three angles? _____

No, because any proportion has two unknowns.

---

44. There are two cases in which we cannot use the Law of Sines for solutions. They are:

    1. When two sides and their included angle are given.
    2. When three sides are given but no angles.

In those cases, we must use the Law of Cosines to solve the triangles. We will introduce the Law of Cosines in the next section.

No, because any proportion has two unknowns.

12-7   THE LAW OF COSINES

When the Law of Sines cannot be used to solve a triangle, we use the Law of Cosines. We will discuss and prove the Law of Cosines in this section.

---

45.   In triangle DFP, we are given two sides ("d" and "f") and their included angle (angle P). We want to find side "p". As we saw in the last section, the Law of Sines cannot be used to find "p" in this triangle.

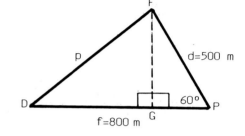

We can find "p" by the right-triangle method. To do so, we draw altitude FG to divide triangle DFP into two right triangles. The steps for finding "p" are:

1. Finding FG. $\sin 60° = \dfrac{FG}{d}$

    FG = (500m)(sin 60°) = 433m

2. Finding GP. $\cos 60° = \dfrac{GP}{d}$

    GP = (500m)(cos 60°) = 250m

3. Finding DG.   DG = DP − GP

    = 800m − 250m

    = 550m

4. Finding "p".

    In right triangle DFG, we know the lengths of both legs (DG = 550m and FG = 433m). Therefore, we can use the Pythagorean Theorem to find the hypotenuse "p".

    $p^2 = (DG)^2 + (FG)^2$

    $p^2 = (550)^2 + (433)^2$

    $p = \sqrt{(550)^2 + (433)^2} = \sqrt{489,989}$

    p = 700m   (This is the solution.)

---

46.   In the last frame, we solved for "p" in the triangle DFP by the right-triangle method. We got: p = 700m .

There is a relationship called the "Law of Cosines" which can be used to solve for "p" in one step. The Law-of-Cosines formula for "p" is:

$$\boxed{p^2 = d^2 + f^2 - 2df \cos P}$$

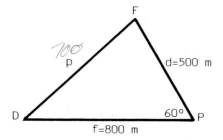

Continued on following page.

46.   Continued

Notice these points:

1. "p" is the unknown side. It is <u>opposite</u> the known angle P.

2. "d" and "f" are the <u>two</u> <u>known</u> <u>sides</u> forming angle P.

3. "2df cos P" is a product of four factors. It can be written (2)(d)(f)(cos P). The factors include the <u>two</u> <u>known</u> <u>sides</u> <u>and</u> <u>the</u> <u>known</u> <u>angle</u>.

Using d = 500, f = 800, and cos P = cos 60° = 0.5, we can substitute in the formula and solve for "p". We get:

$p^2 = (500)^2 + (800)^2 - 2(500)(800)(0.5)$

$p^2 = 250,000 + 640,000 - 400,000$

$p^2 = 490,000$

$p = \sqrt{490,000} = 700m$

Is 700m the same solution we got for "p" in the last frame? _____

47.   For any triangle, we can state three formulas based on the Law of Cosines. The three formulas for triangle BHT are discussed below.

Yes

One formula based on the Law of Cosines is:

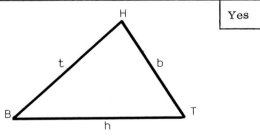

$$t^2 = b^2 + h^2 - 2bh \cos T$$

Note: 1. "t" is the side opposite angle T, and "cos T" is in the formula.
2. "b" and "h" are the other two sides.

A second formula based on the Law of Cosines is:

$$b^2 = h^2 + t^2 - 2ht \cos B$$

Note: 1. "b" is the side opposite angle B, and "cos B" is in the formula.
2. "h" and "t" are the other two sides.

Complete the third formula based on the Law of Cosines:

$h^2 =$ _____

48.   One Law-of-Cosines formula for triangle ACQ is given below. Complete the other two formulas for the triangle.

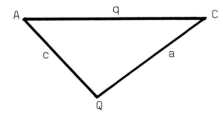

$h^2 = b^2 + t^2 - 2bt \cos H$

$c^2 = a^2 + q^2 - 2aq \cos C$

$a^2 =$ _____

$q^2 =$ _____

a) $a^2 = c^2 + q^2 - 2cq \cos A$ b) $q^2 = a^2 + c^2 - 2ac \cos Q$

49. One Law-of-Cosines formula for triangle MRS is given below. Complete the other two formulas for the triangle.

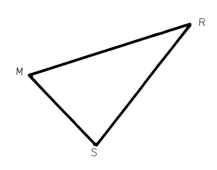

$(MS)^2 = (MR)^2 + (RS)^2 - 2(MR)(RS)(\cos R)$

$(MR)^2 = $ _____

$(RS)^2 = $ _____

50. In this frame we will prove the Law of Cosines. We will not expect you to memorize the proof or reproduce it. The proof is given simply to show that a proof is possible.

a) $(MR)^2 = (MS)^2 + (RS)^2 - 2(MS)(RS)\cos S$

b) $(RS)^2 = (MR)^2 + (MS)^2 - 2(MR)(MS)\cos M$

To understand the proof, you must know the following:

1. The symbol for squaring "sin A" is $\sin^2 A$.
   The symbol for squaring "cos A" is $\cos^2 A$.

   Therefore: $\sin^2 A = (\sin A)(\sin A)$
                $\cos^2 A = (\cos A)(\cos A)$

2. For any angle A, $\underline{\sin^2 A + \cos^2 A = 1}$

   For example, to see that $\sin^2 55° + \cos^2 55° = 1$, follow these steps on your calculator.

   Calculator Steps: 55 $\boxed{\sin}$ $\boxed{x^2}$ $\boxed{+}$ 55 $\boxed{\cos}$ $\boxed{x^2}$ $\boxed{=}$ 1

3. Just as: $(ab)^2 = (ab)(ab) = a^2 b^2$,

   $(a \cos C)^2 = (a \cos C)(a \cos C) = a^2 \cos^2 C$

4. Just as: $(a - b)^2 = (a - b)(a - b) = a^2 - 2ab + b^2$,

   $(b - a \cos C)^2 = (b - a \cos C)(b - a \cos C)$
   $= b^2 - 2ab \cos C + a^2 \cos^2 C$

We will use triangle ABC to prove the Law of Cosines. We will prove the following formula:

$$\boxed{c^2 = a^2 + b^2 - 2ab \cos C}$$

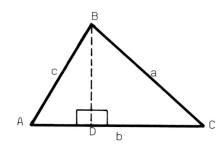

We have drawn altitude BD. Since triangle ABD is a right triangle, the Pythagorean Theorem gives us the following:

$$\boxed{c^2 = (BD)^2 + (AD)^2}$$

Continued on following page.

50.   Continued

We will show that the two boxed equations are equivalent by expressing $(BD)^2$ and $(AD)^2$ in terms of "a", "b" and angle C.

<u>Finding $(BD)^2$</u>:   Since $\sin C = \dfrac{BD}{a}$,

$$BD = a \sin C$$

and $(BD)^2 = \boxed{a^2\sin^2 C}$

<u>Finding $(AD)^2$</u>: $AD = b - DC$

But $DC = a \cos C$, since $\cos C = \dfrac{DC}{a}$

Therefore: $AD = b - a \cos C$

and $(AD)^2 = (b - a \cos C)^2 = \boxed{b^2 - 2ab \cos C + a^2\cos^2 C}$

<u>Substituting the boxed expressions for $(BD)^2$ and $(AD)^2$ into the Pythagorean Theorem</u>, we get:

If $c^2 = (BD)^2 + (AD)^2$: $c^2 = \boxed{a^2\sin^2 C} + \boxed{b^2 - 2ab \cos C + a^2\cos^2 C}$

Rearranging we get: $c^2 = a^2\sin^2 C + a^2\cos^2 C + b^2 - 2ab \cos C$

Factoring the first
two terms we get: $c^2 = a^2(\sin^2 C + \cos^2 C) + b^2 - 2ab \cos C$

Since $\sin^2 C + \cos^2 C = 1$,
          we get: $c^2 = a^2(1) + b^2 - 2ab \cos C$

or

$\boxed{c^2 = a^2 + b^2 - 2ab \cos C}$ (Law of Cosines)

12-8   USING THE LAW OF COSINES TO FIND SIDES AND ANGLES

In this section, we will use the Law of Cosines to find unknown sides and angles in triangles. To simplify the solutions, triangles with obtuse angles will be delayed until a later section.

---

51.   In triangle BDV, two sides and
their included angle are given.
We <u>cannot</u> use the Law of Sines
to find side "d". However, we
<u>can</u> use the Law of Cosines to do so.

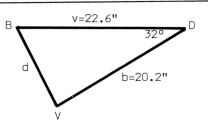

The Law-of-Cosines formula is:

$$d^2 = b^2 + v^2 - 2bv \cos D$$

To solve for "d", we take the square root of each side.

$$d = \sqrt{b^2 + v^2 - 2bv \cos D}$$

Substituting the known values, we get:

$$d = \sqrt{(20.2)^2 + (22.6)^2 - (2)(20.2)(22.6)(\cos 32°)}$$

Continued on following page.

51. Continued

We can then use a calculator to find "d". The steps are:

Calculator Steps: 20.2 $\boxed{x^2}$ $\boxed{+}$ 22.6 $\boxed{x^2}$ $\boxed{-}$ 2 $\boxed{x}$ 20.2 $\boxed{x}$

22.6 $\boxed{x}$ 32 $\boxed{\cos}$ $\boxed{=}$ $\boxed{\sqrt{x}}$ 12.020739

Rounding to tenths, we get: d = _____

---

52. In triangle CFS, two sides and their included angle are given. To find side CS, we must use the Law of Cosines.

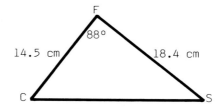

14.5 cm · 88° · 18.4 cm

12.0"

    a) Write the Law-of-Cosines formula.

    $(CS)^2 =$ _____

    b) Solve for CS by taking the square root of each side.

    CS = _____

    c) Substitute the known values in the formula.

    CS = _____

    d) Use a calculator to complete the solution. Round to tenths.

                    CS = _____

---

53. In triangle ATV, three sides and no angles are given. We cannot use the Law of Sines to find angle T. However, we can use the Law of Cosines. The steps are:

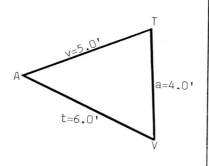

a) $(CS)^2 = (CF)^2 + (FS)^2 - 2(CF)(FS)(\cos F)$

b) $CS = \sqrt{(CF)^2 + (FS)^2 - 2(CF)(FS)(\cos F)}$

c) $CS = \sqrt{(14.5)^2 + (18.4)^2 - 2(14.5)(18.4)(\cos 88°)}$

d) CS = 23.0 cm

Continued on following page.

53. Continued

    1. Write the Law-of-Cosines formula that contains "cos T". It is the one with "t^2" on the left side.

$$t^2 = a^2 + v^2 - 2av \cos T$$

    2. Substitute the known values. Notice that "cos T" is the only unknown.

$$(6.0)^2 = (4.0)^2 + (5.0)^2 - 2(4.0)(5.0)(\cos T)$$

    3. Simplify to find "cos T".

$$36 = 16 + 25 - 40 \cos T$$
$$36 = 41 - 40 \cos T$$
$$-5 = -40 \cos T$$
$$5 = 40 \cos T$$
$$\cos T = \frac{5}{40}$$

Use a calculator to find angle T. Round to a whole number.

Angle T = _____

---

54. In the last frame, we substituted into the formula below and then solved for "cos T".

$$t^2 = a^2 + v^2 - 2av \cos T$$

We could rearrange the formula to solve for "cos T" <u>before</u> substituting. The steps are:

    1. Add $(-a^2)$ and $(-v^2)$ to both sides.

$$t^2 + (-a^2) + (-v^2) = -2av \cos T$$

    2. Take the opposite of each side.

$$(-t^2) + a^2 + v^2 = 2av \cos T$$

or

$$a^2 + v^2 - t^2 = 2av \cos T$$

    3. Divide both sides by "2av", the coefficient of "cos T".

$$\cos T = \frac{a^2 + v^2 - t^2}{2av}$$

With "cos T" solved for, we can use a calculator to find angle T. The known values (a = 4.0', v = 5.0', t = 6.0') are substituted in the formula.

$$\cos T = \frac{(4.0)^2 + (5.0)^2 - (6.0)^2}{2(4.0)(5.0)}$$

83°

Continued on following page.

54.   Continued

The calculator steps are:

Calculator Steps: 4 $\boxed{x^2}$ $\boxed{+}$ 5 $\boxed{x^2}$ $\boxed{-}$ 6 $\boxed{x^2}$ $\boxed{=}$

$\boxed{\div}$ 2 $\boxed{\div}$ 4 $\boxed{\div}$ 5 $\boxed{=}$ $\boxed{\text{2nd}}$ $\boxed{\text{cos}}$ 82.819244

a) Rounded to a whole number, angle T = _____.

b) Is this the same solution we got in the last frame? _____

---

55.   When using the Law of Cosines to find an angle, it is easier to use a calculator if you solve for the cosine of the angle first.

Solve for cos P in this formula.

$$p^2 = b^2 + h^2 - 2bh \cos P$$

a) 83°

b) Yes

---

56.   Let's use the Law of Cosines to find angle V in this triangle.

a) Write the formula that contains "cos V".

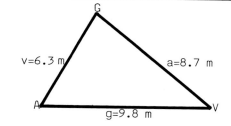

$$\cos P = \frac{b^2 + h^2 - p^2}{2bh}$$

b) Rearrange to solve for cos V.

Continued on following page.

56.   Continued

c) Substitute the known values in the formula.

$$\cos V = \underline{\hspace{5cm}}$$

d) Use a calculator to complete the solution. Rounded to a whole
    number, angle V = _____.

---

57.   Let's use the Law of
       Cosines to find angle M
       in this triangle.

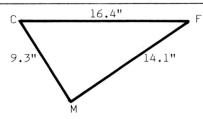

16.4"
C   F
9.3" 14.1"
M

a) Write the formula
    that contains "cos M".

---

b) Rearrange to solve for cos M.

---

c) Substitute the known values in the formula.

$$\cos M = \underline{\hspace{5cm}}$$

d) Use a calculator to complete the solution. Rounded to a whole
    number, angle M = _____.

---

a) $v^2 = a^2 + g^2 - 2ag \cos V$

b) $\cos V = \dfrac{a^2 + g^2 - v^2}{2ag}$

c) $\cos V = \dfrac{(8.7)^2 + (9.8)^2 - (6.3)^2}{2(8.7)(9.8)}$

d) 39°

---

a) $(CF)^2 = (CM)^2 + (MF)^2 - 2(CM)(MF)(\cos M)$

b) $\cos M = \dfrac{(CM)^2 + (MF)^2 - (CF)^2}{2(CM)(MF)}$

c) $\cos M = \dfrac{(9.3)^2 + (14.1)^2 - (16.4)^2}{2(9.3)(14.1)}$

d) M = 86°

SELF–TEST 41 (pages 520-531)

In triangle DEF, find angle E, angle F, and side DE. Round each angle to a whole number. Round the side to tenths.

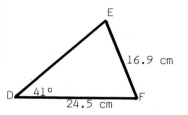

1. Angle E = _____

2. Angle F = _____

3. Side DE = _____

Apply the Law of Cosines to triangle HPT and complete these formulas.

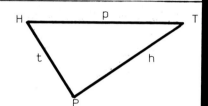

4. $t^2 =$ _____

5. $\cos H =$ _____

---

6. Find side "d". Round to a whole number.

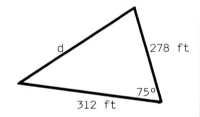

Side "d" = _____

7. Find angle G. Round to a whole number.

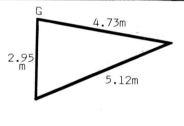

Angle G = _____

ANSWERS:

1. Angle E = 72°
2. Angle F = 67°
3. Side DE = 23.7 cm

4. $t^2 = h^2 + p^2 - 2hp \cos T$

5. $\cos H = \dfrac{p^2 + t^2 - h^2}{2pt}$

   (from: $h^2 = p^2 + t^2 - 2pt \cos H$)

6. d = 360 ft
7. G = 80°

12-9 OBTUSE ANGLES AND THE LAW OF SINES

When using the Law of Sines in earlier sections, we avoided triangles with obtuse angles. We will use the Law of Sines with triangles of that type in this section.

58. Angles between 90° and 180° are called "obtuse" angles. Triangle CDF contains an obtuse angle.

Can a triangle contain more than one obtuse angle? _____

No, because the sum of the angles would then be more than 180°.

59. To find the sine of an obtuse angle on a calculator, we simply enter the angle and press $\boxed{\sin}$. For example, we found the sines of four angles (91°, 127°, 154°, 179°) below.

Calculator Steps: 91 $\boxed{\sin}$ 0.9998477

91 $\boxed{\sin}$ 0.9998477

127 $\boxed{\sin}$ 0.7986355

154 $\boxed{\sin}$ 0.4383711

179 $\boxed{\sin}$ 0.0174524

The sine of an obtuse angle is a positive number between 0 and 1. As the angle increases from 91° to 179°, does the sine increase or decrease? _____

It decreases.

60. Let's use the Law of Sines to find side "a" in triangle ABC. The proportion is:

$$\frac{a}{\sin 59°} = \frac{19.4"}{\sin 102°}$$

$$a = \frac{(19.4")(\sin 59°)}{\sin 102°}$$

The calculator steps are:

Calculator Steps: 19.4 $\boxed{x}$ 59 $\boxed{\sin}$ $\boxed{\div}$ 102 $\boxed{\sin}$ $\boxed{=}$ 17.000548

Rounded to the nearest tenth, a = _____

17.0"

61. Let's use the Law of Sines to find side PR in this triangle. We must find angle R first.

a) Angle R contains _____°.

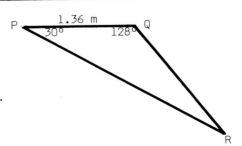

Continued on following page.

61.   Continued

b) Set up the proportion needed to find PR.

c) Use a calculator to find PR. Round to hundredths.

PR = _____

---

62.   Though there is a formal treatment of the "sines" of obtuse angles in <u>Technical</u> <u>Mathematics</u> <u>II</u>, you should know this fact:

The sine of an obtuse angle is the same as the sine of the acute angle obtained by subtracting the obtuse angle from 180°.

That is: sin 150° = sin 30°   (from 180°-150°)
            sin 135° = sin 45°   (from 180°-135°)
            sin 93° = sin 87°   (from 180°-93°)

The facts above can be confirmed with a calculator. For example:

sin 150° = sin 30° = 0.5
sin 135° = sin 45° = 0.7071068

sin 93° = sin 87° = _____

a) 22°

b) $\dfrac{PR}{\sin 128°} = \dfrac{1.36m}{\sin 22°}$

or

$PR = \dfrac{(1.36m)(\sin 128°)}{\sin 22°}$

c) 2.86m

---

63.   When we enter a "sine" between 0 and 1 on a calculator and press ⌐2nd⌐ ⌐sin⌐ to find the angle, we always get an acute angle. For example:

Calculator Steps: 0.9961947 ⌐2nd⌐ ⌐sin⌐   85.000001 (or 85°)

0.8386706 ⌐2nd⌐ ⌐sin⌐   57.000003 (or 57°)

0.3420201 ⌐2nd⌐ ⌐sin⌐   19.999997 (or 20°)

Remember that the above "sines" could also be the "sines" of obtuse angles. To find the obtuse angle, we subtract the acute angle from 180°. That is:

If sin A = 0.9961947, A = 85° or 95°   (from 180°-85°)
If sin A = 0.8386706, A = 57° or 123° (from 180°-57°)

If sin A = 0.3420201, A = 20° or _____

0.9986295

160° (from 180°-20°)

64. When using the Law of Sines to find an angle, a calculator always displays an acute angle even when the angle is obtuse. <u>You</u> <u>have</u> <u>to</u> <u>check</u> <u>the</u> <u>diagram</u> <u>to</u> <u>see</u> <u>whether</u> <u>the</u> <u>angle</u> <u>is</u> <u>acute</u> <u>or</u> <u>obtuse.</u>

Let's apply the principle above to this triangle.

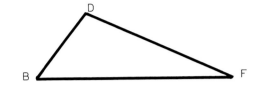

If sin D = 0.9781476 :

   a) Angle D could be either _____°
     or _____°.

   b) Since angle D is an <u>obtuse</u> angle, it must be _____°.

If sin B = 0.809017 :

   c) Angle B could be either _____° or _____°.

   d) Since angle B is an <u>acute</u> angle, it must be _____°.

---

65. Let's use the Law of Sines to find angle T.

   a) Set up the proportion needed to solve for angle T.

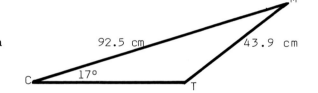

   b) Angle T is an _____ (acute/obtuse) angle.

   c) Rounded to a whole number, angle T = _____

a) 78° or 102°

b) 102°

c) 54° or 126°

d) 54°

---

66. We want to find angle P in this triangle. We cannot use the Law of Sines directly because side AK is unknown.

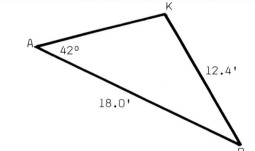

   a) What procedure can we use to find P?

   b) Find angle K. Rounded to a whole number, angle K = _____.

   c) Find angle P. Angle P = _____

a) $\dfrac{\sin T}{92.5 \text{ cm}} = \dfrac{\sin 17°}{43.9 \text{ cm}}$

b) obtuse

c) 142°

---

a) Use the Law of Sines to find angle K° Then use the angle-sum principle to find angle P.

b) 104°   c) 34°

12-10   OBTUSE ANGLES AND THE LAW OF COSINES

When using the Law of Cosines in an earlier section, we avoided triangles with obtuse angles. We will use the Law of Cosines with triangles of that type in this section.

---

67.   To find the <u>cosine</u> of an obtuse angle on a calculator, we simply enter the angle and press $\boxed{\text{cos}}$
For example, we found the cosines of four angles (95°, 112°, 139°, and 174°) below.

Calculator Steps: 95 $\boxed{\text{cos}}$ -0.0871557

112 $\boxed{\text{cos}}$ -0.3746066

139 $\boxed{\text{cos}}$ -0.7547096

174 $\boxed{\text{cos}}$ -0.9945219

The cosine of an obtuse angle is a <u>negative</u> number between 0 and -1. We will discuss why the cosines are negative in <u>Technical Mathematics II</u>.

---

68.   To solve for side FT in this triangle, we must use the Law of Cosines. We get:

$$(FT)^2 = (BT)^2 + (BF)^2 - 2(BT)(BF)(\cos B)$$

$$FT = \sqrt{(3.6)^2 + (4.2)^2 - 2(3.6)(4.2)(\cos 131°)}$$

Use your calculator to complete the solution. Round to tenths. FT = _____

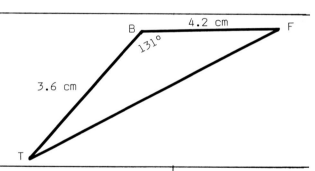

---

69.   There is no confusion between the cosines of obtuse angles and the cosines of acute angles.

Obtuse angles have <u>negative</u> cosines.
Acute angles have <u>positive</u> cosines.

To find the angle corresponding to a negative cosine, we can use a calculator. Some examples are shown. Notice how we use $\boxed{+/-}$ to enter a negative number.

Calculator Steps: 0.1218693 $\boxed{+/-}$ $\boxed{\text{2nd}}$ $\boxed{\text{cos}}$ 96.999997

0.601815 $\boxed{+/-}$ $\boxed{\text{2nd}}$ $\boxed{\text{cos}}$ 126.999998

0.9781476 $\boxed{+/-}$ $\boxed{\text{2nd}}$ $\boxed{\text{cos}}$ 167.999999

Using the above results, complete these. Round to a whole number when necessary.

a) If cos B = -0.1218693 ,   B = _____ .

b) If cos D = -0.601815 ,   D = _____ .

c) If cos F = -0.9781476 ,   F = _____ .

70. To solve for angle G in this triangle, we must use the Law of Cosines. We get:

$$g^2 = d^2 + h^2 - 2dh \cos G$$

a) Rearrange the formula to solve for cos G.

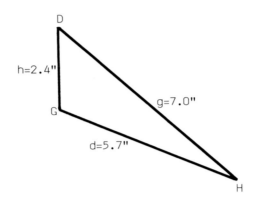

a) 97°

b) 127°

c) 168°

b) Substitute the known values in the formula.

cos G = _____

c) Use a calculator to find angle G. Rounded to a whole number, angle G = _____ .

---

a) $\cos G = \dfrac{d^2 + h^2 - g^2}{2dh}$ b) $\cos G = \dfrac{(5.7)^2 + (2.4)^2 - (7.0)^2}{(2)(5.7)(2.4)}$ c) 113°

12-11   STRATEGIES FOR SOLVING TRIANGLES

In this section, we will review the strategies (or methods) used to solve right and oblique triangles. Exercises requiring strategy-identification are included.

---

71. Except for the <u>angle-sum principle</u> which can be used with all triangles, different strategies (or methods) are used to solve right and oblique triangles.

For <u>right triangles</u>: use one of the <u>three trig ratios</u> or the <u>Pythagorean Theorem</u>.

For <u>oblique triangles</u>: use either the <u>Law of Sines</u> or the <u>Law of Cosines</u>.

When solving oblique triangles, the <u>Law of Sines</u> is used in all cases except the two below for which the <u>Law of Cosines</u> is used.

1. When only <u>two sides and their included angle</u> are known.

2. When only <u>three sides</u> are known.

If you cannot remember the two special cases above, use this strategy.

> WHEN SOLVING OBLIQUE TRIANGLES, TRY THE LAW OF SINES FIRST. IF THE LAW OF SINES DOES NOT WORK, USE THE LAW OF COSINES.

72. For any triangle, check to see whether the angle-sum principle can be used.

In which triangle(s) below can we use the angle-sum principle? _____

a)

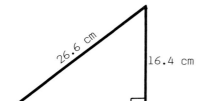

b)

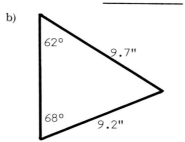

---

73. Check to see whether it is a right or oblique triangle. If it is a <u>right triangle</u>, think: <u>TRIG RATIOS AND PYTHAGOREAN THEOREM.</u>

Only (b)

In right triangle CDE, which trig ratio would you use?

a) To solve for "e"? _____

b) To solve for "d"? _____

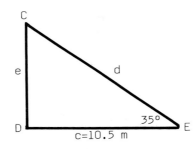

---

74. In triangle BDT, two sides are given.

a) How would you solve for "d"?

_____

b) How would you solve for angle B?

_____

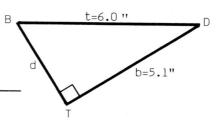

a) tan E or tan 35°

b) cos E or cos 35°

---

75. If it is an oblique triangle, think: <u>LAW OF SINES</u> or <u>LAW OF COSINES.</u>

In this oblique triangle:

a) To solve for side PR, would you use the Law of Sines or the Law of Cosines? _____

b) Write the equation you would use.

_____

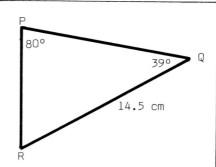

a) Use the Pythagorean Theorem.

b) Use sin B.

76. a) To solve for side CP, would you use the Law of Sines or the Law of Cosines? _____

b) Write the equation you would use.

_____

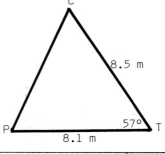

a) Law of Sines

b) $\dfrac{PR}{\sin Q} = \dfrac{QR}{\sin P}$

or

$\dfrac{PR}{\sin 39°} = \dfrac{14.5 \text{ cm}}{\sin 80°}$

---

77. a) To find angle M, which law would you use?

_____

b) Write the equation you would use.

_____

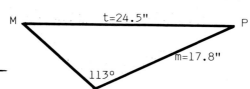

a) Law of Cosines

b) $(CP)^2 = (CT)^2 + (PT)^2 - 2(CT)(PT)(\cos T)$

or

$(CP)^2 = (8.5)^2 + (8.1)^2 - 2(8.5)(8.1)(\cos 57°)$

---

78. a) To find angle A, which law would you use? _____

b) Write the equation you would use.

_____

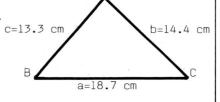

a) Law of Sines

b) $\dfrac{\sin M}{m} = \dfrac{\sin T}{t}$

or

$\dfrac{\sin M}{17.8"} = \dfrac{\sin 113°}{24.5"}$

---

79. To find angle V, two steps are needed. They are:

1. _____

_____

_____

2. _____

_____

a) Law of Cosines

b) $a^2 = b^2 + c^2 - 2bc \cos A$

or

$\cos A = \dfrac{b^2 + c^2 - a^2}{2bc}$

or

$\cos A = \dfrac{(14.4)^2 + (13.3)^2 - (18.7)^2}{2(14.4)(13.3)}$

---

1. Find angle P using the Law of Sines.

2. Find angle V using the angle-sum principle.

80. To find side "q", three steps are needed. They are:

   1. _____

   _____

   2. _____

   3. _____

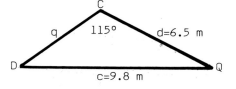

81. To find side CK, two steps are needed. They are:

   1. _____

   _____

   2. _____

   _____

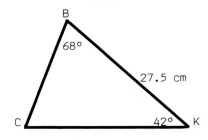

1. Find angle D using the Law of Sines.

2. Find angle Q using the angle-sum principle.

3. Find side "q" using the Law of Sines.

82. To find side PM, would we use the Law of Sines or the Law of Cosines?

   _____

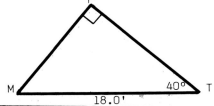

1. Find angle C using the angle-sum principle.

2. Then find CK using the Law of Sines.

83. We want to find two angles, A and T, in this triangle.

   a) Let's find angle A first. Write the equation we should use.

   _____

   b) Having found angle A, we can find angle T in either of two ways. Write the two equations we could use.

   1. _____

   or 2. _____

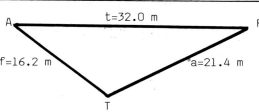

Neither. It is a <u>right</u> triangle. We would use <u>sin T</u>.

a) $a^2 = f^2 + t^2 - 2ft \cos A$

   or

   $\cos A = \dfrac{f^2 + t^2 - a^2}{2ft}$

b) 1. Law of Sines:

   $\dfrac{\sin T}{t} = \dfrac{\sin A}{a}$

2. Law of Cosines:

   $t^2 = a^2 + f^2 - 2af \cos T$

   or

   $\cos T = \dfrac{a^2 + f^2 - t^2}{2af}$

<u>Note</u>: It is preferable to use the Law of Sines because it is simpler.

84.   To find angle V, two steps
      are needed.

      a) What is the first step?

      _____

      b) Write the equation that would
          be used for the first step?

      _____

      c) Knowing side "b", we can find angle V in either of two ways.
          Write the two equations we could use.

              1. _____

          or 2. _____

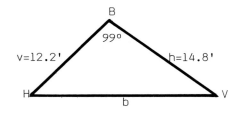

---

a) Find side "b".   b) $b^2 = h^2 + v^2 - 2hv \cos B$ c) 1. Law of Sines: $\dfrac{\sin V}{v} = \dfrac{\sin B}{b}$

2. Law of Cosines: $\cos V = \dfrac{b^2 + h^2 - v^2}{2bh}$

Note: It is preferable to use the
       Law of Sines because it is
       simpler.

12-12   APPLIED PROBLEMS

In this section we will discuss some applied problems that involve solving an oblique triangle.

---

85.   A surveyor has to find the
      distance EF across a river.
      First he laid out line DE.
      Then he used a transit to
      measure angles D and E.

      Find EF. Round to a
      whole number.

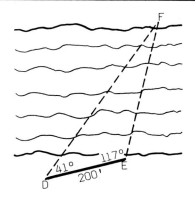

EF = _____

86. The shape at the right is formed by steel girders.

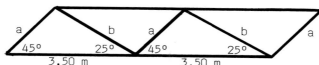

EF = 350 feet

The steps are:

1. Use the angle-sum principle to find angle F.

2. Then use the Law of Sines to find EF.

$$\frac{EF}{\sin D} = \frac{DE}{\sin F}$$

a) Find "a". Round to hundredths.

a = _____

b) Find "b". Round to hundredths.

b = _____

87. Two holes are drilled in a circular metal disc whose center is at point C. Find "d", the center-to-center distance between the two holes. Round to hundredths.

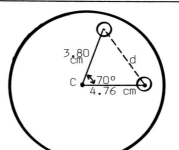

a) a = 1.57m, from:

$$\frac{a}{\sin 25°} = \frac{3.50m}{\sin 110°}$$

b) b = 2.63m, from:

$$\frac{b}{\sin 45°} = \frac{3.50m}{\sin 110°}$$

d = _____

d = 4.97 cm, from:

$$d = \sqrt{(3.80)^2 + (4.76)^2 - 2(3.80)(4.76)(\cos 70°)}$$

88. Three steps are needed to find PT, the length of a lake. They are:

a) Find angle P. Round to a whole number.

P = _____

b) Find angle R. Round to a whole number.

R = _____

c) Find PT. Round to a whole number.

PT = _____

89. Three holes are drilled in a rectangular plate. Their center-to-center distances are given. Find angle A. Round to a whole number.

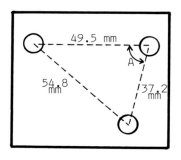

A = _____

a) P = 64°, from:

$$\frac{\sin P}{1429m} = \frac{\sin 39°}{1000m}$$

b) R = 77°, from:

$$180° - (64° + 39°)$$

c) PT = 1,548 meters, from:

$$\frac{PT}{\sin 77°} = \frac{1000}{\sin 39°}$$

---

90. The top and bottom edges of the metal template at the left below are parallel. We want to find side "s". To do so, we draw a dashed line parallel to "s" in the figure at the right. By doing so, we formed an oblique triangle.

A = 77°, from:

$$\cos A = \frac{(49.5)^2 + (37.2)^2 - (54.8)^2}{2(49.5)(37.2)}$$

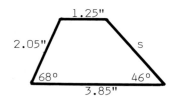

Find side "s". Round to hundredths.

s = _____

---

s = 2.64", from:

$$s^2 = (2.05)^2 + (2.60)^2 - 2(2.05)(2.60)(\cos 68)°$$

or

$$\frac{s}{\sin 68°} = \frac{2.05"}{\sin 46°}$$

Note: Using the Law of Sines is preferable.

SELF-TEST 42   (pages 531-543)

1. Find angle P. Round to a whole number.

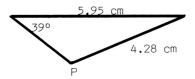

5.95 cm

39°

4.28 cm

P

P = _____

2. Find side "d". Round to tenths.

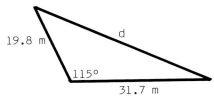

19.8 m

d

115°

31.7 m

d = _____

Which principle at the right (a, b, or c) would be used to find the unknown angle or side represented by a single letter in each triangle below?

> a) Law of Sines
> b) Law of Cosines
> c) Angle-Sum Principle

3.

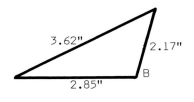

3.62"

2.17"

B

2.85"

_____

4.

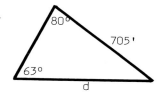

80°

705'

63°

d

_____

5.

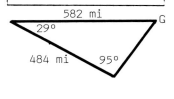

582 mi

G

29°

484 mi

95°

_____

6. Find "h", the height of the cliff shown in the diagram. Round to a whole number.

$\frac{200}{\sin 16} = \frac{x}{\sin 26}$

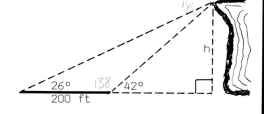

h

26° 138 42°

200 ft

h = _____

ANSWERS: 1. P = 119°   3. b   6. h = 213 ft
             2. d = 43.9m   4. a
                                5. c

SUPPLEMENTARY PROBLEMS - CHAPTER 12

Assignment 40

1. Which of the following are obtuse angles? _____

    a) 100°   b) 96°   c) 87°   d) 90°   e) 24°   f) 172°

Find the third angle in each triangle.

    2. Triangle #1: 58°, 25°, _____   4. Triangle #3: 88°, 44°, _____

    3. Triangle #2: 19°, 71°, _____   5. Triangle #4: 24°, 21°, _____

State whether each triangle above is: a) acute oblique, b) obtuse oblique, or c) right.

    6. #1: _____   7. #2: _____   8. #3: _____   9. #4: _____

In the oblique triangle below: In the oblique triangle below:

    10. The longest side   12. The largest angle
        is side _____.   is angle _____.

    11. The shortest side   13. The smallest angle
        is side _____.   is angle _____.

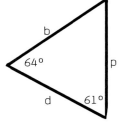
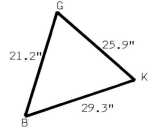

Referring to triangle RST, use the Law of Sines to complete these proportions.

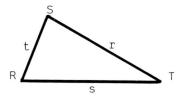

14. $\dfrac{\boxed{}}{t} = \dfrac{\sin R}{\boxed{}}$ 15. $\dfrac{s}{\boxed{}} = \dfrac{\boxed{}}{\sin T}$

In triangle ABC, find side "a", angle B, and side "b". Round each to a whole number.

16. a = _____

17. B = _____

18. b = _____

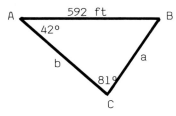

In triangle HKP, find angle H, angle K, and side "k". Round each angle to a whole number. Round the side to tenths.

19. H = _____

20. K = _____

21. k = _____

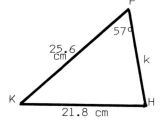

Assignment 41

State whether the <u>Law of Sines</u> or the <u>Law of Cosines</u> would be used to find the unknown angle or side represented by a single letter in each triangle below.

1.

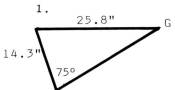

2.

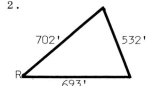

3.

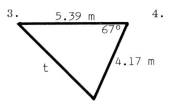

4.

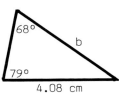

Referring to triangle PRV, complete these formulas using the Law of Cosines.

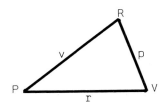

5. $p^2 =$ _____

6. $v^2 =$ _____

Referring to triangle ABC, complete these formulas using the Law of Cosines.

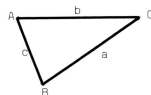

7. $\cos A =$ _____

8. $\cos C =$ _____

In triangle DEF, find side "f", angle D, and angle E. Round the side to tenths. Round each angle to a whole number.

9. f = _____

10. D = _____

11. E = _____

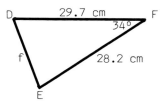

In triangle GPT, find angles G, P, and T. Round each angle to a whole number.

12. G = _____

13. P = _____

14. T = _____

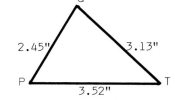

Assignment 42

Using a calculator, find each value. Round to four decimal places.

    1. sin 152°   2. sin 94°   3. cos 103°   4. cos 179°

In Problems 5-7, each angle is <u>obtuse</u>. Find each angle. Round to a whole number.

    5. If sin P = 0.3256,   6. If cos F = -0.6691,   7. If sin A = 0.9945,

          P = _____   F = _____   A = _____

8. In triangle RST, find side "s". Round to tenths.

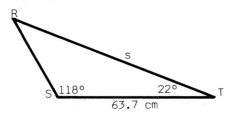

9. In triangle MNP, find angle N. Round to a whole number.

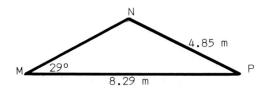

10. In triangle DEF, find side "f". Round to a whole number.

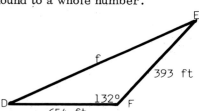

11. In triangle GHK, find angle H. Round to a whole number.

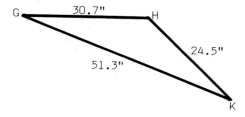

12. To find distance "d" across a swamp, a surveyor made the measurements shown. Find "d". Round to a whole number.

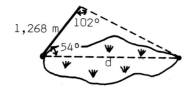

13. The center-to-center distances of three holes in a metal plate are shown. Find angle A and dimension "h". Round A to a whole number. Round "h" to hundredths.

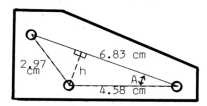

Systems of Two Equations and Formulas

13

In this chapter, we will define a system of two equations and solve systems of that type by the graphing method, the addition method, and the substitution method. Some types of problems involving systems of equations are included. We will also define a system of two formulas and use the substitution method and equivalence method for formula derivations.

13-1   THE GRAPHING METHOD

In this section, we will define a system of two equations and discuss the graphing method for solving systems of equations.

1.   It is sometimes easier to solve a problem if we use two variables and two equations. An example is shown.

> The sum of two numbers is 21. The difference between the larger number and the smaller number is 5. Find the two numbers.

> Using $\underline{x}$ for the larger number and $\underline{y}$ for the smaller number, we can set up two different equations. That is:

> The sum of two numbers is 21.

$$x + y = 21$$

> The difference between the larger and smaller is 5.

$$x - y = 5$$

Continued on following page.

1. Continued

    Therefore, the problem has been translated to the following pair of equations which is called a <u>system</u> <u>of</u> <u>equations</u>.

    $$x + y = 21$$
    $$x - y = 5$$

    The solution of a system of equations is an ordered pair that satisfies <u>both</u> equations. For example, for the system above:

    (15,6) <u>is</u> <u>not</u> <u>a</u> <u>solution</u> because it only satisfies the top equation.

    (10,5) <u>is</u> <u>not</u> <u>a</u> <u>solution</u> because it only satisfies the bottom equation.

    (13,8) <u>is</u> <u>the</u> <u>solution</u> because it satisfies <u>both</u> equations.

    Show that (13,8) satisfies both equations in the system.

    $$x + y = 21 \qquad\qquad x - y = 5$$

---

2. Remember that an ordered pair is a solution of a system of equations <u>only</u> <u>if</u> <u>it</u> <u>satisfies</u> <u>both</u> <u>equations</u>.

    Which ordered pair below is the solution of the system at the right? _____

    $$y = x + 3$$
    $$x + y = 5$$

    (2,5)   (3,2)   (1,4)   (0,3)

    $$x + y = 21$$
    $$13 + 8 = 21$$
    $$21 = 21$$

    $$x - y = 5$$
    $$13 - 8 = 5$$
    $$5 = 5$$

---

3. We graphed both equations in the system below at the right.

    $$x - y = 1$$
    $$x + y = 5$$

    (1,4)

    The point where the two graphed lines cross is called the <u>point</u> <u>of</u> <u>intersection</u>. Since that point lies on both lines, its coordinates satisfy both equations. There-fore, <u>its</u> <u>coordinates</u> <u>are</u> <u>the</u> <u>solution</u> <u>of</u> <u>the</u> <u>system</u>. Let's confirm that fact by answering the questions below.

    a) The coordinates of the point of intersection are _____.

    b) Do those coordinates satisfy both equations? _____

    c) Therefore, the solution of the system is _____.

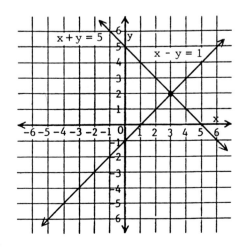

4. When the graphs of both equations are straight lines and the lines intersect, there is only one point of intersection. Therefore, there is only one solution of the system of equations.

We graphed the system of equations below at the right.

$$3x + y = 0$$
$$x - 2y = 7$$

a) Using the coordinates of the point of intersection, write the solution of the system. _____

b) Show that the solution satisfies both equations.

$$3x + y = 0 \qquad\qquad x - 2y = 7$$

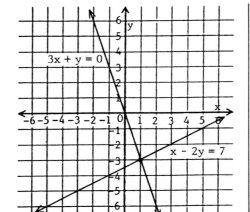

a) (3,2)

b) Yes

c) (3,2)

---

5. Let's use the graphing method to solve the system below.

$$y = 2x$$
$$y - x = 2$$

a) Using the tables provided, find some solutions and graph the equations.

y = 2x   y - x = 2

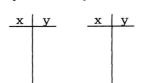

b) The solution of the system is: _____

c) Check the solution in both equations.

$$y = 2x \qquad\qquad y - x = 2$$

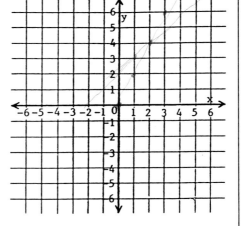

a) (1,-3)

b) 3x + y = 0
   3(1) + (-3) = 0
   3 + (-3) = 0
         0 = 0

   x - 2y = 7
   1 - 2(-3) = 7
   1 + 6 = 7
         7 = 7

6. Graph the following system.

$$2x - y = 3$$
$$x + y = 0$$

2x - y = 3   x + y = 0

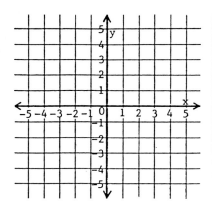

The solution of the system is: _____

a)

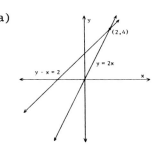

b) (2,4)

c)   y = 2x   y - x = 2
     4 = 2(2)   4 - 2 = 2
     4 = 4   2 = 2

---

7. A system has <u>one</u> solution if its graph is two straight lines that inter- (1,-1)
    sect at one point. A system can have <u>no</u> solution or an <u>infinite num-</u>
    <u>ber</u> of solutions. An example of each type is discussed below.

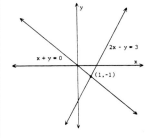

SYSTEM WITH NO SOLUTIONS

The system below is graphed
at the right. Notice that the
straight lines are parallel.
They have the same slope,
m = 3.

$$y = 3x + 2$$
$$y = 3x - 1$$

Since the parallel lines do not
intersect, the equations have <u>no</u>
common solution. Therefore,
the system has <u>no</u> solution.

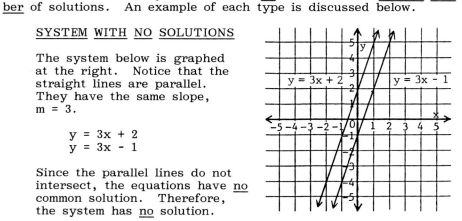

SYSTEM WITH AN INFINITE NUMBER OF SOLUTIONS

In the system below, the bottom
equation can be obtained by
multiplying the top equation by
2. The system is graphed at
the right. Notice that each
equation has the same graph.

$$x + y = 2$$
$$2x + 2y = 4$$

Since the lines are identical, the
equations have <u>an infinite number</u>
of common solutions. Therefore,
the system has <u>an infinite number</u>
of solutions.

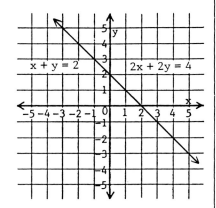

13-2 THE ADDITION METHOD

Because the graphing method for solving systems of equations is time-consuming and only approximate when the solutions are not whole numbers, an algebraic method is ordinarily used. In this section, we will discuss an algebraic method called the <u>addition</u> method for solving systems.

8. The addition method for solving systems is based on the following property of addition.

> If   A = B
> and   C = D,
> then   A + C = B + D.

The property above can be used to add the two equations in a system. For example, we added the two equations and got 4x + 3y = 11. Add the other two equations.

$$x + 5y = 7$$
$$\underline{3x - 2y = 4}$$
$$4x + 3y = 11$$

$$x - 3y = 4$$
$$\underline{5x + y = 0}$$

---

9. In the addition method for solving a system, we add the equations <u>in</u> <u>order</u> <u>to</u> <u>eliminate</u> a <u>variable</u>. For example, we eliminated <u>y</u> by adding the equations below. Add the other two equations.

$$x + 2y = 3$$
$$\underline{x - 2y = 5}$$
$$2x = 8$$

$$a + 4b = 0$$
$$\underline{2a - 4b = 7}$$

6x - 2y = 4

---

10. Let's use the addition method to solve the system at the right. The two steps are described.

$$x + y = 5$$
$$x - y = 1$$

1. <u>Finding</u> <u>the</u> <u>value</u> <u>of</u> <u>x</u>. By adding the two equations, we can eliminate <u>y</u> and solve for <u>x</u>.

$$x + y = 5$$
$$\underline{x - y = 1}$$
$$2x = 6$$
$$x = 3$$

2. <u>Finding</u> <u>the</u> <u>value</u> <u>of</u> <u>y</u>. We can now find the corresponding value of <u>y</u> by substituting 3 for <u>x</u> in either of the original equations.

$$x + y = 5$$
$$3 + y = 5$$
$$y = 2$$

$$x - y = 1$$
$$3 - y = 1$$
$$y = 2$$

The obtained solution is (3,2) or x = 3, y = 2. Check that solution in each original equation below.

a) x + y = 5

b) x - y = 1

3a = 7

11. Let's use the addition method to solve the system at the right.

$s + 2t = 5$
$3s - 2t = 7$

a) $3 + 2 = 5$
$ 5 = 5$

1. <u>Finding the value of s</u>.

By adding the equations, we can eliminate <u>t</u> and solve for <u>s</u>.

$s + 2t = 5$
$\underline{3s - 2t = 7}$
$4s = 12$
$ s = 3$

b) $3 - 2 = 1$
$ 1 = 1$

2. <u>Finding the value of t</u>.

To find the corresponding value of <u>t</u>, we substituted 3 for <u>s</u> in the top equation.

$s + 2t = 5$
$3 + 2t = 5$
$ 2t = 2$
$ t = 1$

<u>Note</u>: We could also have substituted 3 for <u>s</u> in the bottom equation.

The obtained solution is s = 3, t = 1. Show that the solution satisfies each of the original equations.

a) $s + 2t = 5$ b) $3s - 2t = 7$

---

12. Use the addition method to solve each system below.

a) $2x + y = 10$
$2x - y = 6$

b) $p - 4q = 16$
$p + 4q = 0$

a) $3 + 2(1) = 5$
$ 3 + 2 = 5$
$ 5 = 5$

b) $3(3) - 2(1) = 7$
$ 9 - 2 = 7$
$ 7 = 7$

---

a) x = 4, y = 2

b) p = 8, q = -2

13-3   MULTIPLICATION IN THE ADDITION METHOD

When using the addition method to solve a system, we sometimes have to multiply one or both equations by a number before adding the equations. We will discuss solutions of that type in this section.

13.   When using the addition method to solve a system, we sometimes have to multiply one or both equations by a number. When doing so, we must multiply <u>both</u> <u>sides</u> by the number. Two examples are shown.

We multiplied $x - 2y = 4$ by 3 below.

$$3(x - 2y) = 3(4)$$
$$3x - 6y = 12$$

We multiplied $p - 2q = 3$ by -1 below.

$$-1(p - 2q) = -1(3)$$
$$-p + 2q = -3$$

Following the examples, do these.

a) Multiply this equation by 5.   b) Multiply this equation by -2.

$$4x - y = 1 \qquad\qquad 2a + 3b = 4$$

---

14.   If we add the equations in the system at the right, we get $3x + y = 11$. Neither variable is eliminated.

$$x + 2y = 7$$
$$2x - y = 4$$

However, if the $-y$ in the bottom equation were a $-2y$, we could eliminate y by adding. To get a $-2y$ at the right, we multiply both sides of the bottom equation by 2. Then we add and solve for x.

$$\begin{array}{l} x + 2y = 7 \\ \underline{4x - 2y = 8} \quad \text{(Multiplied by 2)} \\ 5x \qquad\; = 15 \\ \qquad x = 3 \end{array}$$

Substituting 3 for x in one of the original equations, we can find the corresponding value of y.

$$x + 2y = 7$$
$$3 + 2y = 7$$
$$2y = 4$$
$$y = 2$$

We get $(3,2)$ as a solution. Check it in each original equation below.

$$x + 2y = 7 \qquad\qquad 2x - y = 4$$

a) $20x - 5y = 5$

b) $-4a - 6b = -8$

15. We could eliminate <u>y</u> in the system at the right if the 3y were -3y in either equation.

$$2x + 3y = 8$$
$$x + 3y = 7$$

$$3 + 2(2) = 7$$
$$3 + 4 \quad = 7$$
$$7 = 7$$

To get a -3y in the bottom equation, we multiply it by -1 at the right and then solve for <u>x</u>.

$$2x + 3y = \ \ 8$$
$$\underline{-x - 3y = -7} \quad \text{(Multiplied by -1)}$$
$$x \qquad = \ \ 1$$

$$2(3) - 2 = 4$$
$$6 \quad - 2 = 4$$
$$4 = 4$$

Substituting "1" for <u>x</u> in the original bottom equation, we find the corresponding value of <u>y</u> at the right.

$$x + 3y = 7$$
$$1 + 3y = 7$$
$$3y = 6$$
$$y = 2$$

Check (1,2) in each original equation below.

a) $2x + 3y = 8$ b) $x + 3y = 7$

---

16. Let's solve the system at the right by eliminating <u>x</u>. To do so, we must multiply the bottom equation by -2.

$$4x + y = \ \ 1$$
$$2x + 3y = 13$$

a) 2(1) + 3(2) = 8
$$2 \ + \ 6 = 8$$
$$8 = 8$$

    a) Write the new system obtained if the bottom equation is multiplied by -2.

b) 1 + 3(2) = 7
$$1 + \ 6 \ = 7$$
$$7 = 7$$

    b) Solve the system.

---

17. Let's solve the system at the right by eliminating <u>p</u>. To do so, we can multiply either equation by -1. Let's multiply the bottom equation by -1.

$$p + q = \ \ 7$$
$$p + 3q = 15$$

a) $4x + y = \ \ 1$
$$-4x - 6y = -26$$

b) (-1,5)

    a) Write the new system obtained if the bottom equation is multiplied by -1.

    b) Solve the system.

13-4 CONVERTING TO STANDARD FORM

To solve a system by the addition method, both equations must be in standard form. Sometimes we have to convert an equation to standard form before using the addition method. We will discuss solutions of that type in this section.

26. In the two systems below, the equations are in standard form $Ax + By = C$. That is, the variables are on the left side and the number is on the right side.

$$x + y = 5$$
$$x - y = 0$$

$$3a + b = 7$$
$$a + 5b = 9$$

An equation in a system is not in standard form if it has a variable on the right side. For example, in the systems below, $y = x - 8$ and $s = 2t$ are not in standard form.

$$y = x - 8$$
$$5x + y = 4$$

$$3s - t = 4$$
$$s = 2t$$

In each system below, circle the equation that is not in standard form.

a) $x + y = 10$
   $y = 1 - 3x$

b) $3p + 2q = 5$
   $4q = 6p - 1$

27. When an equation in a system is not in standard form, we must convert it to standard form before using the addition method. Two examples are shown. Notice how we lined up the variables on the left side.

Original System
$$x + y = 10$$
$$y = 2x + 3$$

Original System
$$a = b + 7$$
$$2a + b = 5$$

Converting 2nd Equation
$$y = 2x + 3$$
$$(-2x) + y = 2x + (-2x) + 3$$
$$-2x + y = 3$$

Converting 1st Equation
$$a = b + 7$$
$$a + (-b) = b + (-b) + 7$$
$$a - b = 7$$

Standard System
$$x + y = 10$$
$$-2x + y = 3$$

Standard System
$$a - b = 7$$
$$2a + b = 5$$

Convert each system below to standard form.

a) $2p + q = 9$
   $q = 4p + 5$

b) $t = 2d - 1$
   $t + d = 10$

---

a) $y = 1 - 3x$

b) $4q = 6p - 1$

28. Two more examples of conversions to standard form are shown below. Notice in each that we lined up the variables on the left side.

<u>Original System</u>

$$p - q = 7$$
$$q = 2p$$

<u>Converting 2nd Equation</u>

$$q = 2p$$
$$(-2p) + q = 2p + (-2p)$$
$$-2p + q = 0$$

<u>Standard System</u>

$$p - q = 7$$
$$-2p + q = 0$$

<u>Original System</u>

$$x = 5 - y$$
$$2x + y = 3$$

<u>Converting 1st Equation</u>

$$x = 5 - y$$
$$x + y = 5 - y + y$$
$$x + y = 5$$

<u>Standard System</u>

$$x + y = 5$$
$$2x + y = 3$$

Convert each system below to standard form.

a) $b = a$
    $2a - 3b = 8$

b) $t + 6m = 1$
   $t = 4 - 3m$

a) $2p + q = 9$
    $-4p + q = 5$

b) $t - 2d = -1$
   $t + d = 10$

---

29. In the system at the right, $y = 3x - 8$ is not in standard form.

To solve the system, we begin by converting $y = 3x - 8$ to standard form.

Then we can use the addition method to solve the standard system at the right.

a) Solve the standard system.

b) Check your solution in each original equation.

$$x - y = 2$$

$$y = 3x - 8$$

$$x - y = 2$$
$$y = 3x - 8$$

$$-3x + y = 3x + (-3x) - 8$$
$$-3x + y = -8$$

$$x - y = 2$$
$$-3x + y = -8$$

a) $-a + b = 0$
   $2a - 3b = 8$

b) $t + 6m = 1$
   $t + 3m = 4$

30. Following the steps in the last frame, solve each system.

    a) 2p + q = 30
         p = 2q

    b) 2a = 7 - 3b
       3a - 2b = 4

a) (3,1)

b) 3 - 1 = 2
    2 = 2

1 = 3(3) - 8
1 = 9 - 8
1 = 1

a) p = 12, q = 6   b) a = 2, b = 1

13-5 DECIMAL COEFFICIENTS

When solving word problems, we sometimes have to solve systems with decimal coefficients. With systems of that type, it is easier to clear the decimals before using the addition method. We will discuss systems of that type in this section.

31. To clear the decimal coefficients in the equation below, we multiplied both sides by 10. Notice how we multiplied by the distributive principle on the left side.

$$.3x + .9y = 4$$
$$10(.3x + .9y) = 10(4)$$
$$10(.3x) + 10(.9y) = 40$$
$$3x + 9y = 40$$

Multiply both sides by 10 to clear the decimals in these.

    a) .5x + .1y = 70   b) .8b - .7d = 1.5

a) 5x + y = 700

b) 8b - 7d = 15

32. To clear the decimal coefficients below, we multiplied both sides by 100.

$$.45x + .8y = 20$$
$$100(.45x + .8y) = 100(20)$$
$$100(.45x) + 100(.8y) = 2,000$$
$$45x + 80y = 2,000$$

Multiply both sides by 100 to clear the decimals in these.

    a) .2x - .65y = 7.5   b) .95s + .35t = 45

33. When multiplying by a power of ten to clear the decimals, the power of ten is determined by the number <u>with</u> <u>the</u> <u>most</u> <u>decimal</u> <u>places</u>. If its last digit is:

in the <u>tenths</u> place, we multiply by 10.

in the <u>hundredths</u> place, we multiply by 100.

Using the facts above, clear the decimals in these.

a) .7x + .15y = 6.5   b) .4a - .1b = 200

a) $20x - 65y = 750$

b) $95s + 35t = 4,500$

---

34. Since it is easier to solve a system without decimal coefficients, we usually clear any decimal coefficients before using the addition method. An example is shown.

<u>Original</u> <u>System</u>

.2x + .5y = 1.6

3x + 2y = 12

<u>Equivalent</u> <u>System</u>

2x + 5y = 16

3x + 2y = 12

Write the equivalent system obtained after clearing the decimals in these.

a) .1x - .25y = 40   b) 1.4p + .9q = 65
   5x + 4y = 1,200   .75p + .2q = 1.5

a) $70x + 15y = 650$

b) $4a - b = 2,000$

---

35. Let's solve the system at the right.

    x + y = 100
  .3x + .7y = 46

a) Clear the decimals and write the equivalent system.

b) Solve the equivalent system.

c) Show that the solution satisfies each original equation.

   x + y = 100   .3x + .7y = 46

a) $10x - 25y = 4,000$
   $5x + 4y = 1,200$

b) $14p + 9q = 650$
  $75p + 20q = 150$

<u>Answer</u> <u>to</u> <u>Frame</u> <u>35</u>:

a)   x + y = 100
    3x + 7y = 460

b) (60,40)

c) 60 + 40 = 100
     100 = 100

  .3(60)+.7(40) = 46
    18 + 28 = 46
        46 = 46

13-6   THE SUBSTITUTION METHOD

The substitution method is a second algebraic method for solving systems of equations.   We will discuss the substitution method in this section.

36.   Instead of adding equations to eliminate a variable, we can <u>substitute</u> to eliminate a variable.   An example is discussed below.

In the system at the right, <u>y</u> is solved for in the top equation.

$$y = 2x$$
$$x + 4y = 45$$

Since $y = 2x$, we can eliminate <u>y</u> by substituting $2x$ for <u>y</u> in the bottom equation.   We get:

$$x + 4y = 45$$
$$x + 4(2x) = 45$$
$$x + 8x = 45$$
$$9x = 45$$
$$x = 5$$

Substituting 5 for <u>x</u> in the top equation, we can solve for <u>y</u>. We get:

$$y = 2x$$
$$y = 2(5)$$
$$y = 10$$

Check (5,10) in each original equation below.

$$y = 2x \qquad\qquad\qquad x + 4y = 45$$

37.   Another example of the substitution method is given below.

In the system at the right, <u>b</u> is solved for in the bottom equation.

$$3a - 2b = 4$$
$$b = a + 2$$

We can eliminate <u>b</u> by substituting $a + 2$ for <u>b</u> in the top equation. We substituted, simplified, and solved for <u>a</u> at the right.

$$3a - 2b = 4$$
$$3a - 2(a + 2) = 4$$
$$3a - (2a + 4) = 4$$
$$3a - 2a - 4 = 4$$
$$a = 8$$

Substituting 8 for <u>a</u> in the bottom equation, we can solve for <u>b</u>. We get:

$$b = a + 2$$
$$b = 8 + 2$$
$$b = 10$$

Check (a = 8, b = 10) in each original equation below.

$$3a - 2b = 4 \qquad\qquad b = a + 2$$

10 = 2(5)
10 = 10

5 + 4(10) = 45
5 + 40 = 45
45 = 45

38. Use the substitution method to solve each system.

    a) $\quad y = 3x$
    $\quad\ \ 5x - y = 4$

    b) $4x + y = 7$
    $\quad\ \ y = x - 8$

$3(8) - 2(10) = 4$
$\quad\ \ 24 - 20 = 4$
$\qquad\qquad 4 = 4$

$10 = 8 + 2$
$10 = 10$

---

39. Neither variable is solved for in the system below.

    $$x - 3y = 0$$
    $$2x + y = 4$$

    To use the substitution method, we must solve for one variable in one equation.

    If we solve for <u>x</u> in the top equation or <u>y</u> in the bottom equation, we get non-fractional solutions.

    $\quad\ x - 3y = 0 \qquad\qquad 2x + y = 4$
    $\quad\ x = 3y \qquad\qquad\quad\ \ y = 4 - 2x$

    If we solve for <u>y</u> in the top equation or <u>x</u> in the bottom equation, we get fractional solutions.

    $\quad\ x - 3y = 0 \qquad\qquad 2x + y = 4$
    $\quad\ x = 3y \qquad\qquad\quad\ \ 2x = 4 - y$
    $\quad\ y = \dfrac{x}{3} \qquad\qquad\qquad x = \dfrac{4 - y}{2}$

    Since it is easier to substitute non-fractional solutions, we would use either x = 3y or y = 4 - 2x for the substitution. <u>Notice</u> <u>that</u> <u>we</u> <u>got</u> <u>the</u> <u>non-fractional</u> <u>solutions</u> <u>by</u> <u>solving</u> <u>for</u> <u>a</u> <u>variable</u> <u>whose</u> <u>coefficient</u> <u>is</u> <u>"1"</u>.

    Which variable in this system would you solve for to get a non-fractional solution?

    $\qquad\qquad\qquad\qquad 4a - 5b = 0$
    $\qquad\qquad\qquad\qquad a - 2b = 8$
    _____

a) (2,6)

b) (3,-5)

---

40. Let's use the substitution method to solve this system.

    $a + 4b = 6$
    $5a - 3b = 7$

    a) Solve for <u>a</u> in the top equation to get a non-fractional solution.

    b) Substitute that solution for <u>a</u> in the bottom equation and then find the value of <u>b</u>.

    c) Substitute in one of the original equations to find the value of <u>a</u>.

    d) The solution is: a = _____, b = _____

<u>a</u> in the bottom equation

41. Let's use the substitution method to solve this system.

$$3x - 4y = 0$$
$$5y - 2x = 7$$

a) $a = 6 - 4b$

b) $5(6-4b) - 3b = 7$
$$30 - 20b - 3b = 7$$
$$30 - 23b = 7$$
$$-23b = -23$$
$$b = 1$$

1. Solve for <u>x</u> in the top equation. (<u>Note</u>: We can't get a non-fractional solution because no variable has "1" as a coefficient.)

$$3x = 4y$$
$$x = \frac{4y}{3}$$

c) $a = 2$

d) $a = 2, \ b = 1$

2. Substitute that solution for <u>x</u> in the bottom equation and then find the value of <u>y</u>.

$$5y - 2\left(\frac{4y}{3}\right) = 7$$

$$5y - \frac{8y}{3} = 7$$

$$3\left(5y - \frac{8y}{3}\right) = 3(7)$$

$$3(5y) - \cancel{3}\left(\frac{8y}{\cancel{3}}\right) = 21$$

$$15y - 8y = 21$$
$$7y = 21$$
$$y = 3$$

3. Substitute 3 for <u>y</u> in the top equation to find the value of <u>x</u>.

$$3x - 4y = 0$$
$$3x - 4(3) = 0$$
$$3x = 12$$
$$x = 4$$

The solution of the system is _____ .

---

42. When using the substitution method, we sometimes get a fraction as a solution for a variable. Another example is discussed.

$(4,3)$

Let's use the substitution method to solve this system.

$$2x - 3y = 9$$
$$4x + 3y = 9$$

1. Solve for <u>x</u> in the top equation.

$$2x = 3y + 9$$
$$x = \frac{3y + 9}{2}$$

2. Substitute that solution for <u>x</u> in the bottom equation and then find the value of <u>y</u>.

$$\overset{2}{\cancel{4}}\left(\frac{3y + 9}{\cancel{2}}\right) + 3y = 9$$
$$6y + 18 + 3y = 9$$
$$9y + 18 = 9$$
$$9y = -9$$
$$y = -1$$

3. Substitute -1 for <u>y</u> in the top equation and find the value of <u>x</u>.

$$2x - 3y = 9$$
$$2x - 3(-1) = 9$$
$$2x + 3 = 9$$
$$2x = 6$$
$$x = 3$$

The solution of the system is _____ .

$(3,-1)$

43. When deciding whether to use the addition method or the substitution method, the following suggestions can be used.

> 1. When a variable is solved for in one equation, the substitution method is probably better.
>
> 2. When neither variable is solved for but one has a coefficient of "1" (so that the solution for it is non-fractional), either method can be used.
>
> 3. Otherwise, use the addition method.
>
> 4. When in doubt, use the addition method.

Identify the method you would use to solve each of these.

a) $3x - y = 10$
$x + 3y = 10$

b) $a + b = 6$
$a = 2b$

c) $2p + 5q = 9$
$3p - 2q = 4$

a) Either method

b) The substitution method

c) The addition method

44. To solve the system at the right, we begin by clearing the fractions in each equation. We did so below.

$$\frac{x}{2} - \frac{y}{3} = \frac{5}{6}$$

$$\frac{x}{5} - \frac{y}{4} = \frac{1}{10}$$

For the top equation, we multiply both sides by 6.

$$6\left(\frac{x}{2} - \frac{y}{3}\right) = 6\left(\frac{5}{6}\right)$$

$$\overset{3}{6}\left(\frac{x}{2}\right) - \overset{2}{6}\left(\frac{y}{3}\right) = \overset{1}{6}\left(\frac{5}{6}\right)$$

$$3x - 2y = 5$$

The resulting system with non-fractional equations is shown at the right.

For the bottom equation, we multiply both sides by 20.

$$20\left(\frac{x}{5} - \frac{y}{4}\right) = 20\left(\frac{1}{10}\right)$$

$$\overset{4}{20}\left(\frac{x}{5}\right) - \overset{5}{20}\left(\frac{y}{4}\right) = \overset{2}{20}\left(\frac{1}{10}\right)$$

$$4x - 5y = 2$$

$$3x - 2y = 5$$
$$4x - 5y = 2$$

a) Which method would you use to solve the system? _____

b) Solve the system.

a) Addition method

b) (3,2)

45. To solve the system at the right, we begin by simplifying the top equation. We get:

$$3(x + y) + y = 38$$
$$3x + 3y + y = 38$$
$$3x + 4y = 38$$

$$3(x + y) + y = 38$$
$$x - 5y = 0$$

The resulting system with the simplified equation is shown at the right.

$$3x + 4y = 38$$
$$x - 5y = 0$$

    a) Which method would you use to solve the system? _____

    b) Solve the system.

---

46. In the system at the right, D is solved for in both equations. Therefore, it is easy to use the substitution method.

$$D = 4(5 - t)$$
$$D = 8(3 - t)$$

a) Either method

b) (10,2)

Substituting $4(5 - t)$ for D in the bottom equation, we can eliminate D and solve for <u>t</u>. We get:

$$4(5 - t) = 8(3 - t)$$
$$20 - 4t = 24 - 8t$$
$$4t = 4$$
$$t = 1$$

Now substituting "1" for <u>t</u> in $D = 4(5 - t)$, we can solve for D. We get:

$$D = 4(5 - t)$$
$$D = 4(5 - 1)$$
$$D = 4(4)$$
$$D = 16$$

Therefore, the solution of the system is D = _____, t = _____.

D = 16, t = 1

SELF-TEST 44 (pages 559 - 568)

1. Solve by the addition method.

$$6x + 6 = 5y$$
$$2y = 3x$$

2. Solve by the substitution method.

$$3p - 2t = 11$$
$$t = 2p - 7$$

3. Solve by any method.

$$9 - 2y = 3(x - y)$$
$$2x - y = 7$$

4. Solve by any method.

$$\frac{r}{2} + \frac{s}{3} = 6$$

$$\frac{r}{4} - 1 = \frac{s}{6}$$

5. Solve by any method.

$$1.5t + 2w = 6.5$$
$$t - 2.5 = 0.5w$$

ANSWERS: 1. $x = 4$ 2. $p = 3$ 3. $x = 2$ 4. $r = 8$ 5. $t = 3$
$\quad\quad\quad\quad\quad\quad y = 6$ $\quad\quad\quad\quad t = -1$ $\quad\quad\quad\quad\quad y = -3$ $\quad\quad\quad\quad s = 6$ $\quad\quad\quad\quad w = 1$

13-7   APPLIED PROBLEMS

In this section, we will discuss some applied problems that can be solved by setting up and solving a system of equations.

47.   The following problem can be solved by setting up and solving a system of equations.

The sum of two numbers is 50. The difference between the larger and smaller numbers is 14. Find the two numbers.

Using x for the larger number and y for the smaller number, we can set up two equations.

The sum of two numbers is 50.

$$x + y = 50$$

The difference between the larger and smaller numbers is 14.

$$x - y = 14$$

Therefore, the problem can be solved by solving the system at the right.

$$x + y = 50$$
$$x - y = 14$$

a) Solve the system.

b) Therefore, the larger number is _____ and the smaller number is _____ .

48.   A system of equations can be used to solve the problem below.

A 10 foot board is cut into two parts. The larger part is 2 feet longer than the smaller. How long is each part?

Using x for the larger part and y for the smaller part, we can set up two equations.

The sum of the two parts is 10 feet.

$$x + y = 10$$

The larger part is 2 feet longer than the smaller part.

$$x = y + 2$$

Continued on following page.

a) $x = 32$, $y = 18$

b) larger number is 32

smaller number is 18

48.   Continued

Therefore, we can solve the problem by solving the system at the right. We will use the substitution method.

$$x + y = 10$$
$$x = y + 2$$

Substituting $y + 2$ for $\underline{x}$ in the top equation, we can find the value of $\underline{y}$.

$$(y + 2) + y = 10$$
$$2y + 2 = 10$$
$$2y = 8$$
$$y = 4$$

Substituting 4 for $\underline{y}$ in the bottom equation, we can find the value of $\underline{x}$.

$$x = y + 2$$
$$x = 4 + 2$$
$$x = 6$$

Therefore, the larger part is _____ feet and the smaller part is _____ feet.

---

49.   The geometric problem below can be solved by means of a system of equations.

larger is 6 feet

smaller is 4 feet

   The perimeter of a rectangle is 180 centimeters. If its length is 20 centimeters longer than its width, what are its length and width?

Using L for length and W for width, we can set up two equations.

   The perimeter of a rectangle is 180 centimeters.

$$2L + 2W = 180$$

   The length is 20 centimeters longer than the width.

$$L = W + 20$$

Therefore, we can solve the problem by solving the system at the right. We will use the addition method.

$$2L + 2W = 180$$
$$L = W + 20$$

Putting the bottom equation in standard form, we get:

$$2L + 2W = 180$$
$$L - W = 20$$

Multiplying the bottom equation by 2 and adding, we can eliminate W and solve for L.

$$2L + 2W = 180$$
$$\underline{2L - 2W = 40}$$
$$4L = 220$$
$$L = 55$$

Substituting 55 for L in the top equation, we get:

$$2(55) + 2W = 180$$
$$110 + 2W = 180$$
$$2W = 70$$
$$W = 35$$

Therefore, its length is _____ centimeters and its width is _____ centimeters.

---

length is 55 cm

width is 35 cm

50. Following the examples in the last two frames, solve these.

a) A piece of wire 54 cm long is cut into two parts. If the larger part is 12 cm longer than the smaller part, how long is each part?

b) The perimeter of a rectangular lot is 144 meters. If the length is twice the width, find the length and width.

---

51. We can use a system of equations to solve this problem.

A small plane flies 120 mph with the wind and 90 mph into the wind. Find the speed of the wind and the speed of the plane in still air.

Letting x equal the speed of the plane in still air and y equal the speed of the wind, we can set up the system at the right.

$$x + y = 120$$
$$x - y = 90$$

Using the addition method, we solved the system at the right.

$$2x = 210$$
$$x = 105$$

$$x + y = 120$$
$$105 + y = 120$$
$$y = 15$$

Therefore, the speed of the plane in still air is _____ mph and the speed of the wind is _____ mph.

a) larger is 33 cm
   smaller is 21 cm

b) L = 48 meters
   W = 24 meters

---

52. We can also use a system of equations to solve this problem.

It takes a boat $2\frac{1}{2}$ hours to go 20 miles downstream and 5 hours to return. Find the speed of the current and the speed of the boat in still water.

Letting x be the speed of the boat in still water and y be the speed of the current, we organized the data in the chart below.

	Distance	Rate	Time
Downstream	20	x + y	$2\frac{1}{2}$
Upstream	20	x - y	5

Continued on following page.

speed of plane in still air = 105 mph

speed of wind = 15 mph

52. Continued

Using the relationship $r = \dfrac{d}{t}$ from $d = rt$, we can set up the following two equations.

$$x + y = \frac{20}{2\frac{1}{2}} \qquad\qquad x - y = \frac{20}{5}$$

$$x - y = 4$$

$$x + y = \frac{20}{\frac{5}{2}}$$

$$x + y = \cancel{20}^{\,4}\left(\frac{2}{\cancel{5}}\right)$$

$$x + y = 8$$

We can now set up and solve the system
of equations at the right.

$$
\begin{aligned}
x + y &= 8 \\
x - y &= 4 \\
\hline
2x \quad\;\; &= 12 \\
x &= 6
\end{aligned}
$$

$$
\begin{aligned}
x + y &= 8 \\
6 + y &= 8 \\
y &= 2
\end{aligned}
$$

Therefore, the speed of the boat in still water is _____ mph and the
speed of the current is _____ mph.

---

53. Following the examples in the last two frames, solve these.

a) If a plane can fly 575 mph with the wind and 435 mph into the wind, find the speed of the wind and the speed of the plane in still air.

b) A plane took 2 hours to fly 800 km against a headwind. The return trip with the wind took $1\frac{2}{3}$ hours. Find the speed of the plane in still air.

---

speed in still water
= 6 mph

speed of current
= 2 mph

---

a) speed in still air = 505 mph
speed of wind = 70 mph

b) speed in still air = 440 km/h

54.   A system of equations can be used to solve this mixture problem.

A chemist has one solution that is 70% acid and a second solution that is 20% acid. She wants to mix them to get 100 liters of a solution that is 50% acid. How many liters of each solution should she use?

Letting $\underline{x}$ represent the number of liters of the 70% acid and $\underline{y}$ represent the number of liters of the 20% acid, we can summarize the given information in the table below.

Liters of solution	Percent	Liters of acid
x	70	.7x
y	20	.2y
100	50	.5(100) = 50

She must mix $\underline{x}$ liters of the first solution and $\underline{y}$ liters of the second solution to get 100 liters. Therefore:

$$x + y = 100$$

The amount of acid in the new solution must be 50% of 100 liters, which is .5(100) = 50 liters. This amount will equal the amounts of acid in the two solutions mixed. These amounts of acid are 70% of $\underline{x}$ or .7x and 20% of $\underline{y}$ or .2y. Therefore:

$$70\%x + 20\%y = 50\%(100)$$
$$.7x + .2y = 50$$

Multiplying by 10 to clear the decimals, we get:

$$10(.7x + .2y) = 10(50)$$
$$10(.7x) + 10(.2y) = 10(50)$$
$$7x + 2y = 500$$

Therefore, the problem can be solved by solving the system at the right.

$$x + y = 100$$
$$7x + 2y = 500$$

Multiplying the top equation by -2 and adding, we can eliminate $\underline{y}$ and solve for $\underline{x}$.

$$-2x - 2y = -200$$
$$\underline{7x + 2y = 500}$$
$$5x = 300$$
$$x = 60$$

Substituting 60 for $\underline{x}$ in the top equation, we can solve for $\underline{y}$.

$$x + y = 100$$
$$60 + y = 100$$
$$y = 40$$

Therefore, she should use _____ liters of the 70% acid and _____ liters of the 20% acid.

---

60 liters of the 70% acid

40 liters of the 20% acid

55. Following the example in the last frame, solve this one.

Solution A is 30% alcohol and solution B is 80% alcohol. How much of each is needed to make 100 milliliters of a solution that is 70% alcohol?

---

56. For the problem below, we get a system in which one equation contains only one variable. Therefore, we don't need the addition method or substitution method to eliminate a variable.

Brine is a solution of salt and water. We want to mix pure water and a solution of brine that is 30% salt to get 100 pounds of a brine solution that is 24% salt. How many pounds of each should be mixed?

Pounds of solution	Percent	Pounds of salt
x	30	.3x
y	0	0y = 0
100	24	.24(100) = 24

We must mix $\underline{x}$ pounds of 30% brine and $\underline{y}$ pounds of water to get 100 pounds. Therefore:

$$x + y = 100$$

The amount of salt in the new solution must be 24% of 100 pounds, which is .24(100) = 24 pounds. This amount is made up of the salt in the 30% brine solution only, because there is no salt in pure water. Therefore:

$$30\%x + 0\%y = 24\%(100)$$
$$.3x = 24$$

Multiplying by 10 to clear the decimal, we get:

$$10(.3x) = 10(24)$$
$$3x = 240$$

Continued on following page.

20 milliliters of solution A

80 milliliters of solution B

56. Continued

Therefore, we can solve the problem by solving the system at the right.

$$x + y = 100$$
$$3x = 240$$

Solving for <u>x</u> immediately in the bottom equation, we get:

$$x = \frac{240}{3} = 80$$

Substituting 80 for <u>x</u> in the top equation, we can solve for <u>y</u>.

$$x + y = 100$$
$$80 + y = 100$$
$$y = 20$$

Therefore, we should mix _____ pounds of the 30% brine and _____ pounds of water.

---

57. Following the example in the last frame, solve this one.

We want to mix some milk containing 6% butterfat with skimmed milk (containing 0% butterfat) to get 100 gallons of milk containing 4.5% butterfat. How many gallons of the 6% milk and the skimmed milk should we use?

Answer column:

80 pounds of the 30% brine

20 pounds of water

75 gallons of the 6% milk, and

25 gallons of the skimmed milk

13-8   MORE APPLIED PROBLEMS

In this section, we will use a system of equations to solve some technical problems.   We will set up the original equations for each problem.   <u>Don't</u> <u>worry</u> <u>if</u> <u>you</u> <u>cannot</u> <u>understand</u> <u>how</u> <u>we</u> <u>got</u> <u>the</u> <u>equations</u>.

---

58.   The following system was set up to calculate the pitch diameter of two gears.   Use any method to solve for d_1 and d_2.

$$d_1 + d_2 = 8.3$$
$$d_2 - d_1 = 3.9$$

---

59.   A steel beam 8 meters long weighs 800 kilograms and is supported at each end.   A load of 1,200 kilograms is applied to the beam 2 meters from its left end.   To find the forces at the ends of the beam (called F_1 and F_2), we can use equilibrium principles to set up the following system of equations.   Solve for F_1 and F_2.

$$F_1 + F_2 = 2,000$$
$$4F_2 + 2,400 = 4F_1$$

$d_1 = 2.2, \quad d_2 = 6.1$

---

60.   To determine the tensile forces F_1 and F_2 in two steel cables holding a 3,400 kilogram weight, we can use equilibrium principles to set up the following system of equations.   Solve for F_1 and F_2.

$$0.8F_1 - 0.5F_2 = 0$$
$$0.4F_1 + 0.6F_2 = 3,400$$

$F_1 = 1,300$ kilograms
$F_2 = 700$ kilograms

61. A voltage of 32 volts is applied to a circuit consisting of two parallel resistors of 10 ohms and 40 ohms connected in series with a 30-ohm resistor. We want to find the currents in the 10-ohm and 40-ohm resistors. If i_1 is the current in the 10-ohm resistor and i_2 is the current in the 40-ohm resistor, we can use basic circuit principles to set up the following system of equations. Solve for i_1 and i_2.

$$30(i_1 + i_2) + 10i_2 = 32$$
$$10i_1 - 40i_2 = 0$$

F_1 = 2,500 kilograms

F_2 = 4,000 kilograms

62. The formula $V = a(t_2 - t_1)$ is used in the study of motion. It contains <u>four</u> variables. If we know the value of <u>the same two variables</u> in two situations, we can use a system of equations to find the value of the other two variables. For example, if $a = 10$ and $t_2 = 40$ is one situation and $a = 20$ and $t_2 = 30$ is a second situation, we can set up the following system. Solve for V and t_1.

$$V = 10(40 - t_1)$$
$$V = 20(30 - t_1)$$

i_1 = 0.8 ampere

i_2 = 0.2 ampere

t_1 = 20

V = 200

SELF-TEST 45 (pages 569-578)

1. A plane took 1 hour to fly 150 miles against the wind. The return trip took $\frac{3}{4}$ hours. Find the speed of the plane in still air and the speed of the wind.

2. We want to add water to a 60% solution of radiator coolant to get 15 liters of a 40% solution of coolant. How many liters of each should be mixed?

3. Using principles of equilibrium, the system below was set up to find the tensile forces F_1 and F_2, in tons, in two steel cables holding a heavy load off the ground. Find F_1 and F_2.

$$3F_2 - 2F_1 = 0$$
$$3F_1 + 2F_2 = 13$$

4. The system below was set up by applying Kirchhoff's laws to an electric circuit. Find the total current I and the branch current i, both in amperes.

$$2I + 2i + I = 8$$
$$2(I - i) - 2i = 0$$

ANSWERS:
1. 175 mph speed in still air

25 mph wind speed

2. 10 liters of 60% coolant
   5 liters of water

3. F_1 = 3 tons
   F_2 = 2 tons

4. I = 2 amperes
   i = 1 ampere

13-9   SYSTEMS OF FORMULAS AND FORMULA DERIVATION

By eliminating a variable or variables from a system of formulas, we can derive a new formula. This process is called <u>formula derivation</u>. In this section, we will define a system of formulas, show an example of formula derivation, and show how formula derivation is used in evaluations.

---

63.   Two formulas form a system only if they contain one or more common variables. For example:

$$Pt = W$$
$$W = Fs$$ form a system. The common variable is W.

$$E = IR$$
$$P = EI$$ form a system. The common variables are E and I.

State whether each pair of formulas below forms a system or not.

a) $H = \dfrac{M}{t}$ b) $S = 0.26DN$ c) $F = ma$

  $at = V$ ____ $v = gt$ ____ $F = \dfrac{GMm}{r^2}$ ____

---

64.   By eliminating a variable or variables from a system of formulas, we can derive a new formula. An example of formula derivation is discussed below.

  In the system of formulas at the right, $W = Pt$
W is solved for in each formula. By $W = Fs$
substituting Pt for W in the bottom
formula, we can eliminate W and obtain
the new formula below.

$$Pt = Fs$$

No variable is solved for in the derived formula. Since a formula is usually written with one of the variables solved for, we solve for one of the variables. Any variable can be solved for. Solve for t and F below.

a) Solve for t.   b) Solve for F.

$$Pt = Fs$$ $$Pt = Fs$$

a) Yes. The common variable is t.
b) No. There is no common variable.
c) Yes. The common variables are F and m.

a) $t = \dfrac{Fs}{P}$

b) $F = \dfrac{Pt}{s}$

65. Formula derivations can be used in evaluations. An example is discussed below.

Suppose we are given the system of formulas at the right and are asked to find P when I = 5 and R = 10. Two methods are possible.

$$V = IR$$
$$P = VI$$

1. Substitute for I and R in the system and then solve for P.

$$V = IR \qquad\qquad P = VI$$
$$V = (5)(10) \qquad P = V(5)$$
$$V = 50 \qquad\qquad P = 5V$$

Since V = 50, we can substitute for V in P = 5V and find P. We get:

$$P = 5V = 5(50) = 250$$

2. Derive a new formula in which P is solved for in terms of I and R. To do so, we can substitute IR for V in the bottom formula. We get:

$$P = VI$$
$$P = (IR)(I)$$
$$P = I^2R$$

If we substitute 5 for I and 10 for R in $P = I^2R$, we also get P = 250. That is:

$$P = (5^2)(10)$$
$$= (25)(10)$$
$$= 250$$

> Note: When we are given a system of formulas and have to perform an evaluation, we usually derive a new formula when needed rather than substitute directly into the system.

13-10 THE SUBSTITUTION METHOD AND FORMULA DERIVATION

In this section, we will show how the substitution method can be used to perform formula derivations.

66. The two major steps in a formula derivation are:

1. Eliminating a common variable or variables from the system.
2. Solving for one of the variables in the new formula.

When the variable we want to eliminate is already solved for in one formula, it is easy to use the substitution method. An example is discussed on the following page.

Continued on following page.

66. Continued

A is already solved for in the top formula in the system at the right.

$$A = DT$$
$$M = AD$$

To eliminate A, we can substitute DT for A in the bottom formula.

$$M = AD$$
$$M = (DT)D$$
$$M = D^2T$$

Solve for T in the new formula.

---

67. Since V is solved for in the top formula at the right, we can eliminate V by substituting IR for V in the bottom formula. The steps are shown.

$$V = IR$$
$$P = VI$$

$$P = VI$$
$$P = (IR)I$$
$$P = I^2R$$

Solve for I in the new formula.

$$T = \frac{M}{D^2}$$

---

68. We eliminated X_L by the substitution method and then solved for L below. Use the same method to eliminate b from the other system and then solve for t.

$$X = X_L - X_c$$
$$X_L = 2\pi fL$$

$$s = b + t$$
$$b = 2s + v$$

$$X = X_L - X_c$$
$$X = 2\pi fL - X_c$$
$$X + X_c = 2\pi fL$$
$$L = \frac{X + X_c}{2\pi f}$$

$$I = \sqrt{\frac{P}{R}}$$

$$t = -s - v$$

69. We used the substitution method to eliminate <u>a</u> below and then solved for V. Use the same method to eliminate <u>t</u> from the other formula and then solve for <u>r</u>.

$$F = ma$$

$$a = \frac{V^2}{r}$$

$$E = dt$$

$$t = \frac{2r^2}{a}$$

$$F = ma$$

$$F = m\left(\frac{V^2}{r}\right)$$

$$F = \frac{mV^2}{r}$$

$$Fr = mV^2$$

$$V^2 = \frac{Fr}{m}$$

$$V = \sqrt{\frac{Fr}{m}}$$

70. In the system below, we substituted to eliminate H and then solved for <u>t</u>. Notice how we reduced the solution to lowest terms. Use substitution to eliminate <u>s</u> from the other system and then solve for V. Be sure to reduce to lowest terms.

$$H = cd$$
$$Hp = ct^2$$

$$K = mas$$

$$s = \frac{V^2}{2a}$$

$$Hp = ct^2$$

$$(cd)p = ct^2$$

$$t^2 = \frac{\cancel{c}dp}{\cancel{c}}$$

$$t^2 = dp$$

$$t = \sqrt{dp}$$

$$r = \sqrt{\frac{aE}{2d}} \text{ , from:}$$

$$E = \frac{2dr^2}{a}$$

71. In the system at the right, we can eliminate both G and H at the same time since GH is solved for in the bottom formula.

$$K = GHv$$
$$R = GH$$

a) Substitute R for GH in the top formula.

b) Solve for <u>v</u> in the new formula.

$$V = \sqrt{\frac{2K}{m}} \text{ , from:}$$

$$K = ma\left(\frac{V^2}{2a}\right)$$

$$K = \frac{m\cancel{a}V^2}{2\cancel{a}}$$

$$K = \frac{mV^2}{2}$$

72. If the variable we want to eliminate is not already solved for in one of the formulas, we must solve for it first. For example, to eliminate D from the system at the right:

$B = DT$
$S = BD$

a) $K = GHv$
   $K = (R)v$
   $K = Rv$

b) $v = \dfrac{K}{R}$

   1. Solve for D in the top formula.

$D = \dfrac{B}{T}$

   2. Substitute that value for D in the bottom formula.

$S = BD$

$S = B\left(\dfrac{B}{T}\right)$

$S = \dfrac{B^2}{T}$

Solve for T in the new formula.

73. Let's eliminate P by the substitution method and then solve for X.

$dK = \dfrac{aT}{P}$

$T = PX$

$T = \dfrac{B^2}{S}$

   a) Solve for P in the top formula.

   b) Substitute that value for P in the bottom formula.

   c) Now solve for X in the new formula. Be sure to reduce to lowest terms.

a) $P = \dfrac{aT}{Kd}$

b) $T = \left(\dfrac{aT}{Kd}\right)X$

$T = \dfrac{aTX}{Kd}$

c) $X = \dfrac{KdT}{aT}$

$X = \dfrac{Kd}{a}$

74. Let's use substitution to eliminate $\underline{m}$ from this system and then solve for $\underline{v}$.

$$F = mg$$
$$Fr = mv^2$$

   a) Solve for $\underline{m}$ in the top formula.

   b) Substitute that value for $\underline{m}$ in the bottom formula.

   c) Now solve for $\underline{v}$. Be sure to reduce to lowest terms.

---

75. We used the substitution method to eliminate V below and then solved for R. Use the same method to eliminate Q from the other system and then solve for M.

$$i = \frac{V}{R + r}$$
$$V = IR$$

$$P = \frac{Q}{M - k}$$
$$Q = aM$$

$$i = \frac{V}{R + r}$$

$$i = \frac{IR}{R + r}$$

$$i(R + r) = IR$$

$$iR + ir = IR$$

$$ir = IR - iR$$

$$ir = R(I - i)$$

$$R = \frac{ir}{I - i}$$

a) $m = \dfrac{F}{g}$

b) $Fr = \left(\dfrac{F}{g}\right)v^2$

   $Fr = \dfrac{Fv^2}{g}$

c) $v = \sqrt{gr}$, from:

   $v^2 = \dfrac{Fgr}{F}$

$M = \dfrac{kP}{P - a}$

76. When using the substitution method, we can get complex fractions. For example, we eliminated D and then solved for A below. Eliminate <u>w</u> from the other system and then solve for <u>s</u>.

$$h = \frac{A}{D + t}$$
$$A = BD$$

$$r = \frac{bw}{s}$$

$$w = \frac{r}{2s}$$

$$D = \frac{A}{B}$$

$$h = \frac{A}{D + t}$$

$$h = \frac{A}{\frac{A}{B} + t}$$

$$h = \frac{A}{\frac{A + Bt}{B}}$$

$$h = \frac{AB}{A + Bt}$$

$$h(A + Bt) = AB$$

$$hA + hBt = AB$$

$$hBt = AB - hA$$

$$hBt = A(B - h)$$

$$A = \frac{Bht}{B - h}$$

$$s \sqrt{\frac{b}{2}}$$

Did you reduce to lowest terms?

13-11   THE EQUIVALENCE METHOD AND FORMULA DERIVATION

The equivalence method is a second method that can be used for formula derivations. When using this second method, we avoid the complex fractions that are sometimes encountered in the substitution methods. We will discuss the equivalence method in this section.

77. The equivalence method for formula derivation is based on the equivalence principle which is stated below.

$$\boxed{\begin{array}{ll} \text{If} & A = B \\ \text{and} & A = C, \\ \text{then} & B = C \end{array}}$$

Continued on following page.

77. Continued

The equivalence principle can be used to eliminate a variable from a system of formulas <u>when the same variable is solved for in both formulas</u>. For example, we eliminated <u>t</u> at the left below and R at the right below.

If $t = pq$ If $R = \dfrac{H}{v}$

and $t = d + s$,   and $R = 0.5M$,

then $pq = d + s$ then $\dfrac{H}{v} = 0.5M$

Use the equivalence principle to eliminate H at the left below and <u>b</u> at the right below.

a) $H = \dfrac{A}{BC}$ b) $b = df$

$H = \dfrac{D}{F}$ $b = \dfrac{t}{0.15G}$

_____ = _____   _____ = _____

---

78. Let's use the equivalence method to eliminate P from the system at the right and then solve for M. The three steps are:

    1. Solve for P in each formula.

    2. Use the equivalence principle to eliminate P.

    3. Now solve for M in the new formula.

$T = 0.15DP$

$M = \dfrac{AL}{P}$

$P = \dfrac{T}{0.15D}$

$P = \dfrac{AL}{M}$

$\dfrac{T}{0.15D} = \dfrac{AL}{M}$

---

a) $\dfrac{A}{BC} = \dfrac{D}{F}$

b) $df = \dfrac{t}{0.15G}$

$M = \dfrac{0.15ADL}{T}$

79. Let's use the equivalence method to eliminate B from the system and then solve for C.

$H = \dfrac{B}{KT}$

$A = BC$

   a) Solve for B in each formula.

   $B =$

   $B =$

   b) Use the equivalence principle to eliminate B.

   $\underline{\hspace{2cm}} = \underline{\hspace{2cm}}$

   c) Now solve for C.

---

80. There are two common variables in the system at the right. At the left below, we eliminated D and then solved for W. Using the same steps, eliminate H and then solve for T.

$D = HT$

$W = DH$

$$D = HT \quad \text{and} \quad D = \dfrac{W}{H}$$

$$HT = \dfrac{W}{H}$$

$$W = H^2T$$

a) $B = HKT$

   $B = \dfrac{A}{C}$

b) $HKT = \dfrac{A}{C}$

c) $C = \dfrac{A}{HKT}$

---

81. Let's eliminate <u>a</u> from this system and then solve for $\overline{V}$.

$F = ma$

$a = \dfrac{V^2}{r}$

   a) Solve for <u>a</u> in each formula.

   $a =$

   $a =$

   b) Use the equivalence principle.

   $\underline{\hspace{2cm}} = \underline{\hspace{2cm}}$

   c) Solve for V.

$H = \dfrac{D}{T}$ and $H = \dfrac{W}{D}$

$$\dfrac{D}{T} = \dfrac{W}{D}$$

$$D^2 = TW$$

$$T = \dfrac{D^2}{W}$$

82. At the left below, we eliminated $\underline{t}$ and then solved for $\underline{r}$. Using the same steps, eliminate $\underline{m}$ from the other system and then solve for $\underline{v}$.

$$E = dt$$
$$t = \frac{2r^2}{a}$$

$$t = \frac{E}{d} \quad \text{and} \quad t = \frac{2r^2}{a}$$

$$\frac{E}{d} = \frac{2r^2}{a}$$

$$aE = 2dr^2$$

$$r^2 = \frac{aE}{2d}$$

$$r = \sqrt{\frac{aE}{2d}}$$

$$F = \frac{mv^2}{r}$$
$$m = \frac{W}{g}$$

a) $a = \dfrac{F}{m}$

$\quad a = \dfrac{V^2}{r}$

b) $\dfrac{F}{m} = \dfrac{V^2}{r}$

c) $V = \sqrt{\dfrac{Fr}{m}}$, from:

$\quad V^2 = \dfrac{Fr}{m}$

83. At the left below, we eliminated V and then solved for I. Using the same steps, eliminate I from the same system and then solve for V.

$$V = IR$$
$$P = VI$$

$$V = IR \quad \text{and} \quad V = \frac{P}{I}$$

$$IR = \frac{P}{I}$$

$$I^2 R = P$$

$$I^2 = \frac{P}{R}$$

$$I = \sqrt{\frac{P}{R}}$$

$$V = IR$$
$$P = VI$$

$m = \dfrac{Fr}{v^2} \quad \text{and} \quad m = \dfrac{W}{g}$

$\dfrac{Fr}{v^2} = \dfrac{W}{g}$

$v = \sqrt{\dfrac{Fgr}{W}}$

84. Sometimes when one variable is eliminated. a second variable is also eliminated. This is true for the system at the right in which both H and $\underline{t}$ are common variables.

We can use the equivalence principle to eliminate H. We get:

If we solve for $\underline{b}$ in the new formula, $\underline{t}$ is also eliminated because we can reduce to lowest terms. That is:

$$H = bt$$
$$H = \frac{GTt}{p^2}$$

$$bt = \frac{GTt}{p^2}$$

$$b = \frac{GT\cancel{t}}{p^2\cancel{t}} = \underline{\hspace{2cm}}$$

$I = \dfrac{V}{R} \quad \text{and} \quad I = \dfrac{P}{V}$

$\dfrac{V}{R} = \dfrac{P}{V}$

$V^2 = PR$

$V = \sqrt{PR}$

$b = \dfrac{GT}{p^2}$

85. Both S and T are common variables in the system at the right. If we eliminate S and solve for $\underline{d}$, we will also eliminate T.

$$Cd = \frac{pT}{S}$$

$$T = SV$$

    a) Solve for S in each formula.

            S =

            S =

    b) Use the equivalence principle.

            =

    c) Now solve for $\underline{d}$ and reduce the fraction to lowest terms.

---

86. We eliminated F from the system below and then solved for $\underline{v}$. Notice that $\underline{m}$ was also eliminated. Eliminate $\underline{s}$ and then solve for $\overline{V}$ in the other system. Notice how $\underline{a}$ is also eliminated.

$$F = mg$$
$$Fr = mv^2$$

$$F = mg \quad \text{and} \quad F = \frac{mv^2}{r}$$

$$mg = \frac{mv^2}{r}$$

$$mgr = mv^2$$

$$v^2 = \frac{\cancel{m}gr}{\cancel{m}}$$

$$v^2 = gr$$

$$v = \sqrt{gr}$$

$$K = mas$$
$$s = \frac{V^2}{2a}$$

a) $S = \dfrac{pT}{Cd}$

   $S = \dfrac{T}{V}$

b) $\dfrac{pT}{Cd} = \dfrac{T}{V}$

c) $d = \dfrac{pV}{C}$, from:

   $pTV = CdT$

   $d = \dfrac{p\cancel{T}V}{C\cancel{T}}$

---

87. Sometimes we can eliminate two variables at one time. For example, we can eliminate both V and $\underline{p}$ at the same time from the system at the right. To do so, we solve for $\dfrac{V}{p}$ in both formulas.

$$H = \frac{V}{p}$$

$$V = \frac{ABp}{t^2}$$

Solving for $\dfrac{V}{p}$ in each, we get:

$$\frac{V}{p} = H$$
$$\frac{V}{p} = \frac{AB}{t^2}$$

Using the equivalence principle, we get:

$$H = \frac{AB}{t^2}$$

Solve for $\underline{t}$ in the new formula.

$V = \sqrt{\dfrac{2K}{m}}$, from:

$$\frac{K}{ma} = \frac{V^2}{2a}$$

$$V^2 = \frac{2\cancel{a}K}{m\cancel{a}}$$

88. Here is another system in which we can eliminate two variables at one time. To do so, we solve for $\underline{as}$ in both formulas.

$W = mas$

$v_f^2 = v_o^2 + 2as$

$t = \sqrt{\dfrac{AB}{H}}$

a) Solve for $\underline{as}$ in both formulas.

$as =$

$as =$

b) Now use the equivalence principle and then solve for W.

---

89. We eliminated E from the system below and then solved for G. Notice that we had to factor by the distributive principle because E appears twice in the top formula. Eliminate $\underline{b}$ from the other system and then solve for $\underline{t}$.

$E = A(N - BE)$
$E = GN$

$b = ct$
$b = d(b + c)$

$E = A(N - BE)$
$E = AN - ABE$
$E + ABE = AN$
$E(1 + AB) = AN$

$$E = \frac{AN}{1 + AB}$$

$$GN = \frac{AN}{1 + AB}$$

$$G = \frac{A\cancel{N}}{(1 + AB)\cancel{N}}$$

$$G = \frac{A}{1 + AB}$$

a) $as = \dfrac{W}{m}$

$as = \dfrac{v_f^2 - v_o^2}{2}$

b) $W = \dfrac{m(v_f^2 - v_o^2)}{2}$

or

$W = \dfrac{mv_f^2 - mv_o^2}{2}$

or

$W = \frac{1}{2}mv_f^2 - \frac{1}{2}mv_o^2$

$t = \dfrac{d}{1 - d}$

90. Though both the substitution method and the equivalence method can be used for formula derivations, the substitution method sometimes leads to a complex fraction. <u>In such cases</u>, <u>it is easier to use the equivalence method</u>. Use either method for the derivations below.

a) Eliminate PV and solve for E.

$$H = E + PV$$

$$\frac{PV}{C(T_1 - T_2)} = R$$

b) Eliminate H and solve for R.

$$\frac{HL}{t_2 - t_1} = AKT$$

$$\frac{H}{0.24} = IR$$

a) $E = H - CR(T_1 - T_2)$

b) $R = \dfrac{AKT(t_2 - t_1)}{0.24IL}$

SELF-TEST 46 (pages 579-592)

1. Eliminate d and solve for t.

$$d = ht$$
$$w = dh$$

2. Eliminate A and solve for r.

$$P = \frac{F}{A}$$
$$A = \pi r^2$$

3. Eliminate B and solve for T.

$$K = \frac{B}{NT}$$
$$N = \frac{BP}{K}$$

4. Eliminate v and solve for s.

$$p = v - ks$$
$$s = a + v$$

ANSWERS: 1. $t = \dfrac{w}{h^2}$ 2. $r = \sqrt{\dfrac{F}{\pi P}}$ 3. $T = \dfrac{1}{P}$ 4. $s = \dfrac{a + p}{1 - k}$

SUPPLEMENTARY PROBLEMS - CHAPTER 13

Assignment 43

Use the graphing method to solve each system.

1. $x + y = 4$
   $x - y = 2$

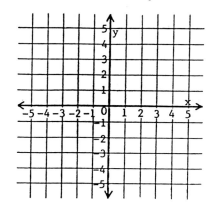

2. $2x - y = -4$
   $x + y = 1$

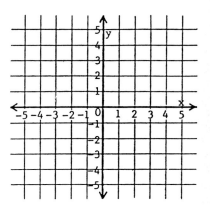

Use the addition method to solve each system.

3. $x - y = 5$
   $x + y = 3$

4. $3h + 2m = 24$
   $h - 2m = 0$

5. $5x - 4y = 23$
   $3x + 4y = 1$

6. $7r - 10v = 40$
   $3r + 10v = 60$

7. $5h - 2k = 35$
   $h + 2k = 7$

8. $3b - m = 20$
   $2b + m = 0$

9. $p + 9r = 65$
   $4p - 9r = 35$

10. $x - y = 2.8$
    $x + y = 5.2$

Use the addition method to solve each system.

11. $x - 3y = 2$
    $2x + y = 11$

12. $4x + y = 3$
    $7x + 3y = 4$

13. $3h + 2k = 28$
    $2h - k = 7$

14. $5r + 2s = 3$
    $3r + 6s = 21$

15. $3c - 4d = 11$
    $7c - 2d = 11$

16. $6v - w = 19$
    $2v - w = 7$

17. $7x + 3y = 26$
    $9x - 2y = 10$

18. $9p - 2r = 11$
    $7p + 5r = 2$

19. $5h + 7k = 32$
    $4h + 5k = 25$

20. $8x - 3y = 10$
    $5x - 2y = 7$

21. $11m - 6n = 8$
    $12m - 5n = 1$

22. $10b + 3d = 15$
    $9b + 5d = 2$

Assignment 44

Convert to standard form and then solve by the addition method.

1. $2x + y = 13$
   $y = x - 2$

2. $b = 3a$
   $2a = b - 3$

3. $t - w = 5$
   $3w + 9 = 2t$

4. $r + 2v = 2$
   $r = v + 5$

5. $h = 4m$
   $2m + 50 = 3h$

6. $s = p + 3$
   $2s - 5 = 3p$

7. $2d = 6 - 5h$
   $2h = 4d - 36$

8. $3k + 13 = 4m$
   $5m - 11 = 2k$

Clear the decimal coefficients and then solve by the addition method.

9. $.3r - .1t = 1.1$
   $.2r + .1t = .9$

10. $.21p + .08w = .5$
    $.17p + .08w = .42$

11. $1.5x - y = 1$
    $3x - 1.5y = 3$

12. $2d = 1.25p + 1$
    $d = 4.4 - .35p$

Use the substitution method to solve each system.

13. $d = 3t$
    $2d + t = 14$

14. $3E - 8R = 12$
    $E = 4R$

15. $5x - y = 8$
    $y = 2x - 5$

16. $v = 2p - 3$
    $7p - 3v = 8$

17. $2b = 5m$
    $3m - b = 2$

18. $4y - 5x = 2$
    $4x = 3y$

19. $3F_1 - 2F_2 = 40$
    $4F_1 - 3F_2 = 50$

20. $7h + 3k = 12$
    $5h - 14 = 6k$

Use any method to solve each system.

21. $7x - 3y = 34$
    $5x + 3y = 14$

22. $p + 4r = 5$
    $p + 5r = 7$

23. $s + 2w = 15$
    $s = 4w$

24. $5a - 4b = 17$
    $2a - 5b = 17$

25. $\dfrac{b}{2} + \dfrac{d}{3} = 4$

    $\dfrac{3b}{4} + \dfrac{d}{6} = 4$

26. $2x + \dfrac{y}{5} = 20$

    $5x - y = 5$

27. $.1b + .2d = 3$
    $.5b - .4d = 8$

28. $2.4r + 3.2 = w$
    $3w - 12 = 4r$

29. $4(x - y) + y = 7$
    $5x - 2y = 0$

30. $s = 3(3 - r)$
    $s = 2(4 - r)$

31. $w + 1 = 2(w - t)$
    $3(t - w) = t - 3$

Assignment 45

Use a system of equations to solve each problem.

1. The sum of two numbers is 218 and their difference is 82. Find the two numbers.

2. The sum of two numbers is 64. The first number is three times the second number. Find the two numbers.

3. The total cost of a camera and calculator is $275. If the price of the camera is four times the price of the calculator, find the price of each.

4. A metal rod 100 centimeters long is cut into two pieces. One piece is 10 centimeters longer than the other. Find the length of each piece.

5. A wire 24.6 meters long is cut into two parts so that the longer part is twice the shorter part. How long is each part?

6. We want to construct a rectangle whose perimeter is 60 centimeters and whose width is 6 centimeters less than its length. Find the length and width.

7. The perimeter of a rectangular room is 66 feet. If the length is twice the width, find the length and width.

8. A plane averages 156 mph with the wind and 130 mph against the wind. Find the speed of the wind and the speed of the plane in still air.

9. It takes a boat 2 hours to go 14 miles downstream and $3\frac{1}{2}$ hours to return. Find the speed of the current and the speed of the boat in still water.

10. Solution A is 30% acid and solution B is 60% acid. How much of each solution should be mixed to get 60 liters of a solution that is 40% acid?

11. How much of a 10% solution of alcohol and a 50% solution of alcohol should be mixed to make 200 milliliters of a 35% solution?

12. Alloy A contains 8% chromium and alloy B contains 5% chromium. How much of each alloy should be mixed to get 300 pounds of an alloy containing 6% chromium?

13. How much pure zinc (100% zinc) and how much of an alloy containing 40% zinc must be melted together to get 15 kilograms of an alloy containing 60% zinc?

14. A plane took 2 hours to fly 720 miles against the wind. The return trip with the wind took 1.8 hours. Find the speed of the plane in still air and the speed of the wind.

15. To find currents i_1 and i_2 in milliamperes in an electric circuit, the system below was set up. Solve the system.

$$3i_2 - 2i_1 = 0$$
$$3(i_1 - i_2) + i_2 = 20$$

16. To find the applied voltage V in volts and the current I in milliamperes in an electric circuit, the system below was set up. Solve the system.

$$2V = 20 - I$$
$$I - 3V = 0$$

17. To find the forces F_1 and F_2 in kilograms in two beam supports, the system below was set up. Solve the system.

$$F_1 + F_2 = 600$$
$$2F_2 + 200 = 2F_1 + 400$$

18. To find the tensile forces T_1 and T_2 in pounds in two steel cables supporting a heavy load, the system below was set up. Solve the system.

$$0.6T_2 + 0.1T_1 = 400$$
$$1.2T_1 - 0.8T_2 = 0$$

19. To find the slope <u>m</u> and the y-intercept <u>b</u> of a straight line through (2,5) and (6,3), the system below was set up. Solve the system.

$$5 = 2m + b$$
$$3 = 6m + b$$

20. To solve a calculus problem involving integration by partial fractions, the system below was set up. Solve the system.

$$2A + 3B = 4$$
$$A - 2B = 9$$

Assignment 46

1. Eliminate <u>m</u> and solve for <u>f</u>.

$$f = ma$$
$$m = kr$$

2. Eliminate W and solve for H.

$$A = LW$$
$$H = CW$$

3. Eliminate <u>v</u> and solve for <u>t</u>.

$$d = vt$$
$$p = hv$$

4. Eliminate G and solve for R.

$$R = GN$$
$$G = NP$$

5. Eliminate <u>h</u> and solve for <u>s</u>.

$$s = bh$$
$$b = hr$$

6. Eliminate T and solve for F.

$$K = FT$$
$$F = TV$$

7. Eliminate <u>a</u> and solve for <u>k</u>.

$$a = ks$$
$$k = \frac{w}{a}$$

8. Eliminate H and solve for B.

$$D = BH$$
$$P = \frac{A}{H}$$

9. Eliminate <u>t</u> and solve for <u>p</u>.

$$rt = pw$$
$$v = \frac{w}{t}$$

10. Eliminate <u>s</u> and solve for <u>f</u>.

$$d - f = s$$
$$s + h = r$$

11. Eliminate P and solve for L.

$$M = L - P$$
$$P = B - L$$

12. Eliminate G and solve for R.

$$F = G + QR$$
$$R = K - G$$

13. Eliminate <u>t</u> and solve for <u>s</u>.

$$r = \frac{p}{t}$$
$$p = \frac{t}{as}$$

14. Eliminate N and solve for A.

$$P = \frac{FN}{A}$$
$$N = \frac{P}{AR}$$

15. Eliminate <u>w</u> and solve for <u>t</u>.

$$d = \frac{bw}{r}$$
$$w = \frac{d}{rt}$$

16. Eliminate G and
solve for V.

$$G = \frac{V^2}{2}$$
$$B = GK$$

17. Eliminate C and
solve for F.

$$K = C + 273$$
$$F = \frac{9C}{5} + 32$$

18. Eliminate $\underline{r}$ and
solve for $\underline{t}$.

$$k = \frac{rt^2}{h}$$
$$h = rs$$

19. Eliminate $\underline{w}$ and
solve for $\underline{p}$.

$$r = w - p$$
$$p = kw$$

20. Eliminate V and
solve for $\underline{t}$.

$$V = at$$
$$V = T + t$$

21. Eliminate $\underline{k}$ and
solve for $\underline{v}$.

$$v + k = r$$
$$mv = kr$$

Systems of Three Equations and Determinants

14

In this chapter, we will discuss the determinant method for solving systems of two equations. We will discuss the addition method and the determinant method for solving systems of three equations. Some applied problems involving systems of three equations are included. Systems of three formulas and formula derivations are also discussed.

14-1 SECOND-ORDER DETERMINANTS

In this section, we will discuss the method for evaluating second-order determinants.

1. A determinant is a square array of numbers enclosed by two vertical lines. Two examples are shown.

$$\begin{vmatrix} 2 & 1 \\ 7 & 5 \end{vmatrix} \qquad \begin{vmatrix} -3 & 8 \\ 6 & -4 \end{vmatrix}$$

The general form for the above determinants is shown below. Since there are 2 rows and 2 columns, the determinant is called a <u>second-order</u> or <u>two-by-two</u> determinant. Each number in a determinant is called an <u>element</u>.

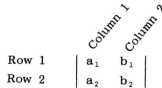

$$\begin{array}{cc} & \text{Column 1} \quad \text{Column 2} \end{array}$$

$$\begin{array}{c} \text{Row 1} \\ \text{Row 2} \end{array} \begin{vmatrix} a_1 & b_1 \\ a_2 & b_2 \end{vmatrix}$$

Continued on following page.

1. Continued

    Any second-order determinant equals one number. To find that number, we use the definition below.

    $$\begin{vmatrix} a_1 & b_1 \\ a_2 & b_2 \end{vmatrix} = a_1 b_2 - a_2 b_1$$

    That is, a determinant equals the difference of the products of the diagonals. Using the definition above, we get:

    $$\begin{vmatrix} 4 & 2 \\ 3 & 5 \end{vmatrix} = (4)(5) - (3)(2) = 20 - 6 = 14$$

    $$\begin{vmatrix} 3 & -5 \\ 2 & 4 \end{vmatrix} = (3)(4) - (2)(-5) = 12 - (-10) = \underline{}$$

---

2. Using the definition in the last frame, find the value of each determinant.

    a) $\begin{vmatrix} 5 & -2 \\ 3 & 4 \end{vmatrix} =$

    b) $\begin{vmatrix} -7 & 3 \\ 2 & 2 \end{vmatrix} =$

    c) $\begin{vmatrix} 6 & -5 \\ -5 & -6 \end{vmatrix} =$

22

---

3. When one element in a determinant is 0, the evaluation is simplified. Evaluate these.

    a) $\begin{vmatrix} 4 & -1 \\ 0 & 2 \end{vmatrix} =$

    b) $\begin{vmatrix} 1 & -3 \\ -2 & 0 \end{vmatrix} =$

    a) 26, from
          20 - (-6)

    b) -20, from -14 - 6

    c) -61, from
          -36 - 25

---

4. The determinant below contains decimal numbers.

    $$\begin{vmatrix} 2.3 & 3.1 \\ 1.7 & 8.6 \end{vmatrix} = (2.3)(8.6) - (1.7)(3.1) = 19.78 - 5.27 = 14.51$$

    We can use a calculator to evaluate the determinant. The steps are:

    Calculator Steps: 2.3 $\boxed{x}$ 8.6 $\boxed{-}$ 1.7 $\boxed{x}$ 3.1 $\boxed{=}$ 14.51

    Use a calculator to evaluate these. Don't forget to press $\boxed{+/-}$ after entering a negative number. Round to tenths.

    a) $\begin{vmatrix} 2.4 & 1.5 \\ -1.7 & 3.7 \end{vmatrix} = \underline{}$ b) $\begin{vmatrix} 1.3 & 3.2 \\ 6.9 & 2.7 \end{vmatrix} = \underline{}$

    a) 8

    b) -6

---

a) 11.4   b) -18.6

14-2 THE DETERMINANT METHOD - CRAMER'S RULE

In this section, we will discuss the determinant method for solving systems of two equations. The determinant method is called <u>Cramer's rule</u>.

---

5. The standard form for a system of two equations is given below.

$$a_1 x + b_1 y = c_1$$
$$a_2 x + b_2 y = c_2$$

The a's and b's are coefficients and the c's are constants. We can use the a's, b's, and c's to set up determinants to solve the system. The three determinants used are D, D_x, and D_y. They are shown.

$$D = \begin{vmatrix} a_1 & b_1 \\ a_2 & b_2 \end{vmatrix} \qquad D_x = \begin{vmatrix} c_1 & b_1 \\ c_2 & b_2 \end{vmatrix} \qquad D_y = \begin{vmatrix} a_1 & c_1 \\ a_2 & c_2 \end{vmatrix}$$

Determinant D is formed by the coefficients of the x's and y's. Notice these points about D_x and D_y.

1) To get D_x from D, we replace the a's (the coefficients of the x's) with the c's (see the arrow).

2) To get D_y from D, we replace the b's (the coefficients of the y's) with the c's (see the arrow).

Using the determinants above, we can set up the following solutions for <u>x</u> and <u>y</u>. They are known as <u>Cramer's rule</u>.

$$x = \frac{D_x}{D} = \frac{\begin{vmatrix} c_1 & b_1 \\ c_2 & b_2 \end{vmatrix}}{\begin{vmatrix} a_1 & b_1 \\ a_2 & b_2 \end{vmatrix}} \qquad\qquad y = \frac{D_y}{D} = \frac{\begin{vmatrix} a_1 & c_1 \\ a_2 & c_2 \end{vmatrix}}{\begin{vmatrix} a_1 & b_1 \\ a_2 & b_2 \end{vmatrix}}$$

---

6. Let's use Cramer's rule to solve this system.

$$5x - 2y = 19$$
$$7x + 3y = 15$$

First we compute D, D_x, and D_y.

$$D = \begin{vmatrix} 5 & -2 \\ 7 & 3 \end{vmatrix} = 15 - (-14) = 15 + 14 = 29$$

$$D_x = \begin{vmatrix} 19 & -2 \\ 15 & 3 \end{vmatrix} = 57 - (-30) = 57 + 30 = 87$$

$$D_y = \begin{vmatrix} 5 & 19 \\ 7 & 15 \end{vmatrix} = 75 - 133 = -58$$

Continued on following page.

6.   Continued

Then using D, D_x, and D_y, we can solve for <u>x</u> and <u>y</u>. We get:

$$x = \frac{D_x}{D} = \frac{87}{29} = 3 \qquad\qquad y = \frac{D_y}{D} = \frac{-58}{29} = -2$$

Show that (3,-2) satisfies each equation in the system below.

5x - 2y = 19 $\qquad\qquad\qquad$ 7x + 3y = 15

---

7.   Let's use the determinant method to solve this system.

$$x + y = 7$$
$$x - y = 3$$

To do so, compute D, D_x, and D_y and then use them to find <u>x</u> and <u>y</u>.

$$D = \begin{vmatrix} 1 & 1 \\ 1 & -1 \end{vmatrix} = -2$$

$$D_x = \begin{vmatrix} 7 & 1 \\ 3 & -1 \end{vmatrix} = -7 - 3 = -10$$

$$D_y = \begin{vmatrix} 1 & 7 \\ 1 & 3 \end{vmatrix} = 3 - 7 = -4$$

$$x = \frac{D_x}{D} = \frac{-10}{-2} = 5 \qquad\qquad y = \frac{D_y}{D} = \frac{-4}{-2} = 2$$

5(3) - 2(-2) = 19
15 +   4   = 19
            19 = 19

7(3) + 3(-2) = 15
21 + (-6) = 15
            15 = 15

---

8.   Let's use the determinant method to solve this system.

$$4a + 3b = 1$$
$$3a - 2b = 22$$

Compute D, D_a, and D_b and then use them to find <u>a</u> and <u>b</u>.

$$D = \begin{vmatrix} 4 & 3 \\ 3 & -2 \end{vmatrix} = -17$$

$$D_a = \begin{vmatrix} 1 & 3 \\ 22 & -2 \end{vmatrix} = -68$$

$$D_b = \begin{vmatrix} 4 & 1 \\ 3 & 22 \end{vmatrix} = 85$$

$$a = \frac{D_a}{D} = 4 \qquad\qquad b = \frac{D_b}{D} = -5$$

$$D = \begin{vmatrix} 1 & 1 \\ 1 & -1 \end{vmatrix}$$
$$= -1 - 1 = -2$$

$$D_x = \begin{vmatrix} 7 & 1 \\ 3 & -1 \end{vmatrix}$$
$$= -7 - 3 = -10$$

$$D_y = \begin{vmatrix} 1 & 7 \\ 1 & 3 \end{vmatrix}$$
$$= 3 - 7 = -4$$

$$x = \frac{-10}{-2} = 5$$

$$y = \frac{-4}{-2} = 2$$

9. The system below is not in standard form. We must put it in standard form before using the determinant method.

$$6.3F_1 + 17 = 1.9F_2$$
$$3.6F_1 + 2.7F_2 = 23$$

Converting the system to standard form, we get:

$$6.3F_1 - 1.9F_2 = -17$$
$$3.6F_1 + 2.7F_2 = 23$$

Now compute D, D_{F_1}, and D_{F_2} and then use them to find F_1 and F_2. Round F_1 to thousandths and F_2 to hundredths.

D =

D_{F_1} =

D_{F_2} =

$$F_1 = \frac{D_{F_1}}{D} = \underline{\hspace{2cm}} \qquad F_2 = \frac{D_{F_2}}{D} = \underline{\hspace{2cm}}$$

$$D = \begin{vmatrix} 4 & 3 \\ 3 & -2 \end{vmatrix} = -17$$

$$D_a = \begin{vmatrix} 1 & 3 \\ 22 & -2 \end{vmatrix} = -68$$

$$D_b = \begin{vmatrix} 4 & 1 \\ 3 & 22 \end{vmatrix} = 85$$

$$a = \frac{-68}{-17} = 4$$

$$b = \frac{85}{-17} = -5$$

D = 23.85	D_{F_1} = -2.2	D_{F_2} = 206.1	F_1 = -0.092	F_2 = 8.64

14-3   SYSTEMS OF THREE EQUATIONS

In this section, we will show how the addition method can be used to solve systems of three equations.

10. Three linear equations with the same three variables form a system of three equations. An example is given below.

$$x + y + z = 4$$
$$x - 2y - z = 1$$
$$2x - y - 2z = -1$$

A solution of a system of three linear equations is an ordered triple that satisfies all three equations. To show that $(2, -1, 3)$ is a solution of the system above, we substituted 2 for $\underline{x}$, -1 for $\underline{y}$, and 3 for $\underline{z}$ in each equation.

$x + y + z = 4$	$x - 2y - z = 1$	$2x - y - 2z = -1$
$2 + (-1) + 3 = 4$	$2 - 2(-1) - 3 = 1$	$2(2) - (-1) - 2(3) = -1$
$4 = 4$	$2 + 2 - 3 = 1$	$4 + 1 - 6 = -1$
	$1 = 1$	$-1 = -1$

Continued on following page.

10. Continued

Is $(2,4,3)$ or $(4,2,-1)$ a solution of the system below? _____

$$x - y + z = 1$$
$$2x - 3y + 4z = -2$$
$$3x - 2y - z = 9$$

$(4,2,-1)$ is a solution

11. For convenience, we numbered the equations in the system below.

$$(1) \quad 4x - y + z = 6$$
$$(2) \quad -3x + 2y - z = -3$$
$$(3) \quad 2x + y + 2z = 3$$

The three steps needed to solve the system by the addition method are discussed below.

  1. Eliminate <u>one</u> variable to get a system of two equations with two variables.

   a) Add a pair of equations to eliminate one variable. We added (1) and (2) below to eliminate <u>z</u>.

$$(1) \quad 4x - y + z = 6$$
$$(2) \quad \underline{-3x + 2y - z = -3}$$
$$(4) \quad x + y \qquad = 3$$

   b) Add a different pair of equations to eliminate <u>the same variable</u>. We added (2) and (3) below to eliminate <u>z</u>. To do so, we had to multiply equation (2) by 2.

$$(2) \quad -3x + 2y - z = -3 \qquad\qquad -6x + 4y - 2z = -6$$
$$(3) \quad 2x + y + 2z = 3 \qquad\qquad \underline{2x + y + 2z = 3}$$
$$\qquad\qquad\qquad\qquad\qquad (5) \quad -4x + 5y \qquad = -3$$

  2. Solve the system of two equations, (4) and (5), to find the values of the two variables <u>x</u> and <u>y</u>.

   a) We multiplied equation (4) by 4 below to eliminate <u>x</u> and solve for <u>y</u>.

$$(4) \quad x + y = 3 \qquad\qquad 4x + 4y = 12$$
$$(5) \quad \underline{-4x + 5y = -3} \qquad\qquad \underline{-4x + 5y = -3}$$
$$\qquad\qquad\qquad\qquad\qquad\qquad 9y = 9$$
$$\qquad\qquad\qquad\qquad\qquad\qquad y = 1$$

   b) We substituted "1" for <u>y</u> in equation (4) to solve for <u>x</u>.

$$(4) \quad x + y = 3$$
$$x + 1 = 3$$
$$x = 2$$

Continued on following page.

11. Continued

   3. Substitute 2 for <u>x</u> and "1" for <u>y</u> in one of the three original
equations to solve for <u>z</u>. We substituted in equation (1).

$$(1)\quad 4x - y + z = 6$$
$$4(2) - 1 + z = 6$$
$$8 - 1 + z = 6$$
$$7 + z = 6$$
$$z = -1$$

The solution is (2,1,-1). We checked it in the three original equations below.

$4x - y + z = 6$	$-3x + 2y - z = -3$	$2x + y + 2z = 3$
$4(2) - 1 + (-1) = 6$	$-3(2) + 2(1) - (-1) = -3$	$2(2) + 1 + 2(-1) = 3$
$8 - 1 - 1 = 6$	$-6 + 2 + 1 = -3$	$4 + 1 + (-2) = 3$
$6 = 6$	$-3 = -3$	$3 = 3$

12. Let's use the addition method to solve the system below.

$$(1)\quad x + y + z = 6$$
$$(2)\quad 2x - y + 3z = 9$$
$$(3)\quad -x + 2y + 2z = 9$$

a) Write the system of two equations obtained if <u>y</u> is eliminated
by adding (1) and (2) and then adding (2) and (3).

b) Solve the system of two equations to find the values of <u>x</u>
and <u>z</u>.

c) Substitute those values for <u>x</u> and <u>z</u> in one of the three original equations to find the corresponding value of <u>y</u>.

d) The solution of the system is: _____

13. In the system below, the variable z does not appear in equation (1). Therefore, the system is easier to solve.

$$(1) \qquad x - y = 4$$
$$(2) \quad x - 2y + z = 5$$
$$(3) \quad 2x + 3y - 2z = 3$$

a) Let's eliminate z to get a system of two equations. One of the equations is (1) above. Find the other equation by adding (2) and (3) to eliminate z.

b) Solve the system of two equations to find the values of x and y.

c) Substitute those values for x and y in either equation (2) or (3) to find the corresponding value of z.

d) The solution of the system is: _____

a) $3x + 4z = 15$
   $3x + 8z = 27$

b) $x = 1$, $z = 3$

c) $y = 2$

d) $(1,2,3)$

---

14. To use the addition method, a system of three equations must be in standard form. If a system is not in standard form, we must convert it to standard form first. For example, we converted one system below to standard form. Convert the other system to standard form.

Original System	Standard Form
$x - 2y + 5 = 3z$	$x - 2y - 3z = -5$
$2x - z - 1 = 3y$	$2x - 3y - z = 1$
$y = x + 2z$	$-x + y - 2z = 0$

$2b - d - 7 = c$

$b - 5c + 3 = 4d$

$c - 3b = 2d + 5$

_____

a) $x - y = 4$
   $4x - y = 13$

b) $x = 3$, $y = -1$

c) $z = 0$

d) $(3,-1,0)$

Answer to Frame 14:

$2b - c - d = 7$

$b - 5c - 4d = -3$

$-3b + c - 2d = 5$

SELF-TEST 47 (pages 597-605)

1. Evaluate: $\begin{vmatrix} -3 & -6 \\ 2 & 5 \end{vmatrix}$

2. Evaluate: $\begin{vmatrix} 1.9 & 2.7 \\ -4.3 & -3.1 \end{vmatrix}$

3. Solve by determinants.

$3x - 2y = 13$ a) $D =$ _____
$4x + 3y = 6$ b) $D_x =$ _____
                    c) $D_y =$ _____
                    d) $x =$ _____
                    e) $y =$ _____

4. Solve by determinants. Round p and r to hundredths.

$2.9p + 1.4r = 10$ a) $D =$ _____
$3.2p - 4.5r = 8$ b) $D_p =$ _____
                    c) $D_r =$ _____
                    d) $p =$ _____
                    e) $r =$ _____

Solve each system by the addition method.

5. $x + y + z = 7$
    $2x + 2y - z = 5$
    $x - y + z = 9$

6. $x - 2y - 2z = 8$
    $2x + y + z = 6$
    $2x + 3z = 2$

ANSWERS: 1. -3

           2. 5.72

           3. a) $D = 17$
              b) $D_x = 51$
              c) $D_y = -34$
              d) $x = 3$
              e) $y = -2$

4. a) $D = -17.53$
   b) $D_p = -56.2$
   c) $D_r = -8.8$
   d) $p = 3.21$
   e) $r = 0.50$

5. $x = 5$
   $y = -1$
   $z = 3$

6. $x = 4$
   $y = 0$
   $z = -2$

14-4   APPLIED PROBLEMS

In this section, we will solve some applied problems that involve a system of three equations.

15.   We can use a system of three equations to solve this problem.

The sum of three numbers is 6. The first number minus the second plus the third is 8. The third minus the first is 2 more than the second. Find the three numbers.

Letting the three numbers be x, y, and z, we can set up three equations.

The sum of three numbers is 6.

$$x + y + z = 6$$

The first number minus the second plus the third is 8.

$$x - y + z = 8$$

The third minus the first is 2 more than the second.

$$z - x = y + 2$$

Therefore, we have the system of three equations below. Notice that we rewrote the system to put the third equation in standard form.

(1) $x + y + z = 6$ (1) $x + y + z = 6$

(2) $x - y + z = 8$ or   (2) $x - y + z = 8$

(3) $z - x = y + 2$ (3) $-x - y + z = 2$

To solve the system, we'll begin by eliminating y. To do so, we added (1) and (2) and then (1) and (3) below.

(1) $x + y + z = 6$ (1) $x + y + z = 6$
(2) $\underline{x - y + z = 8}$ (3) $\underline{-x - y + z = 2}$
(4) $2x \quad\ + 2z = 14$ (5) $\qquad\qquad 2z = 8$

Notice that we also eliminated x when adding (1) and (3). Therefore, solving the system below for x and z is simplified.

(4) $2x + 2z = 14$
(5) $2z = 8$

$z = 4$ (Solving $2z = 8$)

$2x + 2(4) = 14$ (Substituting in $2x + 2z = 14$)
$2x + \ \ 8 \ \ = 14$
$2x = 6$
$x = 3$

Substituting 4 for z and 3 for x in equation (1), we have solved for y below.

$$x + y + z = 6$$
$$3 + y + 4 = 6$$
$$y = -1$$

The solution of the system is (3,-1,4) or x = 3, y = -1, z = 4. Therefore, the first number is _____, the second number is _____, and the third number is _____.

16. We can use a system of three equations to solve this problem.

    In triangle ABC, angle A is 20° more than angle B. Angle C is twice angle B. Find the size of each angle.

Letting the three angles be A, B, and C, we can set up three equations.

    We know that the sum of the angles in a triangle is 180°.

$$A + B + C = 180$$

Angle A is 20° more than angle B.

$$A = B + 20$$

Angle C is twice angle B.

$$C = 2B$$

Therefore, we have the system of three equations below. Notice that we rewrote the system to put the two bottom equations in standard form.

A + B + C = 180		A + B + C = 180
A = B + 20	or	A - B = 20
C = 2B		2B - C = 0

If we solve the system, we get: A = 60, B = 40, and C = 80. Therefore, angle A contains _____°, angle B contains _____°, and angle C contains _____°.

first = 3

second = -1

third = 4

---

17. Following the examples in the last two frames, solve these.

  a) The sum of three numbers is 20. The first minus the second equals the third. The second is 1 more than twice the third. Find the three numbers.

  b) In triangle ABC, angle B is three times angle C. Angle A is 40° more than angle B. Find the size of each angle.

A = 60°

B = 40°

C = 80°

18. Various technical problems involve solving a system of three equations. For example, the system below was set up to find the forces, in kilograms, on a steel beam. Find F_1, F_2, and F_3.

$$3F_1 + 5F_2 - 2F_3 = 1,600$$
$$F_1 + F_2 - 2F_3 = 0$$
$$2F_1 - 3F_2 + 4F_3 = 500$$

a) first = 10
   second = 7
   third = 3

b) A = 100°
   B = 60°
   C = 20°

19. The following system was set up to find three currents, in milliamperes, in an electronic circuit.

$$0.4i_1 - i_2 + 1.2i_3 = 0.4$$
$$0.2i_1 + 0.4i_2 - i_3 = 0.6$$
$$0.2i_1 - 0.2i_2 - 0.1i_3 = 0$$

Before using the addition method, we clear the decimals by multiplying each equation by 10. We get:

$$4i_1 - 10i_2 + 12i_3 = 4$$
$$2i_1 + 4i_2 - 10i_3 = 6$$
$$2i_1 - 2i_2 - i_3 = 0$$

Use the addition method to find i_1, i_2, and i_3.

F_1 = 200 kg
F_2 = 300 kg
F_3 = 250 kg

i_1 = 5 milliamperes
i_2 = 4 milliamperes
i_3 = 2 milliamperes

14-5   SYSTEMS OF THREE FORMULAS AND FORMULA DERIVATION

In this section, we will define a system of three formulas and show how we can derive a new formula from a system of that type by eliminating at least two variables.

20. Three formulas form a system if two of the formulas contain <u>at least</u> <u>one</u> <u>common</u> <u>variable</u> and the third formula contains <u>at least one</u> <u>variable</u> in common with <u>one</u> of the other two.

The three formulas at the right form a system because:

1) Formulas (1) and (2) contain an <u>a</u> in common.

2) Formula (3) contains an <u>r</u> and a <u>v</u> in common with formula (2).

(1) $F = ma$

(2) $a = \dfrac{v^2}{r}$

(3) $v = wr$

The three formulas at the right form a system because:

a) Formulas (1) and (2) contain an _____ in common.

b) Formula (3) contains an _____ in common with formula $\overline{(1)}$.

(1) $B = \dfrac{m}{a}$

(2) $p = m + n$

(3) $q = a - t$

21. To get a new relationship, <u>at least</u> <u>two</u> <u>variables</u> can be eliminated from any system of three formulas. For example, we can eliminate <u>a</u> and <u>r</u> from the system below and then solve for F. The steps are discussed.

(1) $F = ma$

(2) $a = \dfrac{v^2}{r}$

(3) $v = wr$

1) To eliminate <u>a</u>, find two formulas that contain <u>a</u>. The two formulas are (1) and (2).

a) Solve for <u>a</u> in formulas (1) and (2). (<u>Note</u>: <u>a</u> is already solved-for in formula (2).)

$$F = ma \qquad\qquad a = \dfrac{v^2}{r}$$

$$a = \dfrac{F}{m}$$

b) Use the equivalence principle to eliminate <u>a</u>. That is, equate the two solutions above.

$$\dfrac{F}{m} = \dfrac{v^2}{r} \qquad \text{(This is a new formula.)}$$

a) m

b) a

Continued on following page.

21.   Continued

  2) To eliminate r, use the new formula above and the original formula that has not been used. That is, use formula (3).

  a) Solve for r in formula (3) and the new formula above.

$$v = wr \qquad\qquad \frac{F}{m} = \frac{v^2}{r}$$

$$r = \frac{v}{w} \qquad\qquad Fr = mv^2$$

$$\qquad\qquad\qquad r = \frac{mv^2}{F}$$

  b) Use the equivalence principle to eliminate "r".

$$\frac{v}{w} = \frac{mv^2}{F} \qquad \text{(This is a new formula;}$$
$$\qquad\qquad\qquad \underline{a} \text{ and } \underline{r} \text{ are eliminated.)}$$

  3) Solve for F in the new formula in which $\underline{a}$ and $\underline{r}$ are eliminated.

$$\frac{v}{w} = \frac{mv^2}{F}$$

$$Fv = mv^2w$$

$$F = \frac{mv^2w}{v}$$

$$F = \frac{mv^{\cancel{2}}w}{\cancel{v}}$$

$$F = mvw \qquad \underline{\text{Answer}}$$

22.   Let's eliminate $\underline{m}$ and $\underline{a}$ from the system below and then solve for B.

$$(1) \quad B = \frac{m}{a}$$

$$(2) \quad p = m + n$$

$$(3) \quad q = a - t$$

a) Solve for $\underline{m}$ in formulas (1) and (2) and equate the two solutions.

b) Solve for $\underline{a}$ in the new formula and in formula (3), and equate the two solutions.

c) Now solve for B in this new formula in which $\underline{a}$ and $\underline{m}$ are eliminated.

23. Eliminate e_o and $\underline{i}$ from the system below and then solve for A.
    Note: When simplifying at the end, e_i is also eliminated.

    (1) $A = \dfrac{e_o}{e_i}$

    (2) $e_o = iR$

    (3) $i = \dfrac{me_i}{R + r_p}$

a) $aB = p - n$

b) $\dfrac{p - n}{B} = q + t$

c) $B = \dfrac{p - n}{q + t}$

$A = \dfrac{mR}{R + r_p}$

SELF-TEST 48 (pages 606 -612)

1. The sum of three numbers is 24. The first equals the sum of the second and third. The third is 2 more than the second. Find the numbers.

2. To solve a calculus problem involving partial fractions, the system below was set up. Find A, B, and C.

$$A + B - C = 0$$
$$2B + C = -4$$
$$A - 2C = 1$$

Continued on following page.

<u>SELF-TEST 48</u> (<u>pages</u> 606 -612) - Continued

3. The system below involves three forces, in kilograms, present in a structure. Find F_1, F_2, and F_3.

$$F_1 + F_2 - F_3 = 200$$
$$F_1 + 2F_2 - 4F_3 = 0$$
$$3F_1 - F_2 + 2F_3 = 350$$

4. The system below involves three currents, in milliamperes, present in an electronic circuit. Find i_1, i_2, and i_3.

$$3i_1 + 2i_2 - 2i_3 = 10$$
$$i_2 + i_3 = 5$$
$$2i_1 - i_3 = 5$$

5. Eliminate <u>t</u> and <u>s</u> and solve for <u>d</u>.

$$d = rt$$
$$t = s + a$$
$$s = 2d$$

6. Eliminate <u>w</u> and <u>d</u> and solve for <u>r</u>.

$$p = kw$$
$$w = \frac{d}{r}$$
$$k = d - r$$

ANSWERS:

1. first number = 12
   second number = 5
   third number = 7

2. A = 5
   B = -3
   C = 2

3. F_1 = 100 kg
   F_2 = 250 kg
   F_3 = 150 kg

4. i_1 = 4 ma
   i_2 = 2 ma
   i_3 = 3 ma

5. $d = \dfrac{ar}{1 - 2r}$

6. $r = \dfrac{k^2}{p - k}$

14-6   THIRD-ORDER DETERMINANTS

In this section, we will discuss a method for evaluating third-order determinants.

24. The general form for a <u>third-order</u> or <u>three-by-three</u> determinant is shown below. Notice that there are 3 rows and 3 columns.

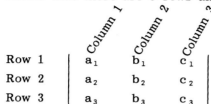

Row 1 $\quad a_1 \quad b_1 \quad c_1$

Row 2 $\quad a_2 \quad b_2 \quad c_2$

Row 3 $\quad a_3 \quad b_3 \quad c_3$

Any third-order determinant equals one number. To find that number, we use the definition below.

$$D = a_1b_2c_3 + b_1c_2a_3 + c_1a_2b_3 - a_3b_2c_1 - b_3c_2a_1 - c_3a_2b_1$$

To find the six terms in the definition, we copy the first two columns to the right of the third column.

$$\begin{vmatrix} a_1 & b_1 & c_1 \\ a_2 & b_2 & c_2 \\ a_3 & b_3 & c_3 \end{vmatrix} \begin{matrix} a_1 & b_1 \\ a_2 & b_2 \\ a_3 & b_3 \end{matrix}$$

The first three terms are the diagonals from left to right starting from a_1, b_1, and c_1. Notice that all three terms $(a_1b_2c_3 + b_1c_2a_3 + c_1a_2b_3)$ are <u>positive</u>.

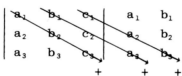

The last three terms are the diagonals from left to right starting from a_3, b_3, and c_3. Notice that all three terms $(-a_3b_2c_1 - b_3c_2a_1 - c_3a_2b_1)$ are <u>negative</u>.

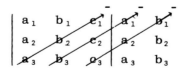

Let's use the preceding definition to find the value of the determinant D below. We recopied the determinant and wrote the first two columns on the right side.

$$D = \begin{vmatrix} 1 & -2 & -3 \\ 4 & -1 & 1 \\ 2 & 0 & 2 \end{vmatrix}$$

$D = (1)(-1)(2) + (-2)(1)(2) + (-3)(4)(0) - (2)(-1)(-3) - (0)(1)(1) - (2)(4)(-2)$

$\quad = \quad (-2) \quad + \quad (-4) \quad + \quad 0 \quad - \quad 6 \quad - \quad 0 \quad - \quad (-16)$

$\quad = \quad 4$

25. Using the definition in the last frame, we found the value of the determinant below. Find the value of the other determinant.

$$D = \begin{vmatrix} 3 & 2 & -1 \\ 1 & 0 & -3 \\ 4 & -2 & 5 \end{vmatrix}$$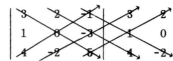

D = 0 + (-24) + 2 - 0 - 18 - 10 = -50

$$D = \begin{vmatrix} 2 & 3 & -5 \\ 1 & 4 & 3 \\ 3 & -2 & 1 \end{vmatrix}$$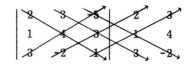

D = __ + __ + __ - __ - __ - __ = _____

---

26. Using the same method, find the value of each determinant.

a)
$$D = \begin{vmatrix} 0 & 2 & 0 \\ 3 & -1 & 1 \\ 1 & -2 & 2 \end{vmatrix}$$

b)
$$D = \begin{vmatrix} 1 & 4 & 1 \\ 2 & -1 & -2 \\ 3 & -2 & 1 \end{vmatrix}$$

D = 8 + 27 + 10 -
(-60) - (-12)
- 3 = 114

---

27. The determinant below contains decimal numbers.

$$D = \begin{vmatrix} 1.2 & 2.1 & 0.6 \\ 0.7 & -1.5 & 0.9 \\ 1.4 & -2.2 & -0.1 \end{vmatrix}$$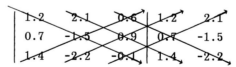

We can use a calculator to find the value of D. The value is 5.685. The steps for the first two terms are:

Calculator Steps:

1.2 $\boxed{\times}$ 1.5 $\boxed{+/-}$ $\boxed{\times}$ 0.1 $\boxed{+/-}$ $\boxed{+}$ 2.1 $\boxed{\times}$ 0.9 $\boxed{\times}$ 1.4 $\boxed{+}$ ...

Use a calculator to complete the evaluation. Remember that the last three terms are subtracted.

D = _____

a) D = -10, from:
0 + 2 + 0 - 0 - 0
- 12

b) D = -38, from:
(-1) + (-24)
+ (-4) - (-3)
- 4 - 8

D = 5.685

28.   Use a calculator to find the value of each determinant.

a)
$$D = \begin{vmatrix} 0.8 & 0.7 & 1.7 \\ 1.2 & 2.4 & 1.4 \\ -1.1 & -3.1 & 0.3 \end{vmatrix}$$

$$D = \underline{\hspace{2cm}}$$

b)
$$D = \begin{vmatrix} 4.3 & 0.8 & -5.1 \\ 0.1 & 6.2 & -0.9 \\ 0.4 & -0.1 & -1.3 \end{vmatrix}$$

$$D = \underline{\hspace{2cm}}$$

a) $D = 0.882$

b) $D = -22.53$

14-7   THE DETERMINANT METHOD - SYSTEMS OF THREE EQUATIONS

In this section, we will discuss the determinant method for solving systems of three equations. The determinant method for systems of three equations is also called <u>Cramer's rule</u>.

29.   Cramer's rule can be extended to solving systems of three equations. To do so, we start with the standard form of a system of three equations.

$$a_1x + b_1y + c_1z = d_1$$
$$a_2x + b_2y + c_2z = d_2$$
$$a_3x + b_3y + c_3z = d_3$$

The a's, b's, and c's are coefficients and the d's are constants. We can use the a's, b's, c's, and d's to set up determinants to solve the system. The four determinants used are D, D_x, D_y, D_z. They are shown below.

$$D = \begin{vmatrix} a_1 & b_1 & c_1 \\ a_2 & b_2 & c_2 \\ a_3 & b_3 & c_3 \end{vmatrix} \qquad D_x = \begin{vmatrix} d_1 & b_1 & c_1 \\ d_2 & b_2 & c_2 \\ d_3 & b_3 & c_3 \end{vmatrix}$$

$$D_y = \begin{vmatrix} a_1 & d_1 & c_1 \\ a_2 & d_2 & c_2 \\ a_3 & d_3 & c_3 \end{vmatrix} \qquad D_z = \begin{vmatrix} a_1 & b_1 & d_1 \\ a_2 & b_2 & d_2 \\ a_3 & b_3 & d_3 \end{vmatrix}$$

Continued on following page.

29. **Continued**

Determinant D is formed by the coefficients of the x's, y's, and z's. Notice these points about D_x, D_y, and D_z.

    1) To get D_x from D, we replace the a's (the coefficients of the x's) with the d's (see the arrow).

    2) To get D_y from D, we replace the b's (the coefficients of the y's) with the d's (see the arrow).

    3) To get D_z from D, we replace the c's (the coefficients of the z's) with the d's (see the arrow).

Using the determinants above, we can set up the following solutions for $\underline{x}$, $\underline{y}$, and $\underline{z}$.

$$x = \frac{D_x}{D} \qquad\qquad y = \frac{D_y}{D} \qquad\qquad z = \frac{D_z}{D}$$

30. Let's use Cramer's rule to solve this system.

$$x - 2y + 3z = 6$$
$$2x - y - z = -3$$
$$x + y + z = 6$$

First we compute D, D_x, D_y, and D_z.

$$D = \begin{vmatrix} 1 & -2 & 3 \\ 2 & -1 & -1 \\ 1 & 1 & 1 \end{vmatrix} = 15 \qquad D_x = \begin{vmatrix} 6 & -2 & 3 \\ -3 & -1 & -1 \\ 6 & 1 & 1 \end{vmatrix} = 15$$

$$D_y = \begin{vmatrix} 1 & 6 & 3 \\ 2 & -3 & -1 \\ 1 & 6 & 1 \end{vmatrix} = 30 \qquad D_z = \begin{vmatrix} 1 & -2 & 6 \\ 2 & -1 & -3 \\ 1 & 1 & 6 \end{vmatrix} = 45$$

Then using D, D_x, D_y, and D_z, we can solve for $\underline{x}$, $\underline{y}$, and $\underline{z}$. We get:

$$x = \frac{D_x}{D} = \frac{15}{15} = 1 \qquad y = \frac{D_y}{D} = \frac{30}{15} = 2 \qquad z = \frac{D_z}{D} = \frac{45}{15} = 3$$

The solution is: x = 1, y = 2, and z = 3. Check the solution in each original equation below.

a) x - 2y + 3z = 6   b) 2x - y - z = -3   c) x + y + z = 6

31. Let's use the determinant method to solve this system.

$$2x - y + z = 2$$
$$x + 3y + 2z = 7$$
$$5x + y - z = -9$$

To do so, compute D, D_x, D_y, and D_z and then use them to find $\underline{x}$, $\underline{y}$, and $\underline{z}$.

D =

D_x =

D_y =

D_z =

$x = \dfrac{D_x}{D} =$ _____ $y = \dfrac{D_y}{D} =$ _____ $z = \dfrac{D_z}{D} =$ _____

a) 1-2(2)+3(3) = 6
      1 - 4 + 9 = 6
                    6 = 6

b) 2(1) - 2 - 3 = -3
        2 - 2 - 3 = -3
                   -3 = -3

c) 1 + 2 + 3 = 6
              6 = 6

---

32. The following system was set up to find the forces (in kilograms) on a steel beam.

$$12F_1 + 25F_2 - 20F_3 = 3,200$$
$$15F_1 - 30F_2 + 30F_3 = 0$$
$$10F_1 + 15F_2 - 12F_3 = 2,200$$

Compute D, D_{F_1}, D_{F_2}, and D_{F_3} and then use them to find F_1, F_2, and F_3. Use a calculator to evaluate the determinants.

D =

D_{F_1} =

D_{F_2} =

Continued on following page.

$D = \begin{vmatrix} 2 & -1 & 1 \\ 1 & 3 & 2 \\ 5 & 1 & -1 \end{vmatrix} = -35$

$D_x = \begin{vmatrix} 2 & -1 & 1 \\ 7 & 3 & 2 \\ -9 & 1 & -1 \end{vmatrix} = 35$

$D_y = \begin{vmatrix} 2 & 2 & 1 \\ 1 & 7 & 2 \\ 5 & -9 & -1 \end{vmatrix} = 0$

$D_z = \begin{vmatrix} 2 & -1 & 2 \\ 1 & 3 & 7 \\ 5 & 1 & -9 \end{vmatrix} = -140$

$x = \dfrac{35}{-35} = -1$

$y = \dfrac{0}{-35} = 0$

$z = \dfrac{-140}{-35} = 4$

32. Continued

$$D_{F_3} = \begin{vmatrix} & & \\ & & \\ & & \end{vmatrix}$$

$$F_1 = \frac{D_{F_1}}{D} = \underline{\hspace{1cm}} \qquad F_2 = \frac{D_{F_2}}{D} = \underline{\hspace{1cm}} \qquad F_3 = \frac{D_{F_3}}{D} = \underline{\hspace{1cm}}$$

Answers to Frame 32:

$$D = \begin{vmatrix} 12 & 25 & -20 \\ 15 & -30 & 30 \\ 10 & 15 & -12 \end{vmatrix} = 420 \qquad D_{F_2} = \begin{vmatrix} 12 & 3,200 & -20 \\ 15 & 0 & 30 \\ 10 & 2,200 & -12 \end{vmatrix} = 84,000$$

$$D_{F_1} = \begin{vmatrix} 3,200 & 25 & -20 \\ 0 & -30 & 30 \\ 2,200 & 15 & -12 \end{vmatrix} = 42,000 \qquad D_{F_3} = \begin{vmatrix} 12 & 25 & 3,200 \\ 15 & -30 & 0 \\ 10 & 15 & 2,200 \end{vmatrix} = 63,000$$

$$F_1 = \frac{42,000}{420} = 100 \text{ kg}$$

$$F_2 = \frac{84,000}{420} = 200 \text{ kg}$$

$$F_3 = \frac{63,000}{420} = 150 \text{ kg}$$

33. The following system was used to find three currents, in milliamperes, in an electronics circuit.

$$1.4i_1 - i_2 + 0.2i_3 = 3$$
$$i_1 + 0.5i_2 + 1.2i_3 = 5$$
$$0.8i_1 - 1.6i_2 - i_3 = 0$$

Find i_1, i_2, and i_3. Use a calculator to evaluate the determinants. Round i_1, i_2, and i_3 to hundredths.

Solving systems of three equations is a lengthy process, even when a calculator is used. Fortunately, programmable calculators and computers can be used to solve systems of that type. A programmable calculator or computer does all the evaluations and prints out the solution. The operator only has to enter the coefficients and constants.

$D = -0.372$

$D_{i_1} = -2.34$

$D_{i_2} = -1.92$

$D_{i_3} = 1.2$

$i_1 = 6.29$ milliamperes

$i_2 = 5.16$ milliamperes

$i_3 = -3.23$ milliamperes

SELF–TEST 49 (pages 612 - 620)

Evaluate each determinant.

1. $\begin{vmatrix} 1 & -3 & 2 \\ -1 & 4 & -3 \\ 2 & -1 & 1 \end{vmatrix}$

2. $\begin{vmatrix} 5 & 0 & 2 \\ 1 & -3 & 0 \\ 0 & 4 & -2 \end{vmatrix}$

3. Solve by determinants.

$$x + 2y - z = 7$$
$$4x + y + 2z = 9$$
$$2x - 3y + z = 1$$

a) $D =$ _____

b) $D_x =$ _____

c) $D_y =$ _____

d) $D_z =$ _____

e) $x =$ _____

f) $y =$ _____

g) $z =$ _____

Continued on following page.

4. Solve by determinants.

$$1.5r + 0.2s + 0.5t = 6$$
$$0.4r + \quad s + 0.3t = 7$$
$$0.6r + 0.6s + 0.2t = 5$$

a) $D =$ _____

b) $D_r =$ _____

c) $D_s =$ _____

d) $D_t =$ _____

e) $r =$ _____

f) $s =$ _____

g) $t =$ _____

ANSWERS: 1. 2   3. a) $D = 21$ e) $x = 3$ 4. a) $D = -0.13$ e) $r = 2$

            2. 38   b) $D_x = 63$ f) $y = 1$ b) $D_r = -0.26$ f) $s = 5$

                   c) $D_y = 21$ g) $z = -2$ c) $D_s = -0.65$ g) $t = 4$

                   d) $D_z = -42$ d) $d_t = -0.52$

SUPPLEMENTARY PROBLEMS - CHAPTER 14

Assignment 47

Find the numerical value of each determinant.

1. $\begin{vmatrix} 2 & 3 \\ 7 & 5 \end{vmatrix}$ -11

2. $\begin{vmatrix} -3 & -5 \\ 6 & 4 \end{vmatrix}$ 18

3. $\begin{vmatrix} 8 & 12 \\ -7 & 0 \end{vmatrix}$

4. $\begin{vmatrix} 8.7 & 3.9 \\ -1.4 & -2.6 \end{vmatrix}$

Solve each system of equations by the determinant method. Record the numerical values of the three determinants used in each solution.

5. $4x + 5y = 10$
$3x + 2y = 11$

   a) D = ____   d) x = ____

   b) D_x = ____   e) y = ____

   c) D_y = ____

6. $3t - 2w = 6$
$4t - 3w = 10$

   a) D = ____   d) t = ____

   b) D_t = ____   e) w = ____

   c) D_w = ____

7. $8Q = 6 - 5P$
$2P = 3Q + 21$

   a) D = ____   d) P = ____

   b) D_P = ____   e) Q = ____

   c) D_Q = ____

8. $a = 2.5b + 3$
$b = 1.5a - 10$

   a) D = ____   d) a = ____

   b) D_a = ____   e) b = ____

   c) D_b = ____

9. $9r - 1 = 5t$
$27 - 6t = 8r$

   a) D = ____   d) r = ____

   b) D_r = ____   e) t = ____

   c) D_t = ____

10. $3.7x - 2.3y = 5$
$1.9x + 5.2y = 8$

   Round x and y to hundredths.

   a) D = ____   d) x = ____

   b) D_x = ____   e) y = ____

   c) D_y = ____

11. Using electrical principles, the following system was set up to calculate currents i_1 and i_2, in milliamperes, in a circuit. Find i_1 and i_2. Round to hundredths.

$$2.8i_1 + 3.4i_2 = 6.3$$
$$4.3i_1 - 1.5i_2 = 6.3$$

12. Using principles of equilibrium, the following system was set up to calculate forces F_1 and F_2, in kilograms, in a structure. Find F_1 and F_2. Round to a whole number.

$$85F_1 - 54F_2 = 8,800$$
$$62F_1 - 15F_2 = 9,400$$

Solve each system of equations by the addition method.

13. $2x - y + z = 9$
$x - 3y - z = 4$
$3x + y + z = 10$

14. $r + s - 2t = 8$
$2r - s + 2t = 4$
$r + s = 2$

15. $a + c + 6 = b$
$2a + b + c = 2$
$a + 2b = c + 7$

16. $h + k + p = 5$
$h + 2k - p = 13$
$h - k - 2p = 3$

17. $t + v + 2w = 4$
$t - 2v + w = 5$
$v + w = 1$

18. $F_1 + F_2 = F_3 + 9$
$F_2 = F_1 + 2$
$F_1 = 8 - F_3$

Assignment 48

1. The sum of three numbers is 7. The first number minus three times the second plus twice the third is 0. The first number plus the second minus the third is -3. Find the three numbers.

2. The sum of three numbers is 3. The first number minus the second plus the third is 7. The difference between the third number and twice the second is 3 less than the first. Find the three numbers.

3. In triangle ABC, angle A is 40° less than angle C. Angle A is three times angle B. Find the size of each angle.

4. In triangle ABC, the sum of angles A and C is 20° more than angle B. Angle B is 25° more than angle C. Find the size of each angle.

5. The system below involves three forces, in tons, in a bridge structure. Find F_1, F_2, and F_3.

$$2F_1 - F_2 + 2F_3 = 1200$$
$$F_1 - 4F_2 + 2F_3 = 0$$
$$F_1 + F_2 + F_3 = 900$$

6. The system below involves three currents, in milliamperes, present in an electronics circuit. Find i_1, i_2, and i_3.

$$2i_1 - i_2 - 2i_3 = 0$$
$$i_1 + i_2 + i_3 = 40$$
$$i_1 + 2i_2 - i_3 = 50$$

Do each formula derivation.

7. Eliminate A and P and solve for V.

$$V = P - A$$
$$B = A + F$$
$$P = KV$$

8. Eliminate w and r and solve for a.

$$t = \frac{w}{d}$$
$$r = w + p$$
$$d = ar$$

9. Eliminate b and r and solve for h.

$$s = bt$$
$$b = \frac{h^2}{r}$$
$$h = kr$$

10. Eliminate s and a and solve for w.

$$p = sw$$
$$r = a + s$$
$$t = aw$$

11. Eliminate s and w and solve for a.

$$p = sw$$
$$r = a + s$$
$$t = aw$$

12. Eliminate P and H and solve for D.

$$P = D^2H$$
$$V = AP$$
$$B = \frac{K}{H}$$

Assignment 49

Evaluate each determinant.

1. $\begin{vmatrix} 2 & 1 & 1 \\ 1 & 4 & 2 \\ 3 & 0 & 1 \end{vmatrix}$

2. $\begin{vmatrix} 1 & -2 & -1 \\ 3 & 0 & 2 \\ 0 & 5 & 4 \end{vmatrix}$

3. $\begin{vmatrix} 0.2 & 1 & -0.5 \\ -1 & 1.5 & 0.1 \\ 2 & -2.4 & 3 \end{vmatrix}$

Solve each system of equations by the determinant method. Record the numerical values of the four determinants used in each solution.

4. $x + y + z = 3$
$2x + y - z = 0$
$x + 2y + z = 6$

a) $D =$ _____
b) $D_x =$ _____
c) $D_y =$ _____
d) $D_z =$ _____
e) $x =$ _____
f) $y =$ _____
g) $z =$ _____

5. $t + 2v + w = 8$
$2t - v - w = 6$
$t - 2v + w = 0$

a) $D =$ _____
b) $D_t =$ _____
c) $D_v =$ _____
d) $D_w =$ _____
e) $t =$ _____
f) $v =$ _____
g) $w =$ _____

6. $2a + b + c = 5$
$a - 3b - c = 2$
$4a - b - c = 1$

a) $D =$ _____
b) $D_a =$ _____
c) $D_b =$ _____
d) $D_c =$ _____
e) $a =$ _____
f) $b =$ _____
g) $c =$ _____

7. $2x + 4y - z = 0$
$x + 2y + 2z = 0$
$x - 3y - 2z = 10$

a) $D =$ _____
b) $D_x =$ _____
c) $D_y =$ _____
d) $D_z =$ _____
e) $x =$ _____
f) $y =$ _____
g) $z =$ _____

8. $1.4r - 0.6s + 2.8t = 5.8$
$0.2r + 1.5s - 0.7t = 0$
$r - s + t = 0$

Round to hundredths.

a) $D =$ _____
b) $D_r =$ _____
c) $D_s =$ _____
d) $D_t =$ _____
e) $r =$ _____
f) $s =$ _____
g) $t =$ _____

9. $2A - B = 10$
$A + C = 2$
$2B - 2C = 6$

a) $D =$ _____
b) $D_A =$ _____
c) $D_B =$ _____
d) $D_C =$ _____
e) $A =$ _____
f) $B =$ _____
g) $C =$ _____

10. The system below involves three currents, in milliamperes, present in an electronic circuit. Find i_1, i_2, and i_3.

$$i_1 - 2(i_2 - i_3) = 2$$
$$i_2 - 3(i_2 - i_1) = 14$$
$$4(i_1 + i_3) - 3i_2 = 25$$

11. The system below involves three forces, in kilograms, on a beam. Find F_1, F_2, and F_3.

$$0.1F_1 + 0.1F_2 - 0.1F_3 = 70$$
$$0.5F_1 - 0.3F_2 - 0.4F_3 = 80$$
$$0.2F_1 - 0.3F_2 + 0.1F_3 = 90$$

15 Powers, Roots, Logarithms

In this chapter, we will discuss powers, roots, common (base 10) logarithms, and natural (base $\underline{e}$) logarithms. The calculator procedures related to those four topics are shown. Evaluations are performed with formulas containing powers of $\underline{e}$, other powers, "log" expressions, and "ln" expressions.

15-1 POWERS AND ROOTS

In this section, we will use a calculator to evaluate powers with integral exponents. We will also define <u>roots</u> and use a calculator to evaluate them.

1. As we saw earlier, any power with a whole-number exponent can be evaluated by converting it to a multiplication of identical factors.
    For example:

    $$4^3 = (4)(4)(4) = 64$$
    $$3^4 = (3)(3)(3)(3) = 81$$
    $$2^5 = (2)(2)(2)(2)(2) = 32$$

Continued on following page

624

1. **Continued**

    We saw earlier that the $\boxed{x^2}$ key and the power key $\boxed{y^x}$ can be used to evaluate any square (second power) or cube (third power). The power key $\boxed{y^x}$ can also be used to evaluate powers with larger exponents. The steps for evaluating 3^4 and 2^5 are shown.

    Calculator Steps: $3 \boxed{y^x} 4 \boxed{=}$ 81

    $2 \boxed{y^x} 5 \boxed{=}$ 32

    Evaluate each power.

    a) $3^9 = $ _____

    b) $15^5 = $ _____

2. The power below has a decimal base. We evaluated it by converting to a multiplication of identical factors.

    $$(1.5)^4 = (1.5)(1.5)(1.5)(1.5) = 5.0625$$

    We can also use the power key $\boxed{y^x}$ to evaluate $(1.5)^4$. The steps are shown.

    Calculator Steps: $1.5 \boxed{y^x} 4 \boxed{=}$ 5.0625

    Evaluate each power.

    a) $(2.2)^5 = $ _____

    b) $(0.9)^7 = $ _____

    a) 19,683

    b) 759,375

3. A power can equal a very large or very small number. In such cases, a calculator gives the number in scientific notation. Do these. Round the first factor to two decimal places.

    a) $(247)^8 = $ _____

    b) $(0.039)^{20} = $ _____

    a) 51.53632

    b) 0.4782969

4. Using the definition for powers with negative exponents, we evaluated each power below.

    $$2^{-1} = \frac{1}{2^1} = \frac{1}{2} = 0.5$$

    $$5^{-2} = \frac{1}{5^2} = \frac{1}{25} = 0.04$$

    We can use the $\boxed{y^x}$ key to evaluate 2^{-1} and 5^{-2}. The steps are shown.

    Calculator Steps: $2 \boxed{y^x} 1 \boxed{+/-} \boxed{=}$ 0.5

    $5 \boxed{y^x} 2 \boxed{+/-} \boxed{=}$ 0.04

    Evaluate each power.

    a) $4^{-1} = $ _____

    b) $20^{-2} = $ _____

    c) $5^{-3} = $ _____

    a) 1.39×10^{19}

    b) 6.63×10^{-29}

    a) 0.25

    b) 0.0025

    c) 0.008

5.   Evaluate these.   Round to millionths.

   a) 12^{-4} = _____   b) $(8.9)^{-5}$ = _____

---

6.   In an earlier chapter, we discussed the square roots and cube roots of numbers.

   To find the <u>square</u> root (or <u>second</u> root) of a number, we must find one of <u>two</u> identical factors whose product is that number.
   Since (7)(7) = 49,   the <u>square</u> root of 49 is 7.

   To find the <u>cube</u> root (or <u>third</u> root) of a number, we must find one of <u>three</u> identical factors whose product is that number.
   Since (5)(5)(5) = 125,   the <u>cube</u> root of 125 is 5.

   Roots beyond the third root are simply called the <u>fourth</u> root, <u>fifth</u> root, <u>sixth</u> root, and so on.

   To find the <u>fourth</u> root of a number, we must find one of <u>four</u> identical factors whose product is that number.
   Since (3)(3)(3)(3) = 81,   the <u>fourth</u> root of 81 is 3.

   To find the <u>sixth</u> root of a number, we must find one of <u>six</u> identical factors whose product is that number.
   Since (4)(4)(4)(4)(4)(4) = 4,096,   the <u>sixth</u> root of 4,096 is _____ .

a) 0.000048

b) 0.000018

---

7.   We have seen that radicals are used for square roots and cube roots. For example:
   $\sqrt{64}$ means: find the <u>square</u> root of 64.
   $\sqrt[3]{27}$ means: find the <u>cube</u> root of 27.

   The small 3 in a cube root is called the <u>index</u>.   A radical with an index is also used for roots beyond cube roots.   That is:
   $\sqrt[4]{81}$ means: find the <u>fourth</u> root of 81.
   $\sqrt[7]{1,089}$ means: find the <u>seventh</u> root of 1,089.

   Though the index of a square root is actually 2, the 2 is not usually written.   That is:
   Instead of $\sqrt[2]{824}$, we simply write $\sqrt{824}$.

   Write each of these in radical form.

   a) the ninth root of 45 _____   b) the square root of 33 _____

4

---

a) $\sqrt[9]{45}$

b) $\sqrt{33}$ $\left(\text{not } \sqrt[2]{33}\right)$

24. In the evaluation below, we divide <u>first</u>. Notice that we press $\boxed{=}$ to complete the division before using $\boxed{y^x}$. We rounded the answer to hundredths.

In $D = \left(\dfrac{a}{b}\right)^{1.8}$, when $a = 72.9$ and $b = 58.6$, $D = 1.48$.

Calculator Steps: $72.9 \boxed{\div} 58.6 \boxed{=} \boxed{y^x} 1.8 \boxed{=} \quad 1.4814733$

Use the same steps for this one. Round to thousandths.

In $F = \left(\dfrac{S}{T}\right)^{0.1}$, when $S = 13.9$ and $T = 47.6$, $F = \underline{\hspace{1cm}}$

a) 3.74

b) 1.63

25. In the evaluation below, we add first. Notice that we press $\boxed{=}$ to complete the addition before using $\boxed{y^x}$. We rounded the answer to a whole number.

In $V = (x + 1)^n$, when $x = 7.8$ and $n = 2.5$, $V = 230.$

Calculator Steps: $7.8 \boxed{+} 1 \boxed{=} \boxed{y^x} 2.5 \boxed{=} \quad 229.72416$

Use the same steps for this one. Round to hundredths.

In $F = (c + d)^{0.2}$, when $c = 11.9$ and $d = 48.3$, $F = \underline{\hspace{1cm}}$

0.884

26. In the formula below, Q^a is a factor in a multiplication. The steps for the evaluation are shown. We rounded the answer to tenths. Note: On some calculators, you have to evaluate Q^a first and then multiply by P.

In $R = PQ^a$, when $P = 1.59$, $Q = 2.24$, and $a = 3.18$, $R = 20.7$

Calculator Steps: $1.59 \boxed{x} 2.24 \boxed{y^x} 3.18 \boxed{=} \quad 20.662641$

Use the same steps for these. Round to ten-thousandths in (a) and to a whole number in (b).

a) In $I = KP^{1.5}$, when $K = 0.001$ and $P = 19$, $I = \underline{\hspace{1.5cm}}$

b) In $b = TV^{0.4}$, when $T = 400$ and $V = 100$, $b = \underline{\hspace{1.5cm}}$

2.27

a) 0.0828

b) 2,524

27. Two methods for the evaluation below are shown. We rounded the answer to tenths.

In $M = T(a + b)^{0.1}$, when $T = 25$, $a = 37$, and $b = 49$,
$$M = 39.0 .$$

   1) Using parentheses symbols, we evaluate $(a + b)$ before pressing $\boxed{y^x}$.

   Calculator Steps:

   25 $\boxed{x}$ $\boxed{(}$ 37 $\boxed{+}$ 49 $\boxed{)}$ $\boxed{y^x}$ 0.1 $\boxed{=}$ 39.029218

   2) Not using parentheses symbols, we evaluate $(a + b)^{0.1}$ first and then multiply by the value of T.

   Calculator Steps:

   37 $\boxed{+}$ 49 $\boxed{=}$ $\boxed{y^x}$ 0.1 $\boxed{x}$ 25 $\boxed{=}$ 39.029218

Do these. Round to a whole number.

   a) $A = P(1 + i)^n$, when $P = 1{,}000$, $i = 0.085$, and $n = 10$,
   $$A = \rule{2cm}{0.4pt}$$

   b) In $D = K(t + 1)^n$, when $K = 250$, $t = 60$, and $n = 0.5$,
   $$D = \rule{2cm}{0.4pt}$$

a) 2,261   b) 1,953

SELF-TEST 50 (pages 624 -633)

Evaluate these powers.

1. 2^{14} = \rule{2cm}{0.4pt} Round to hundreds.

2. $(0.4)^7$ = \rule{2cm}{0.4pt} Round to millionths.

3. $(1.12)^{-5}$ = \rule{2cm}{0.4pt} Round to thousandths.

Evaluate these roots.

4. $\sqrt[5]{260{,}000}$ = \rule{2cm}{0.4pt} Round to tenths.

5. $\sqrt[8]{68.5}$ = \rule{2cm}{0.4pt} Round to hundredths.

6. $\sqrt[4]{0.0973}$ = \rule{2cm}{0.4pt} Round to thousandths.

Convert each radical to a power with a decimal exponent.

7. $\sqrt[5]{8^6}$ = \rule{2cm}{0.4pt}   8. $\sqrt[4]{3^9}$ = \rule{2cm}{0.4pt}

9. $\sqrt[8]{5^2}$ = \rule{2cm}{0.4pt}   10. $\sqrt{17}$ = \rule{2cm}{0.4pt}

Evaluate each power.

11. $(13.8)^{0.9}$ = \rule{1.5cm}{0.4pt} Round to tenths.

12. $(0.45)^{-1.7}$ = \rule{1.5cm}{0.4pt} Round to hundredths.

Evaluate each expression.

13. $\left(\dfrac{36.7}{16.2}\right)^{3.4}$ = \rule{1.5cm}{0.4pt} Round to tenths.

14. $(1 + 0.078)^{10}$ = \rule{1.5cm}{0.4pt} Round to hundredths.

Continued on following page.

SELF-TEST 50 (pages 624-633) - Continued

15. In the formula below, find <u>r</u> when h = 73.4 and p = 12.9 . Round to tenths.

$$r = \left(\frac{h}{p}\right)^{1.8}$$

r = _____

16. In the formula below, find M when K = 5.3, V = 0.86, and a = 2.9 Round to hundredths.

$$M = KV^a$$

M = _____

17. In the formula below, find A when P = 500, i = 0.06, and n = 18. Round to a whole number.

$$A = P(1 + i)^n$$

A = _____

<u>ANSWERS</u>:
1. 16,400	4. 12.1	7. $8^{1.2}$	11. 10.6
2. 0.001638	5. 1.70	8. $3^{2.25}$	12. 3.89
3. 0.567	6. 0.559	9. $5^{0.25}$	13. 16.1
		10. $17^{0.5}$	14. 2.12

15. r = 22.9
16. M = 3.42
17. A = 1,427

15-4   POWERS OF TEN WITH DECIMAL EXPONENTS

Any power of ten with a decimal exponent can be converted to a number, and any number can be converted to a power of ten. We will discuss the conversion methods in this section.

28. To convert powers of ten with decimal exponents to numbers, we enter the exponent and then press $\boxed{10^x}$ or $\boxed{2nd}$ $\boxed{log}$ or $\boxed{INV}$ $\boxed{log}$.
We did so below for $10^{1.7892}$ and $10^{-3.5066}$.

Note: $\boxed{log}$ is an abbreviation for <u>logarithm</u> which means <u>exponent</u>.

Calculator Steps: 1.7892 $\boxed{10^x}$ 61.54602374
or $\boxed{2nd}$ $\boxed{log}$

-3.5066 $\boxed{10^x}$ 0.000311458
or $\boxed{2nd}$ $\boxed{log}$

Therefore: a) Rounding to tenths, $10^{1.7892}$ = _____ .

b) Rounding to millionths, $10^{-3.5066}$ = _____ .

29. $10^1 = 10$ and $10^2 = 100$. Therefore, any power of ten with an exponent between 1 and 2 equals a number between 10 and 100. To show that fact, do these. Round to tenths.

a) $10^{1.2}$ = _____

b) $10^{1.7091}$ = _____

a) 61.5

b) 0.000311

a) 15.8   b) 51.2

30. $10^3 = 1,000$ and $10^4 = 10,000$. Therefore, any power of ten with an exponent between 3 and 4 equals a number between 1,000 and 10,000. To show that fact, do these. Round to a whole number.

    a) $10^{3.25} =$ _____   b) $10^{3.9837} =$ _____

---

31. a) Since $10^{0.7518}$ lies between 10^0 and 10^1, $10^{0.7518}$ equals a number between 1 and 10.

    Rounding to hundredths, $10^{0.7518} =$ _____

b) Since $10^{5.3799}$ lies between 10^5 and 10^6, $10^{5.3799}$ equals a number between 100,000 and 1,000,000.

    Rounding to thousands, $10^{5.3799} =$ _____

a) 1,778   b) 9,632

---

32. Any power of ten with a <u>negative</u> decimal exponent equals a number between 0 and 1. To show that fact, do these.

    a) $10^{-1.3} =$ _____ Round to four decimal places.

    b) $10^{-3.85} =$ _____ Round to six decimal places.

    c) $10^{-5.2075} =$ _____ Round to seven decimal places.

a) 5.65

b) 240,000

---

33. In $y = 10^x$, use the $\boxed{10^x}$ key to find the values of $\underline{y}$ corresponding to the following values of the exponent $\underline{x}$.

    a) If $x = 1.25$, $y =$ _____ Round to tenths.

    b) If $x = 0.4$, $y =$ _____ Round to hundredths.

    c) If $x = -2.5$, $y =$ _____ Round to five decimal places.

a) 0.0501

b) 0.000141

c) 0.0000062

---

34. To convert a number to a power of ten, we use the $\boxed{log}$ key to find <u>the exponent</u>. We did so below for 475 and 0.0936. Remember that $\boxed{log}$ is an abbreviation for <u>logarithm</u> which means <u>exponent</u>.

    Calculator Steps: 475 $\boxed{log}$ 2.6766936

                  0.0936 $\boxed{log}$ -1.0287242

Rounding each exponent to four decimal places and using the base 10, we get:

    $475 = 10^{2.6767}$ $0.0936 =$ _____

a) 17.8

b) 2.51

c) 0.00316

---

35. $100 = 10^2$ and $1,000 = 10^3$. Therefore, the exponent of the power-of-ten form of any number between 100 and 1,000 lies between 2 and 3. To show that fact, convert these to powers of ten. Round each exponent to four decimal places.

    a) $225 =$ _____   b) $879 =$ _____

$10^{-1.0287}$

---

a) $10^{2.3522}$

b) $10^{2.9440}$

36. $10,000 = 10^4$ and $100,000 = 10^5$. Therefore, the exponent of the power-of-ten form of any number between 10,000 and 100,000 lies between 4 and 5. To show that fact, convert these to powers of ten. Round each exponent to four decimal places. a) $35,700 = $ _____ b) $69,699 = $ _____	
37. a) Since 5.49 lies between 1 and 10, the exponent of its power-of-ten form lies between 0 and 1. Rounding the exponent to four decimal places, $5.49 = $ _____. b) Since 7,825 lies between 1,000 and 10,000 , the exponent of its power-of-ten form lies between 3 and 4. Rounding the exponent to four decimal places, $7,825 = $ _____.	a) $10^{4.5527}$ b) $10^{4.8432}$
38. The power-of-ten form of any number between 0 and 1 has a negative exponent. To show that fact, convert these to powers of ten. Round the exponents to four decimal places. a) $0.679 = $ _____ b) $0.000175 = $ _____	a) $10^{0.7396}$ b) $10^{3.8935}$
39. In $y = 10^x$, use the $\boxed{\text{log}}$ key to find the exponent $\underline{x}$ for the following values of $\underline{y}$. Round the exponent to four decimal places. a) If $y = 9,210$, $x = $ _____ b) If $y = 0.008$, $x = $ _____	a) $10^{-0.1681}$ b) $10^{-3.7570}$
	a) 3.9643 b) -2.0969

15-5 COMMON LOGARITHMS

When a number is written in power-of-ten form, the exponent of the 10 is called the <u>common logarithm</u> of the number. We will discuss common logarithms and logarithmic notation in this section.

40. The <u>exponent</u> of the power-of-ten form of a number is called the <u>common logarithm</u> of the number. Since $7.65 = 10^{0.8837}$, the logarithm of 7.65 is 0.8837 . Since $291 = 10^{2.4639}$, the logarithm of 291 is _____.	
41. The logarithm of a number is <u>only the exponent</u> of the power of ten. It does not include the base 10. That is: Since $87.6 = 10^{1.9425}$, the logarithm of 87.6 is 1.9425 (the logarithm <u>is not</u> $10^{1.9425}$) $0.682 = 10^{-0.1662}$. Is -0.1662 or $10^{-0.1662}$ the logarithm of 0.682?	2.4639

42. To find the logarithm of a number on a calculator, enter the number and press the ⌐log⌐ key. Round to four decimal places.

   a) The logarithm of 761,000 is _____.

   b) The logarithm of 0.0829 is _____.

-0.1662

---

43. The logarithm of 27.9 is 1.4456 . Therefore, $27.9 = 10^{1.4456}$

The logarithm of 0.55 is -0.2596 . Therefore, $0.55 = 10^{\boxed{}}$

a) 5.8814

b) -1.0814

---

44. If the logarithm of a number is 2.7959 :

   a) In power-of-ten form, the number is _____.

   b) Rounded to the nearest whole number, the number is _____.

$10^{-0.2596}$

---

45. a) If the logarithm of a number is 0.8549 , the number is _____. Round to hundredths.

   b) If the logarithm of a number is -2.0633, the number is _____. Round to five decimal places.

a) $10^{2.7959}$

b) 625

---

46. The phrase logarithm of is usually abbreviated to log. For example:

   log 1,643 means: the logarithm of 1,643

Using the ⌐log⌐ key, complete these. Round to four decimal places.

   a) log 1,643 = _____   b) log 0.0005 = _____

a) 7.16 , from: $10^{0.8549}$

b) 0.00864 , from: $10^{-2.0633}$

---

47. Only positive numbers have logarithms. Negative numbers and zero do not have logarithms. Therefore, finding the logarithm of a negative number or zero is an IMPOSSIBLE operation. A calculator shows that fact by printing out "Error" or "E" or in some other way. Try these on a calculator.

   log(-155)   log(0)

   Note: When you are using a calculator for a "log" problem and the calculator display shows an IMPOSSIBLE operation, you know that you have made a mistake.

a) 3.2156

b) -3.3010

---

48. Any basic power-of-ten equation can be written as a log equation. For example:

   $56.2 = 10^{1.7497}$ can be written: log 56.2 = 1.7497

   $10^{-0.6925} = 0.203$ can be written: -0.6925 = log 0.203

Write each equation below as a log equation.

   a) $0.0097 = 10^{-2.0132}$ b) $10^{5.8136} = 651,000$

49. When a basic power-of-ten equation contains a variable, it can also be written as a $\underline{\log}$ equation.   For example:

$$y = 10^{1.2514} \quad \text{can be written:} \quad \log y = 1.2514$$

$$10^x = 0.714 \quad \text{can be written:} \quad x = \log 0.714$$

Write each equation as a $\underline{\log}$ equation.

    a) $11.8 = 10^t$ b) $10^{-0.6711} = m$

a) $\log 0.0097$
$$= -2.0132$$

b) 5.8136
$$= \log 651,000$$

---

50. Any basic $\underline{\log}$ equation can be converted to a basic power-of-ten equation.   For example:

$$\log 299 = 2.4757 \quad \text{can be written:} \quad 299 = 10^{2.4757}$$

$$-0.1871 = \log 0.65 \quad \text{can be written:} \quad 10^{-0.1871} = 0.65$$

Write each equation below as a power-of-ten equation.

    a) $\log 0.017 = -1.7696$ b) $4.5846 = \log 38,420$

a) $\log 11.8 = t$

b) $-0.6711 = \log m$

---

51. When a basic $\underline{\log}$ equation contains a variable, it can also be written as a power-of-ten equation.   For example:

$$\log x = 3.6099 \quad \text{can be written:} \quad x = 10^{3.6099}$$

$$y = \log 0.074 \quad \text{can be written:} \quad 10^y = 0.074$$

Write each equation as a power-of-ten equation.

    a) $-1.5277 = \log d$ b) $\log 803 = a$

a) $0.017 = 10^{-1.7696}$

b) $10^{4.5846} = 38,420$

---

a) $10^{-1.5277} = d$ b) $803 = 10^a$

15-6   SOLVING POWER-OF-TEN AND "LOG" EQUATIONS

In this section, we will discuss the methods for solving power-of-ten and $\underline{\log}$ equations that contain a variable.

---

52. In each power-of-ten equation below, the variable stands for $\underline{\text{a number}}$. Use the $\boxed{10^x}$ key to solve each equation.

    a) Round to tenths.   b) Round to four decimal places.

$$10^{1.6294} = x \qquad\qquad y = 10^{-2.1095}$$

$$x = \underline{\hspace{2cm}} \qquad\qquad y = \underline{\hspace{2cm}}$$

53. In each power-of-ten equation below, the variable stands for an exponent. Use the [log] key to solve each equation. Round each exponent to four decimal places.

a) $0.016 = 10^t$

b) $10^b = 81,900$

t = _____

b = _____

a) x = 42.6

b) y = 0.0078

---

54. When a "log" equation contains a variable, we can solve it by writing the equation in power-of-ten form. Let's use that method to solve the equation below.

$$\log V = 2.7427$$

a) Write the equation in power-of-ten form. _____

b) Therefore, rounded to a whole number, V = _____

a) t = -1.7959

b) b = 4.9133

---

55. Let's solve log 0.456 = y.

a) Write the equation in power-of-ten form. _____

b) Therefore, rounded to four decimal places, y = _____

a) $V = 10^{2.7427}$

b) V = 553

---

56. Let's solve -1.7086 = log x.

a) Write the equation in power-of-ten form. _____

b) Therefore, rounded to four decimal places, x = _____

a) $0.456 = 10^y$

b) y = -0.3410

---

57. Let's solve m = log 3,190.

a) Write the equation in power-of-ten form. _____

b) Therefore, rounded to four decimal places, m = _____

a) $10^{-1.7086} = x$

b) x = 0.0196

---

58. When the variable follows log in an equation, it stands for a number.

In log x = -1.53 , x stands for a number since: $x = 10^{-1.53}$

In 2.25 = log y , y stands for a number since: $10^{2.25} = y$

In such cases, convert the log equation to a power-of-ten equation and use [10^x] . Solve these.

a) Round to tenths.

log t = 1.7539

t = _____

b) Round to thousandths.

-0.49 = log m

m = _____

a) $10^m = 3,190$

b) m = 3.5038

59. When the variable is on the opposite side of "log" in an equation, it stands for the exponent (or logarithm). For example:

In log 3.87 = d , d stands for the exponent since: $3.87 = 10^d$

In h = log 0.025 , h stands for the exponent since: $10^h = 0.025$

In such cases, we can simply use the ⌐log⌐ key without converting to a power-of-ten equation. Solve these. Round to four decimal places.

a) log 0.051 = R

R = _____

b) G = log 995,000

G = _____

a) t = 56.7, from: $t = 10^{1.7539}$	
b) m = 0.324, from: $10^{-0.49} = m$	

---

60. Solve: a) Round to hundredths.

log 1,290 = V

V = _____

b) Round to thousandths.

log D = -0.207

D = _____

a) R = -1.2924, from: $0.051 = 10^R$

b) G = 5.9978, from: $10^G = 995,000$

---

61. Solve: a) Round to thousands.

6.37 = log a

a = _____

b) Round to hundredths.

k = log 0.091

k = _____

a) V = 3.11

b) D = 0.621

---

a) a = 2,344,000   b) k = -1.04

15-7 EVALUATING "LOG" FORMULAS

In this section, we will do evaluations with formulas containing log expressions.

---

62. The formula below contains a log expression.

D = 10 log R

To show that there are two factors on the right side, we put them in parentheses below.

D = (10)(log R)

To find D when R = 752, we multiply 10 times log 752. The steps are shown. Notice that we entered 752 and then pressed ⌐log⌐ before pressing ⌐=⌐ . Rounding to tenths, we get D = 28.8 .

Calculator Steps: 10 ⌐x⌐ 752 ⌐log⌐ ⌐=⌐   28.762178

Do these. Round to tenths in (a) and to a whole number in (b).

a) In D = 10 log R , when R = 0.0386 , D = _____ .

b) In H = w log T , when w = 100 and T = 32.5 , H = _____

63. The steps for the evaluation below are shown. Notice again that we entered the value for Q and then pressed [log] . Rounding to tenths, we get P = 71.7 .

In P = A - K log Q , find P when A = 59 , K = 19 , and Q = 0.215 .

Calculator Steps: 59 [-] 19 [x] 0.215 [log] [=] 71.683669

Do this one. Round to tenths.

In P = A - K log Q , when A = 94 , K = 30 , and Q = 3.9 ,

P = _____ .

a) -14.1

b) 151

64. In the formula below, $\log\left(\frac{P_2}{P_1}\right)$ is the log of a fraction or division. Two calculator methods are shown. Rounding to hundredths, we get D = -8.13 .

In $D = 10 \log\left(\frac{P_2}{P_1}\right)$, find D when P_2 = 750 and P_1 = 4,875 .

1) Using parentheses symbols, we evaluate $\frac{P_2}{P_1}$ before pressing [log] .

Calculator Steps: 10 [x] [(] 750 [÷] 4,875 [)] [log] [=]
-8.1291336

2) Not using parentheses symbols, we evaluate $\log\left(\frac{P_2}{P_1}\right)$ first and then multiply by 10. Notice that we pressed [=] to complete the division before pressing [log] .

Calculator Steps: 750 [÷] 4,875 [=] [log] [x] 10 [=]
-8.1291336

Do this one. Round to hundredths.

In $M = 2.5 \log\left(\frac{I_1}{I}\right)$, when I_1 = 79.3 and I = 16.4 , M = _____ .

76.3

65. In the formula below, log(W + H) is the log of an addition. Two calculator methods are shown. Rounding to hundredths, we get B = 4.97 .

In B = K log(W + H) , find B when K = 2.79 , W = 18.3 , and H = 41.9 .

1) Using parentheses symbols, we evaluate (W + H) before pressing [log] .

Calculator Steps: 2.79 x [(] 18.3 [+] 41.9 [)] [log] [=]
4.9650742

1.71

Continued on following page.

65. Continued

    2) Not using parentheses symbols, we must evaluate log(W + H) first and then multiply by the value of K. Notice that we pressed $\boxed{=}$ to complete the addition before pressing $\boxed{\text{log}}$.

    <u>Calculator Steps</u>:   18.3 $\boxed{+}$ 41.9 $\boxed{=}$ $\boxed{\text{log}}$ $\boxed{\text{x}}$ 2.79 $\boxed{=}$

                                                  4.9650742

Do this one. Round to hundredths.

    In G = 1.75 log(a + b) , when a = 3.25 and b = 9.87 ,

                                            G = _____

---

66. In pH = -log aH , -log aH means "the <u>additive inverse</u> of log aH". Therefore, after finding log aH, we press $\boxed{+/-}$ to get its inverse. An example is given. Rounding to hundredths, we get pH = 4.02 .

    In pH = -log aH , find pH when aH = 0.000095 .

    <u>Calculator Steps</u>:   0.000095 $\boxed{\text{log}}$ $\boxed{+/-}$ 4.0222764

Using the same steps, do this one. Round to hundredths.

    In pH = -log aH , when aH = 0.000061 , pH = _____ .

1.96

---

67. A two-step process is needed to find R in the evaluation below. Rounding to tenths, we get R = 56.2 .

    In log R = $\dfrac{D}{10}$, find R when D = 17.5 .

  1) First we find log R by substituting 17.5 for D.

    log R = $\dfrac{D}{10}$ = $\dfrac{17.5}{10}$ = 1.75

  2) Then we use $\boxed{10^{\text{x}}}$ to find R. The steps are:

    <u>Calculator Steps</u>:   1.75 $\boxed{10^{\text{x}}}$ 56.234133

Use the same steps for this one. Round to a whole number.

    In log R = $\dfrac{D}{10}$, when D = 28.9 , R = _____

4.21

776, since: log R = 2.89

SELF-TEST 51 (pages 633-642)

Convert each power of ten to a number.
Round as directed.

1. $10^{5.2354}$ = _____
   Round to thousands.

2. $10^{1.62}$ = _____
   Round to tenths.

3. $10^{-0.1387}$ = _____
   Round to thousandths.

Convert each number to a power of ten.
Round each exponent to four decimal places.

4. 8.59 = _____

5. 596 = _____

6. 0.0204 = _____

Find the numerical value of each of the following. Round to four decimal places.

7. log 93,600 = _____   8. log 5.18 = _____   9. log 0.0072 = _____

Write each power-of-ten equation as a
log equation.

10. $10^{2.8762}$ = 752 _____

11. $4.88 = 10^h$ _____

Write each log equation as a power-of-ten
equation.

12. log 0.018 = -1.7447 _____

13. 1.9148 = log P _____

---

14. Find R. Round to
    tenths.

    log R = 1.8063

    R = _____

15. Find y. Round to
    hundredths.

    $0.00723 = 10^y$

    y = _____

16. Find F. Round to
    thousandths.

    -0.3916 = log F

    F = _____

---

17. In the formula below, find D when
    V_2 = 9.65 and V_1 = 0.038 . Round
    to a whole number.

    $$D = 20 \log\left(\frac{V_2}{V_1}\right)$$

    D = _____

18. In the formula below, find B when
    K = 0.75, W = 23.8, and H = 47.5 .
    Round to hundredths.

    $$B = K \log(W + H)$$

    B = _____

ANSWERS:
1. 172,000
2. 41.7
3. 0.727
4. $10^{0.9340}$
5. $10^{2.7752}$
6. $10^{-1.6904}$
7. 4.9713
8. 0.7143
9. -2.1427
10. 2.8762 = log 752
11. log 4.88 = h
12. 0.018 = $10^{-1.7447}$
13. $10^{1.9148}$ = P
14. R = 64.0
15. y = -2.14
16. F = 0.406
17. D = 48
18. B = 1.39

15-8   POWERS OF e

Some formulas contain powers in which the base is a number called e. The numerical value of e is 2.7182818... . In this section, we will convert powers of e to numbers and numbers to powers of e.

---

68.   To convert $e^{1.5}$ to an ordinary number, one of the two methods below is used. Rounding to two decimal places, $e^{1.5} = 4.48$.

    1) If your calculator has an $\boxed{e^x}$ key, enter 1.5 and press $\boxed{e^x}$.

    <u>Calculator Steps</u>:   1.5 $\boxed{e^x}$ 4.4816891

    2) If your calculator does not have an $\boxed{e^x}$ key, enter 1.5 and press $\boxed{2nd}$ $\boxed{\ln x}$ or $\boxed{INV}$ $\boxed{\ln x}$.

    <u>Calculator Steps</u>:   1.5 $\boxed{2nd}$ $\boxed{\ln x}$ 4.4816891

Convert to numbers. Round to the indicated place.

    a) $e^{2.19} =$ _____ Round to hundredths.

    b) $e^{-3.75} =$ _____ Round to four decimal places.

---

69.   Since $e = e^1$, we can confirm the fact that e = 2.7182818... by entering "1" and pressing either $\boxed{e^x}$ or $\boxed{2nd}$ $\boxed{\ln x}$. Do so.

Any power of e with a positive exponent equals a number larger than "1". To confirm that fact, do these. Round to the indicated place.

    a) $e^{0.5} =$ _____ Round to hundredths.

    b) $e^{3.68} =$ _____ Round to tenths.

    c) $e^{9.757} =$ _____ Round to hundreds.

a) 8.94

b) 0.0235

---

70.   Just as $10^0 = 1$, $e^0 = 1$.   To confirm that fact, enter "0" and press either $\boxed{e^x}$ or $\boxed{2nd}$ $\boxed{\ln x}$.

Any power of e with a negative exponent equals a number between 0 and 1. To confirm that fact, do these. Round to the indicated place.

    a) $e^{-0.545} =$ _____ Round to thousandths.

    b) $e^{-3.37} =$ _____ Round to four decimal places.

    c) $e^{-8.5} =$ _____ Round to millionths.

a) 1.65

b) 39.6

c) 17,300

---

71.   In $y = e^x$:

    a) when x = 6.2 , y = _____ Round to a whole number.

    b) when x = -0.25 , y = _____ Round to thousandths.

a) 0.580

b) 0.0344

c) 0.000203

72. To convert a number to a power of e, we use the $\boxed{\ln x}$ key to find the exponent. We did so below for 67.8 and 0.045 .

    Calculator Steps: 67.8 $\boxed{\ln x}$ 4.2165622

                     0.045 $\boxed{\ln x}$ -3.1010928

    Rounding each exponent to four decimal places, we get:

       67.8 = $e^{4.2166}$ 0.045 = _____

a) 493
b) 0.779

73. Using the $\boxed{\ln x}$ key to find the exponent, convert each number to a power of e. Round each exponent to two decimal places.

       a) 7.95 = _____   b) 0.00635 = _____

    $e^{-3.1011}$

74. When "1" is converted to a power of e, the exponent is 0. That is: 1 = e^0 To confirm that fact, enter "1" and press $\boxed{\ln x}$.

    When a number larger than 1 is converted to a power of e, the exponent is positive. To confirm that fact, convert these numbers to powers of e. Round each exponent to three decimal places.

       a) 1.25 = _____   b) 287 = _____   c) 45,900 = _____

a) $e^{2.07}$
b) $e^{-5.06}$

75. When a number between 0 and "1" is converted to a power of e, the exponent is negative. To confirm that fact, convert these numbers to powers of e. Round each exponent to two decimal places.

       a) 0.979 = _____   b) 0.000408 = _____

a) $e^{0.223}$
b) $e^{5.659}$
c) $e^{10.734}$

76. In y = e^x :

       a) when y = 99.5 , x = _____   Round to four decimal places.

       b) when y = 0.0275 , x = _____   Round to two decimal places.

a) $e^{-0.02}$
b) $e^{-7.80}$

a) 4.6002	b) -3.59

15-9   EVALUATING FORMULAS CONTAINING POWERS OF e

In this section, we will do evaluations with formulas containing powers of e.

77. We can use $\boxed{e^x}$ or $\boxed{2nd}$ $\boxed{\ln x}$ for each evaluation below.

       a) In T = e^x , when x = 2.5 , T = _____   Round to tenths.

       b) In R = e^p , when p = 9.15 , R = _____

                                        Round to a whole number.

78. In the formula below, the -x means "the <u>additive inverse</u> of <u>x</u>". Therefore, we press $\boxed{+/-}$ after entering the value of <u>x</u> and then press either $\boxed{e^x}$ or $\boxed{2nd}$ $\boxed{\ln x}$. Rounding to thousandths, we get S = 0.183 .

    In $S = e^{-x}$, find S when x = 1.7 .

    Calculator <u>Steps</u>:   1.7 $\boxed{+/-}$ $\boxed{e^x}$ .18268352

    or $\boxed{2nd}$ $\boxed{\ln x}$

    Do this one. Round to four decimal places.

    In $V = e^{-p}$, when p = 4.1 , V = _____ .

| a) 12.2 |
| b) 9,414 |

79. To perform the evaluation below, we press $\boxed{+/-}$ after entering 1.4 for <u>x</u> in e^{-x}. Notice that we press $\boxed{=}$ to complete the addition in the numerator before dividing. Rounding to hundredths, we get H = 2.15 .

    In $H = \dfrac{e^x + e^{-x}}{2}$, find H when x = 1.4 .

    Calculator <u>Steps</u>:   1.4 $\boxed{e^x}$ $\boxed{+}$ 1.4 $\boxed{+/-}$ $\boxed{e^x}$ $\boxed{=}$ $\boxed{÷}$ 2 $\boxed{=}$ 2.1508985
    (or $\boxed{2nd}$ $\boxed{\ln x}$ instead of $\boxed{e^x}$)

    Do this one. Round to hundredths.

    In $H = \dfrac{e^x + e^{-x}}{2}$, when x = 0.8 , H = _____.

| 0.0166 |

80. In the formula below, -at means "the <u>additive inverse</u> of <u>at</u>". Therefore, we press $\boxed{+/-}$ after completing the multiplication <u>at</u>. Rounding to five decimal places, we get R = 0.00499 .

    In $R = e^{-at}$, find R when a = 0.5 and t = 10.6 .

    Calculator <u>Steps</u>:   0.5 $\boxed{x}$ 10.6 $\boxed{=}$ $\boxed{+/-}$ $\boxed{e^x}$ .00499159
    (or $\boxed{2nd}$ $\boxed{\ln x}$ instead of $\boxed{e^x}$)

    Use the same steps for this one. Round to thousandths.

    In $R = e^{-at}$, when a = 0.1 and t = 24 , R = _____.

| 1.34 |

| 0.091 |

81.   In the formula below, we must multiply a power of $\underline{e}$ by 14.7 . Two methods are shown. Rounding to tenths, we get $P = 10.3$.

In $P = 14.7e^{-0.2h}$, find P when h = 1.8 .

1) Using parentheses symbols, we evaluate -0.2h before pressing $\boxed{e^x}$ or $\boxed{2nd}$ $\boxed{\ln\ x}$.

Calculator Steps:
14.7 $\boxed{x}$ $\boxed{(}$ 0.2 $\boxed{+/-}$ $\boxed{x}$ 1.8 $\boxed{)}$ $\boxed{e^x}$ $\boxed{=}$ 10.255842

2) Not using parentheses symbols, we have to evaluate the power of $\underline{e}$ first before multiplying by 14.7 .

Calculator Steps:
0.2 $\boxed{+/-}$ $\boxed{x}$ 1.8 $\boxed{=}$ $\boxed{e^x}$ $\boxed{x}$ 14.7 $\boxed{=}$ 10.255842

Do this one. Round to hundredths.

In $P = 14.7e^{-0.2h}$, when   h = 2.1 , P = _____ .

---

82.   Two methods for the evaluation below are shown. Rounding to thousandths, we get A = 0.243 .

In $A = ke^{-ct}$, find A when k = 4, c = 0.02, and t = 140 .

1) Using parentheses symbols, we evaluate $\underline{-ct}$ before pressing $\boxed{e^x}$ or $\boxed{2nd}$ $\boxed{\ln\ x}$. Notice that we press $\boxed{+/-}$ after pressing $\boxed{)}$ to get $\underline{-ct}$.

Calculator Steps:
4 $\boxed{x}$ $\boxed{(}$ 0.02 $\boxed{x}$ 140 $\boxed{)}$ $\boxed{+/-}$ $\boxed{e^x}$ $\boxed{=}$ .24324025

2) Not using parentheses symbols, we must evaluate the power of $\underline{e}$ first before multiplying by $\underline{k}$. Notice that we press $\boxed{+/-}$ to get $\underline{-ct}$ before pressing $\boxed{e^x}$ or $\boxed{2nd}$ $\boxed{\ln\ x}$.

Calculator Steps:
0.02 $\boxed{x}$ 140 $\boxed{=}$ $\boxed{+/-}$ $\boxed{e^x}$ $\boxed{x}$ 4 $\boxed{=}$ .24324025

Do this one. Round to thousandths.

In $A = ke^{-ct}$, when k = 5, c = 0.04, and t = 55, A = _____ .

9.66

0.554

83. Two methods for the evaluation below are also shown. Rounding to tenths, we get i = 11.7 .

$$\text{In} \quad i = I_o e^{-\frac{Rt}{L}} \quad , \text{ find } \underline{i} \text{ when } I_o = 70, \quad R = 15.7 ,$$
$$t = 2.5 , \text{ and } L = 21.9 .$$

1) Using parentheses symbols, we evaluate $-\frac{Rt}{L}$ before pressing $\boxed{e^x}$ or $\boxed{2nd}$ $\boxed{\ln x}$. Notice that we press $\boxed{+/-}$ after $\boxed{)}$ to get $-\frac{Rt}{L}$.

Calculator Steps:
70 $\boxed{x}$ $\boxed{(}$ 15.7 $\boxed{x}$ 2.5 $\boxed{\div}$ 21.9 $\boxed{)}$ $\boxed{+/-}$ $\boxed{e^x}$ $\boxed{=}$ 11.661092

2) Not using parentheses symbols, we must evaluate the power of $\underline{e}$ before multiplying by I_o . Notice that we press $\boxed{+/-}$ to get $-\frac{Rt}{L}$ before pressing $\boxed{e^x}$ or $\boxed{2nd}$ $\boxed{\ln x}$.

Calculator Steps:
15.7 $\boxed{x}$ 2.5 $\boxed{\div}$ 21.9 $\boxed{=}$ $\boxed{+/-}$ $\boxed{e^x}$ $\boxed{x}$ 70 $\boxed{=}$ 11.661092

Do this one. Round to tenths.

$$\text{In} \quad V = Ee^{-\frac{t}{RC}} \quad , \text{ when } E = 200, \; t = 0.75, \; R = 12, \text{ and } C = 0.06,$$
$$V = \underline{\hspace{2cm}} .$$

70.6

15-10 NATURAL LOGARITHMS

When a number is written in power-of-$\underline{e}$ form, the exponent of the $\underline{e}$ is called the <u>natural</u> <u>logarithm</u> of the number. In this section, we will discuss natural logarithms and contrast them with common logarithms. We will also contrast natural logarithmic notation with common logarithmic notation.

---

84. Any positive number can be written in either power-of-ten or power-of-$\underline{e}$ form. For example:
$$325 = 10^{2.5119} \qquad 325 = e^{5.7838}$$

a) The <u>common</u> logarithm of a number is the <u>exponent</u> when the number is written in <u>power-of-ten</u> form.

The <u>common</u> logarithm of 325 is _____ .

b) The <u>natural</u> logarithm of a number is the <u>exponent</u> when the number is written in <u>power-of-e</u> form.

The <u>natural</u> logarithm of 325 is _____ .

85. Since $0.875 = e^{-0.1335}$ and $0.875 = 10^{-0.0580}$:

    a) The common logarithm of 0.875 is _____.

    b) The natural logarithm of 0.875 is _____.

a) 2.5119
b) 5.7838

---

86. a) To find the common logarithm of a number, we enter the number and press log .

    To four decimal places, the common logarithm of 16.7 is _____.

    b) To find the natural logarithm of a number, we enter the number and press ln x .

    To four decimal places, the natural logarithm of 16.7 is _____.

a) -0.0580
b) -0.1335

---

87. The common logarithm of 0.0189 is -1.7235 . The natural logarithm of 0.0189 is -3.9686 . Therefore:

    $$0.0189 = 10^{-1.7235} \qquad 0.0189 = e^{-3.9686}$$

    a) If the common logarithm of a number is 2.5639, the number is $10^{2.5639}$ or _____. Round to a whole number.

    b) If the natural logarithm of a number is 2.5639, the number is $e^{2.5639}$ or _____. Round to tenths.

a) 1.2227
b) 2.8154

---

88. a) If the common logarithm of a number is -1.5, the number is _____. Round to four decimal places.

    b) If the natural logarithm of a number is -1.5, the number is _____. Round to three decimal places.

a) 366
b) 13.0

---

89. The abbreviations log and ln are used for logarithms.

    log 265 means "the common logarithm of 265".

    ln 265 means "the natural logarithm of 265".

    Using the log and ln x keys, complete these. Round each to four decimal places.

    a) log 265 = _____   b) ln 265 = _____

a) 0.0316, from: $10^{-1.5}$
b) 0.223, from: $e^{-1.5}$

---

a) 2.4232
b) 5.5797

90. Any power-of-ten or power-of-$\underline{e}$ equation can be written as a logarithmic equation. For example:

$$27.5 = 10^{1.4393} \quad \text{can be written:} \quad \log 27.5 = 1.4393$$
$$27.5 = e^{3.3142} \quad \text{can be written:} \quad \ln 27.5 = 3.3142$$

Write each equation below as a logarithmic equation.

   a) $10^{-1.5288} = 0.0296$ b) $e^{-2.78} = 0.062$

   _____   _____

91. When a power-of-ten or power-of-$\underline{e}$ equation contains a variable, it can also be written as a logarithmic equation. For example:

$$t = 10^{-2.5684} \quad \text{can be written:} \quad \log t = -2.5684$$
$$1.29 = e^{x} \quad \text{can be written:} \quad \ln 1.29 = x$$

Write each equation below as a logarithmic equation.

   a) $10^{y} = 6.59$ b) $e^{-2.75} = h$

   _____   _____

a) $-1.5288 =$
      $\log 0.0296$

b) $-2.78 = \ln 0.062$

92. Any logarithmic equation can be written as a power-of-ten or power-of-$\underline{e}$ equation. For example:

$$\log 1.33 = 0.1239 \quad \text{can be written:} \quad 1.33 = 10^{0.1239}$$
$$\ln 1.33 = 0.2852 \quad \text{can be written:} \quad 1.33 = e^{0.2852}$$

Write each logarithmic equation as a power-of-ten or power-of-$\underline{e}$ equation.

   a) $2.8209 = \log 662$ b) $4.1759 = \ln 65.1$

   _____   _____

a) $y = \log 6.59$

b) $-2.75 = \ln h$

93. When a logarithmic equation contains a variable, it can also be written as a power-of-ten or power-of-$\underline{e}$ equation. For example:

$$\log 0.589 = d \quad \text{can be written:} \quad 0.589 = 10^{d}$$
$$\ln t = -3.55 \quad \text{can be written:} \quad t = e^{-3.55}$$

Write each logarithmic equation as a power-of-ten or power-of-$\underline{e}$ equation.

   a) $-0.2066 = \log x$ b) $y = \ln 0.265$

   _____   _____

a) $10^{2.8209} = 662$

b) $e^{4.1759} = 65.1$

a) $10^{-0.2066} = x$ b) $e^{y} = 0.265$

15-11 SOLVING POWER-OF-e AND "LN" EQUATIONS

In this section, we will discuss the methods for solving power-of-e and "ln" equations that contain a variable.

---

94. In each power-of-e equation below, the variable stands for a number. Use $\boxed{e^x}$ or $\boxed{2nd}$ $\boxed{\ln x}$ to solve each equation.

    a) Round to tenths.   b) Round to four decimal places.

    $e^{4.25} = t$ $y = e^{-2.685}$

    t = _____   y = _____

---

95. In each power-of-e equation below, the variable stands for an exponent. Use the $\boxed{\ln x}$ key to solve each equation. Round each exponent to three decimal places.

    a) $2{,}750 = e^x$ b) $e^y = 0.0999$

    x = _____   y = _____

| a) t = 70.1 |
| b) y = 0.0682 |

---

96. When a "ln" equation contains a variable, we can solve it by writing the equation in power-of-e form. Let's use that method to solve the equation below.

    $$\ln R = 1.8576$$

    a) Write the equation in power-of-e form. _____

    b) Therefore, rounded to hundredths, R = _____.

| a) x = 7.919 |
| b) y = -2.304 |

---

97. Let's solve $\ln 0.193 = y$.

    a) Write the equation in power-of-e form. _____

    b) Therefore, rounded to four decimal places, y = _____.

| a) $R = e^{1.8576}$ |
| b) R = 6.41 |

---

98. Let's solve $-2.56 = \ln V$.

    a) Write the equation in power-of-e form. _____

    b) Therefore, rounded to four decimal places, V = _____.

| a) $0.193 = e^y$ |
| b) y = -1.6451 |

---

99. Let's solve $t = \ln 6{,}295$.

    a) Write the equation in power-of-e form. _____

    b) Therefore, rounded to hundredths, t = _____.

| a) $e^{-2.56} = V$ |
| b) V = 0.0773 |

| a) $e^t = 6{,}295$ |
| b) t = 8.75 |

100. When the letter follows "ln" in an equation, it stands for a number. For example:

In   ln P = -1.75 , P stands for a number since: $P = e^{-1.75}$

In   3.69 = ln H , H stands for a number since: $e^{3.69} = H$

In such cases, convert the "ln" equation to a power-of-e equation and use $\boxed{e^x}$ or $\boxed{2nd}$ $\boxed{ln\ x}$. Solve these.

a) Round to tenths.

b) Round to thousandths.

ln B = 4.488

-0.36 = ln G

B = _____

G = _____

---

101. When the letter is on the opposite side of "ln" in an equation, it stands for the exponent (or logarithm). For example:

In   ln 10.8 = v ,   <u>v</u> stands for the exponent since: $10.8 = e^v$

In   y = ln 0.725 ,   <u>y</u> stands for the exponent since: $e^y = 0.725$

In such cases, we can simply use the $\boxed{ln\ x}$ key without converting to a power-of-e equation. Solve these. Round to three decimal places.

a) ln 0.087 = F

b) Q = ln 13,500

F = _____

Q = _____

a) B = 88.9 , from:

$\qquad B = e^{4.488}$

b) G = 0.698 , from:

$\qquad e^{-0.36} = G$

---

102. Solve: a) Round to hundredths.

b) Round to thousandths.

ln 966 = m

ln C = -0.199

m = _____

C = _____

a) F = -2.442

b) Q = 9.510

---

103. Solve: a) Round to a whole number.

b) Round to hundredths.

t = ln 0.056

4.95 = ln b

t = _____

b = _____

a) m = 6.87

b) C = 0.820

---

a) b = 141

b) t = -2.88

15-12   EVALUATING "LN" FORMULAS

In this section, we will do evaluations with formulas containing "ln" expressions.

---

104.   The steps for the evaluation below are shown. Notice that we enter 50 and press $\boxed{\ln x}$ before pressing $\boxed{=}$. Rounding to tenths, we get D = 39.1 .

In $D = k \ln P$, find D when k = 10 and P = 50 .

Calculator Steps: 10 $\boxed{x}$ 50 $\boxed{\ln x}$ $\boxed{=}$ 39.12023

Do this one. Round to tenths.

In $D = k \ln P$, when k = 7.5 and P = 18.5 , D = _____ .

---

105.   The steps for the evaluation below are shown. Notice again that we enter the value for P and then press $\boxed{\ln x}$. Rounding to thousandths, we get w = 0.998 .

In $w = \dfrac{d}{\ln P}$, find w when d = 4.24 and P = 70 .

Calculator Steps: 4.24 $\boxed{\div}$ 70 $\boxed{\ln x}$ $\boxed{=}$ .99800041

Do this one. Round to hundredths.

In $w = \dfrac{d}{\ln P}$, when d = 3.99 and P = 80 , w = _____ .

| 21.9 |

---

106.   In the formula below, $\ln\left(\dfrac{A_o}{A}\right)$ is the natural log of a fraction or division. Notice that we press $\boxed{=}$ to complete that division before pressing $\boxed{\ln x}$, and then divide by the value of k. Rounding to hundredths, we get t = 3.54 .

In $t = \dfrac{\ln\left(\dfrac{A_o}{A}\right)}{k}$, find t when $A_o = 60$, A = 20 , and k = 0.31 .

Calculator Steps: 60 $\boxed{\div}$ 20 $\boxed{=}$ $\boxed{\ln x}$ $\boxed{\div}$ 0.31 $\boxed{=}$ 3.5439106

Do this one. Round to hundredths.

In $k = \dfrac{\ln\left(\dfrac{A_o}{A}\right)}{t}$, when $A_o = 38.9$, A = 9.45 , and t = 1.29 ,

k = _____ .

| 0.91 |

| 1.10 |

107. Two calculator methods are shown for the evaluation below. Rounding to hundredths, we get v = 3.57 .

In   v = c ln$\left(\dfrac{M}{m}\right)$, find $\underline{v}$ when c = 1.88 , M = 750 , and m = 112 .

1) Underline parentheses symbols, we evaluate $\left(\dfrac{M}{m}\right)$ before pressing ⎡ln x⎤ .

   Calculator Steps:
   1.88 ⎡x⎤ ⎡(⎤ 750 ⎡÷⎤ 112 ⎡)⎤ ⎡ln x⎤ ⎡=⎤   3.5749598

2) Not using parentheses symbols, we must evaluate ln$\left(\dfrac{M}{m}\right)$ first and then multiply by 1.88 . Notice that we press ⎡=⎤ to complete the division before pressing ⎡ln x⎤ .

   Calculator Steps: 750 ⎡÷⎤ 112 ⎡=⎤ ⎡ln x⎤ ⎡x⎤ 1.88 ⎡=⎤   3.5749598

Do this one. Round to hundredths.

In   v = c ln$\left(\dfrac{M}{m}\right)$, when c = 1.86 , M = 869 , and m = 90 ,

v = _____ .

| | 4.22 |

108. In the formula below, -ln B means "the additive inverse of ln B". Therefore, after finding ln B, we press ⎡+/-⎤ to get its inverse. An example is shown. Rounding to hundredths, we get Q = -1.25 .

In   Q = -ln B , find Q when B = 3.5 .

   Calculator Steps: 3.5 ⎡ln x⎤ ⎡+/-⎤   -1.252763

Do this one. Round to hundredths.

In   Q = -ln B , when B = 120 , Q = _____ .

| | -4.79 |

109. A two-step process is needed to find A in the evaluation below. Rounding to tenths, we get A = 54.6 .

In   ln A = $\dfrac{h}{t}$, find A when h = 80 and t = 20 .

1) First we find ln A by substituting:

   ln A = $\dfrac{80}{20}$ = 4

2) Then we use ⎡e^x⎤ or ⎡2nd⎤ ⎡ln x⎤ to find A. The steps are:

   Calculator Steps: 4 ⎡e^x⎤   54.59815
                       or ⎡2nd⎤ ⎡ln x⎤

Do this one. Round to a whole number.

In   ln A = $\dfrac{h}{t}$, when h = 150 and t = 30 , A = _____ .

| | 148 |

SELF–TEST 52 (pages 643 - 654)

Convert each power of <u>e</u> to a number.

1. $e^{4.25}$ = _____ Round to tenths.

2. $e^{-1.36}$ = _____ Round to thousandths.

Convert each number to a power of <u>e</u>.

3. 915 = _____ Round to hundredths.

4. 0.0827 = _____ Round to thousandths.

---

5. Rounded to seven decimal places, the numerical value of <u>e</u> is _____ .

---

6. Find F when A = 18.7 and t = 9.42 . Round to thousands.

$$F = Ae^t$$

F = _____

7. Find S when x = 2.56 . Round to hundredths.

$$S = \frac{e^x - e^{-x}}{2}$$

S = _____

8. Find W when k = 51.7 , c = 0.065 , and t = 12 . Round to tenths.

$$W = ke^{-ct}$$

W = _____

Find each of the following to four decimal places.

9. The <u>natural</u> logarithm of 46.2 is _____ .

10. The <u>common</u> logarithm of 46.2 is _____ .

---

11. Write this power-of-<u>e</u> equation as a logarithmic equation.

$$314 = e^{5.75}$$ _____

12. Write this logarithmic equation as a power-of-<u>e</u> equation.

$$-1.016 = \ln 0.362$$ _____

---

13. Find G. Round to tenths.

$$\ln G = 2.5181$$

G = _____

14. Find <u>x</u>. Round to hundredths.

$$17.4 = e^x$$

x = _____

15. Find B. Round to thousandths.

$$-1.93 = \ln B$$

B = _____

---

16. Find <u>r</u> when T = 0.727 . Round to thousandths.

$$r = -\ln T$$

r = _____

17. Find W when c = 290 , a = 7,300 , and b = 510 . Round to a whole number.

$$W = c \ln\left(\frac{a}{b}\right)$$

W = _____

18. Find P when b = 81.3 and h = 56.9 . Round to hundredths.

$$\ln P = \frac{b}{h}$$

P = _____

<u>ANSWERS:</u>

1. 70.1
2. 0.257
3. $e^{6.82}$
4. $e^{-2.493}$

5. 2.7182818
6. F = 231,000
7. S = 6.43
8. W = 23.7

9. 3.8330
10. 1.6646
11. ln 314 = 5.75
12. $e^{-1.016}$ = 0.362

13. G = 12.4
14. x = 2.86
15. B = 0.145

16. r = 0.319
17. W = 772
18. P = 4.17

SUPPLEMENTARY PROBLEMS - CHAPTER 15

Assignment 50

Evaluate these powers. Report all digits shown on the display.

1. 3^{10} 2. 12^5 3. $(8.1)^4$ 4. $(0.2)^6$ 5. $(1.35)^3$

Evaluate these powers. Report each answer in scientific notation with the first factor rounded to hundredths.

6. $(480)^8$ 7. $(3.79)^{19}$ 8. $(0.025)^7$ 9. $(0.146)^{18}$

Evaluate these roots. Round as directed.

10. Round to tenths.   11. Round to hundredths.   12. Round to hundredths.

$\sqrt[4]{936,000}$ $\sqrt[7]{350}$ $\sqrt[6]{8.92}$

13. Round to two decimal places.   14. Round to thousandths.   15. Round to four decimal places.

$\sqrt[5]{0.44}$ $\sqrt[3]{0.0617}$ $\sqrt[9]{0.001839}$

Convert each radical to a power with a decimal exponent. Do not evaluate.

16. $\sqrt[5]{18^3}$ 17. $\sqrt[8]{6^4}$ 18. $\sqrt{3^9}$ 19. $\sqrt[4]{542}$

Evaluate these powers. Round as directed.

20. Round to tenths.   21. Round to hundredths.   22. Round to a whole number.

$(263)^{0.75}$ $(3.94)^{1.5}$ $(18.6)^{2.3}$

23. Round to millionths.   24. Round to thousandths.   25. Round to hundredths.

6^{-3} $(2.74)^{-1.2}$ $(0.183)^{-0.5}$

26. In $W = K^r$, find W when $K = 3.9$ and $r = 2.7$. Round to tenths.

27. In $R = PQ^a$, find R when $P = 1.38$, $Q = 3.27$, and $a = 0.6$. Round to hundredths.

28. In $D = K(t + 1)^n$, find D when $K = 80$, $t = 3.9$, and $n = 1.3$. Round to a whole number.

29. In, $H = \left(\dfrac{d}{t}\right)^{0.6}$, find H when $d = 47.6$ and $t = 61.3$. Round to thousandths.

30. Air pressure P, in lb/in^2, and altitude <u>h</u>, in miles, are related by the formula:
$P = (14.7)(10^{-0.087h})$. Find the air pressure at a height of 5 miles. Round to tenths.

Assignment 51

Convert each power of ten to a number. Round as directed.

1. Round to hundredths.   2. Round to thousands.   3. Round to thousandths.   4. Round to millionths.

$10^{0.58}$ $10^{5.3948}$ $10^{-0.17}$ $10^{-3.0259}$

Convert each number to a power of ten. Round each exponent to four decimal places.

    5. 8,250   6. 41.7   7. 0.063   8. 0.000216

Find these common logarithms. Round each to four decimal places.

    9. log 5,320,000   10. log 9.41   11. log 0.78   12. log 0.00115

Write each power-of-ten equation as a log equation.

    13. $185 = 10^{2.2672}$ 14. $10^{-0.7595} = 0.174$ 15. $10^x = 27.4$ 16. $G = 10^{-2.3814}$

Write each log equation as a power-of-ten equation.

    17. 2.4166 = log 261   18. log 0.02 = -1.6990   19. log R = 5.8037   20. y = log 7,190

Find the numerical value of each letter. Round as directed.

    21. Round to hundredths.   22. Round to millionths.   23. Round to thousandths.

        0.8673 = log G   log T = -3.8673 $19,345 = 10^b$

24. In $N = 10 \log K$, find N when $K = 50$. Round to a whole number.

25. In $P = -\log A$, find P when $A = 0.0000037$. Round to hundredths.

26. In $M = 2.5 \log\left(\dfrac{I_1}{I}\right)$, find M when $I_1 = 430$ and $I = 15$. Round to tenths.

27. In $r = 0.95 \log(p + t)$, find r when $p = 25.8$ and $t = 37.6$. Round to hundredths.

28. In an electronic amplifier, the gain D, in decibels, the power input P_1 , in watts, and the power output P_2 , in watts, are related by the formula: $D = 10 \log\left(\dfrac{P_2}{P_1}\right)$. Find D when , $P_1 = 0.018$ watt and $P_2 = 60$ watts. Round to a whole number.

29. In reporting earthquakes, the Richter number R, the earthquake magnitude I, and the comparison magnitude I_c are related by the formula: $I = I_c(10^R)$. (a) When $R = 4.8$ and $I_c = 1$ unit, find I. Round to thousands. (b) For two earthquakes measuring 4.8 and 6.8 on the Richter scale, the second is how many times larger than the first?

Assignment 52

Convert each power of e to a number.

    1. Round to tenths.   2. Round to thousandths.   3. Round to hundreds.

      $e^{3.98}$ $e^{-1.72}$ $e^{9.35}$

Convert each number to a power of e.

    4. Round to hundredths.   5. Round to thousandths.   6. Round to four decimal places.

        116   0.0384   4.72

7. In $H = ae^p$, find H when $a = 2.4$ and $p = 1.72$. Round to tenths.

8. In $y = \dfrac{e^x + e^{-x}}{2}$, find y when $x = 1.36$. Round to hundredths.

9. In $B = ke^{-ct}$, find B when $k = 485$, $c = 7.2$, and $t = 0.096$. Round to a whole number.

Find these natural logarithms. Round each to four decimal places.

     10. ln 29.8   11. ln 7,560   12. ln 2.04   13. ln 0.00192

Convert each power-of-e equation to a logarithmic equation.

     14. $e^{4.15} = 63.4$ 15. $0.0828 = e^{-2.49}$

Convert each logarithmic equation to a power-of-e equation.

     16. ln 9.31 = 2.23   17. -0.851 = ln 0.427

18. Find R.   19. Find y.   20. Find P.
    Round to hundredths.   Round to tenths.   Round to ten-thousandths.

     1.3206 = ln R $e^y = 3,500$ ln P = -3.725

21. In $r = \dfrac{d}{\ln A}$, find r when d = 836 and A = 11.5 . Round to a whole number.

22. In $t = k \ln\left(\dfrac{E}{G}\right)$, find t when k = 3.53 , E = 984 , and G = 127 . Round to hundredths.

23. In $\ln N = \dfrac{a}{w}$, find N when a = 95.7 and w = 23.6 . Round to tenths.

24. In an electronic circuit, current i, in milliamperes, and time t, in seconds, are related by the formula: $i = 500e^{-0.833t}$. Find i when t = 1.26 seconds. Round to a whole number.

25. The final velocity v of a rocket, in mi/sec, its exhaust velocity c, in mi/sec, its initial mass M, in tons, and its final mass m, in tons, are related by the formula: $v = c \ln\left(\dfrac{M}{m}\right)$. Find v when c = 1.8 mi/sec , M = 960 tons , and m = 100 tons. Round to tenths.

ANSWERS FOR SUPPLEMENTARY PROBLEMS

CHAPTER 1 - REAL NUMBERS

Assignment 1

1. 6, 30   2. 0, 6, 30   3. -5, -2, 0, 6, 30   4. $-\sqrt{5}$, $\sqrt{2}$ 5. true   6. false   7. false

8. true   9. false   10. true   11. >   12. <   13. >   14. <   15. >   16. $\leq$ 17. $\geq$

18. 12   19. $\frac{1}{3}$ 20. 0   21. $\frac{9}{4}$ 22. 6.81   23. 5   24. -6   25. -12   26. -25   27. $\frac{2}{3}$

28. -3   29. $\frac{3}{2}$ 30. $-\frac{7}{20}$ 31. $\frac{1}{3}$ 32. $\frac{7}{4}$ 33. $-\frac{13}{24}$ 34. 0   35. -5.5   36. .55

37. 4.04   38. -7.4   39. 0   40. -28   41. 4   42. 17   43. -3.96   44. $\frac{5}{2}$ 45. -10

46. -5   47. -7   48. 6   49. 50   50. 0   51. -7.3   52. -7.49   53. 1.619   54. $-\frac{5}{2}$

55. $\frac{17}{15}$ 56. $\frac{4}{5}$ 57. $\frac{16}{9}$

Assignment 2

1. -56   2. -20   3. 27   4. 0   5. 10   6. -3.9   7. -6.3   8. 1.44   9. $-\frac{3}{8}$ 10. $\frac{1}{2}$

11. $-\frac{5}{2}$ 12. 6   13. 28   14. -30   15. 0   16. 126   17. -6   18. -5   19. 8   20. 0

21. -59   22. -17.6   23. impossible   24. 8   25. $-\frac{1}{6}$ 26. 6   27. $-\frac{4}{7}$ 28. 25

29. identity   30. identity   31. associative   32. commutative   33. associative

34. commutative   35. identity   36. commutative   37. identity   38. 8   39. 256   40. 1

41. -9   42. 0   43. 1   44. $\frac{1}{8}$ 45. $\frac{1}{49}$ 46. $\frac{1}{216}$ 47. $-\frac{1}{32}$ 48. -.125   49. 2.25

50. 1   51. $-\frac{9}{5}$ 52. $\frac{1}{64}$

Assignment 3

1. 6^{-1} 2. 3^2 3. 5^6 4. 8^{-6} 5. 9^{-8} 6. 4^2 7. 10^3 8. 2^3 9. 10^{-3} 10. 2^2

11. 7^0 12. 4^{-3} 13. 6^0 14. a^{m-n} 15. a^{m+n} 16. a^{mn} 17. 10^4 18. 10^{-3} 19. 10^8

20. 10^{-6} 21. 1,000   22. .1   23. 1   24. .0001   25. 168   26. .937   27. 415,000,000

28. .0000062   29. 7.14×10^{-4} 30. 9.53×10^2 31. 3×10^8 32. 1.88×10^{-1} 33. 5,000

34. .00361   35. 9,820,000   36. .00000079   37. 1.8×10^7 instructions per second

38. 1609.344 meters   39. 6.5×10^{-5} centimeter   40. 4.38×10^9 hertz   41. .000000725 second

42. 12,760,000 meters   43. 1.34×10^{-3} horsepower

Assignment 4

1. -6   2. 8   3. -18   4. -11   5. 23   6. 14   7. 0   8. -2   9. 24   10. -4   11. $\frac{5}{4}$

12. $\frac{1}{4}$ 13. 4   14. 1   15. -26   16. 3   17. $-\frac{5}{16}$ 18. $-\frac{3}{2}$ 19. $20 + 2x$ 20. $\frac{30}{x}$

21. $8 - 5x$ 22. $15x^2$ 23. $\frac{50}{x - 10}$ 24. $12x^3 - 6$ 25. $18 - 2xy$ 26. $\frac{x^2 + y^2}{x - y}$

27. $\frac{1}{2}(x^2 - y^2)$ 28. $50 - (x^3 + y^3)$ 29. $\frac{x^3}{y^3} - 9$ 30. $(x + y)^2$ 31. $x > 5$ 32. $x \leq 0$

33. -7   34. -4   35. 2   36. $\frac{5}{4}$ 37. $\frac{1}{2}$ 38. 1   39. 4   40. 16   41. 7   42. A = 120

43. F = 600   44. r = 20   45. F = 95   46. M_r = 6   47. v = 5   48. R_t = 60

CHAPTER 2 - SOLVING EQUATIONS

Assignment 5

1. y = -4   2. t = 11   3. x = 41   4. N = 53   5. G = 1   6. d = -5   7. x = 6.6

8. p = 1.8   9. d = $\frac{2}{3}$ 10. x = $\frac{7}{8}$ 11. V = $-\frac{3}{7}$ 12. m = $-\frac{2}{3}$ 13. p = $\frac{4}{9}$ 14. R = $\frac{15}{8}$

15. F = -1   16. y = $\frac{1}{7}$ 17. w = $-\frac{5}{3}$ 18. h = 0   19. x = -8   20. y = 3   21. p = $\frac{4}{3}$

22. x = $-\frac{2}{5}$ 23. y = $\frac{5}{4}$ 24. H = 0   25. x = 4   26. r = -3   27. m = $-\frac{2}{3}$ 28. s = $\frac{1}{4}$

29. H = $-\frac{3}{2}$ 30. A = -1   31. y = $-\frac{7}{4}$ 32. x = 1   33. c = $-\frac{1}{3}$ 34. p = 0   35. t = -4

36. m = 4.5   37. A = 10   38. y = $\frac{9}{4}$ 39. x = $\frac{3}{16}$ 40. s = 0   41. w = $\frac{2}{7}$ 42. P = -2

Assignment 6

1. 3x + 21   2. 18 + 27y   3. 50R + 10   4. 7x - 14   5. 15 - 20y   6. 8 - 40d   7. -4m - 12

8. -7 + 6P   9. 6x   10. 6y   11. -7m   12. 5d   13. -6r   14. -1 + 5p   15. 5V - 7

16. -3a - 9   17. R = -4   18. y = $-\frac{13}{5}$ 19. P = $\frac{1}{3}$ 20. x = $-\frac{5}{2}$ 21. w = 0   22. h = $\frac{10}{3}$

23. F = $-\frac{3}{4}$ 24. x = 5   25. y = $\frac{4}{3}$ 26. r = $\frac{5}{4}$ 27. b = -2   28. x = $-\frac{2}{3}$ 29. a = $\frac{1}{2}$

30. V = 1   31. t = 3   32. x = $\frac{1}{3}$ 33. N = $\frac{1}{3}$ 34. c = 0   35. E = 1   36. h = -12

37. y = 2   38. w = $-\frac{5}{2}$ 39. k = -1   40. d = $\frac{1}{3}$

Assignment 7

1. a = $-\frac{1}{4}$ 2. R = $\frac{4}{3}$ 3. s = $-\frac{1}{2}$ 4. x = $-\frac{1}{6}$ 5. w = 0   6. b = $\frac{5}{2}$ 7. G = -2   8. h = $-\frac{7}{2}$

9. r = $\frac{2}{5}$ 10. t = $\frac{1}{5}$ 11. x = $-\frac{7}{4}$ 12. y = 10   13. P = $\frac{3}{2}$ 14. F = $\frac{5}{2}$ 15. d = 2

16. E = $-\frac{3}{5}$ 17. x = -2   18. V = -5   19. w = 5   20. d = $-\frac{1}{4}$ 21. x = $-\frac{15}{4}$ 22. y = 15

23. E = $-\frac{1}{2}$ 24. r = 3   25. 2   26. -1   27. ±10   28. -3   29. 1.41   30. ±6.08

31. -8.66   32. 9.95   33. $\frac{5}{4}$ 34. $\pm\frac{1}{3}$ 35. $-\frac{3}{4}$ 36. $\pm\frac{1}{10}$ 37. 1, 25, 64, 81

38. $\frac{1}{100}$, $\frac{16}{49}$, $\frac{100}{81}$, $\frac{25}{64}$

Assignment 8

1. x = 7 and -7   2. y = $\frac{1}{10}$ and $-\frac{1}{10}$ 3. p = $\frac{9}{8}$ and $-\frac{9}{8}$ 4. t = 6 and -6   5. r = 2 and -2

6. a = $\frac{1}{4}$ and $-\frac{1}{4}$ 7. x = $\frac{5}{2}$ and $-\frac{5}{2}$ 8. m = 4 and -4   9. V = ±8.94   10. h = ±3.46

11. x = ±8.25   12. s = ±1.41   13. L = 20 cm   14. R_1 = 90 ohms   15. 100° Celsius

16. an increase of 20°F   17. $V_2 = 500$ in^3 18. R = 40 ohms   19. t = 5 seconds

20. $R_2 = 60$ rpm   21. $d_2 = 3$ meters   22. I = 2.24 amperes   23. a = 2 meters/sec^2

24. s = 2.5 cm   25. $V_2 = 50$ volts   26. The number is 2.   27. The number is 4.   28. 30 cm

and 90 cm   29. 4 amperes and 8 amperes   30. 24 in, 24 in, and 48 in   31. 700 gal/min,

850 gal/min, and 450 gal/min   32. 5.5 hours   33. 18,000 mph   34. 81 mph   35. $\frac{2}{3}$ hour or

40 min   36. 1.5 hours   37. 4:30 p.m.

CHAPTER 3 - FRACTIONAL EQUATIONS

Assignment 9

1. w = 54   2. x = 3   3. y = 25   4. R = 3   5. $G = \frac{12}{5}$ 6. $v = \frac{1}{10}$ 7. $d = \frac{7}{4}$ 8. $p = \frac{8}{3}$

9. t = 0   10. h = -3   11. x = -1   12. $y = \frac{1}{2}$ 13. $w = -\frac{3}{4}$ 14. F = 1   15. a = 6

16. $k = -\frac{3}{4}$ 17. x = -5   18. $t = \frac{3}{2}$ 19. $x = \frac{1}{4}$ 20. H = 5   21. v = -3   22. P = -18

23. y = -1   24. $B = \frac{7}{5}$ 25. t = -3   26. $k = \frac{9}{8}$ 27. s = 0   28. p = -8   29. m = -7

30. $y = \frac{1}{7}$ 31. $R = \frac{1}{10}$ 32. $w = -\frac{1}{4}$ 33. $h = -\frac{1}{3}$

Assignment 10

1. $t = \frac{7}{2}$ 2. $y = \frac{3}{5}$ 3. $w = \frac{6}{5}$ 4. $x = \frac{15}{16}$ 5. x = 1   6. w = -6   7. $r = \frac{15}{8}$ 8. $d = \frac{11}{5}$

9. $a = -\frac{15}{4}$ 10. x = 3 and -3   11. $y = \frac{5}{2}$ and $-\frac{5}{2}$ 12. $t = \pm\sqrt{28} = \pm5.29$ 13. $a = \pm\sqrt{12} = \pm3.46$

14. $h = -\frac{2}{7}$ 15. $P = -\frac{5}{6}$ 16. $b = -\frac{1}{4}$ 17. $E = -\frac{7}{8}$ 18. $t = \frac{13}{21}$ 19. R = 5   20. y = -1

21. $F = \frac{3}{5}$ 22. $A = -\frac{4}{5}$ 23. r = 9   24. N = -5   25. t = -3   26. $h = -\frac{9}{2}$ 27. p = -6

28. $w = \frac{15}{28}$ 29. $m = \frac{2}{3}$ 30. k = 7   31. $m = -\frac{20}{9}$ 32. $h = \frac{14}{13}$ 33. $y = \frac{21}{40}$

Assignment 11

1. $v_2 = 52$ m/sec   2. $V_1 = 120$ liters   3. I = 5 amperes   4. 68°F   5. b = 30 cm   6. $t = \frac{5}{2}$

or $2\frac{1}{2}$ seconds   7. I = 5 amperes   8. $L = 5\frac{1}{2}$ ft   9. r = 40 ft   10. $C_2 = 40$ microfarads

11. $\alpha = \frac{24}{25}$ or 0.96   12. d = 6.4 inches   13. 9 mph   14. 48 mph and 58 mph   15. 240 mph

and 420 mph   16. Wind speed is 20 mph

CHAPTER 4 - PERCENT, PROPORTION, VARIATION

Assignment 12

1. ten-thousands   2. tenths   3. hundreds   4. thousandths   5. millionths   6. 703,015
7. 5,200,080,000   8. 6.009   9. 30.5   10. .0160   11. .000300   12. 569,000   13. 110,000,000
14. 74   15. .000360   16. 35.10   17. 860   18. 60,300,000   19. .01710   20. .0600
21. .600   22. 18.9   23. -114.8   24. -6   25. .511225   26. -.016   27. Not possible
28. 21,700   29. 20.26   30. 6,000,000   31. 310,000   32. 78.0   33. 240   34. 4,600,000

35. .756   36. .034683   37. 193,000   38. 2.24   39. .006273   40. .338   41. 40,000
42. .4684   43. .015

Assignment 13

1. $\frac{3}{4}$ 2. $\frac{1}{5}$ 3. $\frac{1}{3}$ 4. $\frac{9}{10}$ 5. $\frac{1}{4}$ 6. .19   7. .039   8. .0065   9. 1.47   10. 3

11. 81.4%   12. 4.25%   13. .6%   14. 170%   15. 500%   16. 50%   17. 80%   18. 30%

19. $66\frac{2}{3}$%   20. 59%   21. 35%   22. 17%   23. 131%   24. 4.15%   25. .47%   26. 42.02%

27. 40%   28. 300   29. 50   30. $33\frac{1}{3}$%   31. $21.25   32. 700   33. 2   34. 400   35. 4%

36. 17.5 pounds   37. 83%   38. 16 kilograms   39. 60.67%   40. 667 pounds   41. 85.8%

42. .038%   43. 7,300 pounds   44. 24 liters   45. 6 liters

Assignment 14

1. $\frac{5}{12}$ 2. 3.25   3. t = .507   4. x = 2,959,000   5. w = 4.54   6. 644 miles   7. 2.2 pounds

8. 218 grams   9. 8.3 centimeters   10. 8 pounds   11. 214 washers   12. 40 teeth

13. 21.7 centimeters   14. 58.7 kilowatts   15. 5.47 pounds   16. 3.785 liters   17. 396 feet

18. 21 milligrams   19. 322 miles   20. 82.9 tons

Assignment 15

1. w = 18   2. d = 32   3. p = 70   4. F = 4   5. v = 15   6. R = 42   7. d = 425 kilometers
8. P = 6 kg/cm^2 9. X_L = 120 ohms   10. I = 3 microwatts   11. E = 200 joules
12. X_C = 32 ohms   13. F = 20 newtons   14. e = 0.08 inch   15. W = 600 grams
16. P = 250 watts   17. a = 22.5 ft/sec^2 18. 72 minutes   19. s = 320 ft

CHAPTER 5 - CALCULATOR OPERATIONS

Assignment 16

1. R = 332.9   2. V = 6,400   3. Q = 54.7   4. L = 162   5. t = 0.00174   6. v = 6.77
7. x = 29.9   8. P = 0.718   9. C = -40   10. T = 301   11. H = 1.87   12. m = 0.0872
13. E = 31,000,000   14. F = 14.5   15. P = 0.880

Assignment 17

1. P = 806   2. A = 0.697   3. R = 1.90   4. s = 224   5. a = 12.7   6. p = 9.16
7. E = 120   8. b = 480   9. w = 77   10. F = 86.6   11. t = 0.054   12. C = 0.424
13. V = 710,000   14. s = 6.38

Assignment 18

1. T_1 = 295   2. A_2 = 44.1   3. P_1 = 9.96   4. 2.05 x 10^{-10} 5. 4.13 x 10^{14} 6. 3.00 x 10^{-11}
7. 6.03 x 10^{-12} 8. 8.15 x 10^{10} 9. 2.70 x 10^{13} 10. f = 6.66 x 10^{14} 11. R = 2.05 x 10^6
12. .000577   13. 5.52 g/cm^2 14. 0.207 acre   15. 372 miles   16. 28,350 mg   17. 12,400 ft

Assignment 19

1. M = 4,900   2. A = 1.71   3. r = 24.6   4. d = 5,600   5. E = 3,100,000   6. A = 93
7. ΔL = 6.24 x 10^{-2} or 0.0624   8. I = 0.017   9. R = 4.14   10. v = 25.5   11. m = 14,100
12. R = 68.1   13. R_1 = 690   14. 0.10

CHAPTER 6 - MEASUREMENTS

Assignment 20

1. 198 in   2. 25 lb   3. 19,800 ft   4. 3.5 gal   5. 7 mi   6. 108 sec   7. 1.59 tons
8. 26 mi   9. 6.5 km   10. 0.57 g   11. 0.4 m   12. 0.15 hg   13. 820 ml   14. 610 dag
15. 0.0025 sec   16. 2,800 dm   17. 8,000 cm   18. 3.5 cm   19. 2.17 msec   20. 6.2 mi
21. 0.747 hr   22. 3.40 kg   23. 10.65 cm   24. 22 l   25. 17.6 oz   26. 8.11 in
27. 42.2 km   28. 282 gal/min   29. 32.4 km/hr   30. 14 cm/sec   31. 3.8 l/sec
32. 62.1 mi/hr   33. 4 gal/sec   34. 299,800 km/sec or 2.998×10^5 km/sec

Assignment 21

1. 20°C   2. 1,000°C   3. 5°F   4. 212°F   5. -220°F   6. -5°C   7. 27°C   8. 63°K
9. 323°K   10. 10^6 11. 10^{-12} 12. 10^{12} 13. 10^{-3} 14. micro-   15. kilo-   16. nano-
17. giga-   18. centi-   19. hecto-   20. deci-   21. 430 km   22. 55 μsec   23. .098 Mg
24. .549 mm   25. 35,000 km   26. .85 Tg   27. 15,000 nsec   28. .0075 nsec   29. 59,000 Mm
30. 440,000 psec   31. .45 kg   32. .00045 Mg   33. 6 sec   34. 19,000 m   35. 88,000,000
μsec   36. 1,535°C   37. 380 nanoseconds   38. 5,000 picometers

Assignment 22

1. tenths   2. hundred-thousandths   3. thousandths   4. hundredths   5. thousands
6. 0.0195 in   7. 29.6 km   8. 0.00420 g   9. three   10. two   11. four   12. five
13. three   14. 344 cm   15. 0.280 sec   16. 41.93 in   17. 1.5130 sec   18. 1.5130 sec
19. 0.016 sec   20. 349 sec   21. 0.0310 g   22. 725.7 g   23. 18.705 g   24. 0.0310 g
25. 6.34   26. 1.618   27. 18.0   28. 4.50   29. 10.07   30. 2.420   31. 53.9   32. 0.2227
33. 3.7   34. 3.107   35. 0.00043   36. 3,470   37. 0.0480   38. 459.7   39. 0.00462
40. 0.23   41. 88.54   42. 4.69   43. 1.4   44. 0.8080   45. 3.75   46. 15   47. 22.67

CHAPTER 7 - GEOMETRY

Assignment 23

1. 140°   2. 40°   3. 140°   4. AD = 25"   5. AE = 20"   6. 17.8 m²   7. 17.58 m
8. 222 cm²   9. 59.6 cm   10. 24.9 cm   11. 33,200 ft   12. 5.08 in²   13. 118,000 ft²
14. 1,500 cm²   15. 25.7 m²   16. 0.656 in²   17. 27.7 m²   18. 2,996,000 ft²   19. 35.6 in

Assignment 24

1. 900 in²   2. 44.6 m²   3. 908 ft²   4. 1,350 cm²   5. 4.3 m²   6. 7.78 in²   7. 4.5 ft²
8. 29,300 cm²   9. 50 in²   10. 7.18 yd²   11. 72.8 hectares   12. 184 cm²   13. 1,190 in³
14. 18.6 m³   15. 900 cm³   16. 3.5 ft³   17. 6 yd³   18. 3,800,000 cm³   19. 951 cm³
20. 9.51 yd³   21. 1,200 in³   22. 3.52 cm

Assignment 25

1. 4,100 ft³   2. 8.81 m³   3. 4,710 cm³   4. 1,400 ft²   5. 26.0 m²   6. 1,360 cm²
7. 1,350 in²   8. 6,628 cm²   9. 391 cm³   10. 777 cm³   11. 140 in³   12. 14.7 lb
13. 679 g   14. 3.08 lb   15. 25 kg   16. 89 lb

CHAPTER 8 - ALGEBRAIC FRACTIONS

Assignment 26

1. 1   2. $\dfrac{a}{2d}$ 3. $\dfrac{G}{3P}$ 4. $\dfrac{3}{4}$ 5. $\dfrac{1}{c}$ 6. 4T   7. $\dfrac{t}{b}$ 8. $\dfrac{1}{5r}$ 9. $\dfrac{h}{xy}$ 10. $\dfrac{2}{b}$ 11. $\dfrac{p^2 t}{k}$

12. a   13. $\dfrac{E}{G}$ 14. $\dfrac{1}{t}$ 15. 2tx   16. $\dfrac{1}{k-1}$ 17. $h + rw$ 18. $\dfrac{v}{x-y}$ 19. $\dfrac{b}{d}$ 20. $\dfrac{1}{x}$

21. 2P   22. $\dfrac{ks}{t}$ 23. $\dfrac{m}{2pr}$ 24. $\dfrac{F}{T(1-A)}$ 25. $c + k$ 26. $\dfrac{4d}{m}$ 27. $\dfrac{1}{2(y+1)}$ 28. $\dfrac{2}{3}$

29. 2   30. 1

Assignment 27

1. $\dfrac{w}{2}$ 2. $\dfrac{v+w}{r}$ 3. $-2ab$ 4. $\dfrac{3t-2}{w}$ 5. $\dfrac{gh-d}{m}$ 6. $\dfrac{4}{p}$ 7. 0   8. 2R   9. $\dfrac{F-H}{E-2}$

10. $\dfrac{1}{x+y}$ 11. $\dfrac{c+d+p}{h}$ 12. $\dfrac{3}{F}$ 13. $\dfrac{1}{T}$ 14. $\dfrac{d+3}{a+b}$ 15. $\dfrac{P}{6}$ 16. $\dfrac{4r+t}{4t}$ 17. $\dfrac{7r}{8}$

18. $\dfrac{5}{3G}$ 19. $\dfrac{5d+r}{5x}$ 20. $\dfrac{a}{3t}$ 21. $\dfrac{2m-3}{6}$ 22. $\dfrac{3d+4r}{24}$ 23. $\dfrac{kv-pr}{rv}$ 24. $\dfrac{P+Q}{PQ}$

25. $\dfrac{h+sw}{rw}$ 26. $\dfrac{5m-3}{6m}$ 27. $\dfrac{2bw+at}{2ab}$ 28. $\dfrac{c-v}{cpv}$ 29. $\dfrac{x+4}{2x^2}$ 30. $\dfrac{a-d}{ad}$ 31. $\dfrac{rw+st}{sw}$

32. $\dfrac{2vw-bp}{pw^2}$

Assignment 28

1. $\dfrac{E-2}{2}$ 2. $\dfrac{r+w}{r}$ 3. $\dfrac{3P+2}{3}$ 4. $\dfrac{5B-4}{B}$ 5. $\dfrac{4t+1}{t}$ 6. $\dfrac{mk-w}{k}$ 7. $\dfrac{r+2adv}{av}$

8. $\dfrac{1-x}{x}$ 9. $\dfrac{ar-1}{r}$ 10. $\dfrac{h+6}{3}$ 11. $\dfrac{A-B}{A}$ 12. $\dfrac{m+tw}{t}$ 13. $\dfrac{r}{w} + \dfrac{s}{w}$ 14. $\dfrac{2}{3} - \dfrac{1}{2r}$

15. $\dfrac{B}{C} - \dfrac{1}{A}$ 16. $\dfrac{c}{r} - \dfrac{b}{t}$ 17. $\dfrac{1}{H} + \dfrac{1}{F}$ 18. $\dfrac{x}{y+1} + \dfrac{1}{y+1}$ 19. $\dfrac{a}{4} + 2$ 20. $\dfrac{1}{r} - 1$ 21. $\dfrac{2}{d} - \dfrac{4}{3a}$

22. $s + \dfrac{r}{t}$ 23. $\dfrac{V}{3} + 2$ 24. $3a - \dfrac{5}{h}$ 25. True   26. False   27. False   28. $6t(3a+2)$

29. $F(1-P)$ 30. $d(c+h)$ 31. Not possible   32. $4(A_1 - A_2)$ 33. $V(R+1)$

34. Not possible   35. $p(w-2)$

Assignment 29

1. $\dfrac{2x+4y}{3w}$ 2. $\dfrac{t}{p-2}$ 3. $\dfrac{d}{w+1}$ 4. $\dfrac{3s}{s-2v}$ 5. $\dfrac{p+a}{4}$ 6. $\dfrac{1-P}{R}$ 7. $\dfrac{y+1}{1-y}$ 8. $\dfrac{k-2}{2r-1}$

9. $\dfrac{2t+w}{p+3v}$ 10. $\dfrac{1}{A+1}$ 11. $\dfrac{1}{2y-3}$ 12. $b - 1$ 13. True   14. True   15. False

16. (a) and (e)   17. $k(r-a)$ 18. $\dfrac{bd}{t+w}$ 19. $\dfrac{x-y}{2pv}$ 20. $\dfrac{1}{D+H}$ 21. $\dfrac{x+1}{2xy}$ 22. $\dfrac{rt}{1-r}$

23. $\dfrac{R}{R+2}$ 24. $\dfrac{w-1}{bw}$ 25. $\dfrac{3(x+y)}{t}$ 26. $\dfrac{p-t}{c-k}$ 27. $\dfrac{R+1}{R-1}$ 28. $\dfrac{t(a-w)}{w(b+t)}$ 29. $\dfrac{V}{R-V}$

30. $\dfrac{x^2+1}{x^2-1}$ 31. $\dfrac{1}{B}$ 32. $\dfrac{2}{a}$

CHAPTER 9 - FORMULA REARRANGEMENT

Assignment 30

1. $s = \dfrac{P}{4}$ 2. $L = \dfrac{A}{W}$ 3. $r = \dfrac{C}{2\pi}$ 4. $R = \dfrac{E^2}{P}$ 5. $h = \dfrac{V}{\pi r^2}$ 6. $v = \dfrac{d}{t_2 - t_1}$

7. $M = \dfrac{F(p - a)}{2m}$ 8. $V_2 = \dfrac{T_2 V_1}{T_1}$ 9. $W = dF$ 10. $f = \dfrac{1}{t}$ 11. $c = \dfrac{Q}{mT}$ 12. $G = \dfrac{R^2 W}{KN}$

13. $R = \dfrac{E + e}{I}$ 14. $C = \dfrac{2s}{gt^2}$ 15. $h = \dfrac{2dp}{b(1 - a)}$ 16. $L = \dfrac{HK(b_1 + b_2)}{A}$ 17. $A_2 = \dfrac{A_1 P_1}{P_2}$

18. $R_2 = \dfrac{(E_2)^2 R_1}{(E_1)^2}$ 19. $d = \dfrac{hpt}{bw}$ 20. $a = \dfrac{r^2}{m}$

Assignment 31

1. $H = \sqrt{D + T}$ 2. $c = \sqrt{\dfrac{E}{m}}$ 3. $v = \sqrt{2as}$ 4. $t = \sqrt{\dfrac{d}{2k}}$ 5. $h = \sqrt{\dfrac{2M}{b}}$ 6. $P = \sqrt{GN}$

7. $r = \sqrt{\dfrac{2cw}{a}}$ 8. $V = \sqrt{\dfrac{FgR}{W}}$ 9. $F_t = F_1 + F_2$ 10. $W = \dfrac{P - 2L}{2}$ 11. $Z = \sqrt{R^2 + X^2}$

12. $E = e + ir$ 13. $s_2 = \dfrac{-s_1 f_1}{f_2}$ 14. $h = \sqrt{r^2 - v^2}$ 15. $P_1 = P_2 - EI$ 16. $k = \dfrac{ah - w}{s}$

17. $d_2 = d_1 - vt$ 18. $r = \sqrt{R^2 - hN}$ 19. $G_1 = PW - G_2$ 20. $c = \dfrac{1 - 2ad}{m}$

Assignment 32

1. $w = \dfrac{b - dh}{h}$ or $w = \dfrac{b}{h} - d$ 2. $t_1 = \dfrac{at_2 - v}{a}$ or $t_1 = t_2 - \dfrac{v}{a}$ 3. $d = \sqrt{\dfrac{\pi D^2 - 4A}{\pi}}$ or

$d = \sqrt{D^2 - \dfrac{4A}{\pi}}$ 4. $p = \sqrt{\dfrac{k + w}{k}}$ or $p = \sqrt{1 + \dfrac{w}{k}}$ 5. $d_2 = \dfrac{2V - bhd_1}{bh}$ or $d_2 = \dfrac{2V}{bh} - d_1$

6. $P_1 = \dfrac{SW + FTP_2}{FT}$ or $P_1 = \dfrac{SW}{FT} + P_2$ 7. $v = \sqrt{\dfrac{crt - ah^2}{a}}$ or $v = \sqrt{\dfrac{crt}{a} - h^2}$

8. $P_2 = \dfrac{dHP_1 - 3Fms}{dH}$ or $P_2 = P_1 - \dfrac{3Fms}{dH}$ 9. $P = \dfrac{F}{A_1 + A_2 + A_3}$ 10. $x = \dfrac{b}{m - 1}$

11. $f_1 = \dfrac{w + df_2}{d}$ or $f_1 = \dfrac{w}{d} + f_2$ 12. $T = \dfrac{AK}{P - 2}$ 13. $r = \dfrac{a}{h + v}$ 14. $A = \dfrac{G}{D - 1}$

15. $t = \dfrac{sv}{s + v}$ 16. $G_2 = \dfrac{G_1 G_t}{G_1 - G_t}$ 17. $d = \dfrac{b}{s + 1}$ 18. $H = \dfrac{FV}{F - V}$ 19. $N = \dfrac{L}{L + 1}$

20. $v = \sqrt{\dfrac{p(e - k)}{m}}$

Assignment 33

1. $R = \dfrac{A}{1 - H}$ 2. $a = \dfrac{bc}{b - c}$ 3. $F_1 = \dfrac{F_2 F_t}{F_2 - F_t}$ 4. $k = \dfrac{mt}{m - h}$ 5. $P = \dfrac{M + T}{M - B}$ 6. $r = \dfrac{N}{1 + hN}$

7. $M = \dfrac{s - 1}{s + 1}$ 8. $t_1 = T - b(p - p_1)$ 9. $a = t - m(d_1 + d_2)$ 10. $k = \dfrac{v - h}{w - 1}$

11. $T = \dfrac{p_1 - p_2 + bt}{b}$ or $T = \dfrac{p_1 - p_2}{b} + t$ 12. $R_2 = \dfrac{R_t - dR_1}{d + 1}$ 13. $F = \dfrac{h + w}{t + w}$

14. $r_1 = dG(p_1 - p_2) - r_2$ 15. $H = \dfrac{AS(B_1 + B_2)}{V_2 - V_1}$ 16. $m = \dfrac{w(h - t)}{p(a - k)}$ 17. $e_2 = \dfrac{C_1 + C_2 - C_1 e_1 Q}{C_1 Q}$

or $e_2 = \dfrac{C_1 + C_2}{C_1 Q} - e_1$ 18. $d_1 = \dfrac{P - W + Ad_2 P}{AP}$ or $d_1 = \dfrac{P - W}{AP} + d_2$ 19. $r_1 = \dfrac{r_2}{f(w_1 + w_2) + 1}$

20. $d = \dfrac{r(1 - ah)}{1 + ah}$

CHAPTER 10 - FUNCTIONS AND GRAPHS

Assignment 34

1. a, c, d   2. a, d, e   3. (-1,6)   4. (3,-2)   5. (2,0)   6. (-2,7)   7. (2,1)   8. (0,4)

9. Quadrant 2   10. Quadrant 1   11. Quadrant 3   12. Quadrant 4   13. Quadrant 2

14. (10,0)   15. (0,-4)   16. (-2,0)   17. (0,8)

18.   19.   20.

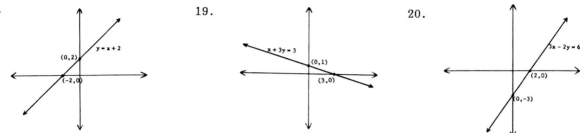

21. $y = 3$ 22. $x = -2$ 23. $x = 1$ 24. $y = -4$ 25. $y = 0$ 26. $x = 0$

Assignment 35

1. 1   2. $-\frac{1}{2}$ 3. -4   4. $\frac{1}{6}$ 5. A, D   6. A   7. B, C, E   8. C   9. 2   10. $-\frac{3}{2}$

11. 1   12. -1   13. $\frac{1}{4}$ 14. $-\frac{3}{5}$ 15. $\frac{1}{2}$ 16. 2   17. -4   18. 0   19. undefined

20. $m = 5$, $(0,2)$ 21. $m = 1$, $(0,-3)$ 22. $m = -\frac{2}{5}$, $\left(0,\frac{3}{2}\right)$ 23. $m = \frac{1}{2}$, $(0,0)$ 24. $y = -4x + 6$

25. $y = 3x + 2$ 26. $y = x - 5$ 27. $y = \frac{1}{4}x - 2$ 28. $y = -\frac{1}{2}x + \frac{5}{3}$ 29. $y = \frac{1}{5}x$

30.   31.   32.

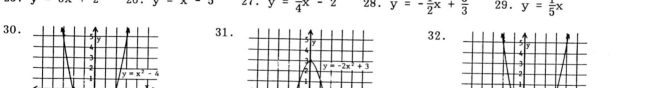

Assignment 36

1. $w = 10p^2$

p	w
0	0
1	10
2	40
3	90
4	160
5	250

2. $E + 2I = 20$

E	I
0	10
4	8
8	6
12	4
16	2
20	0

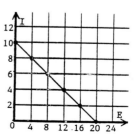

3. 4 amperes   4. 11 amperes   5. 60 volts   6. 35 volts   7. 3 cm   8. 8 cm   9. 12 cm

10. 24 cm or 25 cm   11. 120 cm^2 12. y   13. C   14. $x = 0$ 15. range   16. a, c, d

17. -7   18. 2   19. -6   20. $6a + 1$ 21. $-\frac{3}{2}$ 22. $\frac{10}{t + 2}$ 23. -4   24. -3   25. -20

26. 6   27. all real numbers except $x = 9$ 28. $x \geq 0$ That is, 0 and all positive real numbers.

CHAPTER 11 - RIGHT TRIANGLES

Assignment 37
1. F = 78° 2. P = 28° 3. A = 65° 4. t = 5,600 m 5. a = 32.8 cm 6. p = 5.52"
7. d = 26.2 ft 8. t = 630 in 9. w = 0.463 cm 10. G = 37°, H = 106° 11. A = 30°, B = 30°
12. R = 45°, S = 45° 13. A = 14,500 m² 14. A = 254 cm² 15. A = 20.1 in²

Assignment 38
1. 0.2126 2. 0.9986 3. 0.3256 4. 1. 5. 0.5 6. 0.8660 7. 2.1445 8. 0.0175
9. A = 14° 10. P = 73° 11. H = 89° 12. Q = 72° 13. T = 7° 14. F = 82° 15. A = 45°
16. B = 12° 17. 29.1 18. 11.9 19. 77.6 20. 41.2 21. 347 22. 275 23. 110
24. 516 25. A = 23° 26. G = 59° 27. R = 71° 28. E = 9° 29. $\sin R = \dfrac{r}{s}$ 30. $\cos R = \dfrac{t}{s}$

31. $\tan R = \dfrac{r}{t}$ 32. $\sin T = \dfrac{t}{s}$ 33. $\cos T = \dfrac{r}{s}$ 34. $\tan T = \dfrac{t}{r}$ 35. $\cos B = \dfrac{a}{c}$ 36. $\sin A = \dfrac{a}{c}$

37. $\tan B = \dfrac{b}{a}$ 38. $\tan A = \dfrac{a}{b}$ 39. $\cos A = \dfrac{b}{c}$ 40. $\sin B = \dfrac{b}{c}$

Assignment 39
1. h = 3.63 cm 2. w = 3.31" 3. p = 7.22 m 4. d = 438 ft 5. t = 105 cm 6. b = 268"
7. F = 37°, G = 53° 8. A = 61°, B = 29° 9. R = 41°, S = 49° 10. h = 6.70", v = 5.62"
11. A = 34° 12. w = 11.2 cm

CHAPTER 12 - OBLIQUE TRIANGLES

Assignment 40
1. a, b, f 2. 97° 3. 90° 4. 48° 5. 135° 6. b 7. c 8. a 9. b 10. p

11. d 12. G 13. K 14. $\dfrac{\boxed{\sin T}}{t} = \dfrac{\sin R}{\boxed{r}}$ 15. $\dfrac{s}{\boxed{\sin S}} = \dfrac{\boxed{t}}{\sin T}$ 16. a = 401 ft

17. B = 57° 18. b = 503 ft 19. H = 80° 20. K = 43° 21. k = 17.7 cm

Assignment 41
1. Law of Sines 2. Law of Cosines 3. Law of Cosines 4. Law of Sines
5. $p^2 = r^2 + v^2 - 2rv \cos P$ 6. $v^2 = p^2 + r^2 - 2pr \cos V$ 7. $\cos A = \dfrac{b^2 + c^2 - a^2}{2bc}$
8. $\cos C = \dfrac{a^2 + b^2 - c^2}{2ab}$ 9. f = 17.0 cm 10. D = 68° 11. E = 78° 12. G = 77°
13. P = 60° 14. T = 43°

Assignment 42
1. 0.4695 2. 0.9976 3. -0.2250 4. -0.9998 5. P = 161° 6. F = 132° 7. A = 96°
8. s = 87.5 cm 9. N = 124° 10. f = 962 ft 11. H = 136° 12. d = 3,049 m
13. A = 20°, h = 1.57 cm

CHAPTER 13 - SYSTEMS OF TWO EQUATIONS AND FORMULAS

Assignment 43

1. (3,1)

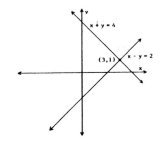

2. (-1,2)

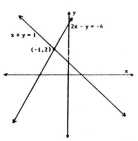

3. $x = 4$, $y = -1$ 4. $h = 6$, $m = 3$ 5. $x = 3$, $y = -2$ 6. $r = 10$, $v = 3$ 7. $h = 7$, $k = 0$
8. $b = 4$, $m = -8$ 9. $p = 20$, $r = 5$ 10. $x = 4$, $y = 1.2$ 11. $x = 5$, $y = 1$ 12. $x = 1$, $y = -1$
13. $h = 6$, $k = 5$ 14. $r = -1$, $s = 4$ 15. $c = 1$, $d = -2$ 16. $v = 3$, $w = -1$ 17. $x = 2$, $y = 4$
18. $p = 1$, $r = -1$ 19. $h = 5$, $k = 1$ 20. $x = -1$, $y = -6$ 21. $m = -2$, $n = -5$ 22. $b = 3$, $d = -5$

Assignment 44

1. $x = 5$, $y = 3$ 2. $a = 3$, $b = 9$ 3. $t = 6$, $w = 1$ 4. $r = 4$, $v = -1$ 5. $h = 20$, $m = 5$
6. $p = 1$, $s = 4$ 7. $d = 8$, $h = -2$ 8. $k = -3$, $m = 1$ 9. $r = 4$, $t = 1$ 10. $p = 2$, $w = 1$
11. $x = 2$, $y = 2$ 12. $d = 3$, $p = 4$ 13. $d = 6$, $t = 2$ 14. $E = 12$, $R = 3$ 15. $x = 1$, $y = -3$
16. $p = -1$, $v = -5$ 17. $b = 10$, $m = 4$ 18. $x = 6$, $y = 8$ 19. $F_1 = 20$, $F_2 = 10$
20. $h = 2$, $k = -\frac{2}{3}$ 21. $x = 4$, $y = -2$ 22. $p = -3$, $r = 2$ 23. $s = 10$, $w = \frac{5}{2}$ 24. $a = 1$, $b = -3$
25. $b = 4$, $d = 6$ 26. $x = 7$, $y = 30$ 27. $b = 20$, $d = 5$ 28. $r = \frac{3}{4}$, $w = 5$ 29. $x = -2$, $y = -5$
30. $r = 1$, $s = 6$ 31. $t = 0$, $w = 1$

Assignment 45

1. 150 and 68 2. 48 and 16 3. $220 for the camera, $55 for the calculator 4. 55 cm and
45 cm 5. 16.4m and 8.2m 6. $L = 18$ cm, $W = 12$ cm 7. $L = 22$ ft, $W = 11$ ft
8. speed of wind is 13 mph, speed in still air is 143 mph 9. speed of current is $1\frac{1}{2}$mph,
speed in still water is $5\frac{1}{2}$ mph 10. 40 liters of solution A, 20 liters of solution B
11. 75 milliliters of 10% solution, 125 milliliters of 50% solution 12. 100 pounds of alloy A,
200 pounds of alloy B 13. 5 kilograms of pure zinc, 10 kilograms of 40% zinc
14. speed in still air is 380 mph, speed of wind is 20 mph 15. $i_1 = 12$ milliamperes,
$i_2 = 8$ milliamperes 16. $V = 4$ volts, $I = 12$ milliamperes 17. $F_1 = 250$ kilograms,
$F_2 = 350$ kilograms 18. $T_1 = 400$ pounds, $T_2 = 600$ pounds 19. $m = -\frac{1}{2}$, $b = 6$
20. $A = 5$, $B = -2$

Assignment 46

1. $f = akr$ 2. $H = \dfrac{AC}{L}$ 3. $t = \dfrac{dh}{p}$ 4. $R = N^2P$ 5. $s = \dfrac{b^2}{r}$ 6. $F = \sqrt{KV}$ 7. $k = \sqrt{\dfrac{w}{s}}$

8. $B = \dfrac{DP}{A}$ 9. $p = \dfrac{r}{v}$ 10. $f = d + h - r$ 11. $L = \dfrac{B + M}{2}$ 12. $R = \dfrac{F - K}{Q - 1}$ or $\dfrac{K - F}{1 - Q}$

13. $s = \dfrac{1}{ar}$ 14. $A = \sqrt{\dfrac{F}{R}}$ 15. $t = \dfrac{b}{r^2}$ 16. $V = \sqrt{\dfrac{2B}{K}}$ 17. $F = \dfrac{9(K - 273)}{5} + 32$ or

$F = \dfrac{9K - 2297}{5}$ 18. $t = \sqrt{ks}$ 19. $p = \dfrac{kr}{1 - k}$ 20. $t = \dfrac{T}{a - 1}$ 21. $v = \dfrac{r^2}{m + r}$

CHAPTER 14 - SYSTEMS OF THREE EQUATIONS AND DETERMINANTS

Assignment 47

1. -11　　2. 18　　3. 84　　4. -17.16　　5. a) $D = -7$　b) $D_x = -35$　c) $D_y = 14$　d) $x = 5$
e) $y = -2$　　6. a) $D = -1$　b) $D_t = 2$　c) $D_w = 6$　d) $t = -2$　e) $w = -6$　　7. a) $D = -31$
b) $D_P = -186$　c) $D_Q = 93$　d) $P = 6$　e) $Q = -3$　　8. a) $D = 2.75$　b) $D_a = 22$　c) $D_b = 5.5$
d) $a = 8$　e) $b = 2$　　9. a) $D = 94$　b) $D_r = 141$　c) $D_t = 235$　d) $r = 1.5$　e) $t = 2.5$
10. a) $D = 23.61$　b) $D_x = 44.4$　c) $D_y = 20.1$　d) $x = 1.88$　e) $y = 0.85$　　11. $i_1 = 1.64$ ma,
$i_2 = 0.50$ ma　　12. $F_1 = 181$ kg, $F_2 = 122$ kg　　13. $x = 3$, $y = -1$, $z = 2$　　14. $r = 4$, $s = -2$,
$t = -3$　　15. $a = -2$, $b = 5$, $c = 1$　　16. $h = 3$, $k = 4$, $p = -2$　　17. $t = 1$, $v = -1$, $w = 2$
18. $F_1 = 5$, $F_2 = 7$, $F_3 = 3$

Assignment 48

1. first is -1, second is 3, third is 5　　2. first is 6, second is -2, third is -1　　3. $A = 60°$,
$B = 20°$, $C = 100°$　　4. $A = 45°$, $B = 80°$, $C = 55°$　　5. $F_1 = 600$ tons, $F_2 = 200$ tons,
$F_3 = 100$ tons　　6. $i_1 = 15$ ma, $i_2 = 20$ ma, $i_3 = 5$ ma　　7. $V = \dfrac{B - F}{K - 1}$ or $V = \dfrac{F - B}{1 - K}$

8. $a = \dfrac{d}{dt + p}$　　9. $h = \dfrac{s}{kt}$　　10. $w = \dfrac{p + t}{r}$　　11. $a = \dfrac{rt}{p + t}$　　12. $D = \sqrt{\dfrac{BV}{AK}}$

Assignment 49

1. 1　　2. -1　　3. 4.448　　4. a) $D = 3$　b) $D_x = -3$　c) $D_y = 9$　d) $D_z = 3$　e) $x = -1$
f) $y = 3$　g) $z = 1$　　5. a) $D = -12$　b) $D_t = -48$　c) $D_v = -24$　d) $D_w = 0$　e) $t = 4$
f) $v = 2$　g) $w = 0$　　6. a) $D = 12$　b) $D_a = 12$　c) $D_b = -24$　d) $D_c = 60$　e) $a = 1$
f) $b = -2$　g) $c = 5$　　7. a) $D = 25$　b) $D_x = 100$　c) $D_y = -50$　d) $D_z = 0$　e) $x = 4$
f) $y = -2$　g) $z = 0$　　8. a) $D = -3.1$　b) $D_r = 4.64$　c) $D_s = -5.22$　d) $D_t = -9.86$
e) $r = -1.50$　f) $s = 1.68$　g) $t = 3.18$　　9. a) $D = -6$　b) $D_A = -30$　c) $D_B = 0$　d) $D_C = 18$
e) $A = 5$　f) $B = 0$　g) $C = -3$　　10. $i_1 = 8$ ma, $i_2 = 5$ ma, $i_3 = 2$ ma　　11. $F_1 = 800$ kg,
$F_2 = 400$ kg, $F_3 = 500$ kg

CHAPTER 15 - POWERS, ROOTS, LOGARITHMS

Assignment 50

1. 59,049　　2. 248,832　　3. 4,304.6721　　4. 0.000064　　5. 2.460375　　6. 2.82×10^{21}
7. 9.87×10^{10}　　8. 6.10×10^{-12}　　9. 9.09×10^{-16}　　10. 31.1　　11. 2.31　　12. 1.44
13. 0.85　　14. 0.395　　15. 0.4967　　16. $18^{0.6}$　　17. $6^{0.5}$　　18. $3^{4.5}$　　19. $542^{0.25}$　　20. 65.3
21. 7.82　　22. 832　　23. 0.004630　　24. 0.298　　25. 2.34　　26. $W = 39.4$　　27. $R = 2.81$
28. $D = 631$　　29. $H = 0.859$　　30. $P = 5.4$ lb/in^2

Assignment 51

1. 3.80　　2. 248,000　　3. 0.676　　4. 0.000942　　5. $10^{3.9165}$　　6. $10^{1.6201}$　　7. $10^{-1.2007}$
8. $10^{-3.6655}$　　9. 6.7259　　10. 0.9736　　11. -0.1079　　12. -2.9393　　13. $\log 185 = 2.2672$
14. $-0.7595 = \log 0.174$　　15. $x = \log 27.4$　　16. $\log G = -2.3814$　　17. $10^{2.4166} = 261$
18. $0.02 = 10^{-1.6990}$　　19. $R = 10^{5.8037}$　　20. $10^y = 7,190$　　21. $G = 7.37$　　22. $T = 0.000136$
23. $b = 4.287$　　24. $N = 17$　　25. $P = 5.43$　　26. $M = 3.6$　　27. $r = 1.71$　　28. $D = 35$ decibels
29. a) $I = 63,000$ units　b) 100 times

Assignment 52

1. 53.5　　2. 0.179　　3. 11,500　　4. $e^{4.75}$　　5. $e^{-3.260}$　　6. $e^{1.5518}$　　7. $H = 13.4$
8. $y = 2.08$　　9. $B = 243$　　10. 3.3945　　11. 8.9306　　12. 0.7129　　13. -6.2554
14. $4.15 = \ln 63.4$　　15. $\ln 0.0828 = -2.49$　　16. $9.31 = e^{2.23}$　　17. $e^{-0.851} = 0.427$
18. $R = 3.75$　　19. $y = 8.2$　　20. $P = 0.0241$　　21. $r = 342$　　22. $t = 7.23$
23. $N = 57.7$　　24. $i = 175$ ma　　25. $v = 4.1$ mi/sec